Progress in Botany

Volume 76

Series Editors
Ulrich Lüttge, TU Darmstadt,
FB Biologie (10), Schnittspahnstraße 3–5,
64287 Darmstadt, Germany
Wolfram Beyschlag, Fakultät für Biologie, Lehrstuhl für
Experimentelle Ökologie und Ökosystembiologie,
Universität Bielefeld, Universitätsstraße 25, 33615 Bielefeld,
Germany

Ulrich Lüttge • Wolfram Beyschlag
Editors

Progress in Botany 76

Springer

Editors
Ulrich Lüttge
TU Darmstadt FB Biologie
Darmstadt
Germany

Wolfram Beyschlag
Universität Bielefeld W4-105 Lehrstuhl
Experimentelle Ökologie
Bielefeld
Nordrhein-Westfalen
Germany

ISSN 0340-4773
ISBN 978-3-319-38438-2 ISBN 978-3-319-08807-5 (eBook)
DOI 10.1007/978-3-319-08807-5
Springer Cham Heidelberg New York Dordrecht London

Printed on acid-free paper

Springer is part of Springer Science+Business Media (www.springer.com)

Contents

List of Contributors

Steven R. Archer School of Natural Resources and the Environment, University of Arizona, Tucson, AZ, USA

Paul W. Barnes Department of Biological Sciences, Loyola University New Orleans, New Orleans, LA, USA

Maik Bartelheimer Institute of Botany, Faculty of Biology and Preclinical Medicine, University of Regensburg, Regensburg, Germany

Sibylle Bauer Institute of Botany, Faculty of Biology and Preclinical Medicine, University of Regensburg, Regensburg, Germany

Jürgen Bender Thünen Institute of Biodiversity, Braunschweig, Germany

Elke Bergmann Thünen Institute of Biodiversity, Braunschweig, Germany

David D. Breshears School of Natural Resources and the Environment, University of Arizona, Tucson, AZ, USA

Katharina Hell Institute of Botany, Faculty of Biology and Preclinical Medicine, University of Regensburg, Regensburg, Germany

Günter Hoch Institute of Botany, University of Basel, Basel, Switzerland

Małgorzata Janicka-Russak Department of Plant Molecular Physiology, Institute of Experimental Biology, Wrocław University, Wrocław, Poland

Marjan Jongen Instituto Superior de Agronomia, Universidade de Lisboa, Lisboa, Portugal

Katarzyna Kabała Department of Plant Molecular Physiology, Institute of Experimental Biology, Wrocław University, Wrocław, Poland

Simcha Lev-Yadun Department of Biology and Environment, Faculty of Natural Sciences, University of Haifa–Oranim, Tivon, Israel

Hartmut K. Lichtenthaler Botanical Institute 2 (Plant Physiology and Plant Biochemistry), Karlsruhe Institute of Technology (KIT), University Division, Karlsruhe, Germany

David Lloyd Cardiff University School of Biosciences, Cardiff, Wales, UK

Ulrich Lüttge Department of Biology, Technical University of Darmstadt, Darmstadt, Germany

Vladislav S. Markin Department of Neurology, University of Texas, Southwestern Medical Center, Dallas, TX, USA

Alejandra Matiz Department of Botany, Institute of Biosciences, University of São Paulo, São Paulo, SP, Brazil

Rebecca L. McCulley Department of Plant and Soil Science, University of Kentucky, Lexington, KY, USA

Helenice Mercier Department of Botany, Institute of Biosciences, University of São Paulo, São Paulo, SP, Brazil

Paulo Tamaso Mioto Department of Botany, Institute of Biosciences, University of São Paulo, São Paulo, SP, Brazil

Ewa Niewiadomska The F. Górski Institute of Plant Physiology Polish Academy of Sciences, Kraków, Poland

Elizabeth A. H. Pilon-Smits Biology Department, Colorado State University, Fort Collins, CO, USA

Maria Aurineide Rodrigues Department of Botany, Institute of Biosciences, University of São Paulo, São Paulo, SP, Brazil

Christoph Schmid Institute of Botany, Faculty of Biology and Preclinical Medicine, University of Regensburg, Regensburg, Germany

Gustavo M. Souza Laboratório de Cognição Ecológica em Plantas – ECOLAB, Presidente Prudente, SP, Brasil

Joana Storf Institute of Botany, Faculty of Biology and Preclinical Medicine, University of Regensburg, Regensburg, Germany

Heather L. Throop Department of Biology, New Mexico State University, Las Cruces, NM, USA

Mark A. Tobler Department of Biological Sciences, Loyola University New Orleans, New Orleans, LA, USA

Stephan Unger Department for Experimental and Systems Ecology, University of Bielefeld, Bielefeld, Germany

Alexander G. Volkov Department of Chemistry and Biochemistry, Oakwood University, Huntsville, AL, USA

Lars H. Wegner Insitute for Pulsed Power and Microwave Technology and Institute of Botany 1, Karlsruhe Institute of Technology, Karlsruhe, Germany

Hans J. Weigel Thünen Institute of Biodiversity, Braunschweig, Germany

Monika Wiciarz Department of Plant Physiology and Biochemistry, Faculty of Biochemistry, Biophysics and Biotechnology, Jagiellonian University, Kraków, Poland

Part I
Review

Fifty-Five Years of Research on Photosynthesis, Chloroplasts, and Stress Physiology of Plants: 1958–2013

Hartmut K. Lichtenthaler

Contents

Abstract In the past 55 years, enormous scientific progress was made in many fields of plant physiology and plant biochemistry. Throughout these years, our knowledge on the photosynthetic light processes, the chemical composition and

H.K. Lichtenthaler (✉)
Botanical Institute 2 (Molecular Biology and Biochemistry of Plants), Karlsruhe Institute of Technology (KIT), University Division, Kaiserstr. 12, 76131 Karlsruhe, Germany
e-mail: hartmut.lichtenthaler@kit.edu

U. Lüttge, W. Beyschlag (eds.), *Progress in Botany*, Progress in Botany 76,
DOI 10.1007/978-3-319-08807-5_1

biosynthesis of the photosynthetic apparatus, the ultrastructure of chloroplasts, and their large adaptation capacity to high-light and low-light was extremely enhanced. The author of this article reviews the substantial scientific evolution in these and other fields in which he was actively involved together with his group. The topics that are reviewed also include forest decline research, the mode of action of herbicides in photosynthesis, and in blocking biosynthetic pathways of chloroplasts, such as de novo fatty acid and isoprenoid biosynthesis, as well as the application of chlorophyll fluorescence imaging in the fast noninvasive determination of photosynthetic activity and early detection of plant stress. Moreover, the detection, elucidation, and metabolic significance of the non-mevalonate chloroplast pathway for isopentenyl diphosphate and isoprenoid biosynthesis, the DOXP/MEP pathway, is reviewed. The author further documents that this extreme progress in plant science was largely due to the continuous development and application of new scientific methods and instruments.

1 Introduction

In the fall of 1958 when I started my Ph.D. thesis performing scientific research in the laboratory of the late Professor Dr. August Seybold, Botanical Institute of the University of Heidelberg, Germany, experimental research in plant physiology and photosynthesis in Europe and elsewhere was still in its infancy. Most of the modern scientific instruments and approaches applied as routine methods today did not yet exist. In fact, in 1958 our knowledge on photosynthetic light reactions, on the development and biosynthesis of functional chloroplasts, and on the genuine biosynthetic pathways of plastids and chloroplasts was very limited. By applying $^{14}CO_2$ Melvin Calvin (Nobel laureate 1961) and his group had elucidated already in 1956 the path of carbon in photosynthesis known today as Calvin–Benson cycle. It was yet unknown that there are two light reactions in photosynthesis with associated photosynthetic electron transport processes, that herbicides can specifically block photosynthesis, and that chloroplasts possess several other unique biosynthetic capacities, such as the non-mevalonate pathway of isoprenoid biosynthesis or the de novo fatty acid biosynthesis. Though the electron microscopy of leaves and other plant materials had started, the fine structure of chloroplasts, the biosynthesis and arrangements of their biomembranes and the processes during biosynthesis and degradation of the photosynthetic apparatus were not known.

Over the past 55 years, however, an enormous, steadily increasing progress has been made in science and particularly in plant science which dramatically increased our knowledge and understanding of the photosynthetic light reactions, of the pigment, prenylquinone, and lipid composition of the photosynthetic biomembrane, as well as of the mode of action of herbicides in blocking either photosynthetic electron transport or special metabolic activities of chloroplasts. Moreover, the

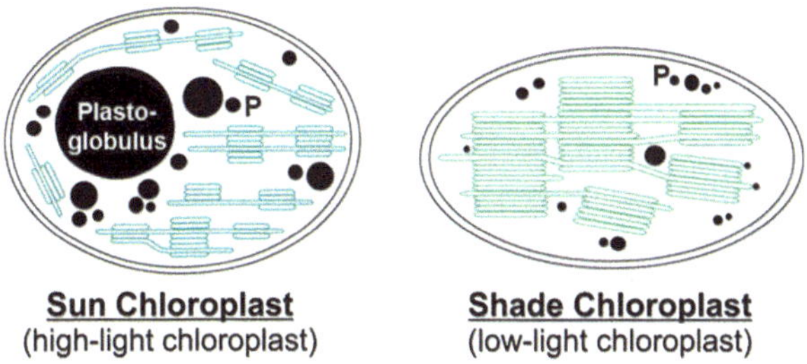

Fig. 1 Scheme showing the differences in the ultrastructure of sun and shade chloroplasts with regard to frequency, width, and stacking degree of thylakoids as well as size and frequency of osmiophilic plastoglobuli. Sun chloroplasts usually contain one or several large starch grains which are not shown here. P = osmiophilic plastoglobuli. The scheme is based on Lichtenthaler (1981), Lichtenthaler et al. (1981a, 1982a) and was presented in a similar form in the review Lichtenthaler (2007)

ultrastructure of functional chloroplasts and their light adaptation, i.e., the arrangement of thylakoids in either high and broad grana stacks (*shade chloroplasts*) or as low and narrow grana stacks (*sun chloroplasts*), has been evaluated (see Fig. 1). In addition, the osmiophilic plastoglobuli were recognized as regular chloroplast structures and as reservoirs for plastohydroquinone-9, α-tocopherol, and other excess plastid lipids. Furthermore, the special genuine, non-mevalonate chloroplast pathway for isopentenyl diphosphate (IPP), isoprenoid, and carotenoid biosynthesis, the DOXP/MEP pathway (see below paragraph 8), was detected between 1995 and 1998 by H. Lichtenthaler, Karlsruhe, in close cooperation with M. Rohmer, Strasbourg, as a special genuine metabolic activity of chloroplasts and all other plastid forms. Besides, the application of the red and far-red chlorophyll fluorescence and the fluorescence imaging technique to leaves and plants as a tool for the investigation of photosynthetic processes and for stress detection in plants were established. The enormous progress in all of these fields, to which my research activities (including those of my students and my cooperation partners from other laboratories) essentially contributed, is briefly summarized and reviewed in this report that also provides information on the general development of the entire field of plant science over the past 55 years.

2 Phylloquinone (Vitamin K_1), Its Localization and Function in Chloroplasts Including Results on Other Prenylquinones

Vitamin K_1, a 2-methyl-1,4-naphthoquinone with a phytyl side chain, had been detected in the late 1920s as a vitamin that is involved in the blood coagulation process and reduces the blood clotting time (see Dam 1942). Biological vitamin tests with animals (chickens) in those years had shown that K_1 is present predominantly in green plant tissue such as leaves, yet even in the late 1950s a chemical analysis of vitamin K_1 in plant tissues had not yet been performed. This then became the topic of my Ph.D. thesis in the fall of 1958 when I started it with Professor August Seybold at the University of Heidelberg. Due to the fact that vitamin K_1 is a typical genuine plant product particularly in green leaves, it was later termed phylloquinone. Since it contains a phytyl side chain, as do the two chlorophylls *a* and *b*, our assumption in 1958 was that it could be associated together with the chlorophylls in chloroplasts and play a role in their photosynthetic process. Upon column chromatography of leaf pigment extracts with the then applied sucrose columns, K_1 showed up in the β-carotene fraction. By a repetition of the column chromatography of the β-carotene fraction with a slowed down elution time I could partially separate K_1 from β-carotene. Via a subsequent paper chromatography the existence of vitamin K_1 in all green leaf tissues was proven, whereas only trace amounts were found in white plant tissues (leucoplasts) and in orange fruit tissue (chromoplasts). K_1 could be well located on the chromatograms because it emits an intense light green color upon illumination with UV light, which is specific for phylloquinone K_1. In contrast, in yeast and various edible fungi I could not detect any K_1. First approaches towards a quantitative determination of K_1 in green leaf extracts of different plants (applying column chromatography followed by a chemical reduction to its hydroquinone) revealed that K_1 was present in leaves at a low concentration of only about one to two molecules per 100 molecules of chlorophyll (Lichtenthaler 1962).

Laboratory research in botanical institutions in Germany and other European countries in the 1950s was rather strenuous and difficult. In those postwar times modern instruments for experimental research were still lacking. The powerful techniques of thin layer chromatography that allows the separation of minor plant lipids (e.g., phylloquinone K_1) from major plant lipids, such as carotenoids or chlorophylls, had not yet been developed. Moreover, recording spectrophotometers were not yet available. In fact, one had to determine the absorbance spectrum of an isolated pigment by measuring the absorbance step by step at each wavelength, first for the blind and then for the sample in order to finally obtain an approximate spectrum of a carotenoid or a chlorophyll in the visible region or of a lipid fraction enriched with phylloquinone in the UV region.

In contrast to Europe, the working conditions for scientific research were completely different in the USA at that time. Thus, in 1962 when I joined, as a postdoctoral research associate, Melvin Calvin's laboratory at the University of California in Berkeley for 2 years, I found there excellent scientific equipment and modern instruments which simplified and advanced scientific research. There I quickly proved that phylloquinone K_1 was, indeed, located in isolated chloroplasts and also in the smallest thylakoid fragments isolated from sonicated chloroplasts, then termed "quantasome aggregates". When I presented M. Calvin a short note to publish these results, he had just received a manuscript in print by F.L. Crane. Already in 1959 the latter had discovered plastoquinone-9 in chloroplasts (Crane 1959), and now also described the location of K_1 in chloroplasts (Kegel and Crane 1962). For this reason, my K_1 results were only published later, together with other observations on the prenylquinone and carotenoid content of thylakoids (Lichtenthaler and Calvin 1964). This information is also found in the paper published with Rod Park where we had summarized the basic lipid and protein composition of thylakoids (Lichtenthaler and Park 1963). This paper in Nature being the first description of the complete lipid composition of a biomembrane strongly stimulated research in other laboratories. Thus, within about 2 years the lipid composition of various other biomembranes in plants and animals was published.

Via further analysis of isolated chloroplasts I detected that their thylakoid membranes also contained α-tocopherol and α-tocoquinone and that plastoquinone-9 was present in its reduced form plastoquinol-9. In addition, I could prove that the thylakoid-free 145,000 × *g* supernatant contained a layer of osmiophilic globuli with high amounts of plastoquinone-9, including its reduced form plastoquinol-9, as well as α-tocopherol (Lichtenthaler and Calvin 1964). For more details on osmiophilic globuli, see below paragraph 3.

After my return to Germany these studies were continued and showed that phylloquinone K_1 was enriched in the photosynthetic photosystem I (PSI), whereas the major part of plastoquinone-9 was bound to photosystem II (PSII) (Lichtenthaler 1969a). α-tocopherol and α-tocoquinone were not specifically bound to one photosystem but were found in both PSI and PSII particle fractions. In further studies on the partition of phylloquinone K_1 between digitonin particles and chlorophyll carotenoid proteins of tobacco we could prove that phylloquinone K_1 is, in fact, exclusively bound to the photosystem I particles (Interschick-Niebler and Lichtenthaler 1981) where it has a function in photosynthetic electron transport (Golbeck 1987). Another observation emphasized the essential requirement of phylloquinone K_1 and β-carotene for a functional photosynthetic apparatus. Etiolated leaf tissue already contained plastoquinone-9 and lutein, yet phylloquinone K_1 and β-carotene in trace amounts only. However, during the first hours of illumination etiolated leaf tissues synthesize and accumulate phylloquinone K_1 and β-carotene at high rates parallel to the formation of the first thylakoids, whereas the de novo accumulation of lutein and plastoquinone-9 that had been formed before in the dark started much later (Lichtenthaler 1969b). Concerning the question which light was responsible for the light-induced biosynthesis of thylakoids and their carotenoids as well as prenylquinones we could show that active phytochrome, P730, is required (Lichtenthaler and Becker 1972) and that this process proceeds also in blue light and red light.

It was also an essential finding that plastoquinone-9 and α-tocopherol are present in the chloroplast envelope membranes and in low levels also phylloquinone K_1 which seem to reflect biosynthetic pool sizes (Lichtenthaler et al. 1981b). Moreover, we addressed the question on the occurrence and concentration of ubiquinone homologues in plants and their mitochondria and found that these contained ubiquinone-9 and ubiquinone-10, also known as coenzyme Q-9 and Q-10 (Schindler 1984; Schindler and Lichtenthaler 1984).

3 Osmiophilic Plastoglobuli: Structure, Composition, and Function

A major part of my early research in the 1960s until the mid-1970s was to establish the fact that osmiophilic plastoglobuli are genuine morphological structures of chloroplasts and other plastid forms and that they function primarily as a lipid store for excess lipids, such as α-tocopherol and plastoquinone-9 + plastoquinol-9, as well as in some plastid stages for glycerolipids and in chromoplasts also for carotenoids. Examples of chloroplasts with many plastoglobuli are shown in Fig. 2. Plastoglobuli are particularly frequent in sun chloroplasts as shown in the scheme of Fig. 1.

In the early 1960s several authors had described the principal structure of chloroplasts as revealed by studies with the electron microscope. Besides the structures of biomembranes partially stapled to grana stacks, which later were termed thylakoids by Menke (1962), usually rather small osmiophilic globuli were found but only after the fixation of leaf tissues with osmium tetroxide. Other authors who had applied $KMnO_4$ as a fixation medium detected only "star-shaped bodies" in the chloroplast stroma. At that time "osmiophilic globuli" and "star-shaped bodies" were regarded as fixation artifacts. Already in 1962 I had analyzed from sonicated chloroplasts a thylakoid-free supernatant of the centrifugation tube (145,000 × *g* supernatant) that contained a yellowish lipid layer

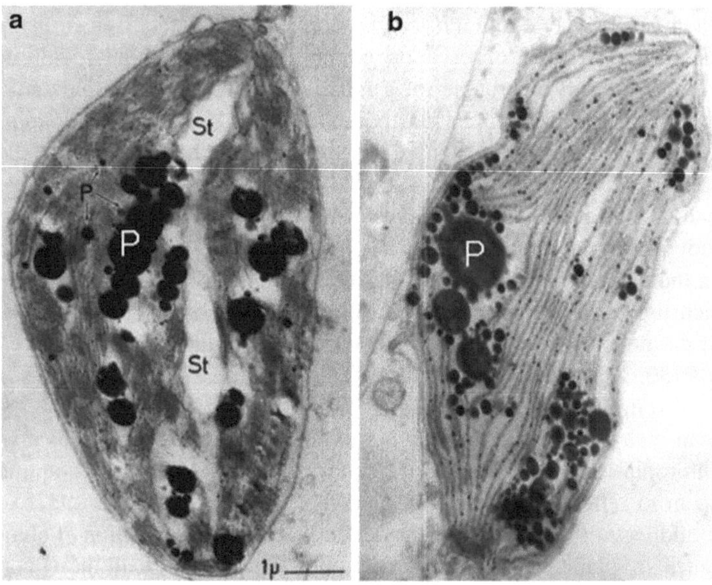

Fig. 2 Chloroplasts with numerous osmiophilic plastoglobuli in green perennial plant tissue. (**a**) From leaves of *Hoya carnosa* R.Br. and (**b**) from the green stem of *Cereus peruvianus* (L.) Mill. *P* plastoglobuli, *St* starch. Fixation of tissues with OsO_4. (Based on Lichtenthaler and Peveling (1966, 1967) and also presented in Lichtenthaler (2013)

consisting of relatively small osmiophilic globuli, first seen by Park and Pon (1961). This layer contained high amounts of α-tocopherol and plastoquinone-9 + plastoquinol-9, apparently excess amounts that were not bound to the photochemically active thylakoids (e.g., Lichtenthaler 1964). This globuli fraction was free of chlorophylls and contained only traces of xanthophylls. Later I showed in a detailed investigation that the "osmiophilic globuli" are regular structural components of the chloroplast stroma and are present in practically all differentiation stages of plastids (see Lichtenthaler 1968). Thus, we termed them "*osmiophilic plastoglobuli*" (Lichtenthaler and Sprey 1966). Since α-tocopherol and plastoquinol-9 are strong reducing agents, the plastoglobuli readily reduce OsO_4 and thus appear osmiophilic. Moreover, we isolated them from several plants, e.g., *Billbergia, Eucharis, Ficus, Spinacia*, and *Tradescantia*, and showed that in chloroplasts of older green leaves the osmiophilic plastoglobuli possess considerably larger diameters than in young spinach leaves (Lichtenthaler and Sprey 1966; Grumbach and Lichtenthaler 1974).

Plastoglobuli predominately function as an extra-thylakoidal store for plant lipids and in particular for excess α-tocopherol and plastoquinone-9 and plastoquinol-9 which are accumulated in high amounts in sun leaves (see review Lichtenthaler 2007) and older green leaves of perennial plant tissues e.g., *Ficus* (Lichtenthaler and Weinert 1970). In chromoplasts plastoglobuli contain also carotenoids and secondary carotenoids that are accumulated together with α-tocopherol and plastoquinone-9 as well as plastoquinol-9 during chromoplast formation. Further, in etioplasts of dark-grown plant seedlings plastoglobuli contain, besides α-tocopherol and plastoquinone-9 + plastoquinol-9, also xanthophylls and possibly also glycerolipids, lipids that are used for the light-induced thylakoid formation. Thus, young chloroplasts are free of osmiophilic plastoglobuli. During chloroplast degeneration and thylakoid breakdown plastoglobuli become more numerous, and usually only a few rather large plastoglobuli remain in the final gerontoplast. Observations by other authors indicate that at a disturbance of normal thylakoid biosynthesis, e.g., by herbicides or other treatments, even triacylglycerides may accumulate and be deposited in plastoglobuli as well, whereby also translucent plastoglobuli can be formed because such lipids are less osmiophilic or not at all [for details and the original references, see the recent plastoglobuli review by Lichtenthaler (2013)]. All data available so far indicate that the interior of osmiophilic plastoglobuli is of pure lipid nature. Due to their more aqueous character proteins cannot be stored inside the plastoglobuli.

Concerning plastoglobuli function, in 1974 we made the highly interesting observation in several plants that the plastoquinol-9 pool in plastoglobuli becomes partially photo-oxidized during the first minutes of the light-induced onset of photosynthetic processes (Grumbach and Lichtenthaler 1974). This photo-oxidation of the plastoquinol-9 pool proceeded in parallel to the photoreduction of violaxanthin to zeaxanthin in the xanthophyll cycle, a process that was reversible in darkness. Thus, it appears that during illumination there occurs an electron flow from plastoglobuli to thylakoids, a process that is partially or fully reversed in the dark (Grumbach and Lichtenthaler 1974). This indicates an interesting regulatory function of the plastoquinol-9 pool of plastoglobuli in the photosynthetic light reactions and the performance of the xanthophyll cycle. More recent observations indicate that in chloroplasts plastoglobuli may contain on their outer surface certain functional chloroplast proteins, which may be coupled to thylakoids and could function in the biosynthesis of chloroplast lipids and possibly also in an active channeling of lipid molecules and lipid breakdown products (Austin et al. 2006; Bréhélin et al. 2007; Bréhélin and Kessler 2008). This opens up an interesting additional aspect concerning plastoglobuli function but requires much further research. More literature and many further details on our research on osmiophilic plastoglobuli as well as references to the parallel observations of other laboratories are found in a recent comprehensive plastoglobuli review (Lichtenthaler 2013).

4 Composition, Structure, and Function of the Photosynthetic Apparatus of Sun and Shade Chloroplasts

An essential part of the research in my group over the last 50 years was to investigate the adaptation of the photosynthetic apparatus and to establish the irradiance-induced differences in pigment composition, photosynthetic quantum conversion, and CO_2 fixation rates of sun and shade chloroplasts of trees. This also included a detailed investigation of the fine structure and in particular of the differential arrangement and stacking of thylakoids in sun and shade chloroplasts as well as in high-light and low-light chloroplasts from leaves of plants grown under high-light and low-light growth conditions (cf. Fig. 1). This research also contained a detailed investigation of the light-induced biosynthesis of the photosynthetic apparatus in dark-grown etiolated leaf tissues. In this very broad field of photosynthesis and plant physiology very little was known in the mid-1960s. We made large progress in our knowledge particularly by applying in parallel various different techniques, including electron microscopy, spectroscopy, and fluorescence analysis, and also by developing new methods, such as reversed phase high-performance liquid chromatography (HPLC) for separation of leaf pigments, gel electrophoresis (PAGE) for the separation of chlorophyll–carotenoid protein complexes of whole chloroplasts, and the superb technique of chlorophyll fluorescence imaging of the photosynthetic quantum conversion of intact leaves. In addition, I redetermined the absorption coefficients of chlorophylls and all individual carotenoids in different solvents and established new equations for the quantitative determination of chlorophyll a and b and the sum of leaf carotenoids $(x+c)$ next to each other in one leaf extract solution (Lichtenthaler 1987). This allows an easy determination of the pigment levels per leaf area or leaf weight unit, including the pigment ratios Chl a/b and total chlorophylls to total carotenoids $(a+b)/(x+c)$. Today this method is applied in most laboratories of photosynthesis and plant physiology.

4.1 Differences in Chlorophyll and Carotenoid Composition

The differentiation between sun and shade leaves as well as sun and shade plants was already made by August Seybold in the 1930s when he analyzed the chlorophyll and carotenoid composition of sun and shade leaves of trees by chromatography of leaf pigment extracts using sugar powder columns. Although he could not yet separate the different xanthophylls from each other, he already demonstrated that sun leaves had higher values for the ratio Chl a/b, and considerably lower values for the ratio xanthophylls to β-carotene, x/c, and also for the ratio of total chlorophylls to total carotenoids, $(a+b)/(x+c)$ (Seybold and Egle 1937). With the establishment of thin layer chromatographic (TLC) techniques for the separation of chlorophylls and individual carotenoids in the 1960s (e.g., Hager and Bertenrath 1962), I reinvestigated the pigment composition of plant leaves in dependence of the incident light and confirmed these particular pigment ratios that are quite different for sun leaves as compared to shade leaves. In addition, the individual levels of the different xanthophylls in sun and shade leaves were determined for the first time. Moreover, we could demonstrate that the same differences in pigment ratios and xanthophyll levels as for sun and shade leaves also existed for leaves of high-light and low-light seedlings that were grown at either high or low irradiances, respectively. Major parts of these results are briefly summarized in the review of Lichtenthaler (2007) and Lichtenthaler and Babani (2004). Reversed phase TLC allowed a distinct separation of zeaxanthin from lutein. Thus, we could clearly demonstrate that sun leaves and leaves of high-light plants had much higher levels of xanthophyll cycle carotenoids (zeaxanthin + antheraxanthin + violaxanthin) as compared to shade leaves or leaves of low-light plants, both on a leaf area as well as on a total carotenoid or on a total chlorophyll $a+b$ level.

After having established a high-performance liquid chromatography (HPLC) technique for fast chlorophyll and carotenoid separation within 20 min (e.g., Schindler et al. 1992, 1994), this strict irradiance dependence of the photosynthetic pigment ratios and the level of xanthophyll cycle carotenoids of chloroplasts during leaf and chloroplast development was further accentuated. With this powerful HPLC method we also determined the kinetics of the light-triggered photoreduction of violaxanthin to zeaxanthin in field-grown maple trees during the course of a sunny and a cloudy day (Schindler and Lichtenthaler 1996) which showed the spontaneous response of the redox state of the xanthophyll cycle carotenoids to transient changes in the irradiance of leaves. Moreover, we could demonstrate in leaves of the tobacco "aurea" mutant Su/su grown at medium irradiance that at high irradiance stress zeaxanthin accumulated in a dynamic biphasic process, i.e., not only via a fast transformation of violaxanthin to zeaxanthin, but by doubling the level of xanthophyll cycle carotenoids within 5 h of high irradiance exposure by de novo biosynthesis and accumulation of new zeaxanthin (Schindler et al. 1992; see also review Lichtenthaler 2007). In the same time period also the β-carotene pool increased by one-third via de novo biosynthesis. These results, which were supplemented by parallel chlorophyll fluorescence measurements and determination of characteristic fluorescence ratios, such as ratios F_v/F_m, dF/F_m' as well as photochemical quenching q_P and non-photochemical quenching q_N (see also Schindler and Lichtenthaler 1996), demonstrated the high flexibility and adaptation capacity of chloroplasts and their photosynthetic pigment apparatus against high-light stress to avoid photo-inhibition and photo-degradation (Lichtenthaler and Schindler 1992). Moreover, additional investigations showed that in fully developed and differentiated leaves the complete photosynthetic pigment apparatus of shade chloroplasts can successively be changed by partial pigment breakdown and de novo pigment accumulation to that of sun chloroplasts and vice versa within a few days.

4.2 *Differences in Photosynthetic Rates of Sun and Shade Leaves*

That sun and shade leaves may have different rates and capacities in photosynthetic quantum conversion had long been assumed, e.g., by Seybold in the 1930s. In the late 1960s and early 1970s it had been shown by some authors that sun plants have higher photosynthetic CO_2 fixation rates than shade plants. Yet, a detailed analysis of sun and shade leaves of the same tree or of plant species grown at different incident light conditions had not yet been performed in a direct comparative way. Some of the early general knowledge of that time was later summarized by Boardman 1977. When we were able to buy one of the (back then still complex) infrared CO_2 gas analyzer systems, I specifically addressed this topic. The small and very handy CO_2 measuring systems (CO_2/H_2O porometers) of today did not yet exist. As expected, our measurements showed that the net CO_2 fixation rates P_N were considerably higher in sun and high-light leaves as compared to shade and low-light leaves of the same plants (Lichtenthaler 1981; Lichtenthaler et al. 1981a), which was correlated with a higher stomata density of the leaves. Further, we could show that sun and high-light leaves exhibited a higher level of soluble sugars. The highly significant differences in photosynthetic quantum conversion between sun and shade leaves are found in different reference systems, not only on a leaf area basis, but also on a total chlorophyll basis. In addition, these differences also showed up in the Hill activity of isolated chloroplasts, which proved to be significantly higher in sun and high-light chloroplasts as compared to shade and low-light chloroplasts.

Furthermore, the same differences were found in the values of the variable chlorophyll fluorescence decrease ratio R_{Fd} that was measured, in parallel, of intact sun and high-light leaves as well as shade and low-light leaves. In the mid-1970s we had established this chlorophyll fluorescence decrease ratio R_{Fd} (originally addressed by us as vF) being based on the measurement of the light-induced slow Chl fluorescence decline during 5 min (slow component of the Kautsky Chl fluorescence induction kinetics) as a valuable indirect measure of the net photosynthetic rates [e.g., Lichtenthaler et al. (1981a, 1984), see also review Lichtenthaler and Babani (2004)]. For details see below paragraph 7.2. The method is much faster than measurements of the photosynthetic CO_2 fixation rates, it can easily be applied in outdoor measurements and was successfully applied as a stress and damage indicator in our forest decline research in the Black Forest between 1983 and 1990 [e.g., Lichtenthaler (1988a, b), Lichtenthaler and Rinderle (1988a)]. The fact that there is a direct correlation between the R_{Fd} values and the photosynthetic net CO_2 fixation rates of sun and shade leaves of trees has been confirmed more recently for various trees at several locations applying the new technique of Chl fluorescence imaging where several ten thousands R_{Fd} values are simultaneously determined for all parts of one leaf (Lichtenthaler et al. 2000a, 2005b, 2007). It has been demonstrated very recently (Lichtenthaler et al. 2013a) that, with respect to their chlorophyll–carotenoid composition and their photosynthetic activity (P_N rates, R_{Fd} values), *blue-shade* and *half-shade leaves* possess an intermediate position between sun and shade leaves.

4.3 Differences in Chloroplast Ultrastructure and Thylakoid Arrangement

Based on the large differences in the chlorophyll–carotenoid composition and photosynthetic activity between sun and shade leaves and the leaves of high-light and low-light plants one could anticipate considerable differences in the fine structure of sun and shade chloroplasts as well as high-light and low-light chloroplasts. In fact, our electron microscopical investigations revealed that the chloroplast ultrastructure of shade and low-light chloroplasts is characterized by a much higher number of thylakoids per granum stack and a significantly higher stacking degree of thylakoids, but also by a significantly broader width of grana thylakoids and grana stacks than in sun and high-light chloroplasts (Lichtenthaler et al. 1981a; Meier and Lichtenthaler 1981) as summarized in Fig. 1 and Table 1. In addition, sun chloroplasts exhibit large starch grains (Fig. 3) which are usually missing in shade and low-light chloroplasts. Moreover, sun and high-light chloroplasts contain more and larger osmiophilic plastoglobuli and consequently higher levels of excess α-tocopherol and plastoquinone-9 that are located in the plastoglobuli as compared to shade and low-light chloroplasts [see the review Lichtenthaler (2007)].

When it had been shown in 1975 that chlorophylls and carotenoids within the photosynthetic membrane are bound to the different chlorophyll–carotenoid proteins CPa, CPI, CPIa and the light-harvesting chlorophyll–xanthophyll proteins LHCPs (Thornber 1975), we adopted the gel electrophoresis techniques (PAGE) for isolated whole chloroplasts and in a quantitative way we studied the presence of the chlorophyll–carotenoid proteins in sun and shade chloroplasts. This way we could show that the higher stacking degree of thylakoids in shade and low-light chloroplasts is, in fact, associated with a significantly higher level of the light-harvesting chlorophyll–carotenoid proteins LHCPs (Lichtenthaler et al. 1982a, b) that are known to be responsible for thylakoid stacking.

In summary, our comparative investigations revealed that leaves and their chloroplasts are highly reactive, adaptive, morphological, and biochemical systems that specifically adapt to the prevailing incident light conditions by forming either sun and high-light chloroplasts or shade and low-light chloroplasts. Thus, sun and high-light leaves with their chloroplasts are adapted for high rates of photosynthetic quantum conversion and CO_2 fixation and contain high amounts of xanthophyll cycle carotenoids to avoid photo-inhibition, whereas the photosynthetic apparatus of shade and low-light leaves primarily "invests" into increasing the light-absorbing pigment cross section in order to catch enough light for performance of

Table 1 Major differences in ultrastructure, thylakoid arrangement, pigment composition, and photosynthetic function of sun and shade chloroplasts

Sun chloroplasts	Shade chloroplasts
Low thylakoid amounts (per chloroplast section)	**High thylakoid amounts** (per chloroplast section)
Narrow grana stacks (width: 0.20–0.26 μm)	**Broad grana stacks** (width: 0.33–0.50 μm)
Few thylakoids per granum	**High grana stacks**
Lower stacking degree (%)	**High stacking degree** (%)
Fagus: 57 ± 6	**82** ± 6
Raphanus: 55 ± 5	**64** ± 4
Triticum: 54 ± 5	**73** ± 3
Zea mays: 55 ± 3	**77** ± 3
Appressed thylakoids: low level	**Appressed thylakoids**: high level
Appressed/exposed thylakoids	**Appressed/exposed thylakoids**
Fagus: 1.3	**4.7**
Raphanus: 1.2	**1.8**
Triticum: 1.2	**2.7**
Zea mays: 1.2	**3.3**
Low levels of LHCPs	**High levels of LHCPs**
Numerous and large plastoglobuli	Few small plastoglobuli
Large starch grains	No starch
High values for Chl a/b	Low values for Chl a/b
3.0–4.3	2.3–2.7
Pigment ratio *x*/*c*	**Pigment ratio *x*/*c***
Low values: 1.7–2.3	High values: **2.6–4.0**
Pigment ratio $(a+b)/(x+c)$	**Pigment ratio $(a+b)/(x+c)$**
Low values: 3.8–4.9	High values: **5.1–6.5**
Xanthophyll cycle carotenoids	Xanthophyll cycle carotenoids
High levels	Low levels
High amounts of excess α-T	Low α-T levels
High level of excess plastoquinone-9 ($PQ\text{-}9 + PQ\text{-}9{\bullet}H_2$)	No excess plastoquinone-9 ($PQ\text{-}9 + PQ\text{-}9{\bullet}H_2$)
High R_{Fd} values	Low R_{Fd} values
3.5–5.5	1.8–2.7
High photosynthetic rates P_N	Low photosynthetic rates P_N
4.6–11.5 μmol $CO_2\,m^{-2}\,s^{-1}$	2.6–3.8 μmol $CO_2\,m^{-2}\,s^{-1}$

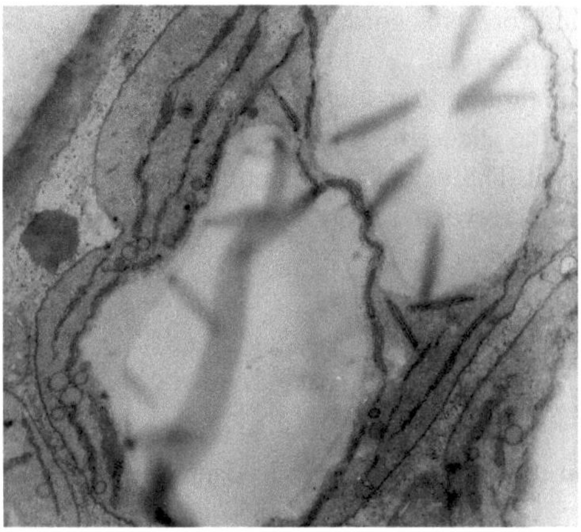

Fig. 3 Sun chloroplast of beech (*Fagus sylvatica*) with large starch grains and several translucent plastoglobuli at the *lower left side* within the chloroplast

photosynthesis. We also checked which other factors besides irradiance control the formation of sun and shade chloroplasts. We could show that the formation of sun chloroplasts is induced by blue light (Lichtenthaler et al. 1980) and enhanced by the phytohormone kinetin (Lichtenthaler and Buschmann 1978), whereas shade chloroplast formation and increased stacking of thylakoids are caused by red light illumination and is further promoted by the application of photosystem 2 herbicides, such as bentazon (Meier and Lichtenthaler 1981). This dependence of the formation of sun and shade chloroplasts on either blue or red light indicates that the phytochrome system—specifically the ratio of red/far-red light—is involved in this adaptation response of chloroplasts. In fact, the incident light in the shade of trees

Table 1 (continued)

Presented are the differential frequency, width, and stacking degree of thylakoids and the level of light-harvesting Chl *a*/*b* proteins LHCPs, which is also documented in the differential ratios of appressed to exposed thylakoid biomembranes. In addition, the size and frequency of osmiophilic plastoglobuli and the differences in the level of total plastoquinone-9 (oxidized and reduced form: PQ-9 and PQ-9•H_2) and α-tocopherol (α-T). Significant differences also exist in the pigment ratios Chl *a*/*b*, xanthophylls to carotenes, *x*/*c*, as well as total chlorophylls to total carotenoids $(a+b)/(x+c)$ that are presented. Moreover, the differences in photosynthetic quantum conversion, i.e., the variable Chl fluorescence decrease ratio R_{Fd} and the net photosynthetic rates P_N yielding high values in sun leaves as compared to shade leaves, are presented

Higher values of individual parameters either present in sun or in shade chloroplasts are shown in bold print. Concerning the stacking degree of thylakoids and width of grana stacks electromicrographs of *Fagus* chloroplasts of sun and shade leaves were investigated and in the case of *Raphanus, Triticum, Zea mays* chloroplast electromicrographs of seedlings grown at high-light and low-light conditions. The ultrastructural and thylakoid arrangement data of chloroplasts in this table are primarily based on Lichtenthaler (1981) and Lichtenthaler et al. (1981a, 1982a, 1984), whereas the pigment ratio data, the level of α-tocopherol and plastoquinone-9, and the differences in R_{Fd} values and photosynthetic CO_2 fixation rates are based on our earlier data reviewed in Lichtenthaler (2007) and Lichtenthaler and Babani (2004); see also Sarijeva et al. (2007) and Lichtenthaler et al. (2013a). The significance levels for the differences between sun and shade leaves and sun and shade chloroplasts in the indicated parameters are ranging from $p<0.05$ to $p<0.001$ as indicated in the original publications cited above.

and forests is enriched with far-red light, whereas sun light and blue skylight only contain relatively low amounts of far-red light. Thus, in sun light and blue skylight the ratio red/far-red light amounts to values of 1.56 and 1.52, respectively, whereas in the shade the red/far-red ratio exhibits a value of 0.21 and in the half-shade of 0.42. The values of the red/far-red ratio presented here were calculated from those of the reverse ratio given by Lichtenthaler et al. (2013a).

5 Forest Decline Research

In the summer of 1982 I became aware of the reports of several German foresters who complained about the unusual decline of spruce (*Picea abies* (L.) H. Karst.) and fir trees (*Abies alba* Mill.) at several locations in the Northern Black Forest, e.g., on the Mauzenberg (altitude 755 m) near the town of Herrenalb. Together with those foresters I examined this decline and the particular damage symptoms. Apparently I was one of the first German plant physiologists who took this threat to our forest trees seriously. The decline started with a considerable loss of the older 3- to 6-year-old needles, a yellowing and bleaching of chlorophyll in younger needles as well as a reduced accumulation of chlorophylls and carotenoids in current and first-year needles, combined with a diminished formation and growth of needles, to just name a few major damage symptoms. In addition, the tree crowns of spruces having shorter branches and fewer side branches became fairly open, whereas the tree tops of firs exhibited a strongly reduced growth of length resulting in compressed tree tops that were termed "stork's nests" and were easily visible from the long distance. In each case tree stands on western exposed slopes and hilltops above 600–1,000 m were affected. The foresters also showed me several unusual damage symptoms on older beech trees (*Fagus sylvatica* L.). In the spring of 1983 more forest sites were affected and the initially more isolated stands were increasing and extended further. In fact, in some places the development was dramatic; thus, from mid-June to the beginning of October 1983 all spruces and firs of a whole mountain top, the Katzenkopf (altitude 900–1,100 m), in the Black Forest had died off.

The causes for such a fast progressing damage and tree decline in the Black Forest initially remained mysterious, although one began to discuss the possible effects of air pollutants, in particular sulfur dioxide, being transported by the predominantly western winds to the western exposed upper tree stands of the Black Forest. At that time high sulfur dioxide levels had already been recognized as the major cause for the forest decline in many mountain areas in Czechoslovakia. The fact that I had early access to the results of fumigation experiments with greenhouse plants of English colleagues (see below) allowed me to be among the first to point out that, in addition to sulfur dioxide, nitrogen oxides—which also provoke the light-induced formation of ozone—were essential causes for the large-scale tree and forest decline.

During a sabbatical in 1981, which I spent at the University of Lancaster, I learned there about the essential research results of Alan Wellburn and Terry Mansfield showing that low atmospheric levels of sulfur dioxide plus nitrogen dioxide had more than additive inhibitory effects on plants cultivated in growth chambers (Wellburn et al. 1981; Mansfield et al. 1982). The cause for this was the fact that sulfur dioxide inactivates the plants' nitrite reductase, which reduces nitrite and starts its successive transformation into ammonia that is incorporated into amino acids. When in 1982 and 1983 I checked the measured levels of nitrogen oxides (NO and NO_2) and sulfur dioxide in the Karlsruhe area and in the Black Forest nearby, it was clear that on a large number of days the levels of these air pollution gases were much higher than the levels used by Mansfield and Wellburn in their growth chamber experiments. Therefore, it was evident that in the Black Forest not only sulfur dioxide but also nitrogen oxides were essentially responsible for the decline of forest trees. In addition, those high levels of nitrogen oxides caused the irradiance-induced formation of ozone which, during sunny and hot summer days, rose to extremely high levels at the Rhine river valley in Karlsruhe and the nearby Black Forest, levels that were not only dangerous to humans but also caused considerable damage to plants and their photosynthetic apparatus. We summarized this information, together with the different damage symptoms of forest trees presented in photos, in a short review "The forest decline: progression, causes and consequences" (Lichtenthaler and Buschmann 1983). This review was sent out to colleagues, to politicians, to people in private industry, as well as to interested laymen. It received great resonance among the public, politicians, and also foresters, and it triggered the general discussions in the public. Yet, at that time many of my German colleagues in plant physiology refused to accept the fact that, except for the particular situation in the Czech mountains, air pollutants would or could cause tree damage and forest decline. In fact, several colleagues postulated infections by unknown fungi and microbes as a primary cause for this large-scale forest decline.

Moreover, we demonstrated via chlorophyll (Chl) fluorescence measurements and also by determining the net CO_2 fixation rates that the photosynthetic quantum conversion of the needles of damaged trees was declining and that the Chl and carotenoid levels of needles declined as well. In addition, together with foresters and the tree physiologist Donald Pigott of the University of Cambridge, England, whom I had invited to Karlsruhe, we checked in the Black Forest the different types of damage symptoms of conifer and broadleaf trees to separate them from symptoms caused by natural stressors, such as heat, cold, or water stress. Then, we reported our findings on the relationship between photosynthesis and tree decline (Lichtenthaler and Buschmann 1984a) as well as on air pollutants as a trigger of the forest decline (Lichtenthaler 1984). For more detailed information, we additionally published a booklet "The Forest decline from a botanical point of view" in German language (Lichtenthaler and Buschmann 1984b). Fortunately, we had the chance of sending several hundred free copies of that booklet to interested colleagues and laymen in the German Democratic Republic, GDR, where any discussion on forest decline was officially forbidden.

With the financial support of the W. & E. Heraeus foundation in Hanau, Germany, I was able to invite, in 1984, 44 foresters and plant physiologists from six European countries to a small workshop in Bad Honnef, Germany. There, German, Swiss, and Austrian foresters involved in forest decline research met for the first time, exchanged their individual observations on forest decline and tree damage, and discussed the causes and consequences with plant physiologists. This workshop strongly stimulated future exchange and research cooperations. In a cooperation with Barry Rock of the NASA, USA, and his team we compared in 1984 and 1985 by means of in situ spectral measurements the forest decline symptoms in Vermont, USA, and the Black Forest, Germany, and found that the symptoms were the same in both locations (Rock et al. 1986). In addition, in cooperation with the German space research center, the Deutsches Zentrum für Luft- und Raumfahrt (DLR) in Oberpfaffenhofen, Germany, and the NASA, USA, we classified the trees of damaged spruce stands in the Northern Black Forest by airborne reflectance and terrestrial Chl fluorescence measurements (Schmuck et al. 1987; Rinderle and Lichtenthaler 1989). At the Mauzenberg forest location (altitude 650–755 m) we analyzed the seasonal variation in photosynthetic activity of healthy and damaged spruce trees over 2 consecutive years, whereby 3 needles years were studied in parallel. In damaged spruce trees we found not only a reduced photosynthetic quantum conversion as detected via Chl fluorescence measurements, but even more reduced rates of net CO_2 fixation P_N, both on a chlorophyll and on a needle area basis, as well as reduced rates of transpiration and stomatal conductivity (e.g., Lichtenthaler et al. 1989). Moreover, needles of damaged spruces (damage class 3–4) could no longer regulate and fully close their stomata, which caused a fast desiccation and dropping down of needles.

At that time the state of Baden-Württemberg started and supported the interdisciplinary European research project PEF (Projekt Europäisches Forschungszentrum), where various research groups studied different aspects of the forest decline in the Black Forest at the same locations, with the Schöllkopf (altitude 840 m) near the town of Freudenstadt being the most important location. Within this program we analyzed the performance and pigment composition of the photosynthetic apparatus of different needle ages of healthy and damaged spruce trees (e.g., Lichtenthaler et al. 1985, 1989; Zimmer-Rinderle and Lichtenthaler 1995). The results of all those investigations (major parts were later published in the book by Bittlingmeier et al. 1995) demonstrated that the large-scale forest decline was caused by a combination of natural environmental stress with air pollution stress (e.g., SO_2, NO_x, ozone), whereby the latter considerably enhanced the natural stress, led to potassium and magnesium deficiency of the soil and trees, caused bleaching of photosynthetic pigments, and reduced the vitality of trees predominantly by an early decline of the photosynthetic function and a progressing damage to the photosynthetic apparatus.

6 Mode of Action of Herbicides in Photosynthesis, Chloroplasts, and the Apicoplast

Besides the Calvin–Benson cycle of CO_2 assimilation and the pigment apparatus catalyzing the photosynthetic light and associated electron transport reactions, chloroplasts possess various other biosynthetic pathways that are potential targets for herbicides and natural antibiotics and inhibitors. The goal of our research was to apply inhibitors and herbicides in order to find out more about the photosynthetic electron transport reactions, to learn more about the special metabolic pathways of chloroplasts, and also to detect the mode of action of new herbicides.

Thus, we clarified that the herbicide bentazon blocks the photosynthetic electron transport by specifically binding to the Q_B protein of the photosystem 2 reaction center (Pfister et al. 1974). In addition, we developed isolated chloroplasts and etioplasts as test systems for inhibitors against de novo fatty acid biosynthesis and proved that the herbicides diclofop and other aryloxy-phenoxy-propionic acids as well as sethoxydim, cycloxydim, and other cyclohexane-1,3-diones, all of them specific graminicides, inhibit the plastidic fatty acid biosynthesis by specifically blocking the acetyl-CoA carboxylase (Kobek et al. 1988a, b; Lichtenthaler 1989). They also block the development and replication of chloroplasts (Lichtenthaler and Meier 1984). Moreover, we detected that the two natural antibiotics cerulenin and thiolactomycin are also inhibitors of de novo fatty acid biosynthesis in chloroplasts (Feld et al. 1989; Golz et al. 1994) where they block the β-ketoacyl-ACP synthases KAS I (cerulenin, thiolactomycin) and KAS III (thiolactomycin). We also showed that ethyl-adenylates inhibit the acetyl-CoA synthetase and acetylmethyl-phosphinates the plastidic pyruvate dehydrogenase complex as shown in Fig. 4 (Golz et al. 1994). Such inhibitors are essential tools for the clarification of the metabolite flow from either acetate or pyruvate into de novo fatty acid biosynthesis and into isoprenoids or branched-chain plastidic amino acids. Later we demonstrated that 6-ketoclomazone is a specific inhibitor of the DOXP synthase (1-deoxy-D-xylulose-5-phosphate synthase) and fosmidomycin a specific inhibitor of the DOXP reductase (1-deoxy-D-xylulose-5-phosphate reductase), i.e., the first and the second enzyme of the plastidic DOXP/MEP pathway of isoprenoid biosynthesis, inhibitors that essentially helped to establish this newly detected non-mevalonate pathway of isoprenoid biosynthesis (see Lichtenthaler 2000a). This plastidic DOXP/MEP pathway for isoprenoid biosynthesis is named after its first and second intermediates: 1-deoxy-D-xylulose-5-phosphate (DOXP) and 2-C-methyl-D-erythritol-4-phosphate (MEP). Details of the DOXP/MEP pathway are found below in paragraph 8.

Another observation of particular interest was the finding by several authors that the malaria inducing parasite *Plasmodium falciparum* had a nongreen, plastid-type cell organelle, the apicoplast that, during evolution, was taken up from either a green or a red alga. Since the malaria parasite is dependent on the metabolic activities of its apicoplast we cooperated with physicians and proved that the apicoplast possesses the DOXP/MEP pathway of isoprenoid biosynthesis which can be blocked by the herbicide fosmidomycin. In fact, in our joint efforts we could show that malaria-

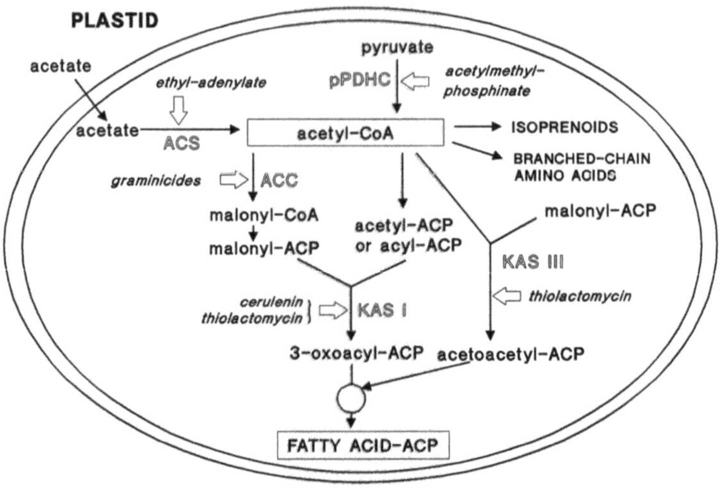

Fig. 4 Scheme of de novo fatty acid biosynthesis in chloroplasts starting from acetate and pyruvate. The enzymes and their specific inhibition by active ingredients and herbicides are indicated (Based on Lichtenthaler 1989, 2000c). *ACC* acetyl-CoA carboxylase, *ACP* acyl carrier protein, *ACS* acetyl-CoA synthetase, *KAS I and KAS III* β-ketoacyl-ACP synthase, *Malonyl-ACP* malonyl-acid carrier protein, *pPDHC* plastidic Pyruvate Dehydrogenase Complex

infected mice were cured by fosmidomycin treatment (Jomaa et al. 1999). Since the DOXP/MEP pathway of isopentenyl diphosphate (IPP) biosynthesis also occurs in pathogenic eubacteria, such as *Mycobacterium tuberculosis* and *Helicobacter pylori* [for a complete list see Lichtenthaler (2000a)], plants with their easy-to-handle DOXP/MEP pathway are very suitable test systems for new drugs against pathogenic bacteria and the malaria parasite (Lichtenthaler et al. 2000b).

7 Chlorophyll Fluorescence and Fluorescence Imaging of Photosynthetic Activity and Plant Stress

Essential progress in our understanding of photosynthetic processes came from the application of chlorophyll fluorescence induction kinetics which several decades later were further promoted by the introduction of laser-induced fluorescence imaging of plant leaves and their photosynthetic activity.

7.1 Chlorophyll Fluorescence

Already in 1931 Hans Kautsky (Fig. 5) had measured in pre-darkened green leaves that upon illumination there is a red Chl fluorescence that initially rises within a few seconds to a maximum and then slowly decreases within a few minutes to a considerably lower steady level (Kautsky and Hirsch 1931). In more than

Fig. 5 Hans Kautsky in Marburg, around 1950

14 subsequent papers [reviewed in Lichtenthaler (1992)] he analyzed and characterized this Chl fluorescence induction kinetics and its dependence on chemicals and environmental factors. Back then already Kautsky had concluded that the photosynthetic light process consisted of two light reactions, one that reduces a substance (e.g., an electron acceptor) and a second light reaction that oxidizes this reduced substance. Therefore, it was Kautsky who first detected that the photosynthetic apparatus consisted of two photosystems cooperating with each other. This knowledge became evident in the photosynthetic community only in the early 1960s when other groups, in particular those of Duysens, Govindjee, and Butler, repeated and advanced Kautsky's Chl fluorescence measurements [for references see Lichtenthaler (1992)]. This was the beginning of the evaluation of the two photosystems and the photosynthetic electron transport chain and the search for its components, whereby plastoquinone-9 was detected by Crane in 1959 and phylloquinone K_1 independently of each other by Crane and by Lichtenthaler in 1962 as mentioned above (see paragraph 2). In subsequent years Chl fluorescence induction kinetics developed to a routine method of photosynthesis research, various Chl fluorescence parameters, ratios, and coefficients were established, e.g., the ratio Fv/Fm and the photochemical and non-photochemical quenching coefficients q_P and q_N. Much of this research on the role of Chl fluorescence in the detection of stress conditions in plants was summarized in the comprehensive review by Lichtenthaler and Rinderle (1988b). Further information is found in the articles by various authors in the two books on Chl fluorescence by Lichtenthaler (1988a) and Papageorgiou and Govindjee (2004). An exact guide of how to measure and correctly apply these Chl fluorescence parameters and ratios was given more recently by Lichtenthaler et al. (2005a).

We also showed that one should be very cautious with the interpretation of such Chl fluorescence parameters when they are solely measured at the upper, i.e., the adaxial, leaf-side. In fact, the values of the Chl fluorescence ratios and quenching coefficients obtained in that way only reflect the responses and reactivity or inhibition of the chloroplasts of the upper leaf-half. The chloroplasts of the lower leaf-half, which are accessible only via Chl fluorescence measurements at the lower leaf-side, may still be fully functional even if the chloroplasts of the upper leaf-half are fully inhibited. Thus, in maple leaves exposed to full sunlight on a hot sunny day the Chl fluorescence ratios F_v/F_m, dF/F_m' as well as photochemical quenching q_P and non-photochemical quenching coefficients q_N, measured at the upper leaf-side, indicated a complete photo-inhibition of the photosynthetic apparatus, yet the CO_2 fixation measurements clearly proved that the leaves still exhibited about 78 % of their maximum CO_2 fixation rates (Schindler and Lichtenthaler 1996). In this respect see also the corresponding results described by Lichtenthaler et al. (2005a). Thus, Chl fluorescence measurements should always be performed on both leaf-sides and be complemented by net CO_2 fixation measurements with a CO_2/H_2O porometer in order to clarify to which degree a presumed photo-inhibition really exists at the whole leaf level.

7.2 *The Chlorophyll Fluorescence Ratios* R_{Fd} *and* F690/F730

Concerning Chl fluorescence I introduced in my research two other Chl fluorescence ratios which are excellent parameters to determine photosynthetic activity and stress effects in plants. Moreover, based on Hans Selye's stress concept for humans I established a general stress concept of plants, a list of stressors and stress responses in order to simplify the discussion on plant stress (Lichtenthaler 1996).

Fluorescence Ratio R_{Fd} The Chl fluorescence decrease ratio R_{Fd}, i.e., the ratio of the slow fluorescence decrease F_d to the steady-state fluorescence F_s reached after 5 min of illumination, ratio F_d/F_s, proved to be an exact indirect indicator of the net photosynthetic CO_2 fixation rates P_N as reviewed in Lichtenthaler and Babani (2004). In the years from 1983 through 1990 we successively applied this Chl fluorescence decrease ratio R_{Fd} in our forest decline research in order to determine the decline in photosynthetic activity and the damage degree of spruces, firs, and deciduous forest trees (e.g., Lichtenthaler 1988b). The ratio R_{Fd} is also a very suitable parameter to sense a decline in photosynthetic quantum conversion due to water stress or other stress events, such as nitrogen deficiency, and also to describe the differential activities of sun and shade leaves of trees as shown in a recent original paper (Lichtenthaler et al. 2013a) and reviewed in Lichtenthaler and Babani (2004).

Fluorescence Ratio *F690/F730* We also introduced another Chl fluorescence ratio, i.e., the ratio of the fluorescence yield in the red (near 690 nm) and far-red (near 730–740 nm) maxima of the Chl fluorescence emission spectra, i.e., the ratio *F690/F730* also known as ratio *F690/F735*. With increasing chlorophyll $a+b$ content of leaves the *F690* maximum decreases, whereas that of *F730* is almost unaffected. Thus, the ratio *F690/F730* is an inverse indicator of the Chl $a+b$ content of leaves. Hence, its increase with decreasing Chl content from low regular values of 0.4–0.6 for green leaves to considerably higher values is an excellent stress indicator (Rinderle and Lichtenthaler 1988; Hák et al. 1990; see also the review of Buschmann 2007). The inclusion of the ratio *F*690/*F*730 opens new possibilities for remote sensing of terrestrial vegetation by a combination of laser-induced Chl fluorescence and reflectance measurements (Lichtenthaler 1989). In further investigations we could retrieve the actually emitted Chl fluorescence emission spectrum as compared to the measurable spectrum of green leaves by evaluating the degree of reabsorption of the emitted red Chl fluorescence by means of absorption and reflectance measurements (Gitelson et al. 1998).

7.3 Fluorescence Imaging of Plants

We also investigated in detail the blue and green fluorescence emission of green leaves and their spectral characteristics together with the red and far-red Chl fluorescence of leaves (Stober and Lichtenthaler 1992; Stober et al. 1994). In contrast to the red and far-red Chl fluorescence, the blue and green fluorescence of plant leaves are also emitted by nongreen plant leaves and they remain constant during the Chl fluorescence induction kinetics known as Kautsky effect (Stober and Lichtenthaler 1993). Thus, the blue fluorescence can be taken as a standard when the red and far-red Chl fluorescence are decreasing due to stress events. In fact, we detected that the ratio of blue to red fluorescence can be taken as stress indicator. The blue fluorescence of plant leaves shows a maximum near 440 nm (*F*440) and the green fluorescence mostly a shoulder (sometimes also a maximum) near 520 nm (*F*520). As the major substance of the blue-green fluorescence emission of plants we identified cell wall bound ferulic acid (Lichtenthaler and Schweiger 1998). All this fluorescence information came from measurements at small individual spots of a leaf. In order to obtain reliable information for the whole leaf several measurements had to be performed at different spots across the leaf surface.

In cooperation with physicists from the CNRS in Cronenbourg near Strasbourg we checked the possibilities for laser-induced imaging of the plants' blue and green fluorescence together with the red and far-red Chl fluorescence. The advantage of fluorescence imaging is that one image contains the information of several 10,000 pixels per leaf, and this is of high statistical significance and reliability. The first fluorescence images were taken of green tobacco leaves (Lang et al. 1994) as shown in Fig. 6, whereby the fluorescence intensity is indicated in false colors. The images clearly indicate that the four fluorescence bands are not homogeneously distributed across the leaf area. The highest blue (*F*440) and green (*F*520) fluorescence emanate from the leaf veins where the chlorophyll content is low. The image also demonstrates that the blue fluorescence is higher than the green fluorescence. In contrast, the red (*F*690) and far-red (*F*740) chlorophyll fluorescence primarily come from the vein-free leaf regions where the Chl density is high. In addition, it can be noticed that the red fluorescence is higher than the far-red fluorescence, and both Chl fluorescences are higher than the blue and green fluorescence. By a pixel-to-pixel division one obtains the corresponding fluorescence ratio images blue/green,

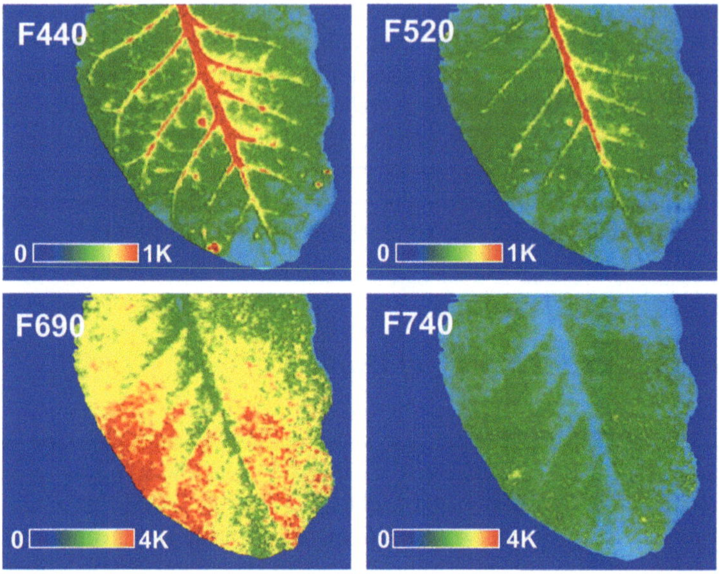

Fig. 6 Fluorescence images of the *upper side* of a green tobacco leaf. The intensity of the blue (*F*440) and green (*F*520) fluorescence as well as the red (*F*690) and far-red (*F*740) chlorophyll fluorescence is shown in false colors, whereby the fluorescence yield in the images increases from blue (no fluorescence) via green and yellow to red as the highest fluorescence. The highest blue and green fluorescence are emitted by the leaf veins, whereas the highest chlorophyll fluorescence comes from the intercostal fields, i.e., the vein-free leaf regions. Note that the scales for the red and far-red chlorophyll fluorescence are different from those of the blue and green chlorophyll fluorescence. K in the scales means kilo (=1,000) counts. [Based on Lang et al. (1994), Lichtenthaler et al. (1996) modified]. Each image consists of several ten thousand pixels over the leaf surface

blue/red, and blue/far-red and red/far-red as indicated in detail in Lang et al. (1996) and Lichtenthaler et al. (1996).

Both fluorescence images and fluorescence ratio images allow the detection of spatial heterogeneities and small local disturbances in fluorescence yield over the leaf surface and also in the values of the individual fluorescence ratios, which are early stress and damage indicators long before a damage can visually be detected. Thus, a high-resolution fluorescence imaging system allows an early detection of vegetation stress (Lichtenthaler et al. 1996). We performed fluorescence imaging of water and temperature stress (Lang et al. 1996); we applied laser-induced fluorescence imaging for monitoring a nitrogen fertilizing treatment (Heisel et al. 1997). This multicolor fluorescence imaging is an excellent diagnostic tool for the detection of plant stress and changes in photosynthetic quantum conversion (Lichtenthaler and Miehé 1997). Its principles and characteristics as well as stress-induced changes of the fluorescence ratios and the application possibilities of this powerful investigation method were summarized by Buschmann and Lichtenthaler (1998) and Buschmann et al. (2000).

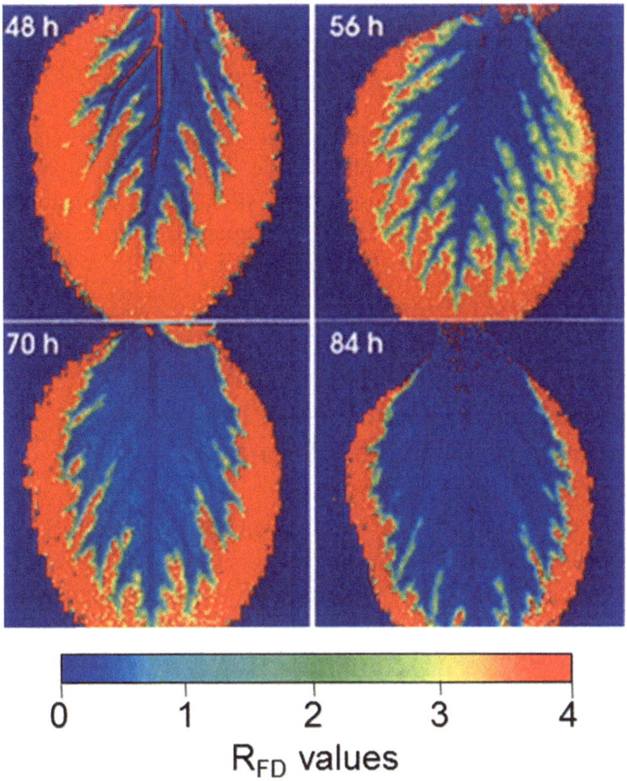

Fig. 7 Successive loss of the photosynthetic activity of intact green leaves of foxglove (*Digitalis purpurea* L.) leaves after the uptake of the photosystem II herbicide diuron as visualized here via images of the Chl fluorescence ratio R_{Fd} that decreases with increasing herbicide uptake. Images of the Chl fluorescence decrease ratio R_{Fd} were taken at different times after application of the herbicide diuron (10^{-5} M) via the root of a young plant. The values of the Chl fluorescence decrease ratio R_{Fd} are given in false colors in absolute values with decreasing values from red (highest R_{Fd} value of 4) via yellow and green to light-blue (low intensity) to dark-blue (zero). (Based on Lichtenthaler and Miehé 1997 and Lichtenthaler et al. 2013b, modified)

Replacing the expensive HeNe laser by a flash lamp UV excitation we developed the much smaller Karlsruhe fluorescence image system which was successfully applied in multicolor fluorescence imaging of sugar beet leaves with different N-status (Langsdorf et al. 2000) and also for imaging the photosynthetic activity of leaves. It was already mentioned above that the Chl fluorescence decrease ratio R_{Fd} (the ratio F_d/F_s) is linearly correlated with the photosynthetic net CO_2 fixation P_N of leaves (e.g., Lichtenthaler and Babani 2004). By imaging the Chl fluorescence (a) in pre-darkened leaves at the fluorescence maximum Fm reached after an illumination time period of ca. 1 s and (b) again at the steady level F_s reached after 5 min of illumination, one can form the R_{Fd} ratio images of leaves providing ample information, e.g., on the differences in the photosynthetic activity between sun and shade leaves of trees including needle twigs of conifers (Lichtenthaler and Babani

2000; Lichtenthaler et al. 2000a, 2005b, 2007). This R_{Fd} imaging technique also allows studying the uptake of the herbicide diuron into green leaves by a progressing decrease of the R_{Fd} values (Lichtenthaler et al. 2013b) as shown in Fig. 7.

The efficient multicolor fluorescence imaging technique is presently the best and superior fluorescence method for plant tissue. In the future it may become an essential method in agriculture, horticulture, silviculture, plant food production, and agro-forestry. It also allows to track the ripening of fruits as shown for the ripening of apples during storage (Lichtenthaler et al. 2012).

8 The Non-mevalonate Chloroplast Pathway for Isopentenyl Diphosphate and Isoprenoid Biosynthesis, the DOXP/MEP Pathway[1]

In the early 1950s it had been shown by the groups of Konrad Bloch and Fjodor Lynen that acetate and acetyl-CoA were the precursors of cholesterol biosynthesis, in 1956 mevalonic acid (MVA) was detected as an intermediate, and in 1958 isopentenyl diphosphate (IPP) and farnesyl diphosphate. In 1958 the well-known plant biochemist and carotenoid specialist T.W. Goodwin was the first to study and to prove the incorporation of ^{14}C-acetate and ^{14}C-MVA yet at low rates, into carotenoids and sterols of higher plants (e.g., Goodwin 1958). Since the typical labeling pattern of the acetate/MVA pathway was found by means of the chemical degradation of ^{14}C-labeled *Euglena* β-carotene (Steele and Gurin 1960), it was generally accepted that the isoprenoids of plants, such as sterols as well as the plastidic carotenoids and chlorophylls (phytol side chain), are all synthesized via the acetate/MVA pathway as is cholesterol in fungi and animals [for original literature of the cited authors see the reviews Lichtenthaler (1999), (2000b)].

There remained, however, doubts and several inconsistencies concerning the biosynthesis of plastidic isoprenoids via the acetate/MVA pathway. Thus, photosynthetically fixed $^{14}CO_2$ was readily incorporated into carotenoids, phytol, and cytosolic sterols, whereas ^{14}C-labeled acetate and MVA were readily incorporated into cytosolic sterols, yet only at very low rates into chloroplast isoprenoids, an observation made by many authors and first by Goodwin (1958). In addition, we found that mevinolin, a specific inhibitor of the acetate/MVA pathway, efficiently blocked the biosynthesis of sterols and the mitochondrial ubiquinones, whereas the accumulation of chlorophylls (phytol side chain), carotenoids, and other plastidic isoprenoids was not affected (Bach and Lichtenthaler 1983). Our attempts to detect a separate plastidic HMG-CoA reductase, which is the key enzyme of the acetate/

[1] The plastidic DOXP/MEP pathway for isoprenoid biosynthesis is named after its first and second intermediates: 1-deoxy-D-xylulose-5-phosphate (DOXP) and 2-C-methyl-D-erythritol-4-phosphate (MEP).

MVA pathway, were negative. These findings suggested that chloroplasts might have their own biosynthesis system for IPP and isoprenoid biosynthesis which should be different from and independent of the cytosolic acetate/MVA pathway.

New progress in this field of chloroplast isoprenoid biosynthesis came from the application of the ^{13}C-labeling technique combined with high-resolution NMR spectroscopy that allows an exact location of the ^{13}C-atoms within the carbon skeleton of carotenoids or any other plant isoprenoid. Using this new evolving technique my chemist colleague Michel Rohmer in Mulhouse, France, had found in 1988 an unusual labeling of hopanoids in two eubacteria. At the beginning of 1993 we started a very close cooperation of our laboratories using this new technique, whereby my group grew the sterile algae and plant cultures on ^{13}C-labeled glucose and Rohmer's group performed the NMR spectroscopy of the ^{13}C- labeled isoprenoid compounds. This way we detected in 1995 the existence of the non-mevalonate plastidic pathway for IPP formation, first in green algae (Lichtenthaler et al. 1995; Schwender et al. 1995, 1996) and later also in higher plants (Lichtenthaler et al. 1997a, b). This pathway starting from pyruvate and glycerinaldehyde-3-phosphate has been termed DOXP/MEP pathway of plastidic IPP and isoprenoid biosynthesis after the first (DOXP, 1-deoxy-D-xylulose-5-phosphate) and second intermediate (MEP, 2-C-methyl-D-erythritol-4-phosphate).

In my group we cloned the genes of the first two enzymes of the DOXP/MEP pathway and found two specific inhibitors: 5-ketoclomazone and fosmidomycin for these two enzymes (cf. Fig. 8). Later I also cooperated with my brother Frieder Lichtenthaler, Darmstadt, a sugar chemist, who provided us with ^{13}C-labeled 1-deoxy-D-xylulose (DOX) that was readily incorporated into phytol, carotenoids, isoprene, and other plastidic isoprenoids by higher plants and algae (Schwender et al. 1997; Zeidler et al. 1997). There exists a cooperation, a cross talk, between both cellular isoprenoid biosynthesis pathways which, at photosynthesis conditions, primarily works via an export of active C_5 units from chloroplasts to the cytosol that are used predominantly for sterol biosynthesis (cf. Fig. 8). An import of short isoprenoid chains from the cytosol into the plastid may occur as well, however, only at extremely low rates if at all. In fact, our investigations with inhibitors demonstrated that the cytosolic acetate/MVA biosynthesis cannot provide the IPP or short chain isoprenyl phosphates required for carotenoid, chlorophyll, and prenylquinone biosynthesis in chloroplast when the plastidic DOXP/MEP pathway has been blocked by fosmidomycin. Once we had detected the plastidic DOXP/MEP pathway of IPP biosynthesis, various other groups jumped into this new research field and detected the following enzymes 3–7 of this pathway. Additional literature references on the detection, establishment and significance of the DOXP/MEP pathway including contributions of other laboratories are found in the extended review articles by Lichtenthaler (1999, 2010).

In cooperation with Rohmers group we also checked evolutionary aspects of the distribution of the DOXP/MEP pathway, e.g., its presence in different algae groups.

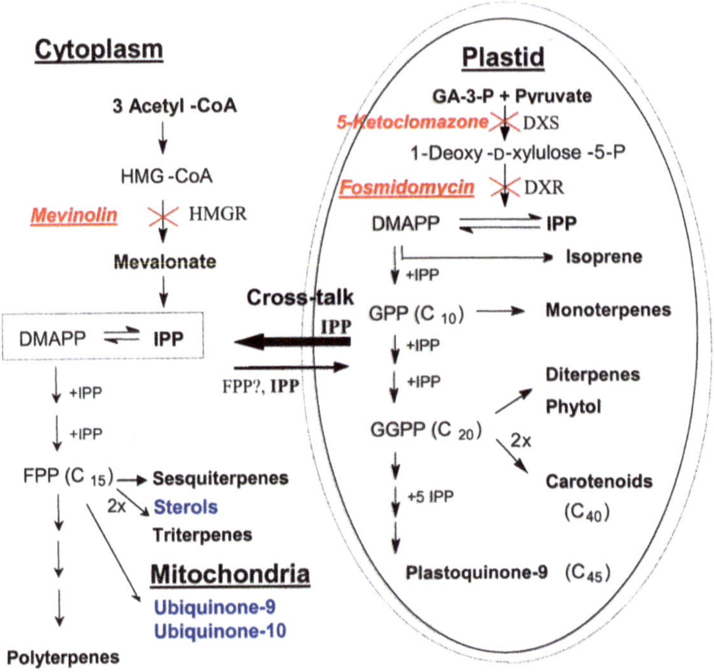

Fig. 8 Scheme showing the two independent isoprenoid biosynthesis pathways in plant cells: (1) the chloroplastidic DOXP/MEP pathway and (2) the cytosolic acetate/mevalonate pathway. The DOXP/MEP pathway provides the active isoprenic C_5 units (IPP, DMAPP) for the biosynthesis of carotenoids, chlorophylls (phytyl side chain), and prenylquinones (phytyl and nonaprenyl side chains). The acetate/mevalonate pathway delivers the isoprenic C_5 units for the biosynthesis of sterols and the prenyl side chain of the mitochondrial ubiquinones. The specific inhibition of the DOXP/MEP pathway by *5-ketoclomazone* (target DOXP synthase, DXS) and *fosmidomycin* (target: DOXP reductase, DXR) and of the acetate/mevalonate pathway by *mevinolin* (target: HMG-CoA reductase = HMGR) is indicated. The indicated "cross talk" between the two cellular biosynthetic isoprenoid pathways primarily consists of an export of IPP from chloroplasts to the cytosol for sterol biosynthesis. Scheme based on Lichtenthaler et al. (1997a) and Lichtenthaler (1999, 2010). *DMAPP* dimethylallyl diphosphate, *DOXP* 1-deoxy-D-xylulose-5-phosphate, *DXR* DOXP reductase, *DXS* DOXP synthase, *IPP* isopentenyl diphosphate, *FPP* farnesyl diphosphate, *GPP* geranyl diphosphate, *GGPP* geranylgeranyl diphosphate, *HMG* 3-hydroxy-3-methyl-glutaryl-CoA, *HMGR* 3-hydroxy-3-methyl-glutaryl-CoA reductase, *MEP* 2-C-methyl-D-erythritol-4-phosphate

It was essential that we could prove the presence of the DOXP/MEP pathway in cyanobacteria, since cyanobacteria-like organisms are regarded as the ancestors of chloroplasts. Like higher plants, Rhodophyta and Heterokontophyta possess both pathways for IPP biosynthesis, whereas Chlorophyta have lost their cytosolic acetate/MVA pathway during the evolution and they not only synthesize plastidic isoprenoids but also cytosolic sterols via the DOXP/MEP pathway (e.g., Lichtenthaler 2004c).

So far *Euglena* (Euglenophyta) is the only exception among all algae groups which, during evolution, has lost its DOXP/MEP pathway of isoprenoid biosynthesis and therefore is dependent on the acetate/MVA pathway for the synthesis of all its isoprenoids including carotenoids and chlorophyll (phytol side chain) [see the reviews Lichtenthaler (2004c) and (2010)]. After Steele and Gurin (1960) had unequivocally shown via the chemical degradation of ^{14}C-labeled *Euglena* β-carotene that it was labeled via the acetate/MVA pathway, nobody doubted anymore that all photosynthetic organisms made their carotenoids via the acetate/MVA pathway. If they had taken a different alga instead of *Euglena* for their ^{14}C-labeling experiments of β-carotene, they and/or Goodwin and his group could have detected already then that chloroplasts possess their own pathway for IPP and isoprenoid biosynthesis.

9 Support by Elder Colleagues

During the start of my research in plant science in 1958 I was essentially supported by professor *August Seybold*, Heidelberg, who accepted me as a Ph.D. student in plant physiology although I had studied pharmacy. He taught me to perform scientific research always on a broad, comparative level. Seybold who since the late 1920s had extensively worked on the transpiration of plants, the photosynthetic pigments in sun and shade plants, and the light perception of plants and algae, also shaped me for ecophysiological research, in particular for questions on the influence of high and low quanta fluence rates on growth and the photosynthetic function of plants. I am grateful to him for his inspiring support.

After my Ph.D. I had the chance, in 1961, to work in the laboratory of *Paul Ozenda* at the Centre d'Etudes Nucleaire, Grenoble, France, where I learned the application of radioisotopes in plant physiology research and studied the kinetics of ion absorption (e.g., ^{32}P) by plant roots. He was very supportive and strongly encouraged the French-German cooperation.

From 1962 to 1964, I had the great privilege to work in the laboratory of *Melvin Calvin* (Nobel laureate 1961) in Berkeley, California (see Fig. 9). This was an exciting time and an atmosphere of departure in plant sciences and photosynthesis research. Melvin Calvin showed much interest in my research on the types of prenylquinones, carotenoids, and lipids in the photosynthetic membrane. He started his day very early, and discussions with him often took place at 7:00 a.m. He was an extremely fast thinker and quick to evaluate consequences of a scientific observation and immediately came up with ideas on the essential steps that should follow. Calvin knew how to stimulate young scientists and gave me the valuable advice to always concentrate fully on new promising research topics and to avoid performing parallel research. The whole scientific environment of Calvin's group was extremely stimulating; he usually had more than 70 individuals (staff members, graduate students and many foreign postdocs) in his laboratories. Among them were chemists, physicists, and plant physiologists, and in the regular Friday morning seminars we had excellent interdisciplinary discussions that often led to scientific cooperation. The scientific exchange with Melvin Calvin continued after

Fig. 9 Melvin Calvin, Berkeley, in his office in 1986

my return to Germany until his passing. He was, indeed, a great and inspiring scientist and academic teacher.

In Berkeley I also had extensive and very stimulating discussions with *Daniel Arnon* on the function of vitamin K_1 in the photosynthetic membrane. He had just shown that vitamin K_3, a methyl-naphthoquinone, catalyzed cyclic photo-phosphorylation in isolated spinach chloroplasts and wanted to learn more on the genuine substance phylloquinone K_1. At that time the idea of two photosynthetic photosystems in series came up, a concept that was based on the early observations of Kautsky (Kautsky and Hirsch 1931). This concept of two photosystems was intensively discussed by Arnon as well as in all the other photosynthesis research groups in Berkeley until it was finally established.

In 1963 I became acquainted with *Andy Benson* who had essentially contributed to the detection of the photosynthetic carbon reduction cycle, today known as Calvin–Benson cycle. Back then he worked in La Jolla, California, on the glycerolipids and the sulfolipid of the photosynthetic membrane. He had invited me to give a lecture on my paper on the total lipid and protein composition of the photosynthetic membrane that had just appeared in Nature (Lichtenthaler and Park 1963). We extensively discussed various possibilities how the photosynthetic pigments and glycero-lipids were arranged in the membrane and finally came up with the conclusion that the sulfo-, galacto-, and phospholipids were arranged in the membrane in a double-layer structure into which the pigments and proteins were integrated or attached to. We developed this concept clearly before the lipid double-layer structure of biomembranes had been established. Various aspects of our discussion together with my just published thylakoid lipid table became an essential

Fig. 10 Andy Benson, La Jolla, here shortly after his 90th birthday in Paris, 2007

part of Andy Benson's review paper "Plant Lipid Membranes" (Benson 1964). This inspiring discussion, followed by many others on the international photosynthesis congresses, was the starting point of a lifelong friendship and scientific exchange. In fact, in 2007 three of us from the photosynthetic community (Bob Buchanan, Roland Douce and myself) celebrated and honored Andy Benson on the occasion of his 90th birthday in a famous restaurant in Paris presenting a special issue of the journal Photosynthesis Research (see Fig. 10) with papers dedicated to him (see Lichtenthaler et al. 2008).

In 1964, shortly after my return to Germany, *Wilhelm Menke* who in the 1930, was the first to isolate chloroplasts from spinach and in 1962 had created the term "thylakoids" invited me to Köln for a lecture and shortly afterwards *André Pirson* invited me to Göttingen. Both photosynthesis researchers accompanied and promoted my further research. I am also grateful to *Hans Reznik*, then at the University of Münster, Westphalia, for his offer to continue my photosynthesis research in Germany and for his promotional support. My research was also inspired by continuous discussions with *Kazuo Shibata*, Tokyo, *Hemming Virgin*, Göteborg, *Kurt Mühlethaler* and *Albert Frey-Wyssling*, Zürich, *Cyrille Sironval*, Liège, and particularly accentuated in repeated discussions, over many years, with *Trevor W. Goodwin*, Liverpool (see Fig. 11), the pioneer and expert of carotenoid and isoprenoid research in plants, whom I first met in 1968 on the first international photosynthesis congress in Freudenstadt, Black Forest, Germany.

Fig. 11 Trevor Goodwin, Liverpool (*right*), with Paul Mazliak, Paris, and Hartmut Lichtenthaler (*left*) on a European plant lipid meeting in September 1993 in Karlsruhe

10 Cooperations with Scientific Colleagues

Extremely essential impulses for photosynthesis research as well as European and international scientific cooperation in plant science came from the first international photosynthesis congress held by *Helmut Metzner* (Tübingen) in Freudenstadt, Black Forest, Germany, in 1968. There, many young as well as older colleagues from Eastern and Western European countries met for the first time after World War II and exchanged their ideas on all aspects of photosynthesis. Among them were physicists, biochemists, classical botanists, plant physiologists, and cytologists. In fact, that congress was the starting point of a broad and interdisciplinary scientific cooperation on a European and worldwide level, which was further promoted by the subsequent international photosynthesis congresses that have been held every 3 years. In this respect I need to emphasize that from the 1960s through the 1990s a much higher percentage of plant scientists worked on particular aspects of photosynthesis than today.

With *Hubert Ziegler* (München) as an essential supporter, *Peter Böger* (Konstanz), *Ulrich Lüttge* (Darmstadt), and other German colleagues I had a close cooperation on the establishment of the "Section Plant Physiology" within the German Botanical Society DBG. On a European level discussions went on with *Paul-Emil Pilet* (Lausanne), *John Dale* (Edinburgh), *Laszlo Erdei* (Szeged), *Anders Kylin* (Lund), *Valentin Kefeli* (Moscow), *Stanislav Procházka* (Brno), *Miloje Saric* (Belgrade), *Ernesto Vieitez* (Santiago de Compostela), *Charles Wittigham* (London), *Zdenek Sestak* (Prague), and many others on a cooperation of European plant physiologists. In 1975 on the XII. International Botanical Congress in Leningrad (today St. Petersburg) we had an informal discussion meeting with the Russian and East European plant physiologists. These discussions, essentially initiated and supported by Hubert Ziegler (München), led to the formation of a Federation of European Societies of Plant Physiology, FESPP, which could finally be founded in 1978 with the participation of several East European countries (for details see the report of Lichtenthaler 2004a). With other colleagues, such as *Paul Stumpf* (Davis, California), *Peter Biacs* (Budapest), *Trevor Goodwin* (Liverpool), *Ernst Heinz* (Köln), *John Harwood* (Cardiff), *Conny Liljenberg* (Göteborg), *Paul Mazliak* (Paris), *Paul-André Siegenthaler* (Neuchatel), *Joseph Wintermans* (Nijmegen), and *Norio Murata* (Japan), I cooperated to establish the International Symposia on Plant Lipids, ISPL which, since 1974, have been held every 2 years all over the world (details are given by Lichtenthaler 2004b). The large and fast progress in photosynthesis, biosynthesis, metabolism, and function of plants lipids as well as in many other topical fields of plant science since the 1970s was, indeed, possible due to the fact that colleagues of different countries got to know each other on such regular international conferences and workshops. They started their scientific cooperation and then an exchange of their graduate students and postdocs. In addition, in 1980 I started with *Peter Böger* (Konstanz), *Aloys Wild* (Mainz), *Manfred Kluge* (Darmstadt), and *Heinrich Fock* (Kaiserslautern) annual photosynthesis workshops where our graduate and Ph.D. students as well as our young scientific staff members could present and discuss their scientific results. Also these workshops provided distinct impulses for progressing with our photosynthesis research towards new horizons.

In 1962 my own scientific cooperation had already started with *Roderic Park* in Berkeley on the lipids and proteins of the photosynthetic unit in thylakoids. Later, this was continued with many colleagues mentioned below in a chronological order, such as *Benno Sprey* and *E. Peveling*, Münster, on osmiophilic plastoglobuli, with *Günter Retzlaff*, BASF, on the mode of action of herbicides in photosynthesis, with *Conny Lilienberg*, Göteborg, on the separation of prenols on TLC plates, with *Alan Wellburn*, Lancaster, on cytosolic and plastidic isoprenoids and their labeling from ^{14}C-mevalonate, with *Pierre Dizengremel*, then Paris, on occurrence and function of different ubiquinone homologues in plants, with *Peter Biacs*, TU Budapest, on saponins in plants, with the biochemist *Janos Retey*, Karlsruhe, on HMG-CoA reductase in plants, with *Roland Douce*, Grenoble, on the localization of prenylquinones and carotenoids in the chloroplast envelope, with *Karl Erismann*, Bern, on ^{14}C-labeling kinetics of prenylquinones in *Chlorella*, with *Barry Rock*, NASA, Pasadena, USA, on forest decline in Germany and the USA including remote sensing and airborne classification of forests, with physicists *Laslo Koscany* and *Peter Richter*, TU Budapest, on creating new instruments for spectroscopy of plant leaves as well as for indoor and outdoor Chl fluorescence measurements, with biochemist *Wilhelm Boland*, Karlsruhe, on the inhibition of fatty acid biosynthesis by cerulenin derivatives, with *Vladimir Saakov*, St. Petersburg, on the effect of gamma ray irradiation on the photosynthetic apparatus, with *Jiri Santrucek* and *Pavel Siffel*, České Budějovice, on photosynthetic activity of green tobacco and aurea mutants, with *Nicola D'Ambrosio*, Napels, on the carotenoid composition of leaf and stem tissue of the CAM plant Cissus, with *Anatoly Gitelson*, Lincoln, Nebraska, on retrieving the actual Chl fluorescence spectra and emissions by plant leaves via simultaneous absorbance and reflectance measurements, with *Fatbardha Babani*, Tirana, Albania, on fluorescence imaging of photosynthetic activity, with *Zoltan Tuba*, Gödöllö, Hungary, on the photosynthetic apparatus of homoio- and poikilochlorophyllous desiccation-tolerant plants, as well as with *Otmar Urban*, Brno, on Chl fluorescence imaging of sun and shade leaves of trees in the Beskydy Mountains. I would also like to mention here the long-term scientific exchange with *Bob Buchanan*, Berkeley, on photosynthetic topics and historical aspects of photosynthetic carbon fixation and the close cooperation with *Tino A. Rebeiz*, Champaign, USA, regarding the organization of the First International Symposium on Chloroplast Bioengineering held at the University of Illinois, Urbana-Champaign in May 2005 and in editing the book "The Chloroplast, Basics and Applications" (Rebeiz et al. 2010).

Particularly close and intensive was the cooperation with physicist *Joseph Miehé* and coworkers, CRNS, Cronenbourg near Strasbourg, from 1994 through 1998, on the development and application of laser-induced fluorescence imaging of plants in the four plant fluorescence emission bands blue, green, red, and far-red as well as with chemist *Michel Rohmer* and his coworkers, University of Strasbourg, in the detection and establishment of the novel DOXP/MEP pathway of chloroplast IPP and isoprenoid biosynthesis from 1993 through 1999. In fact, such international cooperations as mentioned here were, for all those active in plant physiology research, the essential basis for the large progress made in so many fields of plant science in the last five decades. It was a pleasure that I had the chance of contributing to this enormous development.

11 Epilogue

The large progress made in photosynthesis and plant science over the past 55 years was essentially a result of the increasing work with isolated cell organelles and the continuous invention and application of new and advanced instruments as well as investigation techniques that allowed the studies and revelations of details of plant structures, metabolic reactions, and their responses to the environment. Such novel and superior techniques and approaches were the application of electron microscopy, the labeling of cellular metabolites with radioisotopes (e.g., ^{14}C), the application of ^{13}C labeling in combination with high-resolution NMR spectroscopy, the introduction of PAGE and HPLC techniques, the availability of CO_2/H_2O porometer systems, the measurement of laser-induced chlorophyll fluorescence kinetics, the powerful technique of fluorescence imaging of plants and their stress responses as well as the application of inhibitors to specifically block enzymes in metabolic pathways, and of course the use of molecular biology, to just name a few major ones. Thus, progress in science is and has always been dependent on the development of new investigation techniques. In addition to these new techniques there has been the progressing international scientific cooperation and exchange of scientists that have essentially been enhanced and promoted by regular international scientific meetings and workshops, such as the triennial international congresses on Photosynthesis, the biannual meetings of the European plant physiologists, FESPP, or the biannual international symposia on plant lipids, ISPL, which are mentioned above. I hope that also in the future the long-standing, successful international cooperation will continue on a worldwide level and include many more countries. This international cooperation that is so essential for the progress in science is based on the mutual understanding of and the respect for people.

12 Curriculum Vitae of Hartmut K. Lichtenthaler

Hartmut Lichtenthaler was born on June 20, 1934 in Weinheim, Baden, Germany

Education and Professional Experience

1953–1958: Study of pharmacy in Heidelberg and at the University of Karlsruhe

1958: Masters degree (Staatsexamen) in Pharmacy

Spring 1961: Ph.D. in Botany at the University of Heidelberg with August Seybold

1961: Euratom Research Fellow at the Centre d'Etude Nucléaires and the University of Grenoble/France with Paul Ozenda

1962–1964: Research fellow at the University of California, Berkeley, with Melvin Calvin

1964–1970: Botanical Institute, University of Münster/Westphalia, scientific assistant, 1967 Habilitation in Botany, then Dozent and Associate Professor

1970–2001: Full professor in Plant physiology, Pharmaceutical Biology and Plant Biochemistry at the University of Karlsruhe (now: Karlsruhe Institute of Technology, KIT)

since 2001: Professor emeritus

Other Activities

1980–1986: Chairman of the Section Plant Physiology of the German Botanical Society DBG; 1978 Founding member of the Federation of the European Societies of Plant Physiology (FESPP) and FESPP President 1984–1986; Founding Member of the International Symposia on Plant Lipids ISPL 1974–1976. Coordinator and participant in several European research programs, such as OECD, LASFLEUR, QAAFFI, INTERREG, PEF.

1973: Visiting professor at the University of California, Berkeley with Melvin Calvin

1975: Guest professor at the University of Gothenburg, Sweden with Hemming Virgin

1981: Guest professor at the University of Lancaster, England with Alan Wellburn

Honors

1992: Honorary member of the Hungarian Society of Plant Physiology.

Honorary doctoral degrees: 1996 Mendel University of Brno, Czech Republic; 1997 ELTE University, Budapest; and in 2001 St. Istvan University, Gödöllö, Hungary. In 2001 "Bundesverdienstkreuz am Bande" (Cross of Merits) of the Federal Republic of Germany. 2003 Gregor Mendel Medal of the Czech Academy of Sciences; in 2004 Terry Galliard Medal and in 2010 Corresponding Membership Award, American Society for Plant Biology ASPB

Acknowledgements I am very grateful to my entire team of former graduate and Ph.D. students as well as scientific and technical group members who actively participated, over the past 50 years, in performing some truly pioneering research in photosynthesis and plant science. Without their continuous effort, enthusiasm and understanding we could not have made such an enormous progress in our plant science research spanning from biophysics and organic chemistry via pure physiological and structural investigations using the electron microscope to pure biochemical topics including some molecular biology. This interdisciplinary research was also made possible by multiple German and international cooperation partners at universities, research institutions, and the chemical industry and—last but not least—by the continuous support of granting agencies, such as the German Research Council DFG, the German Ministry of Research, the European Community, and various others, all of which is gratefully acknowledged. In addition, I wish to thank Ms Gabrielle Johnson for her long-term English language assistance with many of my scientific publications. Last but not least, my special gratitude goes to my wife for her kind understanding, her patience, and her continuous support of my scientific research over almost five decades now.

References

Austin JR, Frost E, Vidi P-A, Kessler F, Staehelin LA (2006) Plastoglobules are lipoprotein subcompartments of the chloroplast that are permanently coupled to thylakoid membranes and contain biosynthetic enzymes. Plant Cell 18:1693–1703

Bach TJ, Lichtenthaler HK (1983) Inhibition by mevinolin of plant growth, sterol formation and pigment accumulation. Physiol Plant 59:50–60

Benson A (1964) Plant membrane lipids. Annu Rev Plant Physiol 15:1–16

Bittlingmeier L, Reinhardt W, Siefermann-Harms D (eds) (1995) Waldschäden im Schwarzwald: Ergebnisse einer interdisziplinären Freilandstudie zur montanen Vergilbung am Standort Freudenstadt/Schöllkopf. Ecomed Verlagsgesellschaft, Landsberg

Boardman N (1977) Comparative photosynthesis of sun and shade plants. Annu Rev Plant Physiol 28:355–377

Bréhélin C, Kessler F (2008) The plastoglobule: a bag full of lipid biochemistry. Photochem Photobiol 84:1388–1394

Bréhélin C, Kessler F, van Wijk KJ (2007) Plastoglobule: versatile lipoprotein particles in plastids. Trends Plant Sci 12:260–266

Buschmann C (2007) Variability and application of the chlorophyll fluorescence emission ratio red/far-red of leaves. Photosynth Res 92:261–271

Buschmann C, Lichtenthaler HK (1998) Principles and characteristics of multi-colour fluorescence imaging of plants. J Plant Physiol 152:297–314

Buschmann C, Langsdorf G, Lichtenthaler HK (2000) Imaging of the blue, green and red fluorescence emission of plants: an overview. Photosynthetica 38:483–491

Crane FL (1959) Isolation of two quinones with coenzyme Q activity from alfalfa. Plant Physiol 34:546–551

Dam H (1942) Vitamin K, its chemistry and physiology. Adv Enzymol 2:285–324

Feld A, Kobek K, Lichtenthaler HK (1989) Inhibition of fatty-acid biosynthesis in isolated chloroplasts by the antibiotics cerulenin and thiolactomycin. Bright Crop Prot Conf Weeds 4D–10:479–486

Gitelson AA, Buschmann C, Lichtenthaler HK (1998) Leaf chlorophyll fluorescence corrected for re-absorption by means of absorption and reflectance measurements. J Plant Physiol 152:283–296

Golbeck JH (1987) Structure, function and organization of the photosystem I reaction center complex. Biochim Biophys Acta 895:167–204

Golz A, Focke M, Lichtenthaler HK (1994) Inhibitors of *de novo* fatty acid biosynthesis in higher plants. J Plant Physiol 143:426–433

Goodwin TW (1958) Incorporation of $^{14}CO_2$, [2-^{14}C]acetate and [2-^{14}C]mevalonic acid into ß-carotene in etiolated maize seedlings. Biochem J 68:26–27

Grumbach HK, Lichtenthaler HK (1974) Photo-oxidation of the plastohydroquinone-9 pool in plastoglobuli during onset of photosynthesis. In: Avron M (ed) Proceedings of the 3rd international congress on photosynthesis, Weizmann Institute of Science, Rehovot, Israel, vol I. Elsevier, Amsterdam, pp 515–523

Hager A, Bertenrath T (1962) Verteilungschromatographische Trennung von Chlorophyllen und Carotinoiden grüner Pflanzen an Dünnschichten. Planta 58:564–568

Hák R, Lichtenthaler HK, Rinderle U (1990) Decrease of the fluorescence ratio F690/F730 during greening and development of leaves. Radiat Environ Biophys 29:329–336

Heisel F, Sowinska M, Khalili E, Eckert C, Miehé JA, Lichtenthaler HK (1997) Laser-induced fluorescence imaging for monitoring nitrogen fertilizing treatments of wheat. In: SPIE proceedings series: advances in laser remote sensing for terrestrial and oceanographic applications, 21–22 April 1997, Orlando, FL. SPIE 3059:10-21

Interschick-Niebler E, Lichtenthaler HK (1981) Partition of phylloquinone K_1 between digitonin particles and chlorophyll-proteins of chloroplast membranes from *Nicotina tabacum*. Z Naturforsch 36c:276–283

Jomaa H, Wiesner J, Sanderbrand S, Altincicek B, Weidemeyer C, Hintz M, Türbachova I, Eberl M, Zeidler J, Lichtenthaler HK, Soldati D, Beck E (1999) Inhibitors of the nonmevalonate pathway of isoprenoid biosynthesis as antimalarial drugs. Science 285:1573–1576

Kautsky H, Hirsch A (1931) Neue Versuche zur Kohlensäureassimilation. Naturwissenschaften 19:964

Kegel LP, Crane FL (1962) Vitamin K_1 in chloroplasts. Nature 154:1282

Kobek K, Focke M, Lichtenthaler HK, Retzlaff G, Würzer B (1988a) Inhibition of fatty acid biosynthesis in isolated chloroplasts by cycloxydim and other cyclohexane-1,3-diones. Physiol Plant 72:492–498

Kobek K, Focke M, Lichtenthaler HK (1988b) Fatty-acid biosynthesis and acetyl-CoA carboxylase as a target of diclofop, fenoxaprop and other aryloxy-phenoxy-propionic acid herbicides. Z Naturforsch 43c:47–54

Lang M, Lichtenthaler HK, Sowinska M, Summ P, Heisel F (1994) Blue, green and red fluorescence signatures and images of tobacco leaves. Bot Acta 107:230–236

Lang M, Lichtenthaler HK, Sowinska M, Heisel F, Miehé JA, Tomasini F (1996) Fluorescence imaging of water and temperature stress in plant leaves. J Plant Physiol 148:613–621

Langsdorf G, Buschmann C, Babani F, Mokry M, Timmermann F, Lichtenthaler HK (2000) Multicolour fluorescence imaging of sugar beet leaves with different N-status by flash lamp UV-excitation. Photosynthetica 38:539–551

Lichtenthaler HK (1962) Vergleichende Bestimmungen der Vitamin K_1-Gehalte in Blättern. Planta 57:731–753

Lichtenthaler HK (1964) Untersuchungen über die osmiophilen Globuli der Chloroplasten. Ber Deut Bot Ges 78:398–402

Lichtenthaler HK (1968) Plastoglobuli and the fine structure of plastids. Endeavour XXVII:144–149

Lichtenthaler HK (1969a) Localization and functional concentrations of lipoquinones in chloroplasts. In: Metzner H (ed) Photosynthesis research, vol I. ILaupp Press, Tübingen, pp 304–314

Lichtenthaler HK (1969b) Light-stimulated synthesis of plastid quinones and pigments in etiolated barley seedlings. Biochim Biophys Acta 184:164–172

Lichtenthaler HK (1981) Adaptation of leaves and chloroplasts to high quanta fluence rates. In: Akoyunoglou G (ed) Photosynthesis VI. Balaban Internat Science Service, Philadelphia, PA, pp 273–287

Lichtenthaler HK (1984) Luftschadstoffe als Auslöser des Baumsterbens. Naturwissenschaftl Rundschau 37:271–277

Lichtenthaler HK (1987) Chlorophylls and carotenoids, the pigments of photosynthetic biomembranes. In: Douce R, Packer L (eds) Method enzymol, vol 148. Academic, New York, pp 350–382

Lichtenthaler HK (ed) (1988a) Applications of chlorophyll fluorescence. Kluwer, Dordrecht, 1988

Lichtenthaler HK (1988b) In vivo chlorophyll fluorescence as a tool for stress detection in plants. In: Lichtenthaler HK (ed) Applications of chlorophyll fluorescence. Kluwer, Dordrecht, pp 129–142

Lichtenthaler HK (1989) Plant lipid biosynthesis as target for biocides. In: Biacs PA, Gruiz K, Kremmer T (eds) Biological role of plant lipids. Plenum, New York, pp 389–400

Lichtenthaler HK (1992) The Kautsky effect: 60 years of chlorophyll fluorescence induction kinetics. Photosynthetica 27:45–55

Lichtenthaler HK (1996) Vegetation stress: an introduction to the stress concept in plants. J Plant Physiol 148:4–14

Lichtenthaler HK (1999) The 1-deoxy-d-xylulose-5-phosphate pathway of isoprenoid biosynthesis in plants. Annu Rev Plant Physiol Plant Mol Biol 50:47–65

Lichtenthaler HK (2000a) The non-mevalonate isoprenoid biosynthesis: enzymes, genes and inhibitors. Biochem Soc Trans 28:787–792

Lichtenthaler HK (2000b) Discovery of the two parallel pathways for isoprenoid biosynthesis in plants. In: Kung S-D, Yang S-F (eds) Discoveries in plant biology, vol III. World Scientific Publishing, Singapore, pp 141–161
Lichtenthaler HK (2000c) 30 Jahre Botanik II Karlsruhe. Karlsruher Beitr Pflanzenphysiol 37:1–198, ISSN 0173-3133
Lichtenthaler HK (2004a) A history of the Federation of European Societies of Plant Physiology FESPP since its formation in 1978—including notes on events preceding the foundation and following re-naming as the Federation of European Societies of Plant Biology (FESPB) in 2002. J Plant Physiol 161:635–639
Lichtenthaler HK (2004b) Thirty years of international symposia on plant lipids. In: Biacs P, Gerely P (eds) Proceedings of the 16th plant lipid symposium. Mete Publisher, Budapest, pp I–XXIV. https://drive.google.com/file/d/0B9tGkMMkZ8IkQV9qSWhTRHdoWVE/edit?usp=sharing
Lichtenthaler HK (2004c) Evolution of carotenoid and sterol biosynthesis in photosynthetic and non-photosynthetic organisms. In: Biacs P, Gerely P (eds) Proceedings of the 16th plant lipid symposium, Mete Publisher, Budapest, pp 11–24. https://drive.google.com/file/d/0B9tGkMMkZ8IkQV9qSWhTRHdoWVE/edit?usp=sharing
Lichtenthaler HK (2007) Biosynthesis, accumulation and emission of carotenoids, α-tocopherol, plastoquinone and isoprene in leaves under high photosynthetic irradiance. Photosynth Res 92:163–181
Lichtenthaler HK (2010) Chapter 7: The DOXP/MEP pathway of chloroplast isoprenoid biosynthesis. In: Rebeiz C, Bohnert H, Benning C, Daniell H, Hoober K, Lichtenthaler HK, Portis AR, Tripathy BC (eds) The chloroplast: basics and applications, vol 31, Advances in photosynthesis and respiration. Springer, Dordrecht, pp 95–118
Lichtenthaler HK (2013) Chapter 15: Plastoglobuli, thylakoids, chloroplast structure and development of plastids. In: Biswal B, Krupinska K, Biswal UC (eds) Plastid development during leaf growth and senescence, vol 36, Advances in photosynthesis and respiration. Springer Science + Business Media, Dordrecht, pp 337–361
Lichtenthaler HK, Babani F (2000) Detection of photosynthetic activity and water stress by imaging the red chlorophyll fluorescence. Plant Physiol Biochem 38:889–895
Lichtenthaler HK, Babani F (2004) Light adaptation and senescence of the photosynthetic apparatus: changes in pigment composition, chlorophyll fluorescence parameters and photosynthetic activity. In: Papageorgiou GC, Govindjee (eds) Chlorophyll fluorescence: a signature of photosynthesis, Springer, Dordrecht, pp 713–736
Lichtenthaler HK, Becker K (1972) Changes of the plastidquinone and carotenoid metabolism associated with the formation of functioning chloroplast in continuous far-red and white light. In: Proceedings of the 2nd international congress on photosynthesis research, vol 3. Dr. W. Junk, NV Publishers, The Hague, pp 2451–2459
Lichtenthaler HK, Buschmann C (1978) Control of chloroplast development by red light, blue light and phytohormones. In: Akoyunoglou G et al (eds) Chloroplast development in plant biology, vol 2. Elsevier, Amsterdam, pp 801–816
Lichtenthaler HK, Buschmann C (1983) Das Waldsterben, Verlauf, Ursachen und Konsequenzen. Fridericiana 32:38–66, ISSN 0533-0912
Lichtenthaler HK, Buschmann C (1984a) Beziehung zwischen Photosynthese und Baumsterben. Allgemeine Forst Zeitschrift 39:12–16
Lichtenthaler HK, Buschmann C (1984b) Das Waldsterben aus botanischer Sicht. G. Braun, Karlsruhe
Lichtenthaler HK, Calvin M (1964) Quinone and pigment composition of chloroplasts and quantasome aggregates from *Spinacia oleracea*. Biochim Biophys Acta 79:30–40
Lichtenthaler HK, Meier D (1984) Inhibition by sethoxydim of chloroplast biogenesis, development and replication in barley seedlings. Z Naturforsch 39c:115–122
Lichtenthaler HK, Miehé J (1997) Fluorescence imaging as a diagnostic tool for plant stress. Trends Plant Sci 2:316–320

Lichtenthaler HK, Park RB (1963) Chemical composition of chloroplast lamellae from spinach. Nature 198:1070–1072
Lichtenthaler HK, Peveling E (1966) Osmiophile Lipideinschlüsse in den Chloroplasten und im Cytoplasma von *Hoya carnosa* R. Br Die Naturwissenschaften 53:534
Lichtenthaler HK, Peveling E (1967) Plastoglobuli und osmiophile cytoplasmatische Lipideinschlüsse in grünen Blättern von *Hoya carnosa* R. Br. Z Pflanzenphys 56:153–165
Lichtenthaler HK, Rinderle U (1988a) Chlorophyll fluorescence as vitality indicator in forest decline research. In: Lichtenthaler HK (ed) Applications of chlorophyll fluorescence. Kluwer, Dordrecht, pp 143–149
Lichtenthaler HK, Rinderle U (1988b) The role of chlorophyll fluorescence in the detection of stress conditions in plants. CRC Crit Rev Anal Chem 19(Suppl 1):29–85
Lichtenthaler HK, Schindler C (1992) Studies on the photoprotective function of zeaxanthin at high-light conditions. In: Murata N (ed) Research in photosynthesis, vol IV. Kluwer, Dordrecht, pp 517–520
Lichtenthaler HK, Schweiger J (1998) Cell wall bound ferulic acid, the major substance of the blue-green fluorescence emission of plants. J Plant Physiol 152:272–282
Lichtenthaler HK, Sprey B (1966) Über die osmiophilen globulären Lipideinschlüsse der Chloroplasten. Z Naturforsch 21b:690–697
Lichtenthaler HK, Weinert H (1970) Die Beziehungen zwischen Lipochinonsynthese und Plastoglobulibildung in den Chloroplasten von *Ficus elastica* Roxb. Z Naturforsch 25b:619–623
Lichtenthaler HK, Buschmann C, Rahmsdorf U (1980) The Importance of blue light for the development of sun-type chloroplasts. In: Senger H (ed) The blue light syndrome. Springer, Berlin, pp 485–494
Lichtenthaler HK, Buschmann C, Döll M, Fietz H-J, Bach T, Kozel U, Meier D, Rahmsdorf U (1981a) Photosynthetic activity, chloroplast ultrastructure, and leaf characteristics of high-light and low-light plants and of sun and shade leaves. Photosynth Res 2:115–141
Lichtenthaler HK, Prenzel U, Douce R, Joyard J (1981b) Localization of prenylquinones in the envelope of spinach chloroplasts. Biochim Biophys Acta 641:99–105
Lichtenthaler HK, Kuhn G, Prenzel U, Meier D (1982a) Chlorophyll-protein levels and stacking degree of thylakoids in radish chloroplasts from high-light, low-light and bentazon-treated plants. Physiol Plant 56:183–188
Lichtenthaler HK, Prenzel U, Kuhn G (1982b) Carotenoid composition of chlorophyll-carotenoid-proteins from radish chloroplasts. Z Naturforsch 37c:10–12
Lichtenthaler HK, Meier D, Buschmann C (1984) Development of chloroplasts at high and low light quanta fluence rates. Isr J Bot 33:185–194
Lichtenthaler HK, Schmuck G, Döll M, Buschmann C (1985) Untersuchungen über die Funktionsfähigkeit des Photosyntheseapparates bei Nadeln gesunder und geschädigter Koniferen. PEF-Bericht KfK-PEF 2. Kernforschungszentrum Karlsruhe, pp 81–105
Lichtenthaler HK, Rinderle U, Haitz M (1989) Seasonal variations in photosynthetic activity of spruces as determined by chlorophyll fluorescence. Ann Sci Forest 46(suppl):483s–489s
Lichtenthaler HK, Schwender J, Seemann M, Rohmer M (1995) Carotenoid biosynthesis in green algae proceeds via a novel biosynthetic pathway. In: Mathis P (ed) Photosynthesis: from light to biosphere. Kluwer, Amsterdam, pp 115–118
Lichtenthaler HK, Lang M, Sowinska M, Heisel F, Miehé JA (1996) Detection of vegetation stress via a new high resolution fluorescence imaging system. J Plant Physiol 148:599–612
Lichtenthaler HK, Rohmer M, Schwender J (1997a) Two independent biochemical pathways for isopentenyl diphosphate (IPP) and isoprenoid biosynthesis in higher plants. Physiol Plant 101:643–652
Lichtenthaler HK, Schwender J, Disch A, Rohmer M (1997b) Biosynthesis of isoprenoids in higher plant chloroplasts proceeds via a mevalonate independent pathway. FEBS Lett 400:271–274

Lichtenthaler HK, Babani F, Langsdorf G, Buschmann C (2000a) Measurement of differences in red chlorophyll fluorescence and photosynthetic activity between sun and shade leaves by fluorescence imaging. Photosynthetica 38:521–529

Lichtenthaler HK, Zeidler J, Schwender J, Müller C (2000b) The non-mevalonate isoprenoid biosynthesis of plants as a test-system for new herbicides and drugs against pathogenic bacteria and the malaria parasite. Z Naturforsch 55c:305–313

Lichtenthaler HK, Buschmann C, Knapp M (2005a) How to correctly determine the different chlorophyll fluorescence parameters and the chlorophyll fluorescence decrease ratio R_{Fd} of leaves with the PAM fluorometer. Photosynthetica 43:379–393

Lichtenthaler HK, Langsdorf G, Lenk S, Buschmann C (2005b) Chlorophyll fluorescence imaging of photosynthetic activity with the flash-lamp fluorescence imaging system. Photosynthetica 43:355–369

Lichtenthaler HK, Ac A, Marek MV, Kalina J, Urban O (2007) Differences in pigment composition, photosynthetic rates and chlorophyll fluorescence images of sun and shade leaves of four tree species. Plant Physiol Biochem 45:577–588

Lichtenthaler HK, Buchanan BB, Douce R (2008) Honoring Andrew Benson in Paris: a tribut on his 90th birthday. Photosynth Res 96:181–183

Lichtenthaler HK, Langsdorf G, Buschmann C (2012) Multicolor fluorescence images and fluorescence ratio images of green apples at harvest and during storage. Isr J Plant Sci 60:97–106

Lichtenthaler HK, Babani F, Buschmann C (2013a) Chlorophyll fluorescence kinetics, photosynthetic activity and pigment composition of blue-shade and half-shade leaves as compared to sun and shade leaves of different trees. Photosynth Res 117:355–366

Lichtenthaler HK, Langsdorf G, Buschmann C (2013b) Uptake of diuron and concomitant loss of photosynthetic activity in leaves as visualized by imaging the red chlorophyll fluorescence. Photosynth Res 116:355–361

Mansfield TA, Whitmore ME, Law RM (1982) Effects of nitrogen oxides on plants: two case studies. In: Schneider T, Grant L (eds) Air pollution by nitrogen oxides. Elsevier, Amsterdam, pp 511–526

Menke W (1962) Structure and chemistry of plastids. Annu Rev Plant Physiol 13:27–44

Papageorgiou GC, Govindjee (eds) (2004) Chlorophyll a fluorescence: a signature of photosynthesis. Advances in photosynthesis and respiration, vol 19. Springer, Dordrecht

Park RB, Pon NG (1961) Correlation of structure with function in *Spinacia oleracea* chloroplasts. J Mol Biol 3:1–19

Pfister K, Buschmann C, Lichtenthaler HK (1974) Inhibition of the photosynthetic electron transport by bentazon. In: Proceedings of the 3rd international congress on photosynthesis, vol I. Elsevier Scientific Publisher, Amsterdam, pp 675–681

Rebeiz CA, Benning C, Bohnert HJ, Daniell H, Hoober JK, Lichtenthaler HK, Portis AR, Tripathy BC (eds) (2010) The chloroplast: basics and applications. Advances in photosynthesis and respiration, vol 31. Springer, Dordrecht

Rinderle U, Lichtenthaler HK (1988) The chlorophyll fluorescence ratio F690/F735 as a possible stress indicator. In: Lichtenthaler HK (ed) Applications of chlorophyll fluorescence. Kluwer, Dordrecht, pp 189–196

Rinderle U, Lichtenthaler HK (1989) The various chlorophyll fluorescence signatures as a basis for physiological ground truth control in remote sensing of forest decline. In: Internat. Geoscience and remote sensing symposium, IGARSS '89, vol. 2. Vancouver, pp 674–677

Rock BN, Hoshizaki T, Lichtenthaler HK, Schmuck G (1986) Comparison of *in situ* spectral measurements of forest decline symptoms in Vermont (U.S.A.) and the Schwarzwald (F.R.G.). In: Proceedings of the remote sensing symposium (IGARSS), Zürich 1986, vol. III. ESA Publications Division, Noordwijk, pp 1667–1672

Sarijeva G, Knapp M, Lichtenthaler HK (2007) Differences in photosynthetic activity, chlorophyll and carotenoid levels, and in chlorophyll fluorescence parameters in green sun and shade leaves of *Ginkgo* and *Fagus*. J Plant Physiol 164:950–955

Schindler S (1984) Verbreitung und Konzentration von Ubichinon-Homologen in Pflanzen. Karlsr Contrib Plant Physiol 12:1–240

Schindler S, Lichtenthaler HK (1984) Comparison of the ubiquinone homologue pattern in plant mitochondria and their possible prokaryotic precursors. In: Siegenthaler PA, Eichenberger D (eds) Structure, function and metabolism of plant lipids. Elsevier, Amsterdam, pp 273–276

Schindler C, Lichtenthaler HK (1996) Photosynthetic CO_2 assimilation, chlorophyll fluorescence and zeaxanthin accumulation in field-grown maple trees in the course of a sunny and a cloudy day. J Plant Physiol 148:399–412

Schindler C, Burkart S, Lichtenthaler HK (1992) High-light induced accumulation of zeaxanthin in aurea tobacco: a biphasic process. In: Cherif A, Miled-Daoud DB, Marzouk B, Zarrouk M (eds) Metabolism, structure and utilisation of plant lipids. Centre National Pedagogique, Tunis, pp 269–274. ISBN 9973-17-272-8

Schindler C, Reith P, Lichtenthaler HK (1994) Differential levels of carotenoids and decrease of zeaxanthin cycle performance during leaf development in a green and an aurea variety of tobacco. J Plant Physiol 143:500–507

Schmuck G, Lichtenthaler HK, Kritikos G, Amann V, Rock B (1987) Comparison of terrestrial and airborne reflection measurements of forest trees. In: Internat. Geoscience and remote sensing symposium, IGARSS '87, Michigan, vol II. The University of Michigan, Ann Arbor, MI, pp 1207–1212

Schwender J, Lichtenthaler HK, Seemann M, Rohmer M (1995) Biosynthesis of isoprenoid chains of chlorophylls and plastoquinone in *Scenedesmus* by a novel pathway. In: Mathis P (ed) Photosynthesis: from light to biosphere. Kluwer, Amsterdam, pp 1001–1004

Schwender J, Seemann M, Lichtenthaler HK, Rohmer M (1996) Biosynthesis of isoprenoids (carotenoids, sterols, prenyl side-chains of chlorophyll and plastoquinone) via a novel pyruvate/glycero-aldehyde-3-phosphate non-mevalonate pathway in the green alga *Scenedesmus*. Biochem J 316:73–80

Schwender J, Zeidler J, Gröner R, Müller C, Focke M, Braun S, Lichtenthaler FW (1997) Incorporation of 1-deoxy-D-xylulose into isoprene and phytol by higher plants and algae. FEBS Lett 414:129–134

Seybold A, Egle K (1937) Lichtfeld und Blattfarbstoffe I. Planta 26:491–515

Steele JW, Gurin S (1960) Biosynthesis of ß-carotene in *Euglena gracilis*. J Biol Chem 235:2778–2785

Stober F, Lichtenthaler HK (1992) Changes of the laser-induced blue, green and red fluorescence signatures during greening of etiolated leaves of wheat. J Plant Physiol 140:673–680

Stober F, Lichtenthaler HK (1993) Studies on the constancy of the blue and green fluorescence yield during the chlorophyll fluorescence induction kinetics (Kautsky effect). Radiat Environ Biophys 32:357–365

Stober F, Lang M, Lichtenthaler HK (1994) Studies on the blue, green and red fluorescence signatures of green, etiolated and white leaves. Remote Sens Environ 47:65–71

Thornber JP (1975) Chlorophyll-proteins: light-harvesting and reaction center components of plants. Annu Rev Plant Physiol 26:127–158

Wellburn A, Higginson C, Robinson D, Wamsley C (1981) Biochemical exüplanations of more than additive inhibitory effects of low atmospheric leavels of sulphur dioxide plus nitrogen dioxide upon plants. New Phytol 88:223–237

Zeidler JG, Lichtenthaler HK, May HU, Lichtenthaler FW (1997) Is isoprene emitted by plants synthesized via the novel isopentenylpyrophosphate pathway? Z Naturforsch 52c:15–23

Zimmer-Rinderle U, Lichtenthaler HK (1995) Photosyntheseaktivität und Chlorophyllfluoreszenzen geschädigter Fichten. In: Bittlingmeier L et al (eds) Waldschäden im Schwarzwald: Ergebnisse einer interdisziplinären Freilandstudie zur montanen Vergilbung am Standort Freudenstadt/Schöllkopf. Ecomed Verlagsgesellschaft, Landsberg, pp 256–287

Alfred Russel Wallace: Self-Educated Genius and Polymath

David Lloyd

The Childhood shows the man, as morning shows the day. Be famous then by wisdom; as thy empire must extend, So let extend thy mind o'er all the world.
John Milton: Paradise Regained (1671)

Contents

Abstract Alfred Russel Wallace was a Colossus: courageous, heroic, radical, modest, and above all, a man of insatiable curiosity. One hundred years on one can propose that his prescience anticipated many modern scientific developments and that despite relative neglect his far-ranging insight continues to inspire even now.

His earliest memories take us to Usk in South Wales, where he was born in 1823, and many experiences there are fondly recounted as formative influences. Adolescent interest in natural history during apprenticeship to his elder brother, a land-surveyor at the dawn of the railway era in Mid Wales and the Neath valley, blossomed into a lifelong fascination with the living world. The depth and reach of his thinking on the diversity and distribution of species outpaced his contemporaries, and he became the undisputed father of biogeography.

Interaction with the 'poor farmers' of South Wales and exposure to their humble conditions inculcated a concern for the deprivation of the underclasses, and were influential in the shaping of his societal concerns and later activism. After

D. Lloyd (✉)
Cardiff University School of Biosciences, Main Building, Museum Avenue, Cathays Park, Cardiff CF10 3AT, Wales, UK
e-mail: lloydd@cf.ac.uk

U. Lüttge, W. Beyschlag (eds.), *Progress in Botany*, Progress in Botany 76,
DOI 10.1007/978-3-319-08807-5_2

proposing the basic principles of speciation and of selection and arriving at a novel and original concept of evolutionary mechanisms, Wallace daringly pursued several non-scientific interests: phrenology, mesmerism, spiritualism, and the great question of whether we are alone in the cosmos.

Honoured late in a long life, Wallace became regarded as one of the greatest scientists in the world, despite his enthusiasms for supernatural phenomena. Eclipsed after his passing in 1913, a gradual realisation of the depth of his mainstream science as well as premature dismissal of some of his more arcane insights continues beyond his centenary year.

1 Early Years

Alfred Russel Wallace was the son of an unsuccessful lawyer, Thomas Wallace, who having difficulty in making ends meet had moved homes from London to the town of Usk, then as now a rural area in Monmouthshire, South Wales (Hughes 1989). The home where Alfred was born, the eighth of nine children, in 1823 (Fig. 1a) lies on the road leading along the river Usk to Llanbadoc: there in the churchyard two young sisters who died at 8 and 10 years (possibly of scarlet fever) are buried. Of the children born to Thomas and Mary Wallace, only five survived to adulthood. The eldest, William, the surveyor with whom Alfred was to work for 6 years, was to die of pneumonia aged 36, and Herbert, the youngest, died in Pará (Belem) in South America, after following Alfred to help with specimen collecting (Wilson 2000).

Five years after Alfred was born the family moved to Hertford, but his earliest memories of the house and river Usk are recounted in detail in his autobiography (Wallace 1862, 1905, 1910):

> To the time when I was just over three, besides myself, standing on the flat stones and catching lampreys.

His other strong impression was of the fishermen with their coracles:

> An ancient form of boat made of strong wicker-work, somewhat the shape of the deeper half of a cockle-shell, and covered with bullock's hide. Each coracle held one man, and it could be easily carried to and from the river on the owner's back ... this extremely interesting boat, which has been in use from pre-Roman and perhaps even from the Neolithic Age, should continue to be used on several of the Welsh rivers down to the present day. There is probably no other type of vessel now in existence which has remained unchanged for so long a period.

Wallace makes the interesting observation ... "*no doubt common to children of the same age*".... on his ... "*vague shadowy*" ... recollections of his father and mother, brothers, or sisters at that time. On the other hand, the main features and even the details of the outdoor and indoor surroundings then were still clear to

Fig. 1 (**a**) Kensington Cottage the Birthplace of Alfred Russel Wallace in Usk, Monmouthshire, South Wales, a photograph taken in around c. 1900 (Courtesy of A.R. Wallace Memorial Fund and G.W. Beccaloni). (**b**) The Mechanics' Institute in Neath, designed by Alfred and his brother, and still in use. (**c**) Llantwit Cottage, where Wallace and members of his family lived between 1846 and 1848

Wallace in his old age. His hazy memory of having a scalded arm by contact with boiling fat led him to suggest that:

> The sensation of pain does not, probably, reach its maximum till the whole organism is fully developed in the adult individual. This is rather a comforting conclusion in view of the sufferings of so many infants needlessly massacred through the terrible defects of our vicious social system.

In Hertford, Alfred became an ardent reader of books from the local library, and was also exposed to Church of England services and to the "*rare treats*" of the Quaker's chapels, but

> As however, there was no sufficient basis of intelligible fact or connected reasoning to satisfy my intellect, this feeling (of religious fervour) soon left me, and has never returned.

Leaving school at 13, Alfred was sent to London to assist his 19-year-old brother, John, in his apprenticeship as a carpenter. In the evenings they would go to a 'Hall of Science'—a kind of mechanics' institute especially notable for its members, advanced thinkers, and followers of Robert Owen, the Welsh secularist and founder of the English Socialist movement renowned for his munificent management of New Lanark over a period of 26 years. This eighteenth century cotton-mill town, now a World Heritage Site near Glasgow in Scotland, became a model for wise philanthropic and effective community administration and child education.

The "*horrible doctrine of eternal punishment as then commonly taught from thousands of pulpits by both the Church of England and Dissenters*" was thus to give way in Wallace's belief to religious scepticism.

In 1837, Alfred went with his brother, William, to Bedfordshire to begin his training as a land-surveyor:

> It was here too that during my solitary rambles I first began to feel the influence of nature and to wish I knew more of the various flowers, shrubs and trees, I daily met with, but of which for the most part I did not then know the English names. At that time, I hardly realised that there was such a science as systematic botany, that every flower and every meanest and most insignificant weed has been accurately described and classified, and that there was any kind of system or order in the endless variety of plants and animals which I knew existed.

2 Back to South Wales

In Llanbister and Llandrindod Wells in 1839 the Wallace brothers continued their land surveying of the Radnorshire countryside, and Alfred rails against the enclosure of common and wasteland by rich country squires and landowners "*land-robbery*", a lifelong concern, to be reiterated repeatedly (Wallace 1882, 1898, 1913a).

At Trallong, near Brecon, back in South Wales, the flat portions of summits of the Brecon Beacons with their still level sub-strata excited Wallace's attention,

> Here we are able, as it were, to catch Nature at work.

Further up the valleys of the Neath and Usk rivers he stayed at a "*little public-house*" where the people were all thoroughly Welsh, but the landlord of the inn, and a young man who lived with him, spoke English fairly well! More than a year in the Neath valley was "*on the whole very comfortable, although our first experience was a rather trying one*". He describes "*prompt and thorough measures using some ounces of corrosive sublimate* (mercuric chloride) *dissolved in a large pail full of water and liberally painted on the walls, ceiling, bedstead and furniture to eliminate all traces of* Cimex lectularius, *or bedbugs, which attacked us by the hundreds, and altogether banished sleep.*"

Wallace's sojourn in Neath was a time when he had little to do, and he made full use of this opportunity for "*self-education in science and literature*". It was here rather in Leicester that his interest in natural history emerged and developed (Hughes 1989, 1991, 1997; Raby 2001). He became especially fascinated with insects (Claridge 2008) and plant life (Kutschera and Hossfield 2013). His attention to the lines of demarcation of the Welsh language was to find an echo much later in his detailed analyses of word usage amongst the inhabitants of discrete regions in Amazonia and Malaya (Hughes 1997). At 17 or 18 he published his first paper "An essay on the best method of conducting the Kington Mechanics Institute" (Hughes 1989).

Construction of a telescope enabled him:

> to observe the moon and Jupiter's satellites and some of the larger star clusters; but, of course very imperfectly-and it also led me throughout my life to be deeply interested in the grand onward march of astronomical discovery.

But his chief preoccupation was with:

> the variety, the beauty, and the mystery of Nature as manifested in the vegetable kingdom.' I obtained a shilling paper-covered book published by the Society for the Diffusion of Useful Knowledge ... a revelation to me and for a year my constant companion ... Great was my delight when I found I could identify a Crucifer, an Umbellifer and a Labiate, and as one or another, the different orders were recognised. I began to realise for the first time the system that underlay all the variety of nature.

Lindley's "Elements of Botany", a rather mistaken early purchase, was not so useful as a borrowed volume of London's "Encyclopaedia of Plants" for identification of British species, and he annotated Lindley from it (Raby 2001). His brother's disapproval of the time-wasting pastime of construction of a herbarium, although not directly commented upon, came back to Alfred via a letter from their mother:

> Neither he nor I could foresee that it would have any effect on my future life, and I myself only looked upon it as an intensely interesting occupation for time that otherwise would be wasted....
>
> Now I have some reason to believe that this was the turning point of my life, the tide that carried me on, not to fortune but to whatever reputation I have acquired, and which to me has certainly been a never-ending source of much health of body and supreme mental enjoyment.

Wallace then writes about the humble existence that he and his brothers experienced. During the 7 years with his brother, Alfred

> hardly ever had more than a few shillings for personal expenses: but every year or two when I went home, what new clothes were absolutely necessary were provided for me, with perhaps ten shillings or a pound as pocket money till my next visit, and this, I think, was partly or wholly paid out of a small legacy left by my grand-father.... Had my father been a moderately rich man.... I might never have turned to nature.... I should never have even undertaken ... a journey to the almost unknown forests of the Amazon in ... to observe nature and make a living by collecting.

During his stay in the Neath area Wallace became curator of the Museum. Victorian enthusiasm at a time of general intellectual activity and excitement fired new interests on the part of working people who became highly enthusiastic to learn of progress in engineering, the sciences, and natural history. There Wallace also helped collect at first-hand a carefully selected choice of texts (Hughes 1989).

In 1843, Alfred's father died, his mother gave up the family house in Hoddeston, Hertfordshire, and his brother informed him that there was no more work in prospect in Neath; teaching at Leicester became his new occupation. However:

> no amount of teaching or practice would ever have made me a good musician, however much time and study I gave to the subject. I could never have become a good mathematician! ... still the ever-growing complexities of the higher mathematician had a kind of fascination for me as exhibiting powers of the human mind so above my own.

In the town subscription library he read Humboldt, Prescott, and most importantly Malthus's 'Principles of Population', his first introduction to "*philosophical biology*", that volume which he kept in his possession for more than two decades when it was to play a pivotal role in his formulation of his theory of evolution. Also, at Leicester in 1844, Wallace was introduced to the mysteries of mesmerism and the study of phrenology. It was a meeting with Henry Bates, a well-regarded young entomologist that extended Wallace's interests in the taxonomy of plants to that of butterflies and beetles; he was amazed to discover that more than a thousand species could probably be found within a radius of 10 miles of the town. Expeditions took them to Bradgate Park, Kenilworth Castle, and Derbyshire. The death of his brother, William, took him back to Neath, where in the absence of John, he discovered that "*the great railway mania*" had created levelling and surveying work for the Vale of Neath to Merthyr Tydfil line proposed to supplement haulage of coal and iron along the canals.

> Here we had to climb over huge rocks as big as houses, ascend cascades and take cross-levels up steep banks and precipices all densely wooded...

However, not one tenth of the lines proposed that year were ever made and the money wasted upon surveyors, engineers, and law expenses must have amounted to millions.

After brother John had rejoined Alfred, he was persuaded to build a small boat and the local canals made for many pleasant excursions. Collection of payment of tithes from local farmers was the least pleasant part of his occupation, as many were very poor, some could not speak English, didn't understand the system, and even refused to pay.

> This was one of the things that disgusted me with business, and it made me more than ever disposed to give it all up if I could get anything else to do.

That new activity for the brothers was the design and construction of the 'Mechanics Institute' at Neath (Fig. 1b), a sturdy stone building still in use as a museum office near the town centre. During a 2-year period whilst living at Llantwit cottage (Fig. 1c), and when working away from home, the brothers always chose to stay nights in farmhouses where the families spoke Welsh. Lecturing on basic sciences at the Institute, collection of insects, and bird watching with his brother further inspired his interests in the natural world, and of the "*beauties of the Vale of Neath scenery, the old Roman road (Sarn Helen), the caves and waterfalls, and the distant view over the valley to the faint haze of the Bristol Channel*".

3 Four Years in Amazonia

During Wallace's time at Neath he had maintained his correspondence with H.W. Bates and it was during their reunion at South Wales a proposal to go collecting in the tropics was first discussed. At this time in his letters to Bates, Wallace stated that he had read Chambers' (1844) 'Vestiges of the Natural History of Creation' and also mentioned works by Lyell, Darwin's 'Journal', and Humbold's 'Personal Narrative'. Local collecting had become rather inadequate:

> "I should like to take some one family, to study thoroughly principally with a view to the theory of the origin of species. "What decided our going to Pará and The Amazon rather than to any other part of the tropics was the publication in 1847 of 'A Voyage up the Amazon' by Mr. W.H. Edwards. This book was so clearly and brightly written, described so well the beauty and grandeur of tropical vegetation, and gave such a pleasing account of the people, while showing that the expenses of living and travelling were very moderate, that Bates and myself at once agreed that this was the very place to go if there was any chance of paying our expenses by the sale of our duplicate collections ... we set sail on April 28th, 1848." (Wallace 1853a, b).

Bates and Wallace were the only passengers on a 192 t barque, the *Mischief*, "*a very fast sailer*", and landed at Pará (now Belem) 29 days after leaving Liverpool. After 4 months, so as to earn more for their collected specimens sent back to London, they separated so as to get better coverage of the vast area. Bates looked at the Upper Amazon and Wallace went to the Upper Negro. This was after they had met Dr. Richard Spruce, the well-known botanist who had travelled to South America with Wallace's brother, Herbert. Barra (now Manaus) at the mouth of the River Negro and the surroundings turned out to be highly rewarding for botanical studies (Wallace 1853a), but disappointing in terms of insect varieties. Herbert, already having decided to return to England, caught yellow fever at Pará, and tragically died after a very few days. Alfred never fully got over this loss of his

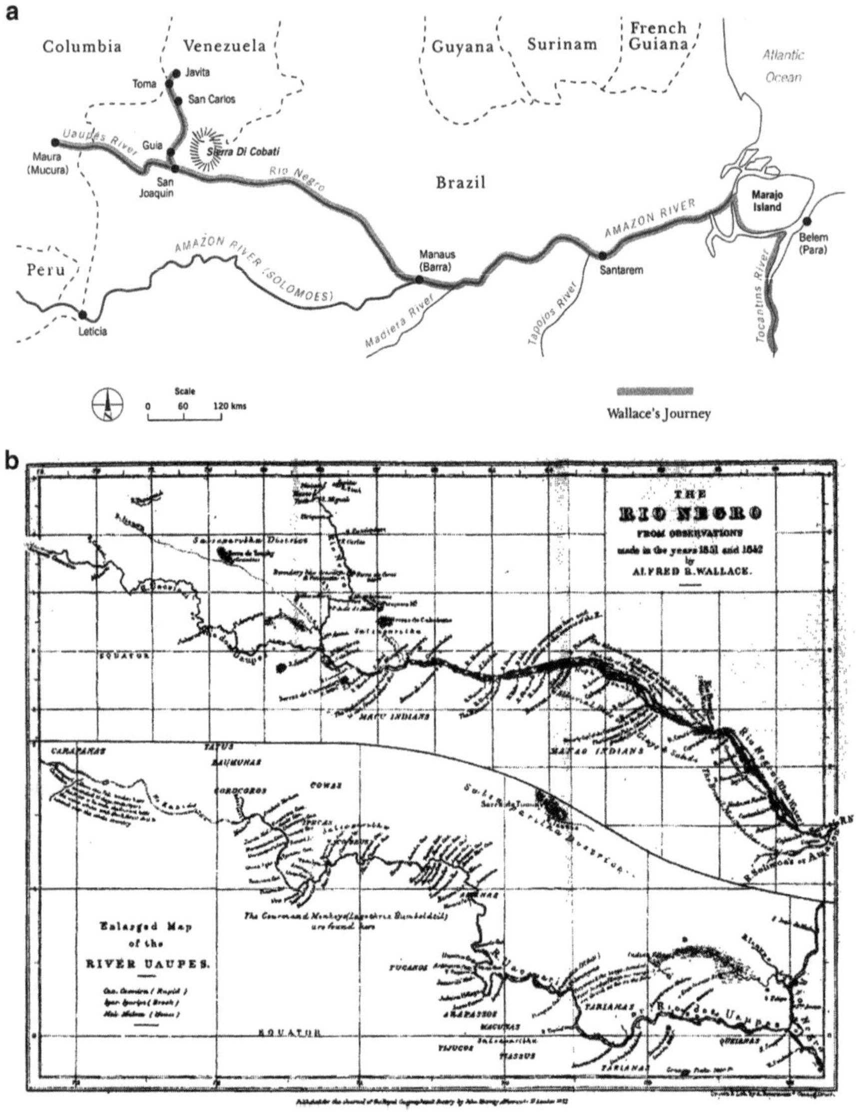

Fig. 2 (**a**) Wallace's journeys along the Amazon and Rio Negro (Courtesy of J.G. Wilson 2000). (**b**) From 1848 to 1852 Wallace charted unexplored territory along these rivers

younger brother and often in later life expressed his great sorrow at this misfortune so far away from home (Wilson 2000).

Two voyages along the Amazon and then the Rio Negro took Wallace beyond Brazil to Javita in Venezuela, and up the Uaupés to Maura (now Macura) in Columbia (Fig. 2a): he made the first detailed map of the area that included much hardly charted territory (Fig. 2b). An enthralling account of these adventures and of

Fig. 3 Palms from the Amazon The Kew palms on loan to National Museum of Wales from the Royal Botanical Gardens: these are some of the oldest surviving specimens collected by Wallace and Bates: (**a**) the Ghost palm, *Mauritiella armarta,* (**b**) the Açai palm *Eutrope oleracae*, and (**c**) the Macaw palm *Acrocomia aculeate*

their unique scientific contributions (Knapp 2013) also provides a fine collection of Wallace's exquisitely delicate pencil line drawings curated at the Natural History Museum at South Kensington in London.

A succession of fevers and dysentery left Wallace physically debilitated; he thought it might have been the dreaded fever that had claimed his brother, but it was probably malaria, only the doses of quinine he was taking were no longer effective. He became so weak that he became anxious for his own life and decided to leave the country. Although botanically extremely interesting, his recent excursions had in any case seemed rather unproductive as far as birds and insects were concerned.

Then another devastating calamity! His latest collections and live animals were destroyed on his ship *Helen* that went on fire Mid-Atlantic. Dried Palm specimens brought back from the Amazon still survive at The Royal Botanical Gardens at Kew (Fig. 3).

His voyage home was "*rather adventurous*": the dangers and discomforts before and after the fire had consumed *The Helen* can hardly be imagined. After 10 days adrift on an open lifeboat, they were picked up in *The Jordeson*, bound for London, and Wallace graphically described their hardship in a letter to Dr. Spruce:

> "we have been for some time on the poorest of fare ... beef or pork of the very worst quality, I have ever eaten or even imagined to exist. This repeated day after day without any variation" ...

The leaky old ship seldom progressed at more than five knots, and an average of two or three, bringing a cargo of heavy timber back from Cuba. Surviving a fearful gale in the English Channel which left four foot of water in the hold, after 80 days out from Pará, Wallace landed in Deal.

4 London and the East

The disastrous shipwreck did not leave Alfred completely destitute as his collection had been insured: presumably it was only this that allowed him to settle with his mother and sister (Fig. 4) in a house conveniently close to Regent's Park and Zoological Gardens. The very few remnants from his adventures in Amazonia that had escaped the shipwreck included a tin box in which he had saved some of his notebooks. One of these included careful pencilled drawings of all the different species of palms he had seen, together with notes on their distribution and uses. He visited scientific society meetings and got to know many London zoologists (including Thomas H. Huxley) and several entomologists; the British Museum (then in Great Russell Street) was close, and at Kew Herbarium he consulted botanical literature to identify his palms; and he also published a popular book on 'Palms of the Amazon and Rio Negro' (Wallace 1853a). During his stay in London, it became evident from Wallace's library and museum researches that it was the Malayan Archipelago that offered the most exciting and biologically unexplored region for investigation.

As the journey would be expensive, Wallace applied for a free passage in a Government ship, and he set out on a Peninsular and Orient steamer, *Bengal*. Visiting Malta, Alexandria (where he commented colourfully on the donkey-drivers, the Nile, Pyramids, and Cairo) and then by the horse-drawn buses across the desert to Suez. This route was a few years later superseded by railway and by the canal. After Aden, Galle, and Penang his ship traversed the Straits of Malacca to Singapore, where he remained for several months collecting birds and insects:

> "to begin eight years of wandering throughout the Malay Archipelago, which constituted the central and controlling incident of my life.". (Fig. 5.)

Fig. 4 Wallace in his late twenties or early thirties with his sister Frances and mother Mary Anne

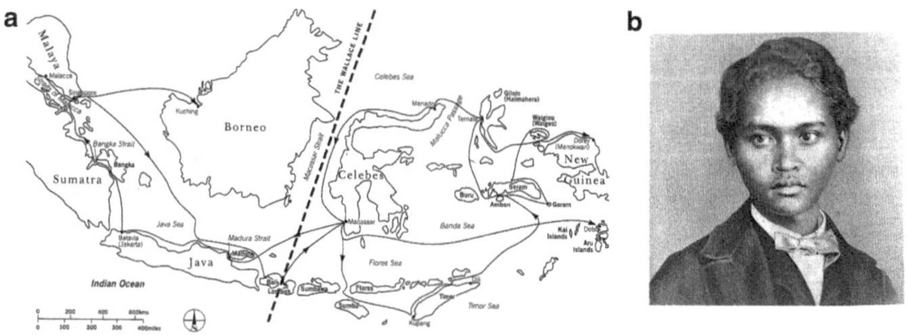

Fig. 5 (**a**) Wallace's voyages in Malaya (Courtesy of J.G. Wilson 2000). (**b**) Ali, Alfred's travelling companion and local expert

Placed immediately upon the equator and surrounded by extensive oceans, it is not surprising that the various islands of the Archipelago should be almost always clothed with a forest vegetation from the level of the sea to the loftiest mountains ... except for a few small and unimportant tracts, due perhaps, in some cases, to an ancient cultivation or accidental fires...

up at half-past five, bath and coffee. Sit down to arrange and put away my insects from the day before, we set them in a safe place to dry. Charles (a boy from London who wished to become a collector), mends our insect-nets, fills our pin-cushions and gets ready for the day. Breakfast at eight; out to the jungle at nine! ... To bed at eight or nine...

The banks of the Saráwak River are everywhere covered with fruit trees, which supply the Dyaks with a great deal of their food. The Mangosteen, Lansat, Rambutan, Jack, Jambou, and Blimbing, are all abundant; but most abundant and most esteemed is the Durian, a fruit about which is little known in England, but which both by natives and Europeans in the Malay Archipelago is reconded superior to all others. The old traveller in Linschott, writing in 1599, says:-"It is of such an excellent taste that it surpasses in flavour all other fruits of the world, according to those who have tasted it." And Doctor Paludanus adds:-"This fruit is of a hot and humid nature. To those not used to it, it seems at first to smell like rotten onions, but immediately they have tasted it they prefer it to all other food. The natives give it honourable titles, exalt it, and make verses on it." ... eating it out of doors, I at once became a confirmed Durian eater ... Poets and moralists, judging from our English trees and fruits, have thought that small fruits always grow on lofty trees, so that their fall should be harmless to man, while the large ones trailed on the ground. Two of the largest and heaviest fruits known however, the Brazil-nut fruit (Bertholletia) and Durian grow on lofty forest trees, from which they fall as soon as they are ripe, and often wound or kill the native inhabitants. From this we may learn two things: first, not to draw general conclusions from a very partial view of nature; and secondly, that trees and fruits, no less than the varied productions of the animal kingdom, do not appear to be organized with exclusive reference to the use and convenience of man... The bamboo is one of the most wonderful and most beautiful productions of the tropics, and one of nature's most valuable gifts to uncivilised man... Pitcher-plants, forming the genus Nepenthes of Botanists, here reach their greatest development. Every mountain-top abounds with them, running along the ground, or climbing over shrubs and stunted trees; their elegant pitchers hanging in every direction. Some of these are long and slender, resembling in form the beautiful Philippine lace-sponge (Euplectella), which has now become so common; others are broad and short. Their colours are green, variously tinted and mottled with red or purple. The finest yet known were obtained on the summit of Kini-balou in North-west Borneo. One of the broad sort, Nepenthes rajah, will hold two quarts of water in its pitcher. Another, Nepenthes Edwardsiania, has a narrow pitcher twenty inches long; while the plant itself grow to a length of twenty feet.

In Sarawak, Wallace wrote his first contribution to the question of the origin of species (Wallace 1855). This contained a concept that became known as the 'Sarawak Law', a milestone publication in the history of evolutionary biology which stated that:

Every species has come into existence coincidently both in space and time with a pre-existing closely-allied species, and thus suggested some type of evolutionary change by a mechanism yet to be discovered.

Wallace's agent back in London, Mr Stevens, who was responsible for selling specimens wrote to express regret that several naturalists having read this paper were concerned that valuable collecting time was being wasted on "*theorising*."

> After this I had in a letter to Darwin expressed surprise that no notice appeared to have been taken of my paper, to which he replied that both Sir Charles Lyell and Mr Edward Blyth, two very good men called his attention to it.

Much later (F. Darwin 1888) it came to light that Huxley too had commented on how little impression the Sarawak paper had made generally (Davies 2008).

Writing to Bates, dated January 4, 1858, Wallace mentions that:

> I have been much gratified by a letter from Darwin, in which he says that he agreed with "almost every word" of my paper. He is now preparing his great work on 'Species and Varieties' for which he has been collecting materials for twenty years. He may save me the trouble of writing more on my hypothesis by proving that the there is no difference between the origin of species and of varieties ... your collections and my own will furnish most valuable material to illustrate and prove the universal application of the hypothesis. The connection between the succession of affinities and the geographical distribution of any group, worked out species by species has never yet been shown as we shall be able to show it.
>
> In this archipelago there are two distinct faunas rigidly circumscribed, which differ as much as do those of Africa and South America, and more than those of Europe and North America, yet there is nothing on the map or on the face of the islands to mark their limits. The boundary line passes between islands closer together than others belonging to the same group. I believe the western part to be a separated portion of continental Asia, while the eastern is a fragmentary prolongation of a former west Pacific continent. In mammalia and birds the distinction is marked by genera, families, and even orders confined to one region; in insects by a number of genera, and little groups of peculiar species, the families of insects having generally a very wide or universal distribution.

Much later in life, Wallace (1905) reminisced that:

> This letter proves that at this time I had not the least idea of the nature of Darwin's proposed work, nor of the definite conclusions he had arrived at, nor had I myself any expectation of a complete solution of the great problem to which my paper was merely the prelude. Yet less than 2 months later that solution flashed upon me and to a large extent marked out a different line of work from that which I had up to this time anticipated.

The processes involved in the gradual change by which one species gave rise to another remained unknown. After arrival on Ternate January 8, 1858, Wallace decided to go to the nearby island of Gilolo (today Helmahera) and while suffering from an intermittent fever, he became too ill to go collecting, but this intermission provided a break and led to a moment of deep inspiration. Malthus's 'Principles of Population' was re-collected from his reading of about 12 years previously:

> I thought of his clear exposition of the positive checks to increase; 'disease, accident, war and famine' – which kept down the population of savage races to a much lower average than most civilized peoples.

Extrapolation to the even more rapid turnover within animal populations led to the question

> *why do some die, and some live? And the answer was clearly, that on the whole the best fitted live"* ... Then it suddenly flashed upon me that this self-activity process would necessarily improve the race, because in every generation the inferior would be *killed* off and the superior would remain – that is, *the fittest would survive*... and as great changes in the environment are always slow, there would be ample time for the change to be affected

by the survival of the best fitted in every generation ... each part of an animal's organisation could be modified exactly as required, and in the very process of this modification the unmodified would die out and thus the definite characteristics and the clear isolation of each new species would be explained. The more I thought over it the more I became convinced that I had at length found the long-sought-for law of nature that solved the problem of the origin of species. For the next hour I thought over the deficiencies in the theories of Lamarck and of the author of the "Vestiges" and I saw that my new theory supplemented these views and obviated every important difficulty. I waited anxiously for the termination of my fit, so that I might at once make notes for a paper on the subject. The same evening I did this pretty fully, and on the two succeeding evenings wrote it out carefully in order to send it to Darwin by the next post, which would leave in a day or two.

I asked him, if he thought it sufficiently important, to show it to Sir Charles Lyell, who had thought so highly of my former paper.

"The effect of my paper upon Darwin was at first almost paralyzing. He had, as I afterwards learnt, hit upon the same idea as my own twenty years earlier, and had occupied himself in all that long period in study and experiment, and sketching out and partly writing a great work"...

Back in London, three men had become frenetically concerned!

Sir Charles Lyell, the foremost geologist, and the distinguished botanist, Sir Joseph Dalton Hooker, Director of the Royal Botanic Gardens at Kew, were Darwin's only confidants for all this time, and the former had frequently suggested that the work be published, even if only in outline lest it be forestalled. When Darwin received Wallace's letter he wrote to Lyell

Your words have come true with a vengeance – that I should be forestalled. I never saw a more striking coincidence...so all my originality whatever it may amount to, will be smashed, though my book if it will have any value, will not be deteriorated, as all the labour consists in the application of the theory...I would far rather burn my whole book than that he (Wallace) or any other man should think I have behaved in paltry spirit.

He therefore left the matter in the hands of his two friends,

and they determined (on their own responsibilities that my essay, together with extracts from Darwins MSS., which they had seen many years before, should be read before the Linnean Society and published in its 'Journal.'

This hasty arrangement by Lyell and Hooker was scheduled for July, 1st 1858; it was conducted in the absence of Darwin (who was arranging for the funeral of his baby son, Charles, who had died of scarlet fever on 28 June). The meeting went ahead without any attempt to request permission from or even inform Wallace (still in Malaysia) of the event.

Long afterwards, Hooker's account of the meeting (Fig. 6) tells that:

the interest was intense, but the subject was too novel and too ominous for the old school to enter into lists before armouring. After the meeting it was talked over with bated breath: Lyell's approval, and perhaps in a small way, mine, as his lieutenant in the affair, rather overawed the Fellows, who would otherwise have flown out against the doctrine.

Wallace reflected long afterwards:

Both Darwin and Dr Hooker wrote to me in the most kind and courteous manner, informing me of what had been done, of which they hoped I would approve. Of course I not only approved, but felt that they had given me more honour and credit than I deserved, by putting

Fig. 6 (**a**) Sir Charles Lyell FRS, Professor of Geology, King's College, London. (**b**) Sir Joseph Hooker FRS, Director Royal Botanic Gardens, Kew. (**c**) A Linnean Society of London meeting room as it looks with the original furniture from the 1858 meeting. The actual room used for that historic meeting is elsewhere (courtesy of Michael Shermer)

> my sudden intuition – hastily written and immediately sent off for the opinion of Darwin and Lyell – on the same level with the prolonged labours of Darwin, who had reached the same point twenty years before me and had worked continuously during that long period, in order that he might be able to present the theory to the world with such a body of systematised facts and arguments as would almost compel conviction.

A letter from Hooker to Wallace informing him of what had happened at the Linnean Society meeting arrived months later with an enclosure from Darwin. It supplied Wallace with a list of the subjects he intended to write about in his book. However, and most significantly, the key item missing from Darwin's list was the subject of '*divergence*'. Nevertheless, and very importantly, it should be especially noted that only when Darwin's 'Origin of Species' book was published more than a year later, in November 1859, '*divergence*' is given its proper recognition as the driving force that Wallace had been exactly discussing and defining for some time as '*descent with modification*'.

When Darwin wrote to Wallace later that year he told him:

> I almost think that Lyell would have proved right and I should never have completed my large work.

It should be noted here that precedence for Darwin was actually contrived by his two friends by their ensuring that Darwin's abstracted works were read first at that meeting, that the proofs of the Linnean Journal were not sent to Wallace before publication, and that the Darwin–Wallace theory of evolution by natural selection rather soon afterwards became attributed solely to Darwin. Davies (2008, 2012, 2013) highlights the clear primacy of Wallace's understanding of the mechanisms of divergence, speciation, and selection. It should also be stressed that although the Darwin–Wallace Theory is mentioned five times in the First Edition of Darwin's Origin of Species (Darwin 1859a, b), "My theory" gets 57 entries (Elin Rhys, personal communication). There was only one belated specific reference to Wallace, on p. 484–5 in the concluding passage of the 2nd Edition (p. 754 in the Peckham edn.). There were many predecessors, and no specific references or acknowledgements were made to any of those who had contributed over the years to various aspects of the eventual synthesis (from Aristotle to Erasmus Darwin, Edward Blyth (Eiseley 1979, 2009), Robert Chambers, and Patrick Matthew (the first to write about 'natural means of selection') (Stott 2012).

During his 8 years and over 60 voyages from 1854 to 1862 in the Malayan Archipelago, Wallace (1869, 1890) had collected and later described thousands of specimens (e.g. 212 new bird species, i.e. 2 % of all the 10,000 bird species now known). He had travelled from the Indo-Malay islands (Singapore to Malacca, Borneo, Java, Sumatra) to the Timor group (Bali, Lombock, Timor), the Celebes group (Celebes, Macassar, Menado), the Moluccas (Banda, Amboyna, Gilolo, the Kaióa Islands, Bouru), and the Papuan group (Ké, Aru, New Guinea, Waigiou, and Ternate), altogether more than 14,000 miles.

On a mountain in Java, Wallace wrote in the 'Malay Archipelago':

> "It is between 2,000 and 5,000 feet that these forests and ravines exhibit the utmost development of tropical luxuriance and beauty. The abundance of noble Tree-ferns, sometimes fifty feet high, contributes greatly to the general effect, since of all the forms of tropical vegetation they are certainly the most striking and beautiful"...
>
> "The splendid foliage of the broad-leaved Musaceae and Zingiberaceae, with their curious and brilliant flowers, and the elegant and varied forms of plants allied to Begonia and Melastoma, continually attract the attention in this region. Filling up the spaces between the trees and larger plants on every trunk and stump and branch, are hosts of Orchids, Ferns and Lycopods which wave and hang and intertwine in ever-ending complexity. At about 5,000 feet I first saw horsetails (Equisetum) very like our own species. At 6,000 feet, Rasberries abound, and thence to the summit of the mountain there are three species of eatable Rubus. At 7,000 feet Cypresses appear, and forest trees become reduced in size, and more covered iwith mosses and lichens. From this point upward these rapidly

increase, soon the blocks of rock and scoria that form the mountain slope are completely hidden in a mossy vegetation. At about 8,000 feet European forms of plant become abundant. Several species of honeysuckle, St. John's wort, and Gelder-rose abound, and at about 9,000 feet we first meet with the rare and beautiful Royal Cowslip (Primula imperialis) which is said to be found nowhere else in the world but on this solitary mountain summit."...

These extensive expeditions with their intensive prodigious collecting work were all meticulously detailed with pencilled sketches and notes in careful handwriting in his field notebooks. His superb observational skills enabled Wallace to propose the boundary between the Oriental and Australian faunal regions, known ever since as the "Wallace Line" (Fig. 7a). Although not the first to address the question of the factors that determine the distribution of animals:

> "An important problem in Natural history, and one that hitherto has been to too little agitated, is that of ascertaining the most natural primary divisions of the earth's surface, taking the amount of similarity or dissimilarity of organised life solely as our guide."....
> (Sclater 1858),

and one of several who had suggested the demarcation of a Eurasian/Australasian boundary, it was however Wallace who exactly defined and redefined its location (Fig. 7b). His proposals still stand, although many investigations highlight differences between different species with respect to their ability to traverse natural barriers. Not until Wegner's hypothesis of continental drift of 1910 was superseded in 1959 by the momentous discovery of plate tectonics, could it be fully realised that the deep chasm separating Lombok and Bali has an enormously significant geophysical explanation. Wallace was later (1876) to lay the foundations of a whole field of investigation on the worldwide distribution of animals and to propose the major sub-continental divisions referred to as 'Wallaces's realms' (Fig. 8).

Wallace's description of the flora in these islands (Figs. 9 and 10) could hardly be bettered today: in 1861 in a letter to his sister, Fanny, he wrote:

> The highest peak is an extinct volcano with the crater nearly filled up forming merely a saucer on the top, in which is a good house built by the government for the old Dutch naturalist who surveyed & explored the mountain. There are a lot of strawberries planted there, wh.[ich] do very well but there were not many ripe. The common weeds & plants of the top were very like English ones such as buttercups, cow-thistle, plantain, wormwood, chickweed, charlock, St. John's wort, violets & many others, all closely allied to our common plants of those names but of distinct species. There was also a honeysuckle & a tall & very pretty kind of cowslip. None of these are found in the low tropical lands & most of them only on the tops of these high mountains. Mr. Darwin supposed them to have come there during a glacial or very cold period when they could have spread over the tropics & as the heat increased, gradually rose up the mountains. They were as you may [written vertically at the left hand side of the page] I also visited a semi-active volcano close by continually sending out steam with a noise like a blast furnace - quite enough to give me a conception of all other descriptions of volcanoes. Imagine most interesting to me, & I am very glad that I have ascended one lofty mountain in the tropics, though I had miserable wet weather & had no view, owing to constant clouds & mist.
>
> The lower parts of the mountains of Java from 3000 to 6000 ft. have the most beautiful tropical vegetation I have ever seen. Abundance of splendid tree ferns, some 50 feet high, & some hundreds of varieties of other ferns, - beautiful leaved plants as Begonias Melastomas

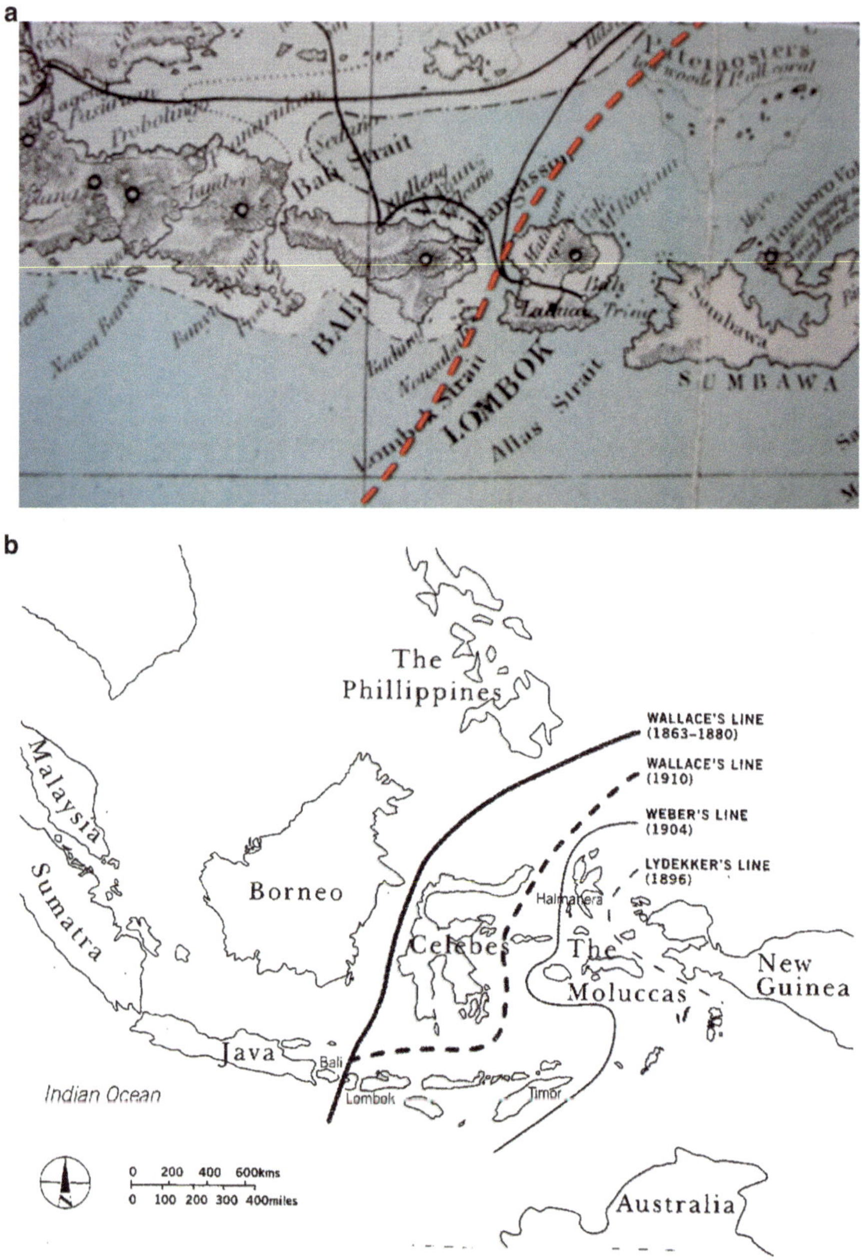

Fig. 7 (**a**) The Wallace Line (s). (**b**) Wallace defined and redefined the demarcation between the Eurasian and Australasian fauna, and was not the only one to be astonished that a 25 mile channel could mark such an enormous faunal divide. Continental drift and plate techtonics now provide the explanation

Fig. 8 Wallace's realms: this example is just one map of the great overall partitioning in faunal distributions over the whole world

& many others & more flowers than are generally seen in the tropics. In fact this region exhibits all the beauty the tropics can produce, but still I consider & will always maintain that our own meadows & woods & mountains are more beautiful. Our own weeds & wayside flowers are far prettier & more varied that those of the tropics. It is only the great

Fig. 9 A clearing in the Rain forest at Tenggarong in Borneo (painting by Wallace)

leaves & the curious looking plants & the deep gloom of the forests, & the mass of tangled vegetation, that astonishes & delights Europeans, & it is certainly grand & interesting & is a certain sense beautiful:- but not the calm sweet warm beauty of our own flowers, - a field of buttercups, a hill of gorse, or of heath, a bank of foxgloves & a hedge of wild roses & purple vetches surpass in beauty any thing I have ever seen in the tropics. - This is a fantastic subject with me but I can not go into it now.

Fig. 10 Some Australasian plants (**a**) *Hibiscus*, possibly *H. rosa-sinensis*. (**b**) *Medinilla myriantha*. (**c**) *Carludovica palmata*. (**d**) *Acalypha hispida*. (**e, f**) *Cycas seemannii*. (**g**) *Cymbidium bicolour*. (Images by Elin Rhys and Ffion Rees; assignments by Lara Jewittt, William Baker, Silke Roch. Courtesy of Dr. Rhian Smith and the Royal Botanic Gardens, Kew)

With the help of Ali (Fig. 5b), his local guide and expert in the ways of the islanders, he collected a huge total of specimens, many of which he sent to his agent, Mr. Stevens, for sale back in London, and the rest he had transported home with him. These totalled 125,660 items, and included 83,200 beetles, 8,050 birds, 13,100 lepidoptera, 13,400 other insects, 7,500 shells, 310 mammals (among those 7 complete hides of orang-utans), and 100 reptiles (Wallace 1869; Fichmann 2004). Of over 5,000 new species Wallace described, 200 bear his name.

In a letter (dated January 4, 1858) replying to one received from Bates the previous summer, Wallace wrote:

> To persons who have not thought much on the subject I fear my paper on the 'Succession of Species' will not appear so clear as it does to you. That paper is, of course, merely the announcement of the theory, not its development. I have prepared the plan and written portions of a work embracing the whole subject, and have endeavoured to prove in detail what I have as yet only indicated. I have been much gratified by a letter from Darwin, in which he says that he agrees with 'almost every word' of my paper. He is now preparing his great work on 'Species and Varieties', for which he has been collecting materials for twenty years. He may save me the trouble of writing more on my hypothesis, by proving that there is no difference in nature between the origin of species and of varieties; or he may give me trouble by arriving at another conclusion; but, at all events, his facts will be given for me to work upon. Your collections and my own will furnish most valuable material to illustrate and prove the universal applicability of the hypothesis. The connection between the succession of affinities and the geographical distribution of a group, worked out species by species, has never yet been shown as we shall be able to show it.

He also wrote of the work,

> I hoped to do myself in describing, cataloguing, and working out the distribution of my insects. I had in fact been bitten by the passion for species and their description and if neither Darwin nor myself had hit upon "Natural Selection," I might have spent the best years of my life in this comparatively profitless work. But the new ideas swept all this away. I have for the most part left others to describe my discoveries, and have devoted myself to the great generalizations which the laborious work of species-describers had rendered possible.

Wallace wrote again to Bates from Ternate on December 24, 1860,

> I know not how, or to whom, to express fully my admiration of Darwin's book. To him it would seem flattery, to others self-praise; but I do honestly believe that with however much patience I had worked and experimented on the subject, I could never have approached the completeness of his book, its vast accumulation of evidence, its overwhelming argument, and its admirable tone and spirit. I really feel thankful that it has not been left to me to give the theory to the world. Mr Darwin has created a new science and a new philosophy; and I believe that never has such a complete illustration of a new branch of human knowledge been due to the labours and researches of a single man. Never have such vast masses of widely scattered and hitherto quite unconnected facts been combined into a system and brought to bear upon the establishment of such a grand and new and simple philosophy.

Singapore, January 20, 1862.

> I cannot write more now. I do not know how long I shall be here; perhaps a month. Then, ho! For England!

Then in *'My Life'*, he writes in 1982,

> While waiting at Singapore for a steamer to take me home I purchased two living specimens of the smaller bird of paradise. They were in a large cage, and the price asked was enormous. As they had never been seen alive in Europe I at once secured them, and had a great deal of trouble with them on my journey home.
>
> Passing through France it was a sharp frost, but they did not seem to suffer; and when we reached London I was glad to transfer them into the care of Mr. Bartlett, who conveyed them to the Zoological Gardens.... Thus ended my Malayan travels.

5 Home Again at Last

Wallace, still the rather shy, polite, modest, and reclusive person having been brought up to show older people respect, and to speak only when spoken to, always maintained a friendly and deferential attitude to Darwin, 15 years his senior. Despite holding him somewhat in awe, Wallace's ideas eventually departed markedly and enhanced those of the older man. In correspondence between them, Wallace writes,

> was always cordial, sympathetic and broad minded .... In 1870 he had written to me:
>
> "I hope it is a satisfaction to you to reflect – and very few things in my life have ever been more satisfactory to me – that we have never felt any jealousy towards each other, though in some sense rivals."

And again Wallace adds:

> This friendly feeling was retained by him to the last, and to have thus inspired and retained it, not with standing our many differences of opinion, I feel to be one of the greatest honours of my life.

Four points of difference between their two final positions may be outlined:

1. Wallace believed that "*some agency other than natural selection, and analogous to that which first produced organic life, had brought into being (man's) moral and intellectual qualities*."
2. The bright colours or ornaments of males had not arisen by selection through female choice as Darwin had suggested, but by independent action of natural selection on each of the two sexes. The females being often more exposed to danger than the males (as in the case of sitting birds) had acquired more subdued coloration, whilst the males had remained bright and comparatively conspicuous.
3. From a study of oceanic islands Wallace concluded that mountain flora had been derived by transmission of seeds by birds or wind, rather than as a result of climate change.
4. Whereas Darwin always believed in the inheritance of acquired characteristics, Wallace was persuaded by the researches of Galton and Weismann that this was not the case.

> It will thus appear that none of my differences of opinion from Darwin imply any real divergence as to the overwhelming importance of the great principle of natural selection, while in several directions I believe that I have extended and strengthened it. The principle

> of "utility", which is one of its chief foundation-stones, I have always advocated unreservedly; while in extending this principle to almost every kind and degree of coloration, and in maintaining the power of natural selection to increase the infertility of hybrid unions, I have considerably extended its range. Hence it is that some of my critics declare that I am more Darwinian than Darwin himself, and in this, I admit, they are not far wrong.

His humble childhood, impecunious condition for most of his life, and his lack of tenured position, even in his mature years and despite a number of job applications, made him sympathetic to the uncertain lives of the vast majority in Victorian and Edwardian times in England. With his Civil List pension he could for the first time relax somewhat about his own financial state. He was still very active lecturing in the 1880s and 1890s throughout the country. In 1886/1887 he completed an arduous 6,000 mile 10-month lecture tour of the United States:

Mr. A.R. Wallace's lectures on Natural History

1. The Darwinian Theory Darwinism, what it is, and how it has been demonstrated. (Illustrated by Diagrams)
2. The Origin and Uses of the Colours of Animals (Illustrated by coloured Stereopticon views)
3. Mimicry and other exceptional modes of Animal Colouration... (Stereopticon pictures)
4. The Origin and Uses of the Colours of Plants. (Stereopticon)
5. The Permanence of the Oceans and the relations of Islands and Continents (Maps and diagrams)
6. Oceanic Islands and their Biological History. (Maps)
7. Continental Islands, their past history and Biological relations. (Maps)
8. The Physical and Biological relations of New Zealand and Australia. (Maps)

He continued publishing well into his 90th year (Wallace 1913a, b). In old age he enjoyed laying out and cultivating his garden (Figs. 11 and 12) and the joys of family life. Eventually tiring of London he moved several times through the home-counties, and finally to Broadstone in Dorset, where he passed away and is buried. He wrote more than 800 papers and 22 books. Not until 1893, when he was 70 was he proposed for Fellowship of the Royal Society, but seemed reluctant to accept:

> I have done so little of what is usually considered scientific work,

although, in 1868, he had already been honoured by its Royal Medal, and in 1890 its (first awarded) Darwin Medal. The Copley Medal followed in 1908, and in the same year the Linnean Society honoured him with its Darwin–Wallace medal. The greatest honour the Nation could bestow, the Order of Merit of the British Empire came also in 1908. In total he received more than 20 awards. Thus, although recognised as one of the most distinguished scientists in the world at the time of death in 1913, his huge achievements were soon forgotten, partly due to his subservient attitude and deference to Darwin throughout, and his reticence and modesty. Now, in his post-centenary year it is an appropriate occasion to reassess his scientific legacy.

Fig. 11 Wallace in his garden at Broadstone standing next to a fully blooming King's spear plant in 1905, the year his 2-volume My Life was published (Courtesy of Richard Milner, Gareth Nelson and Michael Shermer)

6 A Century of Progress

After 1880 Wallace wrote very little on science or natural history, except for retrospective reviews and reassessments of his earlier contributions (Wallace 1858, 1876, 1880) in the light of more recent discoveries (Wallace 1889, 1898). He lectured extensively at home and in the United States. The clarity, depth, breadth, and far-reaching power of his curiosity and intuition come through very clearly in these presentations and publications. The truly extraordinary scope of his experience in field collecting and the popularity of his works are reflected in the multiple editions of his major successes. His "Malay Archipelago" is the most published travel book ever, and has never been out of print. His prodigious output still holds many lessons for the modern world of sociology, natural history, and science. He is the widely acclaimed father of biogeography (Michaux 2008), the first neo-Darwinist (Kutschera and Niklas 2004; Kutschera and Briggs 2009) and an early anthropologist and ethno-geographer. Wallace can also be regarded as the father of astrobiology. Kutschera (2013) furthermore points out that "this unselfish

Fig. 12 Wallace in later life

man in the shadow of Darwin" (Kutschera and Hossfield 2013) was one of the first to abhor "the plunder of the earth" by the "struggle for wealth" (Wallace 1898) by uncaring humans.

He dared to question the assumptions and unfairness of the Victorian attitudes to privilege and class, and was completely against free-enterprise capitalism. He became increasingly concerned about the aggressively competitive, crude, and over-simplistic social interpretations of "Darwinism." His main interest was in land reform, as the laws of those times favoured only the rich landowners. His wide-ranging curiosity led him into activities and involvement with controversial topics. Some of these were clearly mistaken but totally understandable for someone with an inquiring mind and no formal education to speak of (anti-vaccination, phrenology, and spiritualism), but others (mesmerism and the possible existence of extra-terrestrial life) are even now still the subjects of great scientific endeavour. Hypnotherapy and cognitive psychotherapy are beneficial remedies for depressive symptoms and anxiety states, often as effective as medical treatment. What would Wallace have made of the current burgeoning technology that enables us to trace 'brainwaves' as encephalographic traces of neuronal electrical activity in 128 channels, of magnetic resonance imaging of the three dimensional ultrastructure of brain activity in vivo...or of deep magnetic stimulation techniques that benefit the brain or vagus nerve? These developments in the biomedical sciences advance in parallel with molecular studies on intracellular structure and function to provide novel treatments in the clinic.

Ecology has also been revolutionised by modern continuous non-invasive methods of investigation, especially by the availability of rapidly responding and specific sensors, and the ease of automated data collection. For instance, subsurface methane profiles and gas emissions from wetlands and soils are monitored directly by mass spectrometry probes (Lloyd et al. 2002; Beckmann et al. 2004), and photosynthetic activities measured with high spatial and temporal resolution. Performed at various levels in the tropical canopy, studies for over several months using arrays of small light sensors enabled the conclusion that actual photosynthesis of tropical epiphytes was determined by the specific and fluctuating light conditions of their microhabitat and cannot be ascribed to light-adapted ancestors (Rascher et al. 2012).

The ideas of Malthus as expanded by Wallace on the checks and balances in population biology presaged the mathematical theories that underlie oscillatory dynamics (Lotka 1924, 1956). The control of populations by feedback was likened by Wallace to the action of the 'governor' of a steam engine which swings in and out as it controls and steadies the rotational speed of the driving shaft. Novel mathematical and physical principles applied to machines, electronic circuits, as well as metabolic and signalling networks (Thellier and Lüttge 2013; Murray et al. 2013) are now widely employed in understanding the complexity and coherence of living cells and organisms (Lüttge and Beck 1992; Lloyd 2009) as well as the dynamics of populations. Realisation that biological complexity can be explained strictly in terms of physico-chemical mechanisms would really excite Wallace. The vast information storage inherent in the simplest organism and the mode of its transmission down the generations would enable him to understand many of the unsolved aspects of biology in his lifetime.

> We thus find that Darwinian theory, even when carried out to its extreme logical conclusion, not only does not oppose, but lends a decided support to, a belief in the spiritual nature of man.
>
> It shows us how man's body may have been developed from that of a lower animal form under the law of natural selection; but it also teaches us that we possess intellectual and moral facilities which could not have been so developed but must have another origin, and for this origin we can only find an adequate cause in the unseen universe of Spirit.

We now see that Wallace partially and intuitively realised (so much more clearly than Darwin) that:

> 'natural selection is not the all-powerful, all-sufficient and only cause of the development of organic forms'...

The progress of many aspects of evolutionary theory has been delayed as a consequence of the slow appreciation of the details of Wallaceism. The higher cognitive functions are not bestowed by a supra-vital force or principle. Even these functions will eventually become explicable in terms of energy fluxes through open systems so as to allow the emergence of collective properties that are more than the sum of their constituent parts (Lloyd 2009; Lüttge 2012). Oscillatory non-linear behaviour in multiple dimensions can give rise to concentration gradients, controllable chaos, symmetry breaking, and pattern formation in space and time. It is these agencies that explain the generation of increasing complexity in living systems, and it is these discoveries that illuminate the post-Wallace science.

Thus Wallace (2013) has pointed out that there are two major inconsistencies in the neo-Darwinian theory, i.e. that random chromosomal mutations acted upon by natural selection generate new species. Yet natural selection does not depend on increasing complexity, although complexity is a signature of life. Furthermore, human chromosomal DNA sequence variation is either neutral or deleterious, so is insufficient to introduce the variation adequate to explain either speciation or predilection to many common human diseases. It is the ovarian selection mechanism that removes deleterious mutations rapidly generated in the multiple copies of mitochondrial DNA, so that it is only maternally inherited subtle mutations in energy metabolism that are becoming continuously introduced into the species. Thereby this process allows adaptation to regional environmental differences and permits animals to inhabit harsher niches (faster intra-specific variation). There, the more rare nuclear mutations can accumulate, and give rise to slower processes of speciation.

The comparative genomics of microorganisms and viruses (Koonin and Wolf 2012) has revealed mechanisms that lie outside not only the neo-Darwinist refinements of the earlier half of the twentieth century (Dobzhanski 1937; Huxley 1942), but even those being uncovered daily in studies of human genetics. Mutational change is far from random, it often occurs by large-scale increments, natural selection is not the only filtering event, and lateral gene transfer is extremely common. The "Tree of Life" has to be replaced by a rhizome (Raoult 2010; Merhej and Raoult 2012).

From the molecular to the cosmic, and within a century, the bounds of our senses have been extended from the subatomic (large Hadron Collider) and submolecular (X-ray crystallography), to the limits of our Universe (X-ray astronomy), and the speed and facility for communication (world wide web).

7 Eclipse, Reassessment, and Overdue Reappraisal

The contributions of Wallace's genius to Darwin's understanding and written contributions have not only been considerably underestimated, but essentially ignored: McKinney (1966), Beddall (1968), Eiseley (1979, 2009), Brackman (1980), Ospovat (1981), Brooks (1984), Bulmer (2005), Poulton (1913), Marchant (1916), Darlington (1959), Shermer (2002), James (2008), Lloyd et al. (2010), and Kutschera and Nicklas (2013), as well as the many contributors to the fascinating edited volume by Smith and Beccaloni (2008), and that recently available from Glaubrecht (2013), have all contributed to redress this balance.

Most recently Roy Davies, a highly experienced and extremely well-regarded freelance investigative journalist, with a formidable portfolio of historical documentaries for the British Broadcasting Corporation, has summarised 18 years of painstaking work in this field. This research indicates the unfair treatment that has progressively relegated Wallace to second place (Davies 2008, 2012, 2013). Thus the Darwin–Wallace theory of natural selection became the Darwin theory, and shortly after Alfred's death his key role has all but become completely forgotten. In this centenary year of the death of Wallace, worldwide acclaim has been renewed, but hardly on the scale of the 200th anniversary of Darwin's birth and the 150th of the publication of Origin of Species in 2009 (Beccaloni and Smith 2008). As pointed out by Pattison (2009) Darwin's *'extraordinary ancestors, economic advantage*[1]*, a commitment to learning and a desire to serve others were all part of his inheritance and upbringing'*.

Despite opposition from Van Wyhe and Rookmaaker (2012), Davies (2012, 2013) continues to accumulate new details bearing on the precisely controlled and documented voyages between the Malayan Archipelago and the English docks, and the postal service between the English ports of arrival and Darwin's residence at Down House, in the village of Downe, in rural Kent. Details of the exact timing of the postal service with Wallace's Ternate letter are so essential in the debate about Darwin's possible plagiarism of Wallace. Further scrutiny and insightful literary searches indicate the entrenched errors of conventional opinions (Glaubrecht 2013). The inescapable conclusion is that if Wallace had not attempted to evoke the approval of his hero, but instead had submitted his theory directly to a journal as sole author, it would not be Darwin but Alfred Russel Wallace who would now be rightly celebrated for our understanding of divergence, speciation, and natural selection (Glaubrecht 2013).

> "The first and wisest of them all professed
> To know this only, that he nothing knew"
> John Milton: Paradise Regained (1671)

Acknowledgements David Lloyd wishes to thank Professor Dr. Ulrich Lüttge and Dr. John C Cushman for their interest and encouragement, Roy Davies and Dr. W Alfred Venables for useful criticism, Dr John G Wilson for permission to reproduce his maps, Elin Rhys and Ffion Rees for images, and Drusilla Lewis for secretarial help.

References

Beccaloni GW, Smith VS (2008) Celebrations for Darwin downplay Wallace's role. Nature 45:1050

Beckmann M, Sheppard SK, Lloyd D (2004) Mass spectrometric monitoring of gas dynamics in peat monoliths: effect of temperature and diurnal cycles on emissions. Atmos Environ 38:6907–6913

Beddall BG (1968) Wallace, Darwin and the theory of natural selection: a study in the development of ideas and attitudes. J Hist Biol 1:261–324

Brackman AC (1980) A delicate arrangement: the strange case of Charles Darwin and Alfred Russel Wallace. Times Books, New York

[1] In August 2010 the estate values were disclosed as Darwin £146,191 in 1882 (worth more than £6 million today), and Wallace £5,023 in 1913.

Brooks JL (1984) Just before the origin: Alfred Russel Wallace's theory of evolution. Columbia University, New York
Bulmer M (2005) The theory of natural selection of Alfred Russell Wallace FRS. Notes Rec R Soc (Lond) 59:126–136
Chambers R (1844) Vestiges of the natural history of creation. John Churchill, London
Claridge M (2008) Alfred Russel Wallace—a Welsh entomologist! In: Gardinier B, Milner R, Morris M (eds) The Linnean Society. Special Issue No 9: survival of the fittest. The Linnean Society, London, pp 17–24
Darlington CD (1959) Darwin's place in history. Blackwell, Oxford
Darwin C (1859a) On the origin of species. Murray, London
Darwin C (1859b) On the origin of species by means of natural selection, or the preservation of the favoured races in the struggle for life. John Murray, London
Darwin F (1888) The life and letters of Charles Darwin, 3 vols. John Murray, London
Davies R (2008) The Darwin conspiracy: origins of a scientific crime. Golden Square Books, London
Davies R (2012) How Charles Darwin received Wallace's Ternate paper 15 days earlier than he claimed: a comment on van Wyhe and Rookmaaker (2012). Biol J Linn Soc Lond 105:472–477
Davies R (2013) 1 July 1858: what Wallace knew; what Lyell thought he knew; what both he and Hooker took on trust; and what Charles Darwin never told them. Biol J Linn Soc Lond 109:725–736
Dobzhanski T (1937) Genetics and the origin of species. Columbia University Press, New York
Eiseley LC (1979) Darwin and the mysterious Mr. X: new light on the evolutionists. J M. Dent, London
Eiseley LC (2009) Darwin's century: evolution and the men who discovered it. Barnes and Noble, New York (reprint of the first edition of 1958)
Fichmann M (2004) An elusive victorian: the evolution of Alfred Russel Wallace. Chicago University, Chicago, IL
Glaubrecht M (2013) Am Ende de Archipels: Alfred Russel Wallace. Galian Verlag, Berlin
Hughes RE (1989) Alfred Russell Wallace: some notes on the Welsh connection. Br J Hist Sci 22:401–418
Hughes RE (1991) Alfred Russell Wallace (1823-1913): the making of a scientific non-conformist. Proc Roy Inst 63:175–183
Hughes RE (1997) Alfred Russel Wallace: Gwyddonydd Anwyddonol, Gwasg Prifysgol Cymru, Caerdydd CF10 4UP Wales
Huxley JS (1942) Evolution the modern synthesis. Allen and Unwin, London
James K (2008). http://blog.hmsbeagleproject.org/2008/07/guest-post-by-wallaces-rottweiler-on.html
Knapp S (2013) Alfred Russel Wallace in the Amazon: footsteps in the forest. The Natural History Museum, London
Koonin EV, Wolf YI (2012) Evolution in of microbes and viruses: a paradigm shift in evolutionary biology? Front Cell Infect Microbiol 2:119
Kutschera U (2013) The age of man: a father figure. Science 340:1287
Kutschera U, Briggs WR (2009) From Charles Darwin's botanical country-house studies to modern plant biology. Plant Biol (Stuttg) 11:785–795
Kutschera U, Hossfield U (2013) Alfred Russel Wallace (1823-1913): the forgotten co-founder of the new-Darwinian theory of biological evolution. Theory Biosci 132:207–214
Kutschera U, Niklas KJ (2004) The modern theory of biological evolution: an expanded synthesis. Naturwissenschaften 91:255–276
Kutschera U, Niklas KJ (2013) Metabolic scaling theory in plant biology and the three oxygen paradoxa of aerobic life. Theory Biosci 132:277–288
Lloyd D (2009) Oscillations, synchrony and deterministic chaos. Prog Bot 70:69–91

Lloyd D, Thomas KL, Cowie G, Tammam JD, Williams AG (2002) Direct interface of chemistry to microbiological systems using membrane inlet mass spectrometry. J Microbiol Methods 48:289–302

Lloyd D, Wimpenny J, Venables A (2010) Alfred Russel Wallace deserves better. J Biosci 35:339–349

Lotka AJ (1924, 1956) Elements of physical biology mathematical biology. Williams & Wilkins/ Dover, New York

Lüttge U (2012) Modularity and Emergence: biology's challenge in understanding life. Plant Biol (Stuttg) 14:865–871

Lüttge U, Beck F (1992) Endogenous rhythms and chaos in crassulacean acid metabolism. Planta 188:26–38

Marchant J (1916) Alfred Russel Wallace: letters and reminiscences, 2 vols. Cassell & Co, London

McKinney HL (1966) Alfred Russel Wallace and the discovery of natural selection. J Hist Med Allied Sci 21:333–357

Merhej V, Raoult D (2012) Rhizome of life, catastrophes, sequence exchanges, gene creations, and giant viruses: how microbial genomics challenges Darwin. Front Cell Infect Microbiol 2 (113):1–16

Michaux B (2008) Alfred Russel Wallace, Biogeographer. In: Smith CH, Beccaloni G (eds) Natural selection and beyond: the intellectual legacy of Alfred Russel Wallace. Oxford University Press, Oxford

Murray DB, Amariei C, Sasidharan K, Machné R, Aon MA, Lloyd D (2013) Temporal partitioning of the yeast cellular network. In: Aon MA, Saks V, Schlattner U (eds) Systems biology of metabolic and signaling networks: energy, mass and information transfer, 16th edn, Springer series in biophysics. Springer, Heidelberg, pp 323–350

Ospovat D (1981) The development of Darwin's theory: natural history, natural theology, and natural selection, 1838-1859. Cambridge University Press, Cambridge

Pattison A (2009) The Darwins of Shrewsbury. The History Press, Stroud

Poulton EB (1913) Obituary: Alfred Russel Wallace. Nature 92:347

Raby P (2001) Alfred Russel Wallace a life. Chatto & Windus, London

Raoult D (2010) The post-Darwinist rhizome of life. Lancet 375:104–105

Rascher U, Freiberg M, Lüttge U (2012) Functional diversity of photosynthetic light use of 16 vascular epiphyte species under fluctuating light irradiance in the canopy of a giant *Virola michelii* (Myristicaceae) tree in The tropical lowland forest of French Guyana. Front Plant Sci (2012) 2:117. doi:10.3389/fpls.2011.00117.eCollection 2011

Sclater PL (1858) On the general geographical distribution of the members of the class Aves. J Proc Linn Soc Zool 2:130–137

Shermer M (2002) In Darwin's shadow: the Life and science of Alfred Russel Wallace, A biographical study on the psychology of history. Oxford University Press, Oxford

Smith CH, Beccaloni G (2008) Natural selection and beyond: the intellectual legacy of Alfred Russel Wallace. Oxford University Press, Oxford

Stott R (2012) Darwin's ghosts: in search of the first evolutionists. Bloomsbury Publishing, London

Thellier M, Lüttge U (2013) Plant memory: a tentative model. Plant Biol (Stuttg) 15:1–12

Van Wyhe J, Rookmaaker K (2012) A new theory to explain the receipt of Wallace's Ternate essay by Darwin in 1858. Biol J Linn Soc 105:249–252

Wallace AR (1853a) Palm trees of the amazon and their uses. John Van Voorst, London

Wallace AR (1853b). A narrative of travels on the Amazon and Rio Negro, with an account of the native tribes, and observations on the climate, geology, and natural history of the amazon valley, 2nd edn. (1889) Reeve & Co, London

Wallace AR (1855) On the law that has regulated the introduction of a new species. Ann Mag Nat Hist 16:184–196

Wallace AR (1858) Note on the theory of permanent and geographical varieties. The Zool 16:5887–5888

Wallace AR (1862, 1910) My life: a record of event and opinions. Chapman & Hall, London
Wallace AR (1869). The Malay Archipelago; the land of the orang-utan and the bird of paradise; a narrative of travel with studies of man and nature. Vols. 1 and 2, 10th edn. (1891) Macmillan & Co, London
Wallace AR (1876) The geographical distribution of animals, with a study of the relations of living and extinct faunas as elucidating the past changes of the earth's surface, 2 vol. Macmillan, London
Wallace AR (1880) Island Life: or, the phenomena and causes of insular faunas and floras, including a revision and attempted solution of the problem of geological climates, 3rd edn. (1902) Macmillan & Co, London
Wallace AR (1882) Land nationalization; its necessity and its aims; being a comparison of the system of landlord and tenant with that of occupying ownership in their influence of the well-being of the people, 3rd edn. (1883) Trübner & Co, London
Wallace AR (1889) Darwinism; an exposition of the theory of natural selection with some of its applications, 3rd edn. (1905) Macmillan &Co, London
Wallace AR (1898) The wonderful century: its successes and failures, 7th edn (1908) Swan Sonnenschein, London
Wallace AR (1905) My life: a record of events and opinions, vol 1 and 2. Chapman & Hall, London
Wallace AR (1910) The world of life: a manifestation of creative power, directive, mind and ultimate purpose, 5th edn. (1911) Chapman and Hall, London
Wallace AR (1913a) Social environment and moral progress. Cassell and Company Ltd, London
Wallace AR (1913b) The revolt of democracy. Cassell and Company Ltd, London
Wallace DC (2013) Bioenergetics in human evolution and disease: implications for the origins of biological complexity and the missing genetic variation of common diseases. Philos R Trans Soc Lond B Biol Sci 368:20120267. doi:10.1098/rstb.2012.0267.Print 2013
Wilson JG (2000) The forgotten naturalist: in search of Alfred Russel Wallace. Australian Scholarly Publishing Pty Ltd, Kew, VIC

Part II
Physiology

The Role of Plasma Membrane H^+-ATPase in Salinity Stress of Plants

Małgorzata Janicka-Russak and Katarzyna Kabała

Contents

Abstract Plants have always been exposed to various stress factors in wild conditions. A high concentration of salt in the soil, i.e., salt stress, is one of the stressogenic stimuli. Salt stress is a complex abiotic stress in which both ionic and osmotic components are involved. Most plants adapted to salinity maintain a relatively low concentration of Na^+ in the cytosol achieved through the active exclusion of sodium ions in the apoplast and vacuole. Removal of sodium ions out of the cell, catalyzed by the specific plasma membrane Na^+/H^+ antiporter, depends on the electrochemical membrane proton gradient. The only pump which generates an electrochemical proton gradient across the plasma membrane is H^+-ATPase. For this reason, it is believed that plant plasma membrane H^+-ATPase (PM H^+-ATPase) plays a major role in salt stress tolerance.

M. Janicka-Russak (✉) • K. Kabała
Department of Plant Molecular Physiology, Institute of Experimental Biology, Wrocław University, ul. Kanonia 6/8, 50-328, Wrocław, Poland
e-mail: malgorzata.janicka@uni.wroc.pl; katarzyna.kabala@biol.uni.wroc.pl

U. Lüttge, W. Beyschlag (eds.), *Progress in Botany*, Progress in Botany 76,
DOI 10.1007/978-3-319-08807-5_3

1 Introduction

H^+-ATPase is a major enzyme protein of the plant plasma membrane. This protein belongs to a large superfamily of pumps termed P-type ATPases (Arango et al. 2003). By the use of the chemical energy of ATP hydrolysis, PM H^+-ATPase extrudes protons from cells of plants to generate electrochemical proton gradients. The enzyme is a functional single polypeptide chain with mass of about 100 kDa. H^+-ATPase has N- and C-terminal segments, which emerge into the cytoplasm (Duby and Boutry 2009). The C-terminal fragment of the protein acts as an autoinhibitory domain (Pedersen et al. 2007). The PM H^+-ATPase is kept at a low-activity level by its C-terminal domain, interacting with cytoplasmic domains essential for the catalytic cycle (Speth et al. 2010). The catalytic cycle of H^+-ATPase is described by two main conformational states, E_1 and E_2. A hallmark of P-ATPase is the formation of a phosphorylated intermediate during catalysis (Buch-Pedersen et al. 2009). In the E_1 conformation, the transmembrane binding site has high affinity for the proton and for ATP, whereas in E_2 the same site has low affinity for both ligands but high affinity for inorganic phosphate. The states E_1 and E_2 alternate during transport (Buch-Pedersen et al. 2009). The cytoplasmically positioned domains are in charge of ATP hydrolysis. Conformational changes in these domains during catalysis lead to simultaneous movements in the membrane-embedded part that directs the proton transport. In PM H^+-ATPases, a single proton is believed to be transported per hydrolyzed ATP (Palmgren 2001). However, partial uncoupling between ATP hydrolysis and proton transport has been suggested (Buch-Pedersen et al. 2006). The PM H^+-ATPase is stimulated by potassium (Palmgren 2001). K^+ is bound to the proton pump by Asp^{617} in cytoplasmic phosphorylation domain (P-domain). P-domain is located in the large loop, corresponding to the phosphorylation domain, and contains the aspartate phosphorylated during the catalytic cycle (Buch-Pedersen et al. 2009). Binding of K^+ promotes dephosphorylation of the phosphorylated E_1P reaction cycle and it controls the H^+/ATP coupling ratio. It was suggested that potassium acts as an intrinsic uncoupler of the plasma membrane H^+-ATPase (Buch-Pedersen et al. 2006).

The generation of a proton gradient by PM H^+-ATPase has a major role in providing the energy for secondary active transport across the plasma membrane. The PM H^+-ATPase is a proton pump which plays a central role in physiological functions such as nutrient uptake, intracellular pH regulation, stomatal opening, and cell growth (Serrano 1990). Besides regulation of physiological processes, the plasma membrane proton pump also plays a role in adaptation of plants to changing conditions, especially stress conditions. Thus, H^+-ATPase can be a common element for resistance mechanisms that are activated in various stress conditions. Many studies have shown the changes of gene expression of the plasma membrane H^+-ATPase in response to a variety of environmental factors. Moreover, besides the genetic regulation of the proton pump, its activity may undergo fast posttranslational modulation.

2 Regulation of Plant Plasma Membrane H^+-ATPase Activity

The molecular study of plant H^+-ATPase has shown that this enzyme is encoded by a multigene family. In *Arabidopsis thaliana* 12 genes have been identified (Palmgren 2001), in *Oryza sativa* 10 genes (Baxter et al. 2003), in *Nicotiana plumbaginifolia* 9 genes (Ouffatole et al. 2000), and in *Zea mays* 4 genes (Santi et al. 2003). The existence of multiple isoforms of the enzyme might indicate that some pumps could be redundant. Moreover, isoform diversity may also be related to cellular differentiation with individual isoforms exhibiting tissue- and development-specific expression (Palmgren 2001). The existence of multiple isoforms of the enzyme creates the possibility of their role in abiotic stress tolerance, particularly salt stress tolerance. Phylogenetic and gene structure analysis of plant H^+-ATPases divided them into five subfamilies (Arango et al. 2003). Expression of H^+-ATPase subfamilies I and II is not restricted to particular organs. These subfamilies are highly expressed in many cell types. Conversely, expression of genes belonging to subfamilies III, IV, and V is limited to specific organs or cell types (Arango et al. 2003). It has been shown that various genes are expressed in the same organ. Moreover, even within the same cell type at the same developmental stage, at least two H^+-ATPase genes are expressed (Hentzen et al. 1996; Moriau et al. 1999). In *N. plumbaginifolia* two different plasma membrane H^+-ATPase genes, *PMA2* and *PMA4*, are expressed in guard cells (Moriau et al. 1999). This observation suggests that isoforms with distinct kinetics, having slightly different biochemical and regulatory properties, might coexist in the same cell.

In addition to tissue-specific expression, the plasma membrane H^+-ATPases are differentially expressed according to environmental factors. Several studies have indicated that the H^+-ATPase genes might be activated by various abiotic and biotic stresses. With such a phenomenon the amount of H^+-ATPase might be increased under conditions requiring greater transport activity. The external signals which result in changes in plant plasma membrane H^+-ATPase gene expression include salt (Niu et al. 1993; Binzel 1995; Janicka-Russak and Kłobus 2007; Janicka-Russak et al. 2013), low temperature (Ahn et al. 1999, 2000; Janicka-Russak et al. 2012a, b), heavy metals (Janicka-Russak et al. 2008, 2012a, b), dehydration (Surowy and Boyer 1991), and mechanical stress (Ouffatole et al. 2000).

Phosphorylation and dephosphorylation of proteins is a very common example of posttranslational modification that has the potential to alter protein activity. The activity of the enzyme is well known to be regulated by 14-3-3 proteins, the association of which requires phosphorylation of the penultimate H^+-ATPase residues of Thr 947 in the C-terminus (Svennelid et al. 1999). In the low-activity state, the C-terminal tail interacts with the catalytic region of enzyme-limiting pump activity (Portillo 2000). The binding of 14-3-3 regulatory protein displaces the C-autoinhibitory domain, activating the enzyme. 14-3-3 binding to H^+-ATPase is stabilized by the fungal toxin fusicoccin, which decreases the dissociation rate. One 14-3-3 protein dimer binds two C-terminal polypeptides simultaneously, so a high activity state of H^+-ATPase could involve formation of dimers or multimeric complexes. An analysis with cryo-electron microscopy showed that the activated complex consists of six H^+-ATPase molecules and six 14-3-3 molecules (Kanczewska et al. 2005). The protein kinase responsible for this phosphorylation has not yet been identified. However, it was reported that calcium-dependent protein kinase (CDPK) leads to phosphorylation-dependent activation of H^+-ATPase (Yu et al. 2006).

Abiotic stresses lead to changes in the plasma membrane lipid composition altering the fluidity of the membrane. The modulation of the phospholipid environment of the plasma membrane regulates the activity of H^+-ATPase (Kasamo 2003). The activation of H^+-ATPase is dependent on the degree of saturation or unsaturation of the fatty acyl chain and its length. The activity decreased with an increase in the length of the fatty acyl chain and in the degree of unsaturation of fatty acid (Hernandez et al. 2002; Kasamo 2003; Martz et al. 2006). It was found that lysophosphatidylcholine binds to the C-terminal region of the protein and displacing the autoinhibitory domain leads to increase of ATPase activity (Pedechenko et al. 1990; Regenberg et al. 1995).

3 PM-H^+-ATPase in Salt Stress Conditions

Under salt stress, Na^+ is accumulated excessively in the cytoplasm, leading to inhibition of plant growth and development. Salinity tolerance of plants is a complex trait involving adaptation at the level of cells, organs, and the whole plant. The key factor of salinity tolerance, other than osmotic adjustment, is the control of intracellular ion homeostasis (Niu et al. 1993; Munns et al. 2006). To prevent accumulation of toxic Na^+ amounts in the cytosol, active sodium efflux into the apoplast and its compartmentalization inside the vacuole occur. Since Na^+ pumps responsible for sodium extrusion in animals and microorganisms are absent in higher plant cells, secondary sodium/proton antiporters in both the plasma membrane and the tonoplast are needed to translocate sodium ions against their electrochemical gradients (Apse and Blumwald 2007). Removal of sodium ions out of the cell, catalyzed by the specific plasma membrane Na^+/H^+, depends on the electrochemical membrane proton gradient. The only pump which generates an electrochemical proton gradient across the plasma membrane is H^+-ATPase (Palmgren 2001). The importance of the proton pump in plant adaptation to salinity was indicated by the observations carried out on salt-tolerant plants, showing increased activity of the plasma membrane proton pump in both normal and salt conditions (Niu et al. 1993; Vera-Estrella et al. 1994; Chen et al. 2007; Sahu and Shaw 2009). Moreover, the salt treatment of plants induces the activities of the plasma membrane proton pumps in both halophytes and glycophytes (Niu et al. 1993; Perez-Prat et al. 1994; Binzel 1995; Kłobus and Janicka-Russak 2004; Sahu and Shaw 2009; López-Pérez et al. 2009; Shen et al. 2011; Janicka-Russak et al. 2013). A direct role of H^+-ATPase in salt tolerance was confirmed by studies with transgenic tobacco, using a PMA4 mutant, lacking the autoinhibitory domain (ΔPMA4). In the mutant a constitutively activated PMA4 H^+-ATPase isoform was present. The ΔPMA4 plant roots showed better growth in saline conditions than those of untransformed plants (Géwaudant et al. 2007). The importance of PM H^+-ATPase in salinity tolerance is also evident from the Shen et al. (2011) study. The authors demonstrated that in rice phospholipase Dα (PLDα) is involved in salt tolerance by the mediation of H^+-ATPase activity and transcription. When rice suspension-cultured cells were treated with 100 mM NaCl, PLDα activity increased. The knockdown of OsPLDα1 prevented a NaCl-induced increase in transcript levels of the PM H^+-ATPase gene (*OSA2*) as well as ATPase activity.

On the other hand, there are few reports about the inhibition or no effect of NaCl on the PM H^+-ATPase in leaves (Chelysheva et al. 2001; Zörb et al. 2005; Pitan et al. 2009; Wakeel et al. 2010). The authors observed an increase of apoplastic pH in leaf cells during the first phase of salt stress, thus limiting leaf growth. Pitan et al. (2009) found that a salt-sensitive genotype of maize showed a notable decrease in plasma membrane proton pump activity under salinity. Moreover, salt-resistant genotype showed unchanged PM-H^+-ATPase activity under salt stress.

The salt-dependent activation of the plasma membrane proton pump encompasses the transcriptional as well as posttranslational level. The existence of multiple isoforms of the enzyme creates the opportunity for their role in abiotic stress tolerance. NaCl stress induces the expression of PM H^+-ATPase (Niu et al. 1993; Janicka-Russak and Kłobus 2007; Janicka-Russak et al. 2013). Accumulation of mRNAs of H^+-ATPase under salinity and the positive correlation with salt tolerance are well documented (Niu et al. 1993; Perez-Prat et al. 1994; Janicka-Russak and Kłobus 2007; Sahu and Shaw 2009). In *Oryza sativa* salt-tolerant cultivar accumulation of PM H^+-ATPase gene transcript was greater than that in a non-tolerant cultivar of rice treated with NaCl (Sahu and Shaw 2009). The significant role of this plasma membrane protein in salt stress tolerance confirms the emergence of a new isoform of PM H^+-ATPase, not detectable in plants not treated with NaCl. In rice a new isoform of the enzyme (finding maximum homology with *OSA7*) in response to salt treatment was discovered (Sahu and Shaw 2009). Similarly, in cucumber seedling under salt stress there appears the transcript of a new PM H^+-ATPase gene isoform, *CsHA1*. Accumulation of the *CsHA1* transcript is induced by NaCl exposure, and it is not expressed at detectable levels in plants not treated with NaCl (Janicka-Russak et al. 2013). Quite similar observations were made by Kalampanayil and Wimmers (2001), who demonstrated accumulation of the LHA8 transcript induced by NaCl exposure. Appearance of new isoforms of the PM H^+-ATPase transcript in plants, in addition to the increase in enzyme activity, indicates the important role of the enzyme in maintaining ion homeostasis in plants under salt stress. That, consequently, allows the plant to survive stress conditions.

Recently evidence has been presented that NaCl also causes rapid modulation of proton pumps, which is due to the reversible phosphorylation of enzyme proteins (Kerkeb et al. 2002; Kłobus and Janicka-Russak 2004; Janicka-Russak et al. 2013).

In many plant tissues a salt-inducible shift in the cytoplasmic calcium level was observed (Danielsson et al. 1996; Knight et al. 1997; Blumwald et al. 2000; Netting 2000; Xiong et al. 2002), suggesting its involvement in the signaling pathway under NaCl stress conditions. Transient increases in cytosolic Ca^{2+} can induce the phosphorylation of different proteins in cells, improving the salt tolerance (Hasegawa et al. 2000). Evidence suggests that the major role in coupling the calcium signal to specific protein phosphorylation cascade(s) is played by the Ca^{2+}/calmodulin-dependent protein kinases (CDPKs) and the SOS3 family (salt overly sensitive 3) of Ca^{2+} sensors (Xiong et al. 2002; Zhu 2002). The results of Urao et al. (1994) and Saijo et al. (2000) demonstrated that NaCl rapidly induced CDPK in different plant tissues. Furthermore, evidence has been presented that Ca^{2+}/calmodulin-dependent protein kinases are responsible for the phosphorylation of the plasma membrane H^+-ATPase protein (Van der Hoeven et al. 1996; Camoni et al. 2000). Moreover, it was shown that plasma membrane ATPase phosphorylation, as a target of activation by NaCl, a calcium/calmodulin-dependent protein kinase sensitive to staurosporine, is involved (Kłobus and Janicka-Russak 2004). Further research confirmed that in cucumber seedlings under salt stress conditions fast posttranslational modifications take place. Western blot analysis with antibody against phosphothreonine and 14-3-3 proteins showed that under salt stress conditions the level of those elements increased (Janicka-Russak et al. 2013).

Ca^{2+} has been identified as a possible mediator of ABA-induced stimulus-response coupling (Netting 2000). Abscisic acid is known as a stress hormone, which mediates responses to a variety of stresses, including water and salt stress (Skriver and Mundy 1990; Tan et al. 1994; Jia et al. 2001). The endogenous level of ABA increases when plants are stressed with drought or NaCl, and application of ABA to unstressed plants results in the induction of numerous water-deficit-related activities (La Rosa et al. 1985, 1987; Skriver and Mundy 1990; Cowan et al. 1997). They include triggering of stomatal closure to reduce transpirational water loss by posttranslational modulation of ion channels in guard cells (Grabov and Blatt 1998), and alterations in gene expression through induction of ABA-responsive genes coding for structural, metabolic, or transport proteins (Ingram and Bartels 1996; Shinozaki and Yamaguchi-Shinozaki 1996; Barkla et al. 1999). It was reported that ABA treatment of plants, as well as NaCl treatment, increased activity of plasma membrane H^+-ATPases. ABA treatment of plants elevated the level of plasma membrane H^+-ATPase transcript (Janicka-Russak and Kłobus 2007). On the other hand in stomata guard cells the plasma membrane H^+-ATPase activity is diminished by ABA treatments, leading to stomatal closing. However the significance of this phenomenon in relationship to stomatal closure is still debated (Merlot et al. 2007).

Recently, some evidence has indicated that one mode of ABA action is related to oxidative stress in plant cells. ABA can cause increased generation of reactive oxygen species (ROS) (Pei et al. 2000; Zhang et al. 2001; Laloi et al. 2004). Moreover, H_2O_2 generated in response to ABA treatment was detected in the apoplast (Hu et al. 2005, 2006). Furthermore, ABA stimulates H_2O_2 production by plasma membrane NADPH oxidases (Hu et al. 2005). Genetic evidence indicates that increase in expression of NADPH oxidase genes is required for the production of ROS during ABA-induced stomatal closure (Kwak et al. 2003). Increased H_2O_2 accumulation under salt stress conditions was observed (Panda and Upadhyay 2003; Tsai et al. 2004; Hernandez et al. 2010; Janicka-Russak et al. 2013). The reactive oxygen species H_2O_2 is a harmful cellular metabolite, although it also serves as a signaling molecule that mediates responses of acclimation to abiotic stress in plant cells (Suzuki and Mittler 2006; Jubany-Mari et al. 2009). It was shown that H_2O_2 functions as a trigger for induction of many genes encoding enzymes involved in cellular protection under stress conditions (Volkov et al. 2006). Treatment of cucumber seedlings with H_2O_2 markedly elevated the level of PM H^+-ATPase genes (Zhang et al. 2007; Janicka-Russak et al. 2012a, b).

Fatty acids, as the main component of membrane lipids, are considered to be important in salt tolerance of plants, too. Non-tolerant plants subjected to salt stress commonly show decreased levels of 18:3 fatty acid in their membranes (Upchurch 2008). A study with transgenic tobacco showed that overexpression of ω-3 desaturases, which increases 18:3, elevated tolerance to salt stress (Zhang et al. 2005). Additionally, salt-tolerant plants showed an increase of unsaturated fatty acids (Lin and Wu 1996). In broccoli plants subjected to NaCl a high degree of unsaturation in the plasma membrane of roots was observed, as an adaptation mechanism to salinity (López-Pérez et al. 2009). Moreover, the activity of plant PM H^+-ATPase increased with an increase in the degree of unsaturation of fatty acid (Kasamo 2003; Martz et al. 2006). Thus the modulation of the phospholipid environment of the plasma membrane under salt stress regulates the activity of H^+-ATPase, leading to an increase in its activity.

PM H^+-ATPase can also be regulated by other factors. A novel interaction partner of the AHA1 PM H^+-ATPase, named PPI1 (proton pump interactor, isoform 1), was identified in *Arabidopsis thaliana* (Morandini et al. 2002). This protein stimulates the activity of the proton pump in vitro. PPI1 homolog in potato (StPPI1) stimulated H^+-ATPase activity and is induced by salt and cold (Muñiz García et al. 2011).

It has been shown that regulation of the activity of PM H^+-ATPase involves modulation of the H^+/ATP coupling ratio (Kerkeb et al. 2002). Venema and Palmgren (1995) suggested that intrinsic uncoupling of H^+/ATP is an important mechanism for regulation of pump activity. Additionally, it was found that the H^+/ATP coupling ratio in the plasma membrane of cucumber seedlings roots subjected to salt stress significantly increased (Janicka-Russak et al. 2013). Kerkeb et al. (2002) also observed an enhanced H^+/ATP coupling ratio of plasma membrane H^+-ATPase in tomato cells under osmotic stress. Moreover, the correlation between the increase in H^+ pumping activity and the posttranslational modification consisting of phosphorylation of the enzyme protein was demonstrated (Babakov et al. 2000; Kerkeb et al. 2002; Janicka-Russak et al. 2013). In salt-stressed cucumber seedlings, a weak effect of salinity on ATP hydrolytic activity of PM H^+-ATPase and significant stimulation of proton transport through the plasma membrane were observed (Janicka-Russak et al. 2013). It seems that enhancement of the H^+/ATP coupling ratio indicates a more efficient use of ATP hydrolysis to transport protons. This is very beneficial in terms of salinity to generate a proton gradient that can be used by plasma membrane Na^+/H^+ antiporter.

4 SOS Signaling Pathway

Sodium efflux from root cells prevents accumulation of toxic levels of Na^+ ions in the cytosol and their translocation to the shoot. In higher plants, the main mechanism for Na^+ removal from cells is powered by the operation of the plasma membrane H^+-ATPase generating an electrochemical proton gradient. This proton-motive force energizes the plasma membrane Na^+/H^+ antiporter, identified as SOS1 (Salt Overly Sensitive 1), that couples the downhill movement of H^+ into the cell along its electrochemical gradient to the Na^+ extrusion against its electrochemical gradient (Blumwald et al. 2000). From published data it is known that activity of Na^+/H^+ antiporter was observed in glycophytes only after salt stress induction, and was a part of the adaptive mechanism (Wilson and Shannon 1995; Qiu et al. 2003; Kabała and Janicka-Russak 2012). In addition it is known that in halophytes, the antiporter is constitutively expressed and is present in both NaCl-treated plants and plants growing under NaCl-free conditions (Blumwald et al. 2000).

The understanding of how Na^+ is sensed by plant cells is still very limited. It is believed that Na^+ can be sensed either before or after entering the cell, or both. Extracellular Na^+ may be perceived by a receptor or salt sensor present at the plasma membrane, whereas intracellular Na^+ may be sensed by membrane proteins or by Na-sensitive cytosolic enzymes (Conde et al. 2011). Na^+ sensor or Na^+-selective ion channels have not yet been identified in plants. Thus, it is probable that plants contain other proteins with regulatory Na^+ binding sites (Maathuis 2013). On the other hand, some general mechanisms have been described to sense the onset of salinity stress. These include reduction in water delivery and turgor changes, which may be recorded by HK1 kinases, and distortion of cell wall-membrane geometry, which can be relayed by mechanosensitive ion channels (Maathuis 2013).

In *Arabidopsis thaliana*, the SOS signaling pathway, including SOS1, SOS2, and SOS3, is an important regulatory system activated by salt stress to maintain ion homeostasis and salt tolerance. This pathway strictly depends on plasma membrane proton gradient as well as cytosolic calcium signals. The *SOS1* gene has been shown to encode a transmembrane protein with significant sequence similarity to plasma membrane Na^+/H^+ antiporters from bacteria and fungi (Shi et al. 2000). SOS1, with a molecular mass of about 127 kDa, possesses 12 transmembrane domains with a highly hydrophilic regulatory C-terminal region predicted to be cytosolic (Mahajan et al. 2008). Its transcript level is specifically upregulated by NaCl stress but not by drought or cold (Shi et al. 2000). Although the understanding of how Na^+ is sensed in plant cells is very limited, SOS1 antiporter seems to be a good candidate for a membrane protein with dual functions of sodium transport and sensing (Zhu 2003; Conde et al. 2011). In root epidermal cells, the SOS1 transporter is involved in sodium exclusion from the cytosol into the root medium. Moreover, *SOS1* gene is expressed preferentially in cells bordering the xylem vessels throughout the plant and strongly affects long-distance sodium transport. Thus, it was suggested that SOS1 functions in controlling Na^+ translocation between roots and leaves through loading Na^+ into and unloading Na^+ from the xylem (Shi et al. 2002). When expressed in a yeast mutant deficient in endogenous sodium transporters, *SOS1* is able to diminish Na^+ accumulation and improve salt tolerance of yeast cells (Shi et al. 2002). Under salinity, transgenic *Arabidopsis* plants overexpressing *SOS1* accumulate fewer Na^+ ions in tissues, demonstrating that improved salt tolerance could be achieved by limiting sodium content in plant cells (Shi et al. 2003). Reduction in SOS1 expression in *Thellungiella salsuginea* plants adapted to high salinity causes a loss of their halophytic characteristics, indicating that *SOS1* acts as a major salt tolerance determinant (Oh et al. 2009). Analysis of *Arabidopsis sos1-1* mutant plants revealed that SOS1 protein, in addition to its function as a Na^+/H^+ antiporter, mediates vacuolar integrity, membrane trafficking, and pH homeostasis under salt stress (Oh et al. 2010). SOS1 is conserved in higher plants, both monocots and dicots, including rice, tomato, and poplar (Ji et al. 2013).

Sodium transport activity of SOS1 is regulated by other identified components of the SOS pathway, SOS2 and SOS3 (Qiu et al. 2002). It is well known that high salinity (e.g., Na^+) initiates a calcium signal that activates SOS pathway. According to Tuteja (2007), the salt stress signal first activates phospholipase C, which hydrolyzes phosphatidylinositol bisphosphate to generate inositol triphosphate and diacylglycerol resulting in an increased level of Ca^{2+} ions. This change in cytosolic calcium ions is sensed by calcium sensor such as SOS3, a myristoylated Ca^{2+}-binding protein. Some of data suggest the existence of Na^+-dependent Ca^{2+} transport activity in vacuolar membranes (Wang et al. 1994). Recently, *AtNCL* gene encoding a protein similar to animal NCXs has been identified in *Arabidopsis*. It had Na^+/Ca^{2+} exchange activity and was involved in calcium homeostasis under salt stress (Wang et al. 2012). Upon binding with Ca^{2+}, SOS3 changes its conformation and physically interacts with the serine/threonine protein kinase SOS2, which belongs to the SnRK3 family of protein kinases, activating its substrate phosphorylation. The SOS3–SOS2 complex then phosphorylates and activates antiporter SOS1, causing subsequent extrusion of sodium excess from the cytosol (Chinnusamy et al. 2005; Ji et al. 2013). It was shown that SOS1 is maintained in a resting state by an autoinhibitory domain at the C-terminus. The antiporter is relieved from autoinhibition upon phosphorylation of its autoinhibitory region by the SOS2–SOS3 complex (Quintero et al. 2011). Quan et al. (2007) found that an SOS3 homolog, SCaBP8 (SOS3-like Calcium-Binding Protein 8), is an alternative regulator of SOS2 in the shoots of *Arabidopsis*, whereas SOS3 functions mainly in roots.

SOS2 kinase and SOS1 antiporter have emerged as important mediators of salt stress response and signaling through interactions with other regulatory proteins. Additional SOS2-interacting molecules are protein phosphatase ABI2 (Ohta et al. 2003) and H_2O_2 signaling proteins: nucleoside diphosphate kinase NDPK2 and catalases (Verslues et al. 2007) connecting salt stress with ABA signaling and oxidative stress, respectively. Under salt stress, both SOS1 and SOS2 are believed to be involved in pathways controlling ROS production and/or detoxification (Chung et al. 2008; Zhu et al. 2007). Moreover, a long cytoplasmic tail of SOS1 interacts with RCD1 protein, a regulator of oxidative stress responses (Katiyar-Agarwal et al. 2006). Further results have shown that SOS2 plays a critical role in controlling the activities of NHXs, plant vacuolar Na^+/H^+ exchangers, providing evidence that there can be coordination of the activities of the sodium antiporters in the vacuolar and plasma membranes (Qiu et al. 2004; Huertas et al. 2012). Similarly, SOS2 was shown to activate the vacuolar Ca^{2+}/H^+ antiporter CAX1 integrating calcium transport and salt tolerance (Cheng et al. 2004). Batelli et al. (2007) demonstrated that under salinity SOS2 interacts directly with V-ATPase regulatory subunits B1 and B2, stimulating its proton pumping activity. Such results suggest that regulation of V-ATPase is an additional key function of SOS2 in promotion of salt tolerance.

Extensive studies of *SOS* genes and their biochemical functions indicated that the SOS3 calcium sensor also plays an important role in salt tolerance. Plants lacking *SOS3* are hypersensitive to salt stress, but this sensitivity could be partially reversed by the addition of calcium ions. Salt hypersensitivity of the *Arabidopsis sos3* mutant results, at least in part, from disordered actin filaments. Thus, it was suggested that SOS3 links calcium and microfilament dynamics in the salt stress response (Ye et al. 2013). Moreover, phenotypic analysis of the *sos3-1* mutant provided evidence that the SOS signaling pathway is involved in lateral root initiation under low salt. SOS3 plays a regulatory role in lateral root development by mediating auxin gradient formation and auxin polar transport. These results indicate that SOS3 is required for sufficient auxin supply to initiate lateral root development and to maintain their cell division activity under low salt stress (Zhao et al. 2011).

5 Conclusion

The study revealed that NaCl has an activating effect on PM H^+-ATPase in plants. This supports the crucial role of the enzyme in maintaining ion homeostasis and salt tolerance. In plants exposed to abiotic stresses an increase in permeability related to membrane damage is observed. Maintaining ionic balance and replenishing the loss of essential substances in repair processes is an important issue under such conditions. Support of active transport of ions and organic compounds through the plasma membrane requires increased generation of a proton gradient by PM H^+-ATPase. In addition, salinity leads to accumulation of a toxic excess of Na^+ ions. Generation of an electrochemical proton gradient across the membrane results in a proton-motive force that is used by active transport for assimilation of various nutrients, as well as for releasing toxic excess of ions from cells. Thus, plasma membrane H^+-ATPase can be a mutual element for resistance mechanisms that are activated in salt stress conditions.

References

Ahn S, Im Y, Chung G, Cho B (1999) Inducible expression of plasma membrane H^+-ATPase in the roots of fig leaf gourd plants under chilling root temperature. Physiol Plant 106:35–40
Ahn S, Im Y, Chung G, Seong K, Cho B (2000) Sensitivity of plasma membrane H^+-ATPase of cucumber root system in response to low root temperature. Plant Cell Rep 19:831–835
Apse M, Blumwald E (2007) Na^+ transport in plants. FEBS Lett 581:2247–2254
Arango M, Gevaudant F, Oufattole M, Boutry M (2003) The plasma membrane proton pump ATPase: the significance of gene subfamilies. Planta 216:355–365
Babakov AV, Chelysheva VV, Klychnikov OI, Zinyanz SE, Trofimova MS, De Boer AH (2000) Involvement of 14-3-3 proteins in the osmotic regulation of H^+-ATPase in plant plasma membrane. Planta 211:446–448
Barkla B, Vera-Estrella R, Maldonado-Gama M, Pantoja O (1999) Abscisic acid induction of vacuolar H^+-ATPase activity in *Mesembryanthemum crystallinum* is developmentally regulated. Plant Physiol 120:811–819
Batelli G, Verslues PE, Agius F, Qiu Q, Fujii H, Pan S, Schumaker KS, Grillo S, Zhu J-K (2007) SOS2 promotes salt tolerance in part by interacting with the vacuolar H^+-ATPase and upregulating its transport activity. Mol Cell Biol 27:7781–7790
Baxter I, Tchieu J, Sussman M, Boutry M, Palmgren M, Gribskov M, Harper J, Axelsen K (2003) Genomic comparison of P-type ATPase ion pumps in Arabidopsis and rice. Plant Physiol 132:618–828
Binzel M (1995) NaCl induced accumulation of tonoplast and plasma membrane H^+-ATPase message in tomato. Physiol Plant 94:722–728
Blumwald E, Aharon GS, Apse MP (2000) Sodium transport in plant cells. Biochim Biophys Acta 1465:140–151
Buch-Pedersen M, Rudashevskaya E, Berner T, Venema K, Palmgren M (2006) Potassium as an intrinsic uncoupler of the plasma membrane H^+-ATPase. J Biol Chem 281:38285–38292
Buch-Pedersen M, Pedersen P, Veierskov B, Nissen P, Palmgren M (2009) Protons how they are transported by proton pumps. Pflugers Arch Eur J Physiol 457:573–579
Camoni L, Iori V, Marra M, Aducci P (2000) Phosphorylation-dependent interaction between plant plasma membrane H^+-ATPase and 14-3-3 proteins. J Biol Chem 275:9919–9923
Chelysheva V, Zorinyants S, Smolenskaya I, Babakov A (2001) Regulation of H^+ pumping by plants plasmalemma under osmotic stress: the role of 14-3-3 proteins. Russ J Plant Physiol 48:271–280
Chen Z, Pottosin II, Cuin TA, Fuglsang AT, Tester M, Jha D, Zepeda-Jazo I, Zhou M, Palmgren MG, Newman IA, Shabala S (2007) Root plasma membrane transporters controlling K^+/Na^+ homeostasis in salt-stressed barley. Plant Physiol 145:1714–1725
Cheng NH, Pittman JK, Zhu J-K, Hirschi KD (2004) The protein kinase SOS2 activates the *Arabidopsis* H^+/Ca^{2+} antiporter CAX1 to integrate calcium transport and salt tolerance. J Biol Chem 279:2922–2926
Chinnusamy V, Jagendorf A, Zhu J-K (2005) Understanding and improving salt tolerance in plants. Crop Sci 45:437–448
Chung J-S, Zhu J-K, Bressan RA, Hasegawa PM, Shi H (2008) Reactive oxygen species mediate Na^+-induced *SOS1* mRNA stability in *Arabidopsis*. Plant J 53:554–565
Conde A, Chaves MM, Gerós H (2011) Membrane transport, sensing and signaling in plant adaptation to environmental stress. Plant Cell Physiol 52:1583–1602

Cowan A, Richardson G, Maurel J (1997) Stress-induced abscisic acid transients and stimulus-response-coupling. Physiol Plant 100:491–499
Danielsson A, Larsson C, Larsson K, Gustafsson L, Adler L (1996) A genetic analysis of the role of calcineurin and calmodulin in Ca^{2+} dependent improvement of NaCl tolerance of *Saccharomyces cerevisiae*. Curr Genet 30:476–484
Duby G, Boutry M (2009) The plant plasma membrane proton pump ATPase: a highly regulated P-type ATPase with multiple physiological roles. Pflugers Arch Eur J Physiol 457:645–655
Géwaudant F, Duby G, Stedingk E, Zhao R (2007) Expression of a constitutively activated plasma membrane H^+-ATPase alters plant development and increases salt tolerance. Plant Physiol 144:1763–1776
Grabov A, Blatt M (1998) Co-ordination of signaling elements in guard cell ion channel control. J Exp Bot 49:351–360
Hasegawa P, Bressan R, Zhu J, Bohnert H (2000) Plant cellular and molecular responses to high salinity. Annu Rev Plant Physiol Plant Mol Biol 51:463–499
Hentzen A, Smart L, Wimmers L, Fang H, Schroeder J, Bennett A (1996) Two plasma membrane H^+-ATPase genes expressed in guard cells of *Vicia faba* are also expressed throughout the plant. Plant Cell Physiol 37:650–659
Hernandez A, Cooke D, Clarkson D (2002) *In vivo* activation of plasma membrane H^+-ATPase hydrolytic activity by complex lipid-bound unsaturated fatty acids in *Ustilago maydis*. Eur J Biochem 269:1006–1011
Hernandez M, Fernandez-Garcia N, Diaz-Vivancos P, Olmos E (2010) A different role for hydrogen peroxide and antioxidative system under short and long salt stress in *Brassica oleracea* roots. J Exp Bot 61:521–535
Hu X, Jiang M, Zhang A, Lu J (2005) Abscisic acid induced apoplastic H_2O_2 accumulation up-regulates the activities of chloroplastic and cytosolic antioxidant enzymes in maize leaves. Planta 223:57–68
Hu X, Zhang A, Zhang J, Jiang M (2006) Abscisic acid is a key inducer of hydrogen peroxide production in leaves of maize plants exposed to water stress. Plant Cell Physiol 47:1484–1495
Huertas R, Olías R, Eljakaoui Z, Gálvez FJ, Li J, de Morales PA, Belver A, Rodríquez-Rosales MP (2012) Overexpression of *SlSOS2* (*SlCIPK24*) confers salt tolerance to transgenic tomato. Plant Cell Environ 35:1467–1482
Ingram J, Bartels D (1996) The molecular basis of dehydration tolerance in plants. Annu Rev Plant Physiol Plant Mol Biol 47:377–403
Janicka-Russak M, Kłobus G (2007) Modification of plasma membrane and vacuolar H^+-ATPase in response to NaCl and ABA. J Plant Physiol 164:295–302
Janicka-Russak M, Kabała K, Burzyński M, Kłobus G (2008) Response of plasma membrane H^+-ATPase to heavy metal stress in *Cucumis sativus* roots. J Exp Bot 59:3721–3728
Janicka-Russak M, Kabała K, Wdowikowska A, Kłobus G (2012a) Response of plasma membrane H^+-ATPase to low temperature in cucumber roots. J Plant Res 125:291–300
Janicka-Russak M, Kabała K, Burzyński M (2012b) Different effect of cadmium and copper on H^+-ATPase activity in plasma membrane vesicles from *Cucumis sativus* roots. J Exp Bot 63:4133–4142
Janicka-Russak M, Kabala K, Wdowikowska A, Kłobus G (2013) Modification of plasma membrane proton pumps in cucumber roots as an adaptation mechanism to salt stress. J Plant Physiol 170:915–922
Ji H, Pardo JM, Batelli G, Van Oosten MJ, Bressan RA, Li X (2013) The salt overly sensitive (SOS) pathway: established and emerging roles. Mol Plant 6:275–286
Jia W, Zhang J, Liang J (2001) Initiation and regulation of water deficit-induced abscisic acid accumulation in maize leaves and roots: cellular volume and water relations. J Exp Bot 52:295–300
Jubany-Mari T, Munne-Bosch S, Lopez-Carbonell Alegre L (2009) Hydrogen peroxide is involved in the acclimation of the Mediterranean shrub, *Cistus albidus* L., to summer drought. J Exp Bot 60:107–120

Kabała K, Janicka-Russak M (2012) Na^+/H^+ antiport activity in plasma membrane and tonoplast vesicles isolated from NaCl treated cucumber roots. Biol Plant 56:377–382
Kalampanayil B, Wimmers L (2001) Identification and characterization of a salt-stressed-induced plasma membrane H^+-ATPase in tomato. Plant Cell Environ 24:999–1005
Kanczewska J, Marco S, Vandermeeren C, Maudoux O, Rigaud J, Boutry M (2005) Activation of the plant plasma membrane H^+-ATPase by phosphorylation and binding of 14-3-3 proteins convert a dimer into a hexamer. PNAS 102:11675–11680
Kasamo K (2003) Regulation of plasma membrane H^+-ATPase activity by the membrane environment. J Plant Res 116:517–523
Katiyar-Agarwal S, Zhu J, Kim K, Agarwal M, Fu X, Huang A, Zhu J-K (2006) The plasma membrane Na^+/H^+ antiporter SOS1 interacts with RCD1 and functions in oxidative stress tolerance in *Arabidopsis*. PNAS 103:18816–18821
Kerkeb L, Venema K, Donaire J, Rodriguez-Rozales M (2002) Enhanced H^+/ATP coupling ratio of H^+-ATPase and increased 14-3-3 protein content in plasma membrane of tomato cells upon osmotic shock. Physiol Plant 16:37–41
Kłobus G, Janicka-Russak M (2004) Modulation by cytosolic components of proton pump activities in plasma membrane and tonoplast from *Cucumis sativus* roots during salt stress. Physiol Plant 121:84–92
Knight H, Trewavas A, Knight M (1997) Calcium signalling in *Arabidopsis thaliana* responding to drought and salinity. Plant J 12:1067–1078
Kwak JM, Mori IC, Pei ZM, Leonhardt N, Torres MA, Dangl JL, Bloom RE, Bodde S, Jones JD, Schroeder JI (2003) NADPH oxidase *AtrbohD* and *AtrbohF* genes function in ROS-dependent ABA signaling in Arabidopsis. EMBO J 22:2623–2633
La Rosa P, Handa A, Hasegawa P, Bressan R (1985) Abscisic acid accelerates adaptation of cultured tobacco cells to salt. Plant Physiol 79:138–142
La Rosa P, Hasegawa P, Rhodes D, Clithero J (1987) Abscisic acid stimulated osmotic adjustment and its involvement in adaptation of tobacco cells to NaCl. Plant Physiol 85:174–185
Laloi C, Mestres-Ortega D, Marco Y, Meyer Y, Reichheld JP (2004) The Arabidopsis cytosolic h5 gene induction by oxidative stress and its w-box-mediated response to pathogen elicitor. Plant Physiol 134:1006–1016
Lin H, Wu L (1996) Effects of salt stress on root plasma membrane characteristics of salt-tolerant and salt-sensitive buffalograss clones. Environ Exp Bot 36:239–254
López-Pérez L, Martínez Ballesta M, Maurel C, Carvajal M (2009) Changes in plasma membrane lipids, aquaporins and proton pump of broccoli roots, as an adaptation mechanism to salinity. Phytochemistry 70:492–500
Maathuis FJM (2013) Sodium in plants: perception, signalling, and regulation of sodium fluxes. J Exp Bot. doi:10.1093/jxb/ert326
Mahajan S, Pandey GK, Tuteja N (2008) Calcium- and salt-stress signaling in plants: shedding light on SOS pathway. Arch Biochem Biophys 471:146–158
Martz F, Sutinen M, Kiviniemi S, Palta J (2006) Changes in freezing tolerance, plasma membrane H^+-ATPase activity and fatty acid composition in *Pinus resinosa* needles during cold acclimation and de-acclimation. Tree Physiol 26:783–790
Merlot S, Leonhardt N, Fenzi F, Valon C, Costa M, Piette L, Vavasseur A, Genty B, Boivin K, Müller A, Giraudat J, Leung J (2007) Constitutive activation of a plasma membrane H^+-ATPase prevents abscisic acid-mediated stomatal closure. EMBO J 26:3216–3226
Morandini P, Valera M, Albumi C, Bonza M, Giacometti S, Ravera G, Murgia I, Soave C, De Michelis M (2002) A novel interaction partner for C-terminus of *Arabidopsis thaliana* plasma membrane H^+-ATPase (AHA1 isoform): site and mechanism of action on H^+-ATPase activity differ from those of 14-3-3 proteins. Plant J 31:487–497
Moriau L, Michelet B, Bogaerts P, Lambert L, Michel A, Oufattole M, Boutry M (1999) Expression analysis of two gene subfamilies encoding the plasma membrane H^+-ATPase in *Nicotiana plumbaginifolia* reveals the major transport functions of this enzyme. Plant J 19:31–41

Muñiz García M, País S, Téllez-Iñón M, Capiati D (2011) Characterization of StPPI1, a proton pump interactor from *Solanum tuberosum* L. that is up-regulated during tuber development and by abiotic stress. Planta 233:661–674
Munns R, James RA, Läuchli A (2006) Approaches to increasing the salt tolerance of wheat and other cereals. J Exp Bot 57:1025–1043
Netting A (2000) pH, abscisic acid and the integration of metabolism in plants under stressed and non-stressed conditions: cellular responses to stress and their implication for plant water relations. J Exp Bot 343:147–158
Niu X, Narasimhan M, Salzman R (1993) NaCl regulation of plasma membrane H^+-ATPase gene expression in *Glycophyte* and *Halophyte*. Plant Physiol 103:712–718
Oh D-H, Leidi E, Zhang Q, Hwang S-M, Li Y, Quintero FJ, Jiang X, D'Urzo MP, Lee SY, Zhao Y, Bahk JD, Bressan RA, Yun D-J, Pardo JM, Bohnert HJ (2009) Loss of halophytism by interference with SOS1 expression. Plant Physiol 151:210–222
Oh D-H, Lee SY, Bressan RA, Yun D-J, Bohnert HJ (2010) Intracellular consequences of SOS1 deficiency during salt stress. J Exp Bot 61:1205–1213
Ohta M, Guo Y, Halfter U, Zhu J-K (2003) A novel domain in the protein kinase SOS2 mediates interaction with the protein phosphatase 2C ABI2. PNAS 100:11771–11776
Ouffatole M, Arango M, Boutry M (2000) Identification and expression of tree new *Nicotiana plumbaginifolia* genes which encode isoforms of a plasma-membrane H^+-ATPase, and one of which is induced by mechanical stress. Planta 210:715–722
Palmgren M (2001) Plant plasma membrane H^+-ATPase: powerhouse for nutrient uptake. Annu Rev Plant Physiol Plant Mol Biol 52:817–845
Panda SK, Upadhyay RK (2003) Salt stress injury induces oxidative alterations and antioxidative defence in the roots of *Lemna minor*. Biol Plant 48:249–253
Pedechenko V, Nasirova G, Palladina T (1990) Lysophosphatidylcholine specifically stimulates plasma membrane H^+-ATPase from corn roots. FEBS Lett 275:205–208
Pedersen BP, Buch-Pedersen MJ, Morth JP, Palmgren MG, Nissen P (2007) Crystal structure of the plasma membrane proton pump. Nature 450:1111–1114
Pei ZM, Murata Y, Benning G, Thomine S, Klüsener B, Allen GJ et al (2000) Calcium channels activated by hydrogen peroxide mediate abscisic acid signaling in guard cells. Nature 406:731–734
Perez-Prat E, Narasimhan M, Niu X, Botella M, Bressan R, Valupesta V, Hasegawa P, Binzel M (1994) Growth cycle stage dependent NaCl induction of plasma membrane H^+-ATPase mRNA accumulation in de-adapted tobacco cells. Plant Cell Environ 17:327–333
Pitan B, Schubert S, Mühling K (2009) Decline in leaf growth under salt stress is due to an inhibition of H^+ pumping activity and increase in apoplastic pH of maize leaves. J Plant Nutr Soil Sci 172:535–543
Portillo F (2000) Regulation of plasma membrane H^+-ATPase in fungi and plants. Biochim Biophys Acta 1469:31–42
Qiu Q-S, Guo Y, Dietrich M, Schumaker KS, Zhu J-K (2002) Regulation of SOS1, a plasma membrane Na^+/H^+ exchanger in *Arabidopsis thaliana*, by SOS2 and SOS3. PNAS 99:8436–8441
Qiu QS, Barkla BJ, Vera-Estrella R, Zhu JK, Schumaker KS (2003) Na^+/H^+ exchange activity in the plasma membrane of *Arabidopsis*. Plant Physiol 132:1041–1052
Qiu Q-S, Guo Y, Quintero FJ, Pardo JM, Schumaker KS, Zhu J-K (2004) Regulation of vacuolar Na^+/H^+ exchange in *Arabidopsis thaliana* by the salt-overly-sensitive (SOS) pathway. J Biol Chem 279:207–215
Quan R, Lin H, Mendoza I, Zhang Y, Cao W, Yang Y, Shang M, Chen S, Pardo JM, Guo Y (2007) SCaBP8/CBL10, a putative calcium sensor, interacts with the protein kinase SOS2 to protect *Arabidopsis* shoots from salt stress. Plant Cell 19:1415–1431
Quintero FJ, Martinez-Atienza J, Villalta I, Jiang X, Kim W-Y, Ali Z, Fujii H, Mendoza I, Yun D-J, Zhu J-K, Pardo JM (2011) Activation of the plasma membrane Na/H antiporter salt-overly sensitive 1 (SOS1) by phosphorylation of an auto-inhibitory C-terminal domain. PNAS 108:2611–2616

Regenberg B, Villalba J, Lanfermeijer F, Palmgren M (1995) C-terminal deletion analysis of plant plasma membrane H^+-ATPase: yeast as a model system for solute transport across the plant plasma membrane. Plant Cell 7:1655–1666

Sahu B, Shaw B (2009) Salt-inducible isoform of plasma membrane H^+-ATPase gene in rice remains constitutively expressed in natural halophyte, *Suaeda maritima*. J Plant Physiol 166:1077–1089

Saijo Y, Hata S, Kyozuka J, Shimamoto K, Izui K (2000) Over-expression of a single Ca^{2+}-dependent protein kinase confers both cold and salt/drought tolerance on rice plants. Plant J 23:319–327

Santi S, Locci G, Monte R, Pinton R, Varanini Z (2003) Induction of nitrate uptake in maize roots: expression of a putative high-affinity nitrate transporter and plasma membrane H^+-ATPase isoforms. J Exp Bot 54:1851–1864

Serrano R (1990) Plasma membrane ATPase. In: Larsson C, Moller IM (eds) The plant plasma membrane. Springer, Berlin, pp 127–154

Shen P, Wang R, Zhang W (2011) Rice phospholipase Dα is involved in salt tolerance by the mediation of H^+-ATPase activity and transcription. J Integr Plant Biol 53:289–299

Shi H, Ishitani M, Kim C, Zhu J-K (2000) The *Arabidopsis thaliana* salt tolerance gene *SOS1* encodes a putative Na^+/H^+ antiporter. PNAS 97:6896–6901

Shi H, Quintero FJ, Pardo JM, Zhu J-K (2002) The putative plasma membrane Na^+/H^+ antiporter SOS1 controls long-distance Na^+ transport in plants. Plant Cell 14:465–477

Shi H, Lee B, Wu S-J, Zhu J-K (2003) Overexpression of a plasma membrane Na^+/H^+ antiporter gene improves salt tolerance in *Arabidopsis thaliana*. Nat Biotechnol 21:81–85

Shinozaki K, Yamaguchi-Shinozaki K (1996) Molecular responses to drought and cold stress. Curr Opin Biotechnol 7:161–167

Skriver K, Mundy J (1990) Gene expression in response to abscisic acid and osmotic stress. Plant Cell 2:503–512

Speth C, Jaspert N, Marcon C, Oecking C (2010) Regulation of the plant plasma membrane H^+-ATPase by its C-terminal domain: what we know for sure? Eur J Cell Biol 89:145–151

Surowy T, Boyer J (1991) Low water potentials affect expression of genes encoding vegetative storage proteins and plasma membrane proton ATPase in soybean. Plant Mol Biol 16:252–262

Suzuki N, Mittler R (2006) Reactive oxygen species and temperature stresses: a delicate balance between signaling and destruction. Physiol Plant 126:45–51

Svennelid F, Olsson A, Piotrowski M, Rosenquist M, Ottman C, Larsson C, Oecking C, Sommarin M (1999) Phosphorylation of Thr-948 at the C-terminus of the plasma membrane H^+-ATPase creates a binding site for regulatory 14-3-3 protein. Plant Cell 11:2379–2391

Tan B, Schwartz S, Zeevaart J, Mc Carty R (1994) Genetic control of abscisic acid biosynthesis in maize. PNAS 94:12235–12240

Tsai YC, Hong CY, Liu LF, Kao CH (2004) Relative importance of Na^+ and Cl^- in NaCl-induced antioxidant systems in roots of rice seedlings. Physiol Plant 122:86–94

Tuteja N (2007) Mechanisms of high salinity tolerance in plants. Methods Enzymol 428:419–438

Upchurch RG (2008) Fatty acid unsaturation, mobilization, and regulation in the response of plants to stress. Biotechnol Lett 30:967–977

Urao T, Katagiri T, Mizoguchi T, Yamaguchi-Shinozaki K, Hayashida N, Shinozaki K (1994) Two genes that encode Ca^{2+}-dependent protein kinases are induced by drought and high-salt stresses in Arabidopsis thaliana. Mol Gen Genet 244:331–340

Van der Hoeven P, Siderius M, Korthout H, Drabkin A, De Boer A (1996) A calcium and free fatty acid modulated protein kinase as putative effector of the fusicoccin 14-3-3 receptor. Plant Physiol 111:857–865

Venema K, Palmgren MG (1995) Metabolic modulation of transport coupling ratio in yeast plasma membrane H^+-ATPase. J Biol Chem 270:19659–19667

Vera-Estrella R, Barkla B, Higgins V, Blumwald E (1994) Plant defense response to fungal pathogens (activation of host-plasma membrane H^+-ATPase by elicitor-induced enzyme dephosphorylation). Plant Physiol 104:209–215

Verslues PE, Batelli G, Grillo S, Agius F, Kim Y-S, Zhu J, Agarwal M, Katiyar-Agarwal S, Zhu J-K (2007) Interaction of SOS2 with nucleoside diphosphate kinase 2 and catalases reveals a point of connection between salt stress and H_2O_2 signaling in *Arabidopsis thaliana*. Mol Cell Biol 27:7771–7780

Volkov RA, Panchuk II, Mullineaux PM, Schöffl F (2006) Heat stress-induced H_2O_2 is required for effective expression of heat shock genes in *Arabidopsis*. Plant Mol Biol 61:733–746

Wakeel A, Hanstein S, Pitann B, Schubert S (2010) Hydrolytic and pumping activity of H^+-ATPase from leaves of sugar beet (*Beata vulgaris* L.) as affected by salt stress. J Plant Physiol 167:725–731

Wang YZ, Li RL, Xu XZ, Zhu WC (1994) Ca^{2+} transport across the tonoplast of wheat roots. J Wuhan Univ (Nat Sci Ed) 5:111–115

Wang P, Li Z, Wei J, Zhao Z, Sun D, Cui S (2012) A Na^+/Ca^{2+} exchanger-like protein (AtNCL) involved in salt stress in Arabidopsis. J Biol Chem 287:44062–44070

Wilson C, Shannon MC (1995) Salt-induced Na^+/H^+ antiport in root plasma membrane of a glycophytic and halophytic species of tomato. Plant Sci 107:147–157

Xiong L, Schumaker K, Zhu J-K (2002) Cell signaling during cold, drought, and salt stress. Plant Cell 14:165–183

Ye J, Zhang W, Guo Y (2013) *Arabidopsis* SOS3 plays an important role in salt tolerance by mediating calcium-dependent microfilament reorganization. Plant Cell Rep 32:139–148

Yu X, Li M, Gao G, Feng H, Geng X, Peng C, Zhu S, Wang X, Shen Y, Zhang D (2006) Abscisic acid stimulates a calcium-dependent protein kinase in grape berry. Plant Physiol 140:558–579

Zhang X, Zhang L, Dong F, Gao J, Galbraith D, Song C (2001) Hydrogen peroxide is involved in abscisic acid-induced stomatal closure in *Vicia faba*. Plant Physiol 126:1438–1448

Zhang M, Barg R, Yin M, Gueta-Dahan Y, Leikin-Frenkel A, Salts Y, Shabtai S, Ben-Hayyim G (2005) Modulated fatty acid desaturation via over-expression of two distinct ω-3 desaturases differentially alters tolerance to various abiotic stresses in transgenic tobacco cells and plants. Plant J 44:361–371

Zhang F, Wang Y, Wang D (2007) Role of nitric oxide and hydrogen peroxide during the salt resistance response. Plant Signal Behav 2:473–474

Zhao Y, Wang T, Zhang W, Li X (2011) SOS3 mediates lateral root development under low salt stress through regulation of auxin redistribution and maxima in *Arabidopsis*. New Phytol 189:1122–1134

Zhu J-K (2002) Salt and drought stress signal transduction in plants. Annu Rev Plant Biol 53:247–273

Zhu J-K (2003) Regulation of ion homeostasis under salt stress. Curr Opin Plant Biol 6:441–445

Zhu J, Fu X, Koo YD, Zhu J-K, Jenney FE Jr, Adams MWW, Zhu Y, Shi H, Yun D-J, Hasegawa PM, Bressan RA (2007) An enhancer mutant of *Arabidopsis salt overly sensitive 3* mediates both ion homeostasis and the oxidative stress response. Mol Cell Biol 27:5214–5224

Zörb C, Stracke B, Tramnitz B, Denter D, Sümer A, Mühling KH, Yan F, Schubert S (2005) Does H^+ pumping by plasmalemma ATPase limit leaf growth of corn (*Zea mays* L.) during the first phase of salt stress? J Plant Nutr Soil Sci 168:550–557

Selenium in Plants

Elizabeth A.H. Pilon-Smits

Contents

Abstract Selenium (Se) and sulfur (S) are chemically similar. Most plants cannot discriminate between the two, with the exception of Se hyperaccumulator species, which preferentially accumulate Se over S. Genetic engineering of various genes from the S/Se assimilation pathway has successfully enhanced plant Se tolerance, accumulation, and volatilization, in both laboratory and field. Results from genomic studies are beginning to shed better light on Se tolerance and (hyper)accumulation mechanisms, pointing to particular growth regulators (jasmonic acid, salicylic acid, ethylene) and constitutive upregulation of S/Se uptake and assimilation pathways. Selenium accumulation in plants profoundly affects ecological interactions. It protects plants from herbivores via both deterrence and toxicity, as well as from microbial pathogens. High-Se plants do not deter pollinators. Selenium hyperaccumulators enhance Se levels in neighboring plants, which can have a negative (allelopathic) effect if these are Se sensitive, but a positive effect if they are Se tolerant, via protection from herbivores. Thus, in seleniferous ecosystems Se hyperaccumulators may favor Se-resistant ecological partners while selecting against Se-sensitive partners. In this way, hyperaccumulators may affect species composition at multiple trophic levels, as well as Se cycling.

E.A.H. Pilon-Smits (✉)
Biology Department, Colorado State University, Fort Collins, CO 80523-1878, USA
e-mail: epsmits@lamar.colostate.edu

U. Lüttge, W. Beyschlag (eds.), *Progress in Botany*, Progress in Botany 76,
DOI 10.1007/978-3-319-08807-5_4

1 Introduction to Selenium in Biology and Natural Systems

The element selenium (Se) is chemically similar to sulfur (S). Selenium occurs naturally in soils, typically at very low levels, but up to 100 mg Se kg^{-1} in seleniferous soils, such as Cretaceous shale (Beath et al. 1939a, b). Selenium is essential for mammals and many other animals, as well as for many bacteria and some algae (Fu et al. 2002). Selenium is part of so-called selenoproteins which contain selenocysteine (SeCys) in their active site. Some of the most prominent examples are glutathione peroxidases and thioredoxin reductases (Zhang and Gladyshev 2009). These proteins have various redox functions, including the scavenging of free radicals. Thus, Se supplementation has been reported to reduce the probability of developing cancer (Rayman 2005). Higher plants may have lost essential Se metabolism during their evolution (Zhang and Gladyshev 2009). While not essential for higher plants, Se is considered a beneficial element, stimulating growth at low levels (Pilon-Smits et al. 2009). The mechanisms for this beneficial effect are still largely unknown but may be associated with enhanced antioxidant activity (Hartikainen 2005). At elevated levels Se is toxic to most plants, due to nonspecific incorporation of Se into sulfur (S) compounds, and to oxidative stress (Stadtman 1996; Van Hoewyk 2013). Compared to most elements, the window between benefit and toxicity is particularly small for Se, and both deficiency and toxicity are problems worldwide. Selenium deficiency occurs in areas where soil Se is low, including parts of Europe, China, North America, Australia, New Zealand, and Southern Africa (Sors et al. 2005). Selenium toxicity occurs in areas where soil Se is naturally high, including areas of China, India, and the United States. Toxicity from naturally occurring Se may be exacerbated by irrigation of seleniferous soil, mining, and use of Se-rich fossil fuels (Terry et al. 2000). At the basis of the food chain, plants collect Se from the soil and provide it to higher trophic levels. Using this principle, plants may be used to remove Se from natural or polluted Se-rich areas and as a food source to alleviate Se deficiency in humans or animals. The first process is called phytoremediation (Pilon-Smits 2005), and the second biofortification (White and Broadley 2009). The two may even be combined: plants that have accumulated Se from polluted soil may be used as fortified food. These management practices benefit from a thorough understanding of the mechanisms of plant Se uptake and the fate of Se in different plant species. In addition, it is important for these technologies to have insight into the ecological effects of plant Se accumulation.

2 Plant Se Physiology: What Can Plants Do with Se?

Figure 1 summarizes Se metabolism in plants (for other reviews see Lauchli 1993; Terry et al. 2000; Sors et al. 2005). Environmental selenate (prevalent in oxic environments) and selenite (prevalent in reducing environments) are typically taken up non-specifically by plants using transporters for S analogues. Following

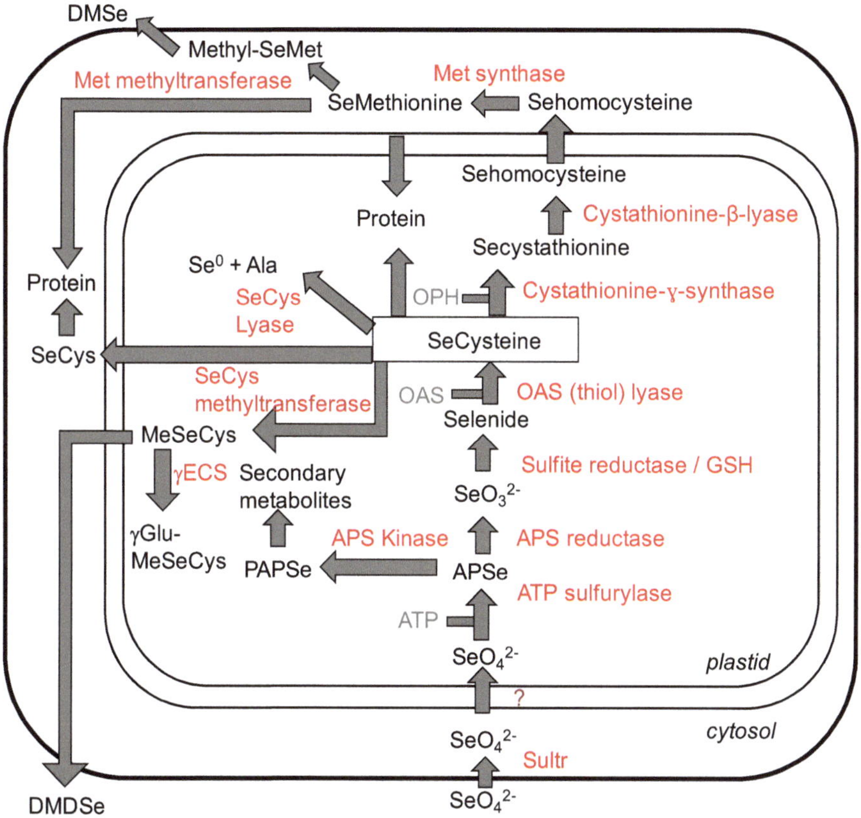

Fig. 1 Proposed model for Se assimilation in plants. Enzymes are shown in *red* and metabolites in *black* or *gray*. *Sultr* sulfate/selenate transporter, *APSe* adenosine phosphoselenate, *PAPSe* phospho adenosine phosphoselenate, *OAS* *O*-acetylserine, *OPH* *O*-phosphohomoserine, *SeCys* selenocysteine, *(Se)Met* (seleno)methionine, *Ala* alanine, *MeSeCys* methyl-SeCys, *gGlu-MeSeCys* g-glutamyl MeSeCys, *gECS* g-glutamylcysteine synthetase, *GSH* glutathione, *DMSe* dimethylselenide, *DMDSe* dimethyldiselenide

uptake, these inorganic forms of Se may be assimilated via the sulfate assimilation pathway into selenocysteine (SeCys), selenomethionine (SeMet), and other organic S compounds. This process can happen in both root and shoot, but is thought to be more prominent in the shoot. When seleno amino acids inadvertently get incorporated into proteins, replacing Cys and Met, this impairs protein function and thus results in toxicity (Stadtman 1990). Most plants can metabolize SeMet into volatile dimethylselenide (DMSe), which may help avoid toxicity (Terry et al. 2000). Another potential Se detoxification mechanism in plants is the breakdown of SeCys into elemental Se and alanine (Van Hoewyk et al. 2005). Both volatilization and breakdown of SeCys are nonspecific, using enzymes that function in S metabolism (Fig. 1; Terry et al. 2000; Van Hoewyk et al. 2007).

In addition to these general mechanisms by which plants metabolize Se inadvertently, some plants may be able to discriminate between Se and S analogues, and thus may be said to have Se-specific metabolism. These plants can, for instance, specifically methylate SeCys into methyl-SeCys, which serves as a very effective Se detoxification mechanism since methyl-SeCys does not get incorporated into proteins (Neuhierl and Böck 1996). This methylation process is mediated by the enzyme SeCys methyltransferase (SMT). The best-known plant taxa that contain this enzyme are the so-called Se hyperaccumulator plants, which can accumulate up to 1.5 % of their dry weight as Se (15,000 mg kg^{-1} DW, Beath et al. 1939a, b). However, SMT has also been found in broccoli (*Brassica oleracea*) (Lyi et al. 2005), and methyl-SeCys has been found in several *Allium* species such as garlic (Ge et al. 1996). These species are known to be sulfur-loving; they are not hyperaccumulators but do accumulate appreciable amounts of Se non-specifically due to high levels of sulfate uptake. They are sometimes referred to as Se accumulators.

True Se hyperaccumulation occurs in 4–5 genera in the Brassicaceae, Fabaceae, and Asteraceae. They occur predominantly or even exclusively on seleniferous soils (Beath et al. 1939a, b). Hyperaccumulators of Se have several properties that set them apart from other species. They accumulate ~100-fold higher Se levels and have higher tissue Se/S levels than surrounding vegetation (Lauchli 1993). While most plants accumulate inorganic Se, hyperaccumulators accumulate organic forms like methyl-SeCys and selenocystathionine (Anderson 1993). Since these forms of Se do not interfere with S metabolism, hyperaccumulators are completely tolerant to their extreme Se levels, and often even grow better under high-Se conditions than without Se (Broyer et al. 1972; El Mehdawi et al. 2012). Like other plants, hyperaccumulators can volatilize Se, but mostly in the form of dimethyldiselenide (DMDSe), which originates from methyl-SeCys (Terry et al. 2000). Selenium hyperaccumulators also show tissue-specific and organ-specific sequestration patterns that are different from other plants. Relative to non-accumulators, a larger fraction of the Se in hyperaccumulators is translocated from root to shoot; also, a larger fraction is remobilized from aging leaves to young leaves and reproductive organs, particularly pollen and ovules (Quinn et al. 2011a, b). Within leaves, hyperaccumulators store most of their Se in the vacuoles of epidermal cells, which may include the trichomes (leaf hairs) (Freeman et al. 2006a, 2010). For comparison, the non-hyperaccumulators *Arabidopsis thaliana* and *Brassica juncea* were found to store most of their Se in the form of selenate in the vascular bundles, and they contained higher Se levels in leaves than in floral tissues (van Hoewyk et al. 2005; Freeman et al. 2006a; Quinn et al. 2011a). Interestingly, selenate uptake in Se hyperaccumulators is not inhibited by sulfate, suggesting they have a selenate-specific transporter; this is in stark contrast to the non-hyperaccumulator *B. juncea* (Harris, Schneberg, and Pilon-Smits, unpublished results) and may explain the elevated Se/S ratios that are typical for hyperaccumulators (White et al. 2007). Similarly, Se and S remobilization in hyperaccumulators follows different patterns, both developmentally and seasonally (Galeas et al. 2007; Quinn et al. 2011a). Selenium levels are highest in young leaves and reproductive tissues, while S levels are highest in mature leaves. Leaf Se levels in the field peak in early spring, while leaf S levels peak in midsummer. In non-hyperaccumulators both Se and S levels peaked in midsummer (Galeas et al. 2007).

3 Genetic and Evolutionary Aspects of Se Tolerance and Accumulation

Plant species vary several orders of magnitude in their capacity to accumulate and tolerate Se (Beath et al. 1939a, b; White et al. 2007). Based on maximum Se levels in shoot tissues in the field, species may be classified as Se hyperaccumulator (>1,000 mg $Se.kg^{-1}$ DW), Se accumulator (100–1,000 mg $Se.kg^{-1}$ DW), or non-Se accumulator (<100 mg $Se.kg^{-1}$ DW). True Se hyperaccumulation is found in 4–5 genera from three different families, and probably evolved independently in each lineage. Even within the genus *Astragalus*, Se hyperaccumulation may have evolved multiple times, judged from its occurrence in more derived taxa that do not form a natural group with a common ancestor. The polyphyletic origin of Se hyperaccumulation suggests this trait can evolve relatively easily, and may be controlled by relatively few genes. It is possible that there is a key gene such as a transcription factor that controls a suite of genes that together bring about the syndrome of hyperaccumulation and hypertolerance. Several studies have been carried out to obtain better insight into the molecular mechanisms involved in Se tolerance and accumulation, both in non-hyperaccumulators and hyperaccumulators.

Making use of model species *A. thaliana*, a non-Se accumulator, a comparative study was performed with recombinant inbred lines (RILs). Several quantitative trait loci (QTL) were identified that co-segregated with the higher selenate tolerance in accession Columbia compared to Landsberg erecta (Zhang et al. 2006a). Several S-related genes are present in the identified QTL regions, including a selenocysteine methyl transferase (SMT) homologue, an ATP sulfurylase, and a serine acetyl transferase (SAT). In another study by Zhang et al. (2006b) comparing nineteen different ecotypes of Arabidopsis with variable tolerance and accumulation of Se, the traits tolerance and accumulation were found to be not correlated. Also, tolerance to selenate and selenite appeared to be controlled by different loci.

In another study with *A. thaliana*, a transcriptome analysis was performed on plants grown with or without selenate (Van Hoewyk et al. 2008). It was found that genes involved in ethylene and jasmonic acid pathways were upregulated by Se. In agreement with a role for these hormones, Arabidopsis mutants with a defect in genes involved in ethylene synthesis, ethylene signaling, and jasmonic acid signaling were shown to have reduced tolerance to selenate, and overexpression of a protein involved in ethylene signaling resulted in increased selenate resistance (Van Hoewyk et al. 2008). A similar study into selenite resistance by Tamaoki et al. (2008) also pointed to the involvement of ethylene and jasmonic acid and also suggested reactive oxygen species (ROS) may have a signaling role. Perhaps as a result of their elevated levels of these hormones, the resistant accessions showed

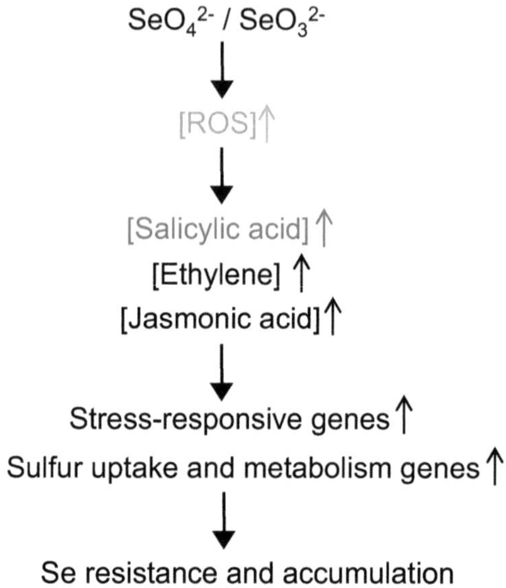

Fig. 2 Model for cellular plant responses to selenate or selenite that may play a role in Se resistance and accumulation [from Tamaoki et al. (2008), Freeman et al. (2010)]. Taxa with enhanced Se resistance were shown to have constitutively higher levels of the plant hormones and gene transcripts indicated. This was found both for moderately Se-resistant ecotypes of non-accumulator *Arabidopsis thaliana* and for Se hyper-tolerant hyperaccumulator *Stanleya pinnata*. Note: enhanced reactive oxygen species (ROS) levels were observed in *A. thaliana* but not in *S. pinnata*, while enhanced salicylic acid (SA) levels were observed in the hyperaccumulator but not in *A. thaliana*

enhanced expression of genes involved in sulfate uptake and reduction, as well as higher levels of total S and of reduced antioxidant S compounds. This may help the plants prevent Se from replacing S in proteins and other S compounds, and prevent Se-induced oxidative stress. The simple model shown in Fig. 2 summarizes these responses in plants to selenate or selenite, and how they may lead to Se resistance and accumulation.

The hyperaccumulator *Stanleya pinnata*, which is in the same family as *A. thaliana* (Brassicaceae), may in part use similar mechanisms for Se tolerance as *A. thaliana* (Freeman et al. 2010). The plant hormones JA and ethylene, as well as the hormone salicylic acid (SA), were constitutively elevated in *S. pinnata*, as compared to non-hyperaccumulator *S. albescens*. Probably as a response to the elevated levels of these hormones, the hyperaccumulator had constitutively upregulated expression of several sulfate transporters and S assimilatory enzymes, and higher levels of total S and of reduced S compounds (including the antioxidant glutathione), as well as higher levels of total Se. The mechanisms underlying some of the other attributes of *S. pinnata* remain to be elucidated, such as its tissue- and organ-specific Se sequestration patterns, in vacuoles of leaf epidermal cells and in young leaves and reproductive organs (Freeman et al. 2006a, 2010). Also, the key mechanisms responsible for its capacity to store Se as methyl-SeCys and selenocystathionine remain to be elucidated. More genome-wide and genus-wide studies are needed to reveal key genes for Se hyperaccumulation in *Stanleya*.

The driving force for the evolution of increasing plant capacity for Se accumulation and ultimately Se hyperaccumulation may be both physiological and ecological benefits. Many plant species show a positive growth response to Se, perhaps due to enhanced antioxidant capacity (Cartes et al. 2005; Djanaguiraman et al. 2005; Hartikainen 2005; Kong et al. 2005). There are also several ecological benefits of the accumulation of Se, since it can protect the plant from a variety of herbivores and pathogens, and maybe also have allelopathic effects on neighboring plants. These ecological effects of plant-accumulated Se are described in more detail in Sect. 5.

4 Genetic Engineering of Plant Se Metabolism

Several genes from the sulfate assimilation pathway were manipulated in order to study the effect on plant Se tolerance and accumulation (see Fig. 1 for the function of each of these enzymes). Overexpression in *B. juncea* of the key enzyme for the reduction of selenate to selenite, ATP sulfurylase (from *Arabidopsis thaliana*), resulted in enhanced production of organic Se as well as enhanced Se tolerance and accumulation (Pilon-Smits et al. 1999). Therefore, this enzyme appears to be a rate-limiting step for selenate assimilation in this Se accumulator. The ATP sulfurylase overexpressors showed three- to fivefold enhanced Se accumulation not only under controlled laboratory conditions but also when grown on naturally seleniferous soil and on polluted sediment in the field, making them interesting candidates for both biofortification and phytoremediation (Van Huysen et al. 2004; Bañuelos et al. 2005).

Overexpression of an *A. thaliana* cystathionine-γ-synthase (CγS) in *B. juncea* resulted in threefold enhanced Se volatilization from either selenate or selenite (van Huysen et al. 2003), showing that this enzyme is a limiting factor for Se volatilization. Probably as a result of their enhanced Se volatilization rate, the CγS plants accumulated 40 % less Se in their tissues, and were more tolerant to selenate compared to untransformed plants. Similar results were obtained under laboratory conditions and in a greenhouse pot experiment using naturally seleniferous soil (van Huysen et al. 2004).

Another transgenic approach overexpressed SeCys methyltransferase (SMT) from the Se hyperaccumulator *Astragalus bisulcatus* in *A. thaliana* and *B. juncea* (Ellis et al. 2004; LeDuc et al. 2004). The SMT transgenics showed enhanced Se volatilization and tolerance, as well as enhanced Se accumulation in the form of nontoxic methyl-SeCys (Montes-Bayón et al. 2002; Meija et al. 2002). These effects were more pronounced when supplied with selenite than selenate, suggesting that conversion of selenate to selenite was a rate-limiting step for the selenate assimilation pathway. Indeed, when double transgenic APSxSMT plants were created by crossing APS and SMT transgenics, they accumulated around eight times more MeSeCys than the wild type and about twofold more than the SMT transgenics (LeDuc et al. 2006).

In another genetic engineering approach, selenocysteine lyase (SL) from mouse or from *A. thaliana* was expressed in *A. thaliana* and *B. juncea*. The SL enzyme breaks down SeCys into alanine and insoluble elemental Se, which was predicted to reduce the nonspecific incorporation of SeCys into proteins. Indeed, the SL transgenics showed reduced Se incorporation into protein (Pilon et al. 2003; Garifullina et al. 2003; Van Hoewyk et al. 2005). The SL transgenics also showed enhanced Se accumulation (up to twofold) compared to wild-type plants, both in controlled laboratory conditions and in the field on Se-polluted sediment (Bañuelos et al. 2007).

The results obtained with the various transgenics in the laboratory, greenhouse, and field were similar. The different transgenics showed enhanced Se accumulation, volatilization, and/or tolerance. These are all promising traits for phytoremediation or biofortification. Before Se-accumulating plants can be grown on a large scale, however, whether they are wild type or transgenic, a careful consideration needs to be done regarding the potential ecological implications (Bañuelos et al. 2002). Studies on the ecological effects of plant Se accumulation also help shed light on the functional significance of Se hyperaccumulation, and the selection pressures that may have driven the evolution of Se hyperaccumulation. These will be discussed in the next section.

5 Ecological Aspects of Se (Hyper)Accumulation

Plant Se (hyper)accumulation generally has a negative effect on Se-sensitive ecological partners; this was found to be the case for plant interactions with herbivores, neighboring plants, microbes, and perhaps also pollinators (El-Mehdawi and Pilon-Smits 2012, and references therein). Selenium accumulation offers plants a broad protection against herbivores, due to a combination of deterrence and toxicity. This protective effect can already be observed at tissue levels as low as 10 mg Se kg^{-1} DW (Hanson et al. 2004). Selenium may also protect plants against microbial pathogens (Hanson et al. 2003). There is also evidence that hyperaccumulator plants use their accumulated Se in their competition with neighboring plants: soil around hyperaccumulators can be tenfold higher in Se due to litter deposition and root exudation, and this may negatively affect germination and growth of Se-sensitive competing plants (El Mehdawi et al. 2011a).

While Se in or around hyperaccumulators deters or is toxic to Se-sensitive neighboring organisms, it also offers a potential exclusive niche for Se-tolerant partners, which may actually benefit from this interaction via enhanced growth or stress resistance. This was found for various hyperaccumulator-associated herbivores, microbes, and neighboring plants (for a review see El-Mehdawi and Pilon-Smits 2012). The two Se hyperaccumulators *S. pinnata* and *A. bisulcatus* both harbored Se-resistant herbivores in their natural habitat, which included Lepidoptera (moths), Coleoptera (beetles), and Hymenoptera (wasps). The Se resistance mechanisms of these herbivores either consisted of an inability to break down methyl-SeCys (protecting them from toxic SeCys incorporation into protein) or of excluding Se from their tissues (Freeman et al. 2006b, 2012; Valdez Barillas et al. 2012). There also is evidence of Se-resistant endophytic bacteria and fungi in Se hyperaccumulators (Lindblom et al. 2013; Valdez Barillas et al. 2012) and Se-resistant litter-decomposing microbes and micro-arthropods (Quinn et al. 2011b). In addition, several Se-tolerant plant species were found to often occur in the vicinity of Se hyperaccumulators, where they can accumulate an order of magnitude higher Se levels compared to when growing away from hyperaccumulators. Interestingly, these Se-tolerant plant species profit from their enhanced tissue Se levels, both physiologically (enhanced growth) and ecologically (reduced herbivory) (El Mehdawi et al. 2011b). Selenium did not deter honeybees and other potential pollinators (Quinn et al. 2011a; Hladun et al. 2012). The Se levels in pollen and nectar of Se hyperaccumulators are extremely high; it remains to be determined whether ingestion of Se-rich pollen and nectar poses a health hazard to pollinators. In a preliminary survey the honey from seleniferous areas had slightly elevated Se levels that would make it suitable as a Se-fortified food source for Se-deficient consumers (Quinn et al. 2011a).

Through the combined negative and positive effects of Se hyperaccumulators on their Se-sensitive and Se-resistant ecological partners they may affect plant, microbial, and animal species composition and species richness in the area that is under their influence. It appears that specialized Se-tolerant herbivores, detritivores, microbial symbionts, and perhaps also pollinators tend to evolve under the influence of Se hyperaccumulators, to occupy the extreme symbiotic niche offered by these plants. Owing to these co-evolved symbionts, hyperaccumulator plants may experience no net evolutionary cost of hyperaccumulation. The only apparent limitation of Se hyperaccumulation is that it limits the geographic distribution to seleniferous soils: hyperaccumulators appear to be physiologically or ecologically dependent on Se for their competitive strength, perhaps because they have lost other mechanisms to defend themselves against biotic or abiotic stress. There is also to date no evidence for a physiological cost of Se hyperaccumulation (Prins et al. 2011).

Selenium hyperaccumulators may play an important role in Se cycling through seleniferous ecosystems. They transform inorganic forms of Se to organic forms, concentrate this Se, and then disperse it back into their local environment and up the food chain via Se-tolerant ecological partners. In Fig. 3 an overview is presented of the processes in and by plants with respect to Se movement and transformation. It also summarizes the effects of plant Se and plant processes on the biogeochemistry of Se in the local ecosystem.

The ecological effects of plant Se accumulation have relevance for the management of seleniferous habitats, as well as applications in agriculture and phytoremediation. Selenium is an essential trace element for most animals, but also a toxin at higher level, with a very narrow window between Se deficiency and

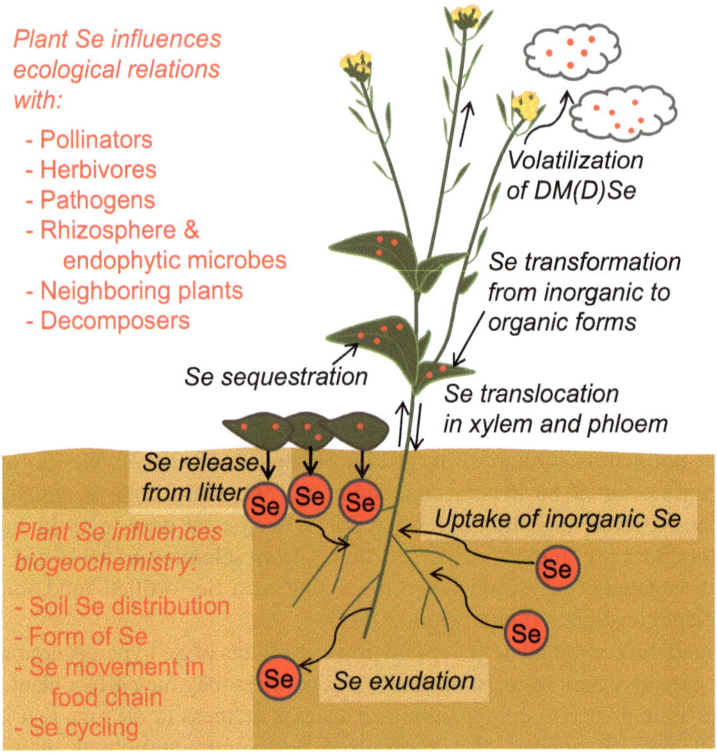

Fig. 3 Plant effects on Se cycling and transformation, and potential effects of plant Se processes on ecological partners. Inorganic Se: selenate, selenite; organic Se: methylselenocysteine or selenocystathionine; volatile DM(D)Se: dimethyl(di)selenide. Typically, Se from plants has a negative effect on Se-sensitive ecological partners, which may protect plants from pathogens and herbivores, and have allelopathic effects on neighboring plants. Conversely, Se-rich plants offer a niche to specialized Se-resistant partners (both mutualists and pathogens/herbivores), and may select for their evolution

toxicity. Selenium from phytoremediation crops can be transferred biologically to insects and mammals (Bañuelos et al. 2002). This may be beneficial if it reduces herbivory (higher productivity, reduced need for pesticides) or if it accumulates in low enough levels in the consumer (e.g., honeybee or other mutualist symbiont) to have a beneficial health effect. If, however, Se in flowers should negatively affect honeybee health this may have serious consequences for honeybee populations and agricultural productivity (Hladun et al. 2012). The other ecological observation that Se hyperaccumulators enhance Se levels in neighboring plants may be utilized in co-cropping or intercropping, to boost Se levels in biofortified crops like wheat or rice. Moreover, Se-tolerant endophytic or rhizosphere microbes isolated from Se hyperaccumulators may have applications in bio- or phytoremediation, or in biofortification, either in association with plants or by themselves. De Souza et al. have already demonstrated the potential of bacteria isolated from seleniferous areas in enhancing plant Se accumulation and volatilization (1999). If Se-rich plants affect Se cycling through their local ecosystem, this may have complex ecological implications as well. It is advisable to consider these multifaceted ecological implications when using Se-accumulating plants in agriculture or environmental restoration.

6 Conclusions and Future Prospects

Combined physiological, biochemical, and genetic/genomic approaches have shown plant Se metabolism to largely follow S uptake and metabolic pathways. Most plants cannot discriminate between Se and S and incorporate Se into all S compounds. An exception to this rule are the Se hyperaccumulator plants, which appear to be able to discriminate between Se and S, and to preferentially take up Se over S. These plants also show different spatial and temporal patterns of Se and S translocation and sequestration in organs and tissues. Selenium responses in plants appear to involve the growth regulators JA and ethylene, and in hyperaccumulators also SA. In response to higher levels of these growth regulators, the S assimilation pathway is upregulated, as well as a variety of stress-related proteins. Se-resistant taxa tend to have higher levels of these growth regulators and also a constitutively upregulated S assimilation pathway. More extensive genomic studies are needed to shed more light on Se hypertolerance and hyperaccumulation mechanisms. The Brassicaceae hyperaccumulator *S. pinnata* and its non-Se accumulator, non-Se-tolerant relatives likely will be a good model system for such studies (El Mehdawi et al. 2012; Feist and Parker 2001). Genetic engineering of genes from the S/Se assimilation pathway has been successful in enhancing Se tolerance, accumulation, and volatilization in plants, both under controlled laboratory conditions and in the field on contaminated soil. These transgenics may be suitable for phytoremediation or biofortification. Ecological studies have shown that Se accumulation protects plants from a wide variety of herbivores, via both deterrence and toxicity, as well as from some microbial pathogens. Thus, cultivation of high-Se plants may require less herbicides. High-Se plants do not deter pollinators, and thus their effects on pollinator health warrant further study. Selenium hyperaccumulators enhance Se levels in neighboring plants, which may have a negative (allelopathic) effect if these are Se sensitive, but a positive effect if they are Se tolerant, since it can protect the neighbors from herbivores as well. Thus, co-cropping or intercropping of popular phytoremediation/biofortification crops with hyperaccumulators may enhance crop Se accumulation. In natural seleniferous ecosystems Se hyperaccumulators likely play an important role in Se biogeochemistry and Se movement in the food chain. Moreover, through their positive effects on Se-resistant ecological partners and negative effects on Se-sensitive partners, hyperaccumulators may affect species composition at multiple trophic levels, and thus may be keystone species. This will be an interesting area of further research.

Acknowledgments Funding for Se research by Elizabeth A. H. Pilon-Smits was provided by the National Science Foundation (grant # IOS-0817748).

References

Anderson JW (1993) Selenium interactions in sulfur metabolism. In: De Kok LJ (ed) Sulfur nutrition and assimilation in higher plants—regulatory, agricultural and environmental aspects. SPB Academic, The Netherlands, pp 49–60

Bañuelos GS, Vickerman DB, Trumble JT, Shannon MC, Davis CD, Finley JW, Mayland HF (2002) Biotransfer possibilities of selenium from plants used in phytoremediation. Int J phytoremediation 4:315–331

Bañuelos G, Terry N, LeDuc DL, Pilon-Smits EAH, Mackey B (2005) Field trial of transgenic Indian mustard plants shows enhanced phytoremediation of selenium contaminated sediment. Environ Sci Technol 39:1771–1777

Bañuelos G, LeDuc DL, Pilon-Smits EAH, Tagmount A, Terry N (2007) Transgenic Indian mustard overexpressing selenocysteine lyase, selenocysteine methyltransferase, or methionine methyltransferase exhibit enhanced potential for selenium phytoremediation under field conditions. Environ Sci Technol 41:599–605

Beath OA, Gilbert CS, Eppson HF (1939a) The use of indicator plants in locating seleniferous areas in the Western United States. I. General. Am J Bot 26:257–269

Beath OA, Gilbert CS, Eppson HF (1939b) The use of indicator plants in locating seleniferous areas in the Western United States. II. Correlation studies by states. Am J Bot 26:296–315

Broyer TC, Huston RP, Johnson CM (1972) Selenium and nutrition of Astragalus. 1. Effects of selenite or selenate supply on growth and selenium content. Plant Soil 36:635–649

Cartes P, Gianfreda L, Mora ML (2005) Uptake of selenium and its antioxidant activity in ryegrass when applied as selenate and selenite forms. Plant Soil 276:359–367

de Souza MP, Chu D, Zhao M, Zayed AM, Ruzin SE, Schichnes D, Terry N (1999) Rhizosphere bacteria enhance selenium accumulation and volatilization by Indian mustard. Plant Physiol 119:565–574

Djanaguiraman M, Durga Devi D, Shanker AK, Sheeba JA, Bangarusamy U (2005) Selenium—an antioxidative protectant in soybean during senescence. Plant Soil 272:77–86

El Mehdawi AF, Quinn CF, Pilon-Smits EAH (2011a) Effects of selenium hyperaccumulation on plant–plant interactions: evidence for elemental allelopathy. New Phytol 191:120–131

El Mehdawi AF, Quinn CF, Pilon-Smits EAH (2011b) Selenium hyperaccumulators facilitate selenium-tolerant neighbors via phytoenrichment and reduced herbivory. Curr Biol 21:1440–1449

El Mehdawi AF, Cappa JJ, Fakra SC, Self J, Pilon-Smits EAH (2012) Interactions of selenium and non-accumulators during co-cultivation on seleniferous or non-seleniferous soil—the importance of having good neighbors. New Phytol 194:264–277

Ellis DR, Sors TG, Brunk DG, Albrecht C, Orser C, Lahner B, Wood KV, Harris HH, Pickering IJ, Salt DE (2004) Production of Se-methylselenocysteine in transgenic plants expressing selenocysteine methyltransferase. BMC Plant Biol 4:1–11

El-Mehdawi AF, Pilon-Smits EAH (2012) Ecological aspects of plant selenium hyperaccumulation. Plant Biol 14:1–10

Feist LJ, Parker DR (2001) Ecotypic variation in selenium accumulation among populations of *Stanleya pinnata*. New Phytol 149:61–69

Freeman JL, Zhang LH, Marcus MA, Fakra S, McGrath SP, Pilon-Smits EAH (2006a) Spatial imaging, speciation and quantification of selenium in the hyperaccumulator plants *Astragalus bisulcatus* and *Stanleya pinnata*. Plant Physiol 142:124–134

Freeman JL, Quinn CF, Marcus MA, Fakra S, Pilon-Smits EAH (2006b) Selenium tolerant diamondback moth disarms hyperaccumulator plant defense. Curr Biol 16:2181–2192

Freeman JL, Tamaoki M, Stushnoff C, Quinn CF, Cappa JJ, Devonshire J, Fakra S, Marcus MA, McGrath S, Van Hoewyk D, Pilon-Smits EAH (2010) Molecular mechanisms of selenium tolerance and hyperaccumulation in *Stanleya pinnata*. Plant Physiol 153:1630–1652

Freeman JL, Marcus MA, Fakra SC, Devonshire J, McGrath SP, Pilon-Smits EAH (2012) Seeds of Selenium Hyperaccumulators *Stanleya pinnata* and *Astragalus bisulcatus* are colonized by Se-resistant, Se-excluding wasp and beetle herbivores. PLoS One 7(12):e50516

Fu L-H, Wang X-F, Eyal Y, She Y-M, Donald LJ, Standing KG, Ben-Hayyim G (2002) A selenoprotein in the plant kingdom: mass spectrometry confirms that an opal codon (UGA) encodes selenocysteine in *Chlamydomonas reinhardtii* glutathione peroxidase. J Biol Chem 277:25983–25991

Galeas ML, Zhang LH, Freeman JL, Wegner M, Pilon-Smits EAH (2007) Seasonal fluctuations of selenium and sulfur accumulation in selenium hyperaccumulators and related non-accumulators. New Phytol 173:517–525

Garifullina GF, Owen JD, Lindblom S-D, Tufan H, Pilon M, Pilon-Smits EAH (2003) Expression of a mouse selenocysteine lyase in *Brassica juncea* chloroplasts affects selenium tolerance and accumulation. Physiol Plant 118:538–544

Ge HH, Cai X-J, Tyson JF, Uden PC, Denover ER, Block E (1996) Identification of selenium species in selenium-enriched garlic, onion and broccoli using high-performance ion chromatography with inductively coupled plasma mass spectrometry detection. Anal Commun 33:279–281

Hanson BR, Garifullina GF, Lindblom SD, Wangeline A, Ackley A, Kramer K, Norton AP, Lawrence CB, Pilon-Smits EAH (2003) Selenium accumulation protects *Brassica juncea* from invertebrate herbivory and fungal infection. New Phytol 159:461–469

Hanson BR, Lindblom SD, Loeffler ML, Pilon-Smits EAH (2004) Selenium protects plants from phloem-feeding aphids due to both deterrence and toxicity. New Phytol 162:655–662

Hartikainen H (2005) Biogeochemistry of selenium and its impact on food chain quality and human health. J Trace Elem Med Biol 18:309–318

Hladun KR, Smith BH, Mustard JA, Morton RR, Trumble JT (2012) Selenium toxicity to honey bee (*Apis mellifera* L.) pollinators: effects on behaviors and survival. PLoS One 7(4):e34137

Kong LA, Wang M, Bi DL (2005) Selenium modulates the activities of antioxidant enzymes, osmotic homeostasis and promotes the growth of sorrel seedlings under salt stress. Plant Growth Regul 45:155–163

Lauchli A (1993) Selenium in plants: uptake, functions, and environmental toxicity. Bot Acta 106:455–468

LeDuc DL, Tarun AS, Montes-Bayon M, Meija J, Malit MF, Wu CP, AbdelSamie M, Chiang C-Y, Tagmount A, deSouza MP, Neuhierl B, Bock A, Caruso JA, Terry N (2004) Overexpression of selenocysteine methyltransferase in arabidopsis and Indian mustard increases selenium tolerance and accumulation. Plant Physiol 135:377–383

LeDuc DL, AbdelSamie M, Montes-Bayón M, Wu CP, Reisinger SJ, Terry N (2006) Overexpressing both ATP sulfurylase and selenocysteine methyltransferase enhances selenium phytoremediation traits in Indian mustard. Environ Pollut 144:70–76

Lindblom SD, Valdez-Barillas JR, Fakra SC, Marcus MA, Wangeline AL, Pilon-Smits EAH (2013) Influence of microbial associations on selenium localization and speciation in roots of *Astragalus* and *Stanleya* hyperaccumulators. Environ Exp Bot 88:33–42

Lyi SM, Heller LI, Rutzke M, Welch RM, Kochian LV, Li L (2005) Molecular and biochemical characterization of the selenocysteine Se-methyltransferase gene and Se-methylselenocysteine synthesis in broccoli. Plant Physiol 138:409–420

Meija J, Montes-Bayón M, LeDuc DL, Terry N, Caruso J (2002) Simultaneous monitoring of volatile selenium and sulfur species from Se accumulating plants (wild-type and genetically modified) by GC-MS and GC-ICP-MS using SPME for sample introduction. Anal Chem 74:5837–5844

Montes-Bayón M, LeDuc DL, Terry N, Caruso J (2002) Selenium speciation in wild-type and genetically modified Se accumulating plants with HPLC separation and ICP-MS/ES-MS detection. J Anal At Spectrom 17:872–879

Neuhierl B, Böck A (1996) On the mechanism of selenium tolerance in selenium accumulating plants. Purification and characterization of a specific selenocysteine methyltransferase from cultured cells of *Astragalus bisulcatus*. Eur J Biochem 239:235–238

Pilon M, Owen JD, Garifullina GF, Kurihara T, Mihara H, Esaki N, Pilon-Smits EAH (2003) Enhanced selenium tolerance and accumulation in transgenic *Arabidopsis thaliana* expressing a mouse selenocysteine lyase. Plant Physiol 131:1250–1257

Pilon-Smits EAH (2005) Phytoremediation. Annu Rev Plant Biol 56:15–39

Pilon-Smits EAH, Hwang S, Lytle CM, Zhu Y, Tai JC, Bravo RC, Chen Y, Leustek T, Terry N (1999) Overexpression of ATP sulfurylase in Indian mustard leads to increased selenate uptake, reduction, and tolerance. Plant Physiol 119:123–132

Pilon-Smits EAH, Quinn CF, Tapken W, Malagoli M, Schiavon M (2009) Physiological functions of beneficial elements. Curr Opin Plant Biol 12:267–274

Prins CN, Hantzis LJ, Quinn CF, Pilon-Smits EAH (2011) Effects of Selenium accumulation on reproductive functions in *Brassica juncea* and *Stanleya pinnata*. J Exp Bot 247(62):5633–5640

Quinn CF, Prins CN, Freeman JL, Gross AM, Hantzis LJ, Reynolds RJB, Yang S, Covey PA, Bañuelos GS, Pickering IJ, Pilon-Smits EAH (2011a) Selenium accumulation in flowers and its effects on pollination. New Phytol 192:727–737

Quinn CF, Wyant K, Wangeline AL, Shulman J, Galeas ML, Valdez JR, Paschke MW, Pilon-Smits EAH (2011b) Enhanced decomposition of selenium hyperaccumulator litter in a seleniferous habitat—evidence for specialist decomposers. Plant Soil 341:51–61

Rayman MP (2005) Selenium in cancer prevention: a review of the evidence and mechanism of action. Proc Nutr Soc 64:527–542

Sors TG, Ellis DR, Salt DE (2005) Selenium uptake, translocation, assimilation and metabolic fate in plants. Photosynth Res 86:373–389

Stadtman TC (1990) Selenium biochemistry. Annu Rev Biochem 59:111–127

Stadtman TC (1996) Selenocysteine. Annu Rev Biochem 65:83–100

Tamaoki M, Freeman JL, Pilon-Smits EAH (2008) Cooperative ethylene and jasmonic acid signaling regulates selenite resistance in *Arabidopsis thaliana*. Plant Physiol 146:1219–1230

Terry N, Zayed AM, de Souza MP, Tarun AS (2000) Selenium in higher plants. Annu Rev Plant Physiol Plant Mol Biol 51:401–432

Valdez Barillas JR, Quinn CF, Freeman JL, Lindblom SD, Marcus MS, Fakra SC, Gilligan TM, Alford ER, Wangeline AL, Pilon-Smits EAH (2012) Selenium distribution and speciation in hyperaccumulator *Astragalus bisulcatus* and associated ecological partners. Plant Physiol 159:1834–1844

Van Hoewyk D (2013) A tale of two toxicities: malformed selenoproteins and oxidative stress both contribute to selenium stress in plants. Ann Bot 112:965–972

Van Hoewyk D, Garifullina GF, Ackley AR, Abdel-Ghany SE, Marcus MA, Fakra S, Ishiyama K, Inoue E, Pilon M, Takahashi H, Pilon-Smits EAH (2005) Overexpression of AtCpNifS enhances selenium tolerance and accumulation in Arabidopsis. Plant Physiol 139:1518–1528

Van Hoewyk D, Abdel-Ghany SE, Cohu C, Herbert S, Kugrens P, Pilon M, Pilon-Smits EAH (2007) The Arabidopsis cysteine desulfurase CpNifS is essential for maturation of iron-sulfur cluster proteins, photosynthesis, and chloroplast development. Proc Natl Acad Sci U S A 104:5686–5691

Van Hoewyk D, Takahashi H, Hess A, Tamaoki M, Pilon-Smits EAH (2008) Transcriptome and biochemical analyses give insights into selenium-stress responses and selenium tolerance mechanisms in Arabidopsis. Physiol Plant 132:236–253

Van Huysen T, Abdel-Ghany S, Hale KL, LeDuc D, Terry N, Pilon-Smits EAH (2003) Overexpression of cystathionine-γ-synthase in Indian mustard enhances selenium volatilization. Planta 218:71–78

Van Huysen T, Terry N, Pilon-Smits EAH (2004) Exploring the selenium phytoremediation potential of transgenic *Brassica juncea* overexpressing ATP sulfurylase or cystathionine γ-synthase. Int J Phytoremediation 6:111–118
White PJ, Broadley MR (2009) Biofortification of crops with seven mineral elements often lacking in human diets—iron, zinc, copper, calcium, magnesium, selenium and iodine. New Phytol 182:49–84
White PJ, Bowen HC, Marshall B, Broadley MR (2007) Extraordinarily high leaf selenium to sulfur ratios define 'Se-accumulator' plants. Ann Bot 100:111–118
Zhang Y, Gladyshev VN (2009) Comparative genomics of trace elements: emerging dynamic view of trace element utilization and function. Chem Rev 109:4828–4861
Zhang L, Byrne PF, Pilon-Smits EAH (2006a) Mapping quantitative trait loci associated with selenate tolerance in Arabidopsis thaliana. New Phytol 170:33–42
Zhang L-H, Ackley AR, Pilon-Smits EAH (2006b) Variation in selenium tolerance and accumulation among nineteen Arabidopsis ecotypes. J Plant Physiol 164:327–336

Interplay of Water and Nutrient Transport: A Whole-Plant Perspective

Lars H. Wegner

Contents

Abstract This review aims to summarise the many facets of how water flow in higher plants affects nutrient transport and vice versa. Initially, some theoretical background is given on physico-chemical concepts to describe fluxes and their (in) direct coupling, followed by a brief overview on some of the relevant methods (pressure probes, ZIM probe, MIFE technique, radioactive and stable isotopes, MRT flow imaging, heat balance technique, modelling of nutrient fluxes). This essay focusses on roots, on vascular tissues and on the whole-plant level, with only occasional in-depth reference to the molecular scale. Radial water and nutrient transport in roots are discussed in analogy to processes in mammalian epithelia,

L.H. Wegner (✉)
Institute for Pulsed Power and Microwave Technology and Institute of Botany 1, Karlsruhe Institute of Technology, Postbox 3640, 76021 Karlsruhe, Germany
e-mail: lars.wegner@kit.edu

U. Lüttge, W. Beyschlag (eds.), *Progress in Botany*, Progress in Botany 76,
DOI 10.1007/978-3-319-08807-5_5

including a possible role of salt/water cotransporters for generating the "non-osmotic" component of root pressure. Moreover, the significance of low reflection coefficients for the coupling of water and solute transport in roots is critically addressed. Separate sections deal with interactions of water and nutrient transport in vascular tissue (xylem, phloem). Finally, a whole-plant perspective is taken; the significance of transpiration for plant nutrition in general, and for the nutrients N, Ca and K in particular, is discussed.

1 Introduction

In the preceding volume of Progress in Botany, Lüttge (2013) pleaded for a revival of an organismic approach to (whole) plant physiology. Examples for a holistic view on plant science treating vascular plants as a highly integrated functional unit were presented. The author recurred to earlier work from the 1970s and 1980s when whole-plant physiology flourished, and also referred to more recent literature standing in this tradition. Systemic responses of the plant, e.g. to stress or to nutrient availability are highlighted that are brought about by electric, hydraulic and chemical signaling. The opposing concept that currently seems to dominate plant sciences considers the plant as an assembly of more or less autonomously acting "modules". According to this view, it is sufficient to study physiological process at the molecular or cellular scale, since the key processes shaping plant life are supposed to take place at this level: Gene expression can be seen entirely as a cellular process with each cell being fully equipped with the required biochemical machinery. The same is true for chemical interaction of a cell with its environment, mediated by membrane proteins like receptors and ion channels; the information is subsequently transduced, e.g. by transcription factors, leading, via an impact on gene expression, to a modification of the chemical properties of cells (and, in turn, of plant organs and the whole organism). Frequently, it is tacitly implied that features of the organism can largely be extrapolated by simply scaling up cellular processes. A separate discipline, systems biology, has evolved that is concerned with the integration of molecular process at a higher organisational level. However, rather than analysing data obtained by hypothesis-driven research, this discipline is largely concerned with treating datasets obtained by "omics" approaches, using statistical tools to find correlations among them. As pointed out by Lüttge, these concepts of physiology tend to undervalue and neglect synergistic emergence as a key feature of living organisms.

Here I want to elaborate on Lüttge's view, referring to an aspect of emergence that was given less attention in his preceding essay: Integration is not only an issue with respect to different parts of the plant (e.g. root–shoot) or different tissues or cell types. Another aspect is the integration of fluxes, namely those of water, ions, energy and metabolites. Fluxes of energy and matter are key features of living organisms, essentially operating away from thermodynamic equilibrium. These fluxes are highly interactive. The focus of this paper, following my own expertise, will be on the interaction of water and ion (nutrient) fluxes, with occasional references to the other two.

The choice to focus on water and nutrient *fluxes* in this review article, rather than on processes at the molecular level, was also guided by the following more specific considerations:

- Several excellent reviews have been published in the last 5 years that provide a comprehensive overview on our knowledge of different classes of transport proteins, including ion channels (Hedrich 2012) and aquaporins (Maurel et al. 2008). In order to avoid just being repetitive, it was preferable to take a different viewpoint in this essay. By contrast, integrative features of the plant, such as nutrient and water fluxes, received comparatively little attention recently.
- Physiological processes are frequently organised in a highly redundant way with respect to the contribution of various types of proteins. This contributes to the plasticity of the organism and serves as a protection against a complete loss of function by mutations. Hence, it may be difficult to quantify the contribution of a particular protein to a physiological process (even though there are also striking counter-examples; see, e.g. Gajdanowicz et al. (2011)).
- Fluxes are quantitative features of a biological system based on a firm physico-chemical concept (see below) that are relatively easy to integrate into computational modelling of physiological processes. Theoretically, trans-membrane fluxes can be traced back to the contribution of individual membrane proteins, but in practice this link is still out of reach in many cases, among other things because of the plasticity of transport processes (see previous bullet point). Generally speaking, extreme environmental conditions provide the best opportunity to link physiological processes to the activity of a limited number of proteins, since the plant's response to a stress situation clearly dominates all other processes. But such a situation may be a rare event in the life of a plant.
- Not all transport phenomena are mediated by proteins! Although membranes (and membrane proteins) are major sites of control, nutrients and water on their way from root to shoot pass through the bulk of various cellular and extracellular compartments. (Electro)diffusion and mass-flow-driven transport of ions, e.g. are important transport mechanisms.
- Fluxes can readily be interpreted in an ecological context, e.g. when the soil–plant continuum is considered.

Of course, focussing on fluxes and their interaction does by no means preclude highlighting the contribution of individual proteins in a transport process where this is appropriate. Rather, the integrative approach that is favoured here implies that processes at the higher organisation level (plant tissues, organs and the whole organism) and emergence phenomena are given at least equal attention.

In this essay, I will focus on the macronutrients K, N (provided either as NO_3^- or as NH_4^+) and Ca and ion-specific aspects of their interaction with water transport [for their physiological roles, see Marschner (1995)]. Moreover, I will discuss aspects of flux coupling relevant for all nutrients in a more general way.

2 Physico-Chemical Concepts to Describe Interactions of Water and Nutrient Transport

Generally speaking, a flux, J_i, denotes the number of molecules n of a species i moving across a boundary area, A, (e.g. a membrane surface) per time increment:

$$J_i = -\frac{1}{A}\frac{dn_i}{dt} \quad (1)$$

In order to quantify the transport of water, it is more convenient from an experimental point of view to consider volume flow (J_v) rather than transport of water molecules:

$$J_v = -\frac{1}{A}\frac{dn_{H_2O}}{dt}\bar{V}_w \quad (2)$$

$\bar{V}_w$ = molar volume of water (Note that for a dilute aqueous solution, the contribution of the solutes to volume flow can be neglected).

A flux cannot be treated as a singular event, but is part of a complex pattern of interconnected processes. In most textbooks dealing with this matter, coupling of water and nutrient transport is solely considered with respect to the osmotic force that solutes dissolved in water exert on water transport. Net transport of nutrients across a membrane will alter concentration gradients and, via the concomitant effect on the osmotic balance, will induce passive water transport. In plant cells, water flow between a cell and its environment also affects the cell turgor P, i.e. the hydrostatic overpressure inside the cell. In order to take the interplay of hydrostatic and osmotic pressure into account, the water potential, Ψ, was defined that is usually considered as the key parameter for water transport:

$$\Psi = P - \pi \quad (3)$$

with the osmotic pressure, π, calculated according to van't Hoff's law: $\pi = RT$ c (R being the gas constant and T the absolute temperature in Kelvin). Where appropriate, this equation has to be extended by a gravitational term $\rho*g*h$ (ρ being the density of the media, g the gravitational constant and h the height) and a matrix potential τ (with a negative sign) taking into account the capillary forces exerted by the cell wall material that have to be overcome when this material is (partly) dehydrated.

Water is thought to follow the Ψ gradient, flowing from higher to lower potential. However, this approach is not unambiguous, and some notes of caution are required:

- Water flow is not induced by an osmotic gradient per se, but only between compartments differing in osmotic pressure *and* separated by a semi-permeable membrane. This second, mandatory prerequisite is fulfilled when water exchange between cells and their environment is considered, but, e.g. usually not for water transport within the apoplast. Osmotic gradients within a compartment (in the absence of a semi-permeable transport barrier) dissipate by diffusion, not by volume flow!

- Most importantly, direct interaction of water and solute transport during membrane passage is neglected by this approach. A number of studies published in the 1980s of the last century seemed to justify this approach (e.g. Palta and Stadelmann 1980). Evidence was provided that plant membranes are ideally semi-permeable for physiologically relevant salts and sugars, implying that direct interaction of water and solute transport is negligible, and solutions exert their full osmotic pressure as a driving force for water transport. However, recently Wegner (2014) has argued that water-solute cotransport across plant cellular membranes needs to be re-considered in the light of previous advances in our understanding of water secretion in mammalian epithelia [for a review, see Zeuthen and MacAulay (2012)]. In specialised cell types like xylem parenchyma, transport proteins, e.g. those of the CCC class that carry a stoichiometrically fixed number of water molecules together with a salt (e.g. KCl) could account for the generation of root pressure and refilling of embolised xylem vessels (for a more detailed discussion of those physiological processes, see below). Gradients in chemical potential of solutes could drive water secretion across a membrane in the absence of a water potential gradient, or even against it. Salt or sugar-driven water transport is not easy to detect, since it is necessarily associated with a subsequent retrieval of the "lost" solutes by the cell to "recharge the battery" and thus keep the process going. This will, at some step, require metabolic energy. Only constant cycling of solutes can drive sustained water secretion, and net solute gradients may even not be affected by this process. Seemingly futile cycling of solutes has indeed been demonstrated to occur between symplast and apoplast in plant tissues, using sophisticated radiaoactive labelling techniques (Britto and Kronzucker 2006). Although osmotica are involved in this process, water flow is not driven by (local) osmotic forces as hypothesised by Diamond and Bossert (1967). In their "standing osmotic gradient" hypothesis that is fundamentally different from the one proposed here, local cavities play a crucial role that are in free exchange with the bulk via one end, whereas the other end is closed. Salt is actively transported into the cavity at the closed end by adjacent cells. Due to passive water flow into the cavity, the osmotic pressure is decreasing along its length; progressive dilution towards the open end results in the constant export of a medium of low osmotic pressure into the bulk. A similar mechanism was also proposed to operate in plants, e.g. in salt glands (Pate and Gunning 1972; Lüttge 1975) and to drive root pressure exudation (Katou et al. 1987), but neither in mammalian epithelia nor in plant tissues clear experimental evidence in favour of this hypothesis could be obtained.

Co-transport of water and solutes may, in fact, fundamentally change our view of many processes in plants that involve a shift of water. Inside the water-filled bladder traps of carnivorous *Utricularia* species, e.g. a hydrostatic underpressure of up to 17 kPa is generated with respect to the ambient pond water (Bentrup 1979). Water efflux occurs against an osmotic gradient across the bladder walls consisting of two cell layers. Hence, water is secreted against the water potential gradient that favours water influx into the lumen of the trap. This could be brought about by water-solute cotransport across the cellular membrane of the wall cells facing the ambient medium, thus overcoming passive water influx. Consistently, generation of underpressure is correlated with the export of K^+, Na^+ and Cl^- from the trap lumen (Bentrup 1979). Fast movements, e.g. stomatal closure, as well as nastic responses of leaves and leaflets to touch that are associated with water shift in pulvini, e.g. in *Mimosa pudica*, may also operate by a water secretion mechanism involving cotransport (Zeuthen 1996 p. 109; Morillon et al. 2001).

It should be noted that direct coupling of water and solute flow also occurs in other transport proteins. Ion channels allow passage of ion *and* water, and it has been demonstrated, e.g. for K^+ channels that water flow can drive K^+ transport through the pores (Homblé and Véry 1992). Moreover, some aquaporins are not only permeable to water but also to other small solutes such as urea or CO_2 (Uehlein et al. 2003).

A discussion of fundamental aspects of the coupling of water and nutrient transport would be incomplete without mentioning unstirred layer effects. When a transport barrier is more permeable to water than to solutes (as is the case with semipermeable membranes), volume flow across that barrier will generate a local concentration gradient across the membrane (and, in turn, an osmotic gradient that counteracts volume flow). Solutes accumulate at the membrane in the "upstream compartment" (concentration-polarisation) whereas in the "downstream compartment" solutes are diluted close to the membrane (sweep-away effect). Unstirred layer effects are most pronounced in compartments that equilibrate by comparatively slow diffusion processes rather than by convection. For membrane transport processes, actual concentrations at the membrane surface are more relevant than bulk concentrations, but are also more difficult to measure! Unstirred layer effects are frequently undervalued. They can have a strong impact on transport processes and may be the actual cause of many interactions of water and solute transport reported in the literature that are ascribed to membrane proteins.

A formalism taking co-transport of water and solutes across membranes into account is provided by the following equation. (For the derivation of this equation and the theoretical framework provided by the thermodynamics of irreversible processes, the reader is referred to Zimmermann and Steudle (1978)):

$$J_v = L_P(\Delta P - \sigma \Delta \pi) \qquad (4)$$

with L_P being the hydraulic conductance of the transport barrier. Direct interaction of water and solute transport is accounted for by the reflection coefficient σ. If the membrane is ideally semipermeable, σ is close to unity and water and solutes move independently across the membrane; σ values <1 provide evidence for an interaction of solute and water fluxes, e.g. by solute–water cotransport (Wegner 2014). Solute-driven water flow counteracts water flow driven by osmotic forces (that is largely mediated by aquaporins) and, hence, reduces the apparent impact of the osmotic pressure gradient (whereas turgor-driven water flow remains unaffected).

Correspondingly, solute transport, J_s, can be described as a sum of two components, one of them comprising the interaction with water flow ("solvent drag"), whereas the other is equivalent to diffusion, i.e. flux driven by a concentration gradient across the membrane (other forces are not taken into account):

$$J_s = \underbrace{J_v \bar{c}_s (1 - \sigma)}_{\text{solvent drag}} + \underbrace{\varpi \Delta c_s}_{\text{diffusion}} \tag{5}$$

$\bar{c}_s$ and ω denote the averaged concentration of the solute s in both compartments separated by a transport barrier (membrane) and the permeability for this solute, respectively. These equations are unsuitable to describe the complex transport phenomena at membrane level in all details (e.g. they neglect electrical forces). However, they provide a simple mathematical framework for taking the interaction of water and nutrient transport into account and can potentially be extended on the basis of the simple formalism provided by the thermodynamics of irreversible processes (Zimmermann and Steudle 1978). Interestingly, Eq. (5) is also valid in the absence of a semipermeable membrane, e.g. when mass flow in the capillaries of the cell wall matrix and in xylem vessels is considered. Under these conditions, σ attains a value of zero, and J_s is equal to $J_v \times c_s$ (ignoring the relatively slow diffusion process that will make a small contribution to J_s under these conditions).

Further refinement of our physico-chemical concepts will be required in the future, e.g. to take non-linearities in the relationships between fluxes and driving forces into account (Ciancio and Verhás 1994).

3 A Brief Survey of Techniques Used to Measure Nutrient and Water Fluxes in Plants

In experimental sciences, progress is bound to the available methods, their advantages and limitations. Hence, it appears appropriate to introduce some of the methods that are routinely used for recording nutrient and water fluxes and their interactions. This summary and the information given on each of these methods cannot be comprehensive, and suggestions for further reading will also be provided.

Pressure Probe Techniques In plants, hydrostatic pressures can vary over a wide range, with overpressures of up to about 1 MPa prevailing in the cells (the "turgor pressure") to values below vacuum in functional xylem vessels [for a review, see Zimmermann et al. 2004)]. Pressure gradients provide a major driving force for volume flow; frequently, hydrostatic pressure data can be used to obtain information on changes in osmotic pressure and to determine the hydraulic conductivity of a biological structure. The basic concept of the turgor pressure probe was initially introduced by Zimmermann et al. (1969); a fine-tipped glass capillary suitable for impaling a cell is attached to a microbaric chamber carrying a miniature pressure sensor. The interior of the chamber and the capillary are filled with silicone oil forming—once being inserted—a meniscus with the cells sap that is kept at a constant position close to the cell surface to eliminate artefacts related to the elasticity of the probe during pressure recording (Tomos and Leigh 1999). Later, this tool was modified in order to measure root pressure in excised roots ("root pressure probe"; Steudle and Jeschke 1983), and to record xylem pressure in individual vessels ("xylem pressure probe"; Balling and Zimmermann 1990). The latter is filled with degassed water (or electrolyte solution) instead of oil (for experimental details on this type of probe, see Zimmermann et al. (2004); Wegner (2012)). Still later, an advanced version of the xylem pressure probe, the multifunctional xylem probe, was developed (Fig. 1a). In addition to measuring xylem pressure, this tool allows simultaneous recording of the electrical potential in a xylem vessel with respect to a reference electrode outside the plant (the so-called trans-root potential) and of ion concentrations (or rather, activities) in the xylem. For multifunctional xylem probes, double-barreled electrodes are used; one barrel serves for measuring xylem pressure and electrical potential. This barrel is attached to the body of the microbaric probe (now additionally containing an Ag/AgCl electrode), whereas the other one is prepared as an ion-selective electrode. For more technical details, the reader is referred to a recent review by Wegner (2012). Note that this instrument is particularly well suited to measure interactions of water and nutrient transport, since the main driving forces for water and nutrient transport are accessible at a particular site in the vascular system; probes measuring K^+ and pH have been published; moreover, a nitrate selective probe has been designed (S. Scherzer and L.H. Wegner, unpublished).

The types of pressure probes introduced so far were predominantly devised for use in the laboratory or greenhouse. They are hardly suitable for field studies, though, and their use requires some effort and skill. Recently, however, a novel type of pressure probe, the ZIM probe, has been introduced by Zimmermann and coworkers, which offers a simple and inexpensive way for measuring relative values of turgor pressure (changes) both in the field and the laboratory in real time and with high precision. The ZIM probe makes use of a miniature pressure sensor embedded in a polymeric matrix that is pressed to a leaf by magnetic force (Fig. 1b); a magnetic counterpad is required at the opposite side of the leaf. Turgor pressure of cells in the leaf patch covered by the probe tends to oppose the magnetic force exerted upon the clamped leaf area. As a consequence, the pressure sensed by the pressure sensor (the "output pressure") is relieved, i.e. turgor pressure and output pressure are inversely related [for details on the measuring principle, see Zimmermann et al. (2013); Westhoff et al. (2009); Zimmermann et al. (2008); Rüger et al. (2010)]. The ZIM probe has great potential for application, e.g. in plant phenotyping and irrigation scheduling (Bramley et al. 2013), but it is also a very useful tool for basic research. Rüger et al. (2010) could, e.g. monitor water shifting in the canopy of Avocado and Eucalyptus trees when branches were differently

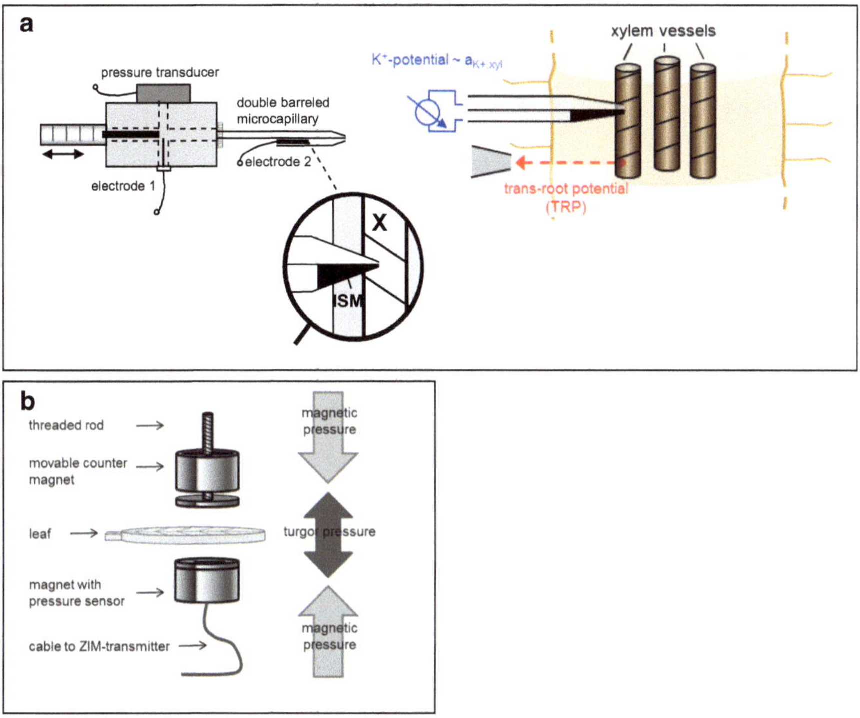

Fig. 1 Diagrams showing advanced pressure probe techniques. (**a**) Multifunctional xylem pressure probe (*left*), consisting of a microbaric Perspex chamber with miniaturised pressure sensor attached to it. An Ag/AgCl electrode (electrode 1) is integrated into the probe for simultaneous recording of hydrostatic pressure and electrical potential in an individual xylem vessel (x). A double-barreled microcapillary is used. One barrel connects (in the impaled state) the lumen of the xylem conduit with the probe for recording pressure and electrical potential, the second one is designed as an ion-selective electrode (electrode 2) to measure, e.g. K^+ or H^+. The very tip of this barrel is filled with an ion-selective matrix (ISM; under the "magnification glass"). An Ag/AgCl electrode is inserted into this barrel to record the K^+ potential. The experimental design for roots is shown in the scheme on the right side. The potential difference of the two barrels corresponds to the K^+ potential; this potential is used to calculate the xylem K^+ activity ($a_{K+,xyl}$) based on pre-/post-calibration of the ion-selective electrode. Moreover, the electrical potential difference between a xylem vessel and an external electrode, the trans-root potential (TRP), is recorded. For further details, see Wegner (2012). (**b**) Schematic diagram of the measuring principle of the magnetic ZIM-probe. For more details, see the text

exposed to light; water was re-directed to those branches that suffered from temporary drought stress. Uneven allocation of water to different branches was also observed in oak trees during the course of the day (Fig. 2).

The MIFE Technique An elegant way to quantify ion fluxes at the surface of plant organs or isolated cells is provided by the Microelectrode Ion Flux Estimate (MIFE) technique originally designed by Ian Newman (for reviews, see Newman 2001; Shabala et al. 2012). The basic concept relies on the fact that ion fluxes across

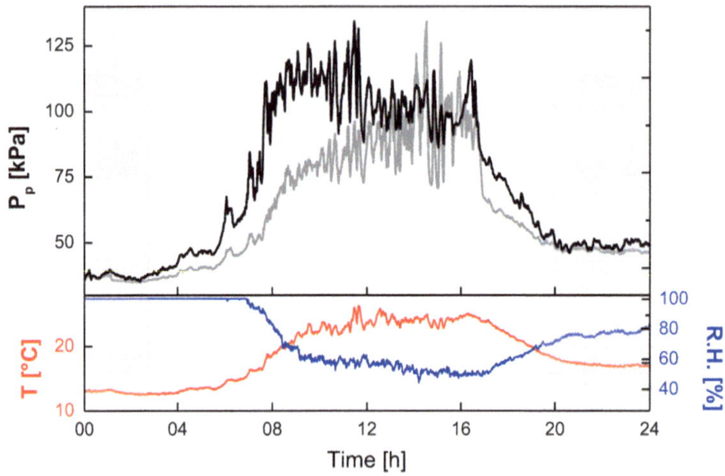

Fig. 2 Multiple probe readings with the ZIM probe (Fig. 1b) on leaves of a 30-m tall oak tree for one day (02 July 2013, Germany) together with the corresponding profiles of local air temperature (T; *lower panel*; *red line*) and relative humidity (R.H.; *lower panel*, *blue line*; unpublished data, U. Zimmermann and S. Rüger). P_p values (that are inversely proportional to leaf turgor pressures) were normalised to the P_p range of the respective probes located on the west side. Nocturnal hours are marked by *grey bars*. Probes were clamped in the east (*black line*) and in the west (*grey line*). Please note the P_p oscillations due to stomatal oscillations during the day. For details see also Zimmermann et al. (2008) and Rüger et al. (2010)

a membrane or the surface of a plant organ (such as the root) will establish an ion gradient in a dilute medium bordering on that surface. This gradient is experimentally accessible by using ion-selective microelectrodes that constantly migrate between two positions (at a distance of a few μm). From the known diffusion coefficient of the dilute medium, ion fluxes can directly be calculated at a high temporal and spatial resolution. An important prerequisite of this approach is, however, that volume flow across the membrane is negligible, since volume flow tends to build up local unstirred layers [see above, sect. 2. In fact, Pohl and co-workers even made use of this fact to quantify volume flow across an ideally semi-permeable membrane; Pohl et al. (1997)]. These volume-flow induced ion gradients would tend to be misinterpreted in the theoretical framework of the MIFE theory.

Use of Isotopes to Label H_2O or Certain Nutrients An early boost in plant transport physiology in the 1960s and 1970s profited from the availability of radioactive tracers for chemical elements relevant for plant nutrition, particularly H (^3H, to label water), K (^{86}Rb), N (^{13}N), S (^{35}S) and ^{32}P. When fed, e.g. to the root, distribution of these tracers in the plant could easily be followed by local recording of radiation, e.g. by scintillation. Moreover, techniques for measuring and interpreting washout kinetics of radioactive tracers (Metzner et al. 2008) were developed (Britto and Kronzucker 2013). A particular challenge is provided by short-lived radioisotopes such as ^{42}K (half life about 12 h). In the meantime, rare stable isotopes also gained popularity as tracers (e.g. ^{41}K, ^{26}Mg, ^{15}N, ^{44}Ca, ^{18}O and ^{2}H (=deuterium) for labelling water); these isotopes can be identified by mass spectrometry, e.g. with high spatial resolution in combination with cryo-microscopy (*S*econdary *I*on *M*ass *S*pectrometry = SIMS-technique; Metzner et al. 2008). A definite advantage of using isotopes as tracers is that unidirectional fluxes can (initially) be measured. This can provide information, e.g. about rapid cycling of ions across membranes (Britto and Kronzucker 2006) that is unavailable by any other technique.

Isotope labelling proved to be ineffective for measuring water flow in plants because of the extremely fast exchange kinetics of water, but other techniques are available to measure volume flow.

Flow Imaging by MRT Nuclear magnetic resonance tomography (MRT) provides a convenient way to monitor, among other things, volume flow with a high temporal and spatial resolution. It is a big advantage of this technique that it is truly non-invasive. The technique has been used, among other things, to monitor xylem and phloem flow simultaneously (Rokitta et al. 1999; Peuke et al. 2001). Flow imaging takes advantage of the fact that non-aligned H-spins move into the measuring plane during scanning. From the data, a flow velocity can be calculated; by multiplication with the conducting area (that can be obtained from flow images) volume flow becomes accessible (e.g. Schulze-Till et al. 2009).

Sapflow Measurements Based on Local Heating While the MRT technique provides very detailed information on volume flow, but is challenging from a technical point of view and requires sophisticated equipment, the heat dissipation or heat balance techniques provide a simpler alternative suitable for field studies (Smith and Allen 1996; Renninger and Schäfer 2012). Local, constant heating of a shoot (e.g. a tree trunk) leads to subsequent convection in the tissue until a constant temperature (measured with respect to a second thermocouple some 10–15 cm below the first one, serving as a reference) is attained. At this point of time energy input equals dissipation. Sap flow passing the electrode will lower this temperature, and the temperature drop is a direct measure of volume flow (also called thermal dissipation or Granier-style probe). Alternatively, sapwood is heated locally and the asymmetrical temperature increase below and above that site (upstream and downstream, respectively) is monitored that is related to volume flow (heat balance technique in the strict sense, or heat field deformation). Because of their technical simplicity these as well as related methods have frequently been used especially to measure transpirational flow in trees, but there is some debate on the reliability of volume flow data obtained this way (Shackel et al. 1992; Renninger and Schäfer 2012).

Flow Modelling According to the Method of Pate and Jeschke An amazingly simple and efficient way of *ex post* modelling of fluxes of nutrients between root and shoot is provided by tapping xylem and phloem sap from individual plants and subsequently harvesting root and shoot biomass to analyse them with respect to the particular nutrient of interest (Jeschke and Pate 1991). From these data, xylem and phloem fluxes as well as total net uptake by the root can be estimated for the whole lifetime of the plant [summarised, e.g. by Peuke (2010)]. A critical point is that the composition of vascular saps may be highly variable with time and data just represented a—somewhat arbitrary—snapshot, whereas the nutrient content of root and shoot tissue has evolved over the whole lifetime of that plant.

4 Radial Transport of Water and Nutrients in Roots: Transpiring Plants

Uptake of water and nutrients by roots and subsequent radial transport into xylem vessels provides a good case study to illustrate the more general considerations outlined in Sects. 1 and 2. The cellular membrane of cortex cells is frequently considered to be the primary interface separating the plant from its local environment, the adjacent soil (Schroeder et al. 2013; Wang et al. 2012). A plethora of transport proteins co-located in this membrane has been studied in much detail. However just focusing on the membrane level would be too reductionist a view on transport processes in this highly complex organ. It is more adequate to adopt a concept well known from animal physiology and to consider root tissue as an epithelium consisting of several cell layers that separate two extracellular compartments: The lumens of the (dead) xylem vessels in the root centre, and the soil solution at the periphery of the root. In analogy to mammalian epithelia, transfer of water and nutrients across this barrier can occur symplastically—via the cells and plasmodesmata connecting them—or apoplastically, i.e. via the cell walls. Apoplastic transport is limited by the Casparyian band, with suberin depositions in the cell wall of the endodermis in its mature state, separating the cortex from the stele in the root centre. In some species the exodermis forms an additional, more peripheral transport barrier in the root apoplast. Some research effort has been invested into quantifying the relative contributions of the two transport pathways—symplastic and apoplastic—to radial water and nutrient transport. Many researchers favoured the "composite transport model" that was originally proposed and promoted by E. Steudle and coworkers (Steudle and Peterson 1998; Steudle 2000a), based on results obtained with the root pressure probe. When the root is challenged with a hyper-osmotic shock by adding various osmotica to the ambient medium, in many cases the root pressure drop did not match the change in external osmotic pressure, indicating that the radial reflection coefficient (calculated from the ratio $\sigma_r = \Delta P/\Delta\pi$) is significantly lower than unity. Following the "composite membrane" model, the root reflection coefficients were considered a measure to quantify the contribution of the cellular pathway (for which $\sigma_r = 1$ was assumed for many osmotica) to overall volume flow; consistently, a σ_r value of 0 was ascribed to the apoplastic pathway (Steudle 2000b). Steudle and co-workers concluded from their root pressure probe measurements that the apoplastic pathway contributed significantly to radial water transport in many species. However, more recently the composite transport model and the experimental approach on which it was based were heavily critisised for various reasons. Bramley et al. (2007) identified technical flaws in the use of the root pressure probe and the interpretation of the probe data when performing experiments with two probes each attached to one cut surface of a root segment. Moreover, Bramley et al. insisted that the experimental procedures to determine hydraulic conductivity and reflection coefficient of the root were prone to produce artefacts, since they tend to affect local gradients in water potential (and gradients in solute concentrations) and give rise to unstirred layers (see also the rebuttal by Steudle and coworkers; Knipfer et al. (2007)). Unstirred layer effects could also be responsible for low values of radial root reflection coefficients. Most likely the significance of the apoplastic pathway is erroneously over-estimated when these effects are neglected.

The same conclusion was drawn, about a decade before, by Schneider et al. (1997a, b) from their osmotic experiments on intact seedlings of maize, barley and wheat using the xylem pressure probe. Schneider et al. observed that the radial reflection coefficient in roots challenged with an osmotic shock strongly depended on the transpiration rate. In accordance with Steudle (2000a, b), apparent σ values were significantly smaller than one in roots of non-transpiring plants and at low transpiration rates, but increased to unity when transpiration was stimulated by the light regime. Only at peak photon densities prevailing in greenhouses on Hawaii at noon, σ values passed through a maximum and tended to decrease again. Schneider et al. (1997a) argued that transpiration-driven flow would tend to abolish unstirred layer effects to some extent, leading to an increase of apparent σ values. Consistently, it could be shown that the transpiration rate remained unaffected by the osmotic challenges in those experiments, with the exception of seedlings exposed to extremely high irradiation rates. It was concluded that radial water flow was predominantly symplastic and that low σ values were due to unstirred layer effects, but not to an apoplastic bypass for water transport, at least in the cereals tested in those studies. The same conclusion was later drawn by Knipfer and Fricke (2010) when repeating earlier experiments on barley with the root pressure probe while stirring the external medium. Evidence against the composite transport model was also obtained by Fritz and Ehwald (2011). For maize, they investigated radial transport into the root xylem of mannitol and other test solutes that are known to be almost membrane-impermeable. Hence, radial root transport of these solutes is necessarily predominantly apoplastic. In contrast to the predictions of the composite transport model, no evidence for a solvent drag effect in the transport rates of these solutes was obtained and the radial root reflection coefficient was close to unity; transport into the xylem did occur, however, but it was largely diffusive, leading Fritz and Ehwald to the conclusion that the endodermis prevented radial water transport in the root, but retained some permeability to solutes like mannitol. The same conclusion may be true for Ca^{2+} (White 2001). Suberins and waxes are generally thought to form an effective diffusion barrier. Therefore, apoplastic diffusion of solutes across the endodermis is most likely restricted to the developing root zone where cell walls are not fully suberised, especially at passage cells, and at sites of lateral root formation. However, no real quantitative information is available on the permeability of suberin layers to solutes as pointed out, e.g. by White (2001).

While radial reflections coefficients were apparently misinterpreted in the past, this is no reason to dismiss these results. On the contrary, apparent reflection coefficients provide insight into the response of the plant root to an osmotic challenge. Values significantly lower than one imply that the impact of soil osmotic pressure fluctuations on xylem pressure is damped. This is highly beneficial for the plant, since a drop in xylem pressure to very negative values could entail cavitation.

The previous discussion of radial root transport at the organ level has revealed severe conceptual as well as methodological problems of techniques to measure interaction of water and nutrient transport at the organ level. Even though there are no "artefact-free" methods in physiology, it is important to remain aware of potential sources of error and to minimise them as far as possible. A major problem of the root pressure probe (and some other techniques) is that work is done on excised roots and relevance for intact, transpiring plants is questionable (Shabala et al. 2009). Moreover, transport parameters and flux data are extracted under rather non-physiological conditions. It was vigorously debated, e.g. whether pressure clamp or pressure pulse relaxation protocols would be more suitable to determine the hydraulic conductance of the root, but the reader is left with the conclusion that eventually both approaches are inadequate. Physiologically meaningful data on root hydraulic conductance of transpiring plants can only be obtained with steady-state volume flow and at a "free-running" xylem pressure under transpirational control. While volume flow can be assessed gravimetrically (Wegner and Zimmermann 2009), using the heat balance technique or by gas exchange measurements, xylem probes are the only instruments to provide us with relevant data on hydrostatic xylem pressure. Therefore, Wegner and Zimmermann (2009) revisited the problem, using a multifunctional xylem probe that allowed to record xylem K^+ in addition to xylem pressure (Fig. 3). With this approach, root hydraulic conductance could be calculated for intact maize seedlings according to Eq. (4) using steady-state data. All parameters were experimentally accessible, assuming the osmolarity of the sap to be four times the K^+ activity. Interestingly, both hydraulic conductance and radial K^+ transport depended in a non-linear way on radial volume flow (Fig. 4a–c), but not directly on hydrostatic xylem pressure, the main driving force for radial volume flow. However, in the light of the context described above this was not interpreted in terms of a solvent drag effect in the strict sense. Rather, it was argued that radial volume flow into the xylem leads to an accumulation of K^+ in xylem parenchyma cells, building up a steep K^+ concentration gradient across the membrane of these cells. This will lead to an enhanced rate of K^+ efflux because of an increase in the driving force; moreover, the open probability of the relevant ion channel, the K^+ outward rectifier SKOR (Liu et al. 2006), is enhanced by an increase in cytosolic K^+, providing an additional feedforward effect. Non-linear dependence of volume flow on the driving force is probably due to the regulation of aquaporins in inner cortex and endodermis cells that serve as a bottleneck for radial water transport (Wegner and Zimmermann 2009). Evidence for an enhancement of

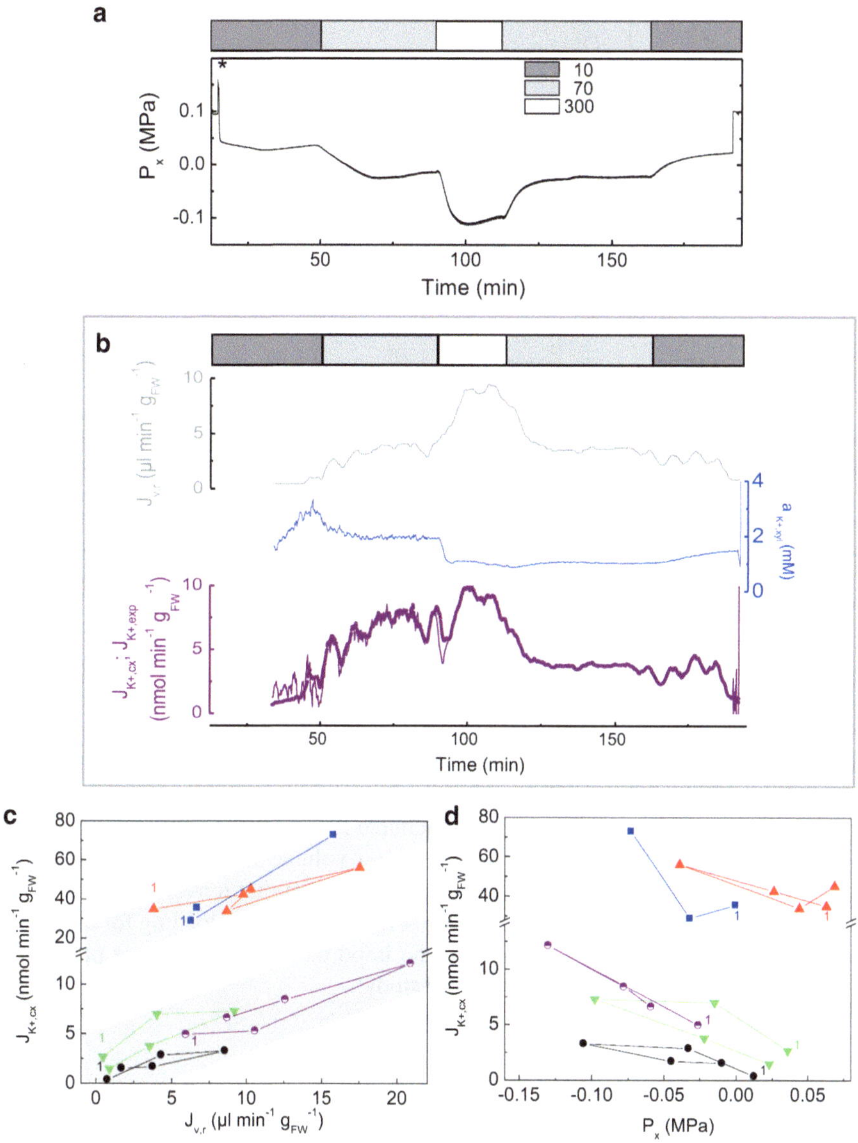

Fig. 3 Dependence of various water and K^+ transport parameters on a varying light regime in a maize seedling. Response of xylem pressure (P_x), radial volume flow ($J_{v,r}$), xylem K^+ activity ($a_{K+,xyl}$), K^+ flux into the xylem ($J_{K+,cx}$) and K^+ export to the shoot ($J_{K+,exp}$) to a stepwise increase and subsequent decrease of light irradiation was recorded with a multifunctional xylem probe (see Fig. 1a) and simultaneous gravimetrical recording of water uptake by the seedling. *Bars* on *top* of the figures indicate the time schedule of light regime changes (in µmol m^{-2} s^{-1}); Fig. 1a shows the time course of xylem pressure. Impalement of a xylem vessel (indicated by *asterisk*) at laboratory light (10 µmol m^2 s^{-1}) was associated with a pressure drop from atmospheric to 0.44 MPa. In (**b**), the radial volume flow (*top trace*), the xylem K^+ activity (*middle trace*) as well as the K^+ flux into the root xylem (*bottom, thin line*) and, almost identically, the K^+ export to the shoot (*bottom, thick line*) are plotted with time for the same experiment. Note that for technical reasons, recording of $J_{v,r}$ started with a delay of about 15 min with respect to impalement. For this (*filled inverted*

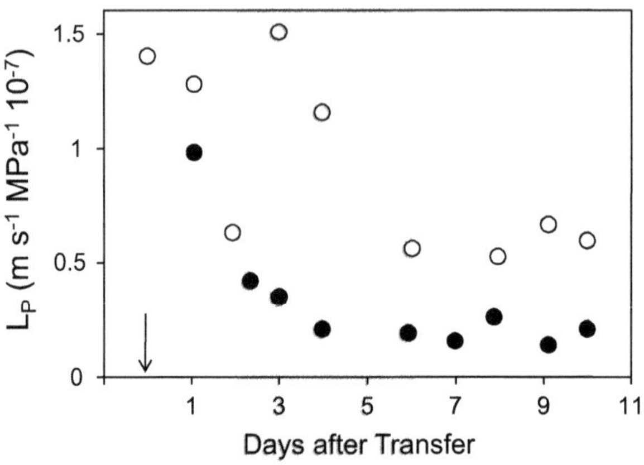

Fig. 4 The hydraulic conductance recorded on individual root cortex cells of cotton depends on the nutritional status of the plant. The experiment was performed with a cell pressure probe. Seeds were allowed to germinate on moist vermiculite and transferred at day 3 (*arrow*) either to a full nutrient solution (*open symbols*) or to a –N medium composed in the same way except for nitrate being replaced by chloride (*solid symbols*). Note that hydraulic conductance dropped on a daily basis in root cells deprived of N, whereas the drop in fully supplied roots was less pronounced. After Radin and Matthews (1989), with modifications

K^+ and nitrate loading into the xylem by volume flow was also obtained by Schurr and Schulze (1995) on intact castor bean plants (but not on detopped root systems!). Xylem sap was sampled locally at a site of incision of the shoot while pressurising the root system. This approach was retrospectively justified by the observation of Wegner and Zimmermann (2009) that xylem pressure has no direct effect on the rate of xylem loading of K^+. Schurr and Schulze found little dependence of the K^+ and nitrate concentrations in the xylem sap on volume flow in the range tested experimentally (in contrast to the detopped root system that rendered a hyperbolic dependence). This pinpoints to an increase in the rate of xylem loading for both ions with the radial volume flow. Moreover, the importance of using intact plants for those studies is again highlighted by their study!

Fig. 3 (continued) triangle) as well as four other seedlings, dependence of $J_{K+,cx}$ on $J_{v,r}$ and P_x is shown in (**c**) and (**d**), respectively. The sequences of light regimes for these experiments were (irradiance in µmol m^{-2} s^{-1}): 10–70–300–70–10 (*filled circle, filled inverted triangle, filled triangle*); 10–70–300 (*filled square*). The index "1" indicates the starting value for the respective experiment. Despite some hysteresis, plants could be divided into two groups with respect to the dependence of $J_{K+,cx}$ on $J_{v,r}$ as indicated by shaded areas in (**c**): In three plants, K^+ flux was close to zero in the absence of volume flow, whereas in two further plants, extrapolation yielded a considerable rate of xylem loading at zero volume flow (about 15 nmol min^{-1} $g_{FW}{}^{-1}$). A weak correlation was also found between $J_{K+,cx}$ and P_x (**d**), but separate experiments revealed that varying P_x at a constant volume flow would not affect xylem loading of K^+ (not shown). For more details, see Wegner and Zimmermann (2009)

5 Radial Transport of Water and Nutrients in Roots: Root Pressure

The xylem pressure probe and its further developments proved to be very useful tools to study water and nutrient transport in transpiring plants. However, since a below -atmospheric pressure is required to locate the probe tip in a vessel with certainty, these probes are not suited to impale roots of non-transpiring plants building up root pressure. From a physiological point of view, root pressure is a special case; it dominates long distance water transport, e.g. in very young seedlings and at a water-saturated atmosphere. According to most textbooks, radial water flow is driven by an osmotic overpressure of the xylem sap with respect to the ambient medium. However, this explanation is at least insufficient to describe the phenomenon adequately; there are many reports in the literature that water secretion into the vessels prevails in the absence of an osmotic gradient between xylem sap and ambient medium, or even against such a gradient. The latter observation has puzzled researchers for decades (Oertli 1966; Enns et al. 2000). Recently Wegner (2014) has reviewed the available experimental evidence and suggested a new hypothesis to explain root pressure [and root pressure exudation; i.e. constant "bleeding" from the cut surface(s) of excised roots, root segments and even cortex sleeves after the removal of the stele (Volkov and Zholkevich 1993)] by a "non-osmotic" mechanism. It was suggested that water secretion across the plasma membrane of xylem parenchyma cells is driven by a cotransport of water and solutes as previously shown for mammalian epithelia; solute concentration gradients across the cellular membrane of xylem parenchyma cells are supposed to provide the free energy to drive water secretion into the xylem vessels, even against a gradient in the chemical potential of water (or "water potential"). For various mammalian epithelia, T. Zeuthen and his co-workers provided multiple evidence for the existence of membrane transporters that co-translocate solutes and water at a fixed stoichiometry (Zeuthen and MacAulay 2012; Zeuthen 2010). A key role is apparently played by transporters of the CCC type that transport either KCl, NaCl or both salts simultaneously together with a fixed number of 150–500 water molecules. In order to maintain the ionic gradient across the membrane that is dissipated by this transport step, ions have to be retrieved again from the extracellular compartment at the expense of metabolic energy.

Interestingly, homologues of the CCC family have also been discovered in plants. Kong et al. (2011) cloned a cation-chloride cotransporter in rice that seems to translocate K^+, but not Na^+. These authors tested subcellular localisation of a CCC–GFP fusion protein and could demonstrate that the protein was predominantly allocated to the plasma membrane. The only CCC-type transporter found in *Arabidopsis* showed also highest homology with the subfamiliy of KCl cotransporters (the KCCs), but reconstitution in oocytes showed that $^{86}Rb^+$ (as a tracer for K^+) was only translocated in the presence of Na^+, indicating that this transporter functions like cotransporters that transport Na^+ and K^+ together with two Cl^- ions (the NKCCs; Colmenero-Flores et al. (2007)). Interestingly, the cotransporter found in *Arabidopsis* turned out to be prominently expressed in vascular tissue. It is well conceivable that these transporters are involved in the directed, radial transport of water into xylem vessels by which root pressure is built up, although availability of Na^+ and $Cl^-$and the gradients of these ions are a critical factor. It is unknown whether the plant transporters can also work with other anions like nitrate that are of no relevance for the animal system, nor is any information on water permeability of the plant transporters available yet. A major challenge to the "water cotransport hypothesis" both in mammalian and plant tissues is also provided by the presence of aquaporins in the cellular membranes of the cells that tend to dissipate any water potential gradient; however, it could be shown by model calculations (Wegner 2014) that cotransporters could operate against a moderate hydraulic conductance of the membrane. Re-absorption of K^+ and Na^+ (required for "keeping the battery charged") would be brought about by inward-rectifying K^+ channels in xylem parenchyma cells (Wegner and Raschke 1994; Wegner et al. 1994) and HKT transporters, respectively. For Cl^-, this role could be played by $Cl^-/2H^+$ symporters. Note that salt release by cotransporters is an electroneutral process (Zeuthen and MacAulay 2012) and would not interfere with K^+ re-uptake by ion channels that requires a membrane potential more negative than the Nernst potential of K^+, which is maintained by proton pump activity. Evidence for "simultaneous" uptake and release of K^+ has indeed been obtained for root tissue, using refined radioactive tracer techniques (Britto and Kronzucker 2006). Rapid, seemingly "futile cycling" of ions is apparently a common phenomenon at root membranes that was found for K^+, Na^+ and Cl^- and becomes more prominent at elevated concentrations of these ions.

Co-transport of water together with one or more substrates is not a unique property for the CCC transporters. The Na^+-glucose co-transporter SGLT1 and the glucose transporter GLUT1 also translocate water at a fixed stoichiometry (Loo et al. 1999; Zeuthen 2010), and evidence was obtained that the same is true for a range of amino acid transporters, including those from plants. Even ion channels, that co-transport 4–12 water molecules together with one ion, could be involved in water secretion (Wegner 2014), provided that ion release into the xylem and re-absorption occur via different pathways and differ in ion/water stoichiometry, and that the membrane potential oscillates continuously. Further research into this direction is required in the near future. For more details on this matter, the reader is referred to the original publication (Wegner 2014).

6 Long-Distance Transport of Water and Nutrients in the Xylem

Axial transport of water and nutrients from roots to shoot occurs via the xylem, more precisely—in angiosperms—via both xylem vessels (trachees) and tracheids that form a continuum of highly interconnected pipelines extending from the fine roots to the leaves. Nutrients dissolved in the xylem sap are transported upwards by the transpiration-driven mass flow, i.e. nutrient transport is proportional to volume flow. Volume flow depends on vessel anatomy, i.e. on the geometric properties of the conduits, and on local pressure gradients and is, to a first approximation, well described by Hagen–Poiseuille's law (Nobel 1991). Mechanisms of mass flow in the xylem against gravity have been debated violently during the last decades; the more than 100-year-old cohesion-tension theory has been questioned repeatedly and obviously needs at least modifications and extension (Zimmermann et al. 2004; Wegner 2014), but this will not be discussed in more detail here.

While coupling of nutrient and water transport in the xylem conduits follows quite simple physical principles, some aspects require special attention. Those are (1) the ion exchanger properties of the matrix of the xylem walls that can buffer changes in the ionic composition of the xylem sap by selective de- and resorption of ions; (2) the effect of K^+ and other cations dissolved in the xylem sap on the hydraulic conductance of the interconnecting pits and, hence, the entire xylem and (3) the role of adjacent cells in changing the composition of the xylem sap.

The "chromatographic effect" of the xylem walls for ions results mainly from fixed negative charges of polygalacturonic acids that are part of the pectic matrix; this effect is highly dependent on the protonation of carboxylic groups and, hence, on xylem sap pH (Wolterbeek 1987). The interaction, preferentially of divalent cations with xylem walls and the role of these processes in translocation in the xylem, was investigated in a range of classical papers (Bell and Biddulph 1963; van Ieperen et al. 2000; de Geijn and Petit 1979; Wolterbeek 1987). Xylem walls were found to have a fixed cation exchange capacity (CEC) and tend to bind divalent cations (Ca^{2+}, Cd^{2+}) tightly (but less than chelators like EDTA). Hence, an increase in divalent cation concentration in the sap would be buffered by the xylem wall. This buffering effect would be even more pronounced with respect to ion exchange between xylem and adjacent cells and the phloem.

While in the past the xylem conduits were considered to have a fixed hydraulic conductivity mostly resulting from the length and diameter of the vessels and being invariant to short-term adjustment, this view started to change gradually during the last decade. Ion-mediated regulation of xylem conductivity (frequently short-termed the "the ionic effect") was originally demonstrated by M. Zimmermann (1978) and later revisited by van Ieperen et al. (2000) and Zwieniecki et al. (2001). These reports caused much excitement and since then numerous follow-up studies have dealt with this topic (summarised, e.g. by Nardini et al. 2011) even though the physiological significance of the effect is still under debate (van Ieperen 2007) and may be highly species-dependent (Herbette and Cochard 2010). In short, it was observed that the axial hydraulic conductance of shoot segments and individual conduits increased with an increase in the electrolyte concentration in the xylem fluid (mostly monovalent and divalent cations). Other osmolytes had no comparative effects. Conductivity increased by 1.9 up to 58 % in 35 species tested (Nardini et al. 2011). However, these results were, in the vast majority, obtained with artificial perfusion media that frequently did not reflect the natural composition of the xylem sap (van Ieperen 2007). As pointed out by this author, it is inadequate to use distilled water as a reference solution, since this medium is non-physiological and will affect the mechanical properties of cell walls and in particular the pectin structure of inter-conduit pit membranes that were identified as the major axial resistance for longitudinal water flow. These are highly susceptible to the ionic composition of the ambient medium. Swelling and shrinking of pectin matrix, and concomitant changes in pit membrane porosity, or rather membrane thickness (Lee et al. 2012), was identified as the most likely molecular mechanism underlying the ionic effect (Zwieniecki et al. 2001; Nardini et al. 2011). Van Ieperen has argued that in the presence of Ca^{2+}-free media not matching the natural composition of the xylem sap properties of the pectin gel matrix may change in a non-physiological way. Moreover, the ionic effect tends to saturate with increasing Ca^{2+} and K^+ concentration in the xylem. The most dramatic effect occurs when distilled water is exchanged for (artificial) xylem sap containing the usual background ionic concentrations (but see also Nardini et al. (2007)). Despite this ambiguity with respect to the physiological relevance of the ionic effect, it was discussed as an important mechanism of short-term adjustment of xylem conductivity at sites where a large fraction of vessels is blocked by embolisms (Trifilò et al. 2008, 2011), for water allocation to branches receiving sunlight (Sellin et al. 2010; Nardini et al. 2011), for seasonal adjustments in xylem conductivity, or at fluctuating environmental conditions (Nardini et al. 2012). Maximum increase in xylem hydraulic conductivity was observed when adjacent vessels became dysfunctional by embolism (Trifilò et al. 2008); at those sites, increase in xylem K^+ under natural conditions appears to be most pronounced (Trifilò et al. 2011), and evidence for a physiological relevance is most compelling. Increased local K^+ concentration may also reflect an elevated osmotic pressure that may be part of mechanisms to repair embolism (Wegner 2014) and to circumvent embolised vessels with water passing through ray cells (Zimmermann et al. 2004). More research on this "ionic effect", its variability and its molecular basis are required to establish its physiological significance unambiguously.

This discussion on the interplay of xylem ionic composition and volume flow in the xylem would be incomplete without a few words on the role of xylem parenchyma cells in controlling the composition of the xylem sap. This topic would merit a separate review paper because of tremendous recent progress on this issue. Unfortunately, it can only be covered in the form of case studies here. Since the role of xylem parenchyma ion channels in controlling composition of the xylem sap was highlighted for the first time 20 years ago (Wegner and Raschke 1994), a great number of transport proteins located at the plasma membrane of cells bordering on xylem vessels have been identified and characterised with respect to their role in long-distance transport. In the methodological and conceptual context discussed here, the elegant work of Metzner et al. (2010b) on lateral water and nutrient exchange between the xylem and adjacent cells in *Phaseolus vulgaris* deserves particular attention. These authors used cryo-mircroscopy in combination with time-of-flight secondary ion mass spectrometry (SIMS) to trace the distribution of stable isotopes in tissues (Metzner et al. 2008) that were fed via the xylem conduits across the cut surface of the excised stem. A very high spatial resolution could be realised (<1 μm in some of the images), allowing to separate apoplastic and symplastic transport. First of all, this work highlighted again the importance of the exchange of water and nutrients (K, Mg, Ca) between xylem vessels and their environment. Complete equilibration of labelled water between lumens of the vessel and adjacent tissues occurred very fast [with the exception of lignified cell walls in the vicinity of the vessels Metzner et al. (2010b)]. Unexpectedly, however, the investigated rare isotopes of K^+ and even Ca^{2+} and Mg^{2+} supplied via the transpirational stream also equilibrated rapidly with both apoplastic and symplastic pools of xylem parenchyma, and to a lesser extent also with other tissues, indicating that these ions were highly mobile and permeability of cellular and vacuolar membrane were comparatively high. In a separate study (Metzner et al. 2010a) evidence was presented that ion exchange occurred by diffusion and that solvent drag was not likely to play a major role. A modelling approach to these datasets would probably be rewarding, since it potentially offers the possibility to quantify individual fluxes at the single-cell level and possibly relate them, e.g. to ion channel activity.

The control of relative ion concentrations seems to be of major importance for the plant, rather than adjusting absolute xylem sap concentrations that are subject to perpetual fluctuations. A good example is maintenance of the K^+/Na^+ ratio that is optimised in salt-tolerant barley cultivars when exposed to mild salt stress, whereas absolute concentrations seem to be less important (Shabala et al. 2010). Much progress has recently been made in unraveling the mechanisms of how xylem sap Na^+ load is adjusted and Na^+ accumulation in the shoot is reduced, e.g. by retrieval of Na^+ from the xylem sap by xylem parenchyma cells at root and shoot base (Jacoby 1979), and by re-circulation via the phloem (Davenport et al. 2007; for more details, see Sect. 8). Circulation of ions between xylem and phloem seems to be a general mechanism in higher plants contributing to ion homeostasis (Lüttge 2013).

7 Xylem Unloading and Water and Nutrient Transport in Leaf Tissues

While nutrients allocated to leaves remain there serving various physiological functions, or are, to a varying extent, recirculated via the phloem, the water is mostly lost to the atmosphere by evaporation via stomatal pores and the epidermis. This implies a constant mass flow from the vascular tissue to substomatal cavities and to the epidermis, passing through mesophyll tissue. Water could move either apoplastically via cell walls, or through the symplast. Like in the root, this is of relevance for the coupling of flows. Fluorescent dyes have been used as tracers to explore pathways of water and solutes in leaves, but it was not before the landmarking review of Canny (1990) that a firm basis for the correct interpretation of these data was established. Canny argued that local apoplastic accumulation of a dye (called "swamps") marks sites where partitioning of water and solutes occurs, and water passes cellular membranes to enter the symplast (termed "flumes"). Visible traces of the dye extending through cell walls from the vasculature to the epidermis result from diffusive transport starting at swamps. From the evaluation of a large series of micrographs, he concluded that water transport in the leaf is predominantly symplastic. This was later confirmed by other techniques, e.g. measurements of turgor pressure (Ye et al. 2008). It is well known that nutrients are unevenly distributed among different cell types (and sites) in the leaf, but little attention was paid so far to the impact of water flow. An exception is the work of Fricke (2004), who observed for barley leaves that Cl and Ca were preferentially accumulated in the epidermis cells, while P was primarily found in the mesophyll. Changes in the transpiration rate affected this distribution significantly and increased Ca levels close to the substomatal cavity. Basic patterns were, however, not affected.

8 Phloem Transport

When discussing the interplay of water and nutrient transport, the phloem seems, at first glance, to be the least obvious candidate. Nutrient transport is usually not associated with phloem function nor with the transport mechanism. According to a general consensus, volume flow in the phloem is thought to be driven by a pressure gradient supported by phloem loading of mainly sucrose at source tissues (and subsequent passive water uptake by sieve tubes) and phloem unloading (and, in turn, water release) at the sinks. This so-called pressure flow ("Druckstrom") hypothesis first introduced by (Münch 1930) is based on tight coupling of water and solute flow, but corresponding to the main function of the phloem associated with assimilate transport, major osmolytes are supposed to be sugars rather than salts. However, this textbook scenario may oversimplify the real situation; K^+ seems to play a previously undervalued role in phloem transport, particularly when photosynthesis is reduced and sucrose loading at the sink is limiting, or when the H^+ ATPase activity is insufficient to energise transport across the sieve tube membrane. First evidence for this was obtained by Hartt (1970), and later a detailed model was worked out by Lang (1983). K^+ gradients in phloem sap tapped along the shoot provided direct evidence for K^+ loading at the source and K^+ release associated with the sink (Vreugdenhil and Koot-Gronsveld 1989). Consistently, Deeken et al. (2002) demonstrated that in mutants that lacked the phloem-located K^+ channel AKT2, phloem transport was strongly affected. More recently, Gajdanowicz et al. (2011) undertook a comprehensive study on the role of this channel in phloem transport. Mutants expressing an inward-rectifying version of AKT2 (in the wild type, the channel mediated both K^+ influx and efflux across the sieve tube membrane) were deficient with respect to phloem transport, especially at low rates of photosynthesis. Combining experimental work with extensive modelling, the authors came to the conclusion that maintenance of a K^+ gradient across the sieve tube membrane serves as an energy source for loading assimilates into the phloem even at low H^+ ATPase activity.

A final remark on the phloem refers to its role in re-circulation of nutrients from the shoot to the root that was already discussed in detail by Lüttge (2013). Particularly excess K^+ is transported back to the root via the phloem and may serve as a shoot-to-root signal on the K^+ status of the shoot (Wegner and De Boer 1997).

9 The Whole-Plant Perspective: Macronutrients and Transpirational Flow

While the previous paragraphs deal with coupling of water and nutrient transport in plant organs, such as the root, and in vascular tissues representing "functional units" within the plant, the remaining part of this essay is dedicated to the main macronutrients K, N and Ca, and their interaction with long-distance volume flow (that is to a great extent identical with transpirational flow). Interaction is truly mutual, since, on the one hand, nutrients (and nutrient availability) regulate hydraulic properties of plant tissue, e.g. via gating of aquaporins and by affecting stomatal function. On the other hand, evidence has been presented for transpirational flow having impact on the allocation of nutrients in the shoot and among parts of it (e.g. younger and older leaves) in a nutrient-specific way.

The latter aspect touches a long-standing, more fundamental debate on the significance of transpiration for nutrient supply to the shoot. Tanner and Beevers (1990, 2001) provided evidence that transpiration is essentially not required to provide the shoot of sunflower plants with the full spectrum of nutrients. No evidence for nutrient deficiencies were found in sunflower plants that had been grown on hydroponics in a climate chamber, and that received mineral nutrients only during the dark period when the shoot was exposed to nearly 100 % humidity. Tanner and Beevers argued that water circulating between xylem and phloem and growth water would induce xylem flow sufficient for supplying the shoot with nutrients. It should be noted, though, that some residual transpiration was retained (about 7 % compared to the rate of control plants), that still contributed about 50 % to the total water flow from root to shoot in the humidity-exposed plants. Hence, a complete uncoupling of nutrient supply to the shoot from transpirational flow could not be achieved by their experimental approach. The case of Tanner and Beevers is supported by findings on aquatic higher plants growing in a submerged state. These plants maintain acropetal xylem water flow in the complete absence of transpiration, most likely to supply leaves with mineral nutrients like P, Fe and Mn that are hardly available from the ambient water at the leaf surface. These nutrients have to be taken up by the roots from sediments and are transported to the shoot by mass flow via xylem conduits (Pedersen and Sand-Jensen 1993, 1997). Independence of nutrient transport on transpiration as advocated by Tanner and Beevers contrasts, however, with other reports that established a link between down-regulation of transpiration (e.g. as a consequence of elevated ambient CO_2 partial pressures) and reduced nutrient supply to the shoot (Conroy and Hocking 1993). Considerable transpiration rates at night were also hypothesised to serve the function of supplying the shoot with nutrients. The issue may be solved by stating that transpiration is, strictly speaking, not required to provide the shoot with nutrients, but since it is there and unavoidable under most conditions, plants have "learned" to make use of it. In habitats that do not require strict optimisation of water use efficiency, part of the transpirational flow being in excess of water requirements of the shoot may serve other purposes, such as optimising nutrient supply (Cramer et al. 2009).

Interactions of nutrient and water at the whole-plant level cannot exclusively be described in a mechanistic way, since various indirect effects have to be taken into account, e.g. regulation of stomatal conductance and photosynthesis (Cramer et al. 2009). Therefore, it is more adequate in some cases to talk rather about trade-offs.

9.1 Nitrogen (Nitrate and Ammonium)

The prime candidate for considering interactions and trade-offs of transport of water and inorganic ions in plants is certainly nitrate. Transport of water and nitrate interacts in various ways that have been a matter of extensive research since the 1980s of the last century. Radin and Boyer (1982) were among the first to detect that root hydraulic conductance was strongly affected by the availability of nitrate. In low nitrate medium, hydraulic conductance would be about half that of roots well supplied with nitrate. Transpiration was affected in the same way. Consistently, turgor pressure probe experiments revealed that the hydraulic conductance of cortex cells in roots deprived of nitrate was significantly lower compared to roots grown in full medium (Fig. 4; Radin and Matthews 1989). Later, this effect could be ascribed to a regulation of aquaporin activity in those cells. Nitrate complementation to roots grown in N free medium was shown to induce an up-regulation of aquaporin expression in fava bean roots (Guo et al. 2007), and N deprivation would suppress aquaporin expression (Clarkson et al. 2000). Moreover, regulation of aquaporin gating by (intracellular) nitrate independent of aquaporin expression was reported for maize (Gorska et al. 2008a, b). Both effects would contribute to an improved water supply to the shoot of plants well provided with nitrate and, in turn, an increase in transpiration and up-regulation of photosynthesis. Gorska et al. (2008a) and Cramer et al. (2009) have argued that an increased water flow would facilitate nitrate acquisition in soil by solvent drag. This may be particularly important in soils with local differences in nitrate availability, a situation that appears to be quite common in natural soils. The effect would favour effective exploitation of local nitrate resources over uptake of ammonium that does not induce a similar effect. However, when linking this effect to the response of the whole plant, complexity increases. In split-root experiments on bean with part of the root supplied with nitrate and the other provided with ammonium, Schulze-Till et al. (2009) using the MRT technology, observed higher rates of water flow in the nitrate-fed roots due to a larger number of vessels per root contributing to flow than in those provided with ammonium. Flow velocity and xylem pressure of conducting xylem elements did not differ much, though, and anatomical properties were also unaffected by the N-form. Schulze-Till et al. hypothesised that part of the vessels remained non-functional in the ammonium-fed roots and thus were "switched off" to prevent cavitation.

Consistent with a facilitated water supply in the presence of nitrate, stomatal conductance (g_s) was found to increase when N-deficient plants received nitrate (Wilkinson et al. 2007). However, the dependence of g_s on soil nitrate was found to follow an optimum curve. High nitrate concentrations would tend to induce partial stomatal closure. Cramer et al. (2009) hypothesised that high nitrate delivery to the shoot (that would not only depend on nitrate uptake, but also on the nitrate assimilation rate in the root tissues) would lead to an increased NO production in the leaf and, in turn, to stomatal closure. Note that CO_2 uptake is also coupled to nitrate reduction in the leaves for adjustment of malate synthesis. Malate produced in the leaves neutralises OH^- formed as a by-product of nitrate reduction; moreover, malate replaces NO_3^- as a counter-ion for excess K^+ that is transported from root to shoot in the xylem conduits and is subsequently recirculated back to the roots via the phloem [see also Sect. 9.3 and Lüttge (2013)].

From this survey of various tight interactions of N fluxes and transpirational flow, it is not surprising that N supply was found to have strong impact on transport and accumulation of other nutrients, including K^+ and Ca^{2+} (Matimati et al. 2014).

9.2 Calcium

Calcium distribution in the plant is predominantly shaped by apoplastic water flow, since symplastic mobility of this divalent ion is low. Cytosolic Ca^{2+} concentrations are kept at extremely low levels of up to about 500 nM—values exceeding this low regime are sensed as a stress signal. Vacuolar Ca^{2+} concentrations are much higher, but this Ca^{2+} pool is rather immobile and does not contribute to Ca^{2+} transport. Clarkson (1993) reported that radial transport of Ca^{2+} into the xylem was restricted to apical parts of the root and correlated with the maturation of the endodermis; he concluded that root Ca^{2+} transport was mainly apoplastic and only for circumventing the suberin barrier of the Casparian strip, Ca^{2+} was taken up and subsequently released into the stelar apoplast. By contrast, White (2001) hypothesised that Ca^{2+} may be transported into xylem vessels by a purely apoplastic pathway, and that Ca^{2+} in a complexed form may also be mobile in the symplast. Ca^{2+} transport in the phloem is also supposed to be negligible (and hence, root-to-shoot net Ca^{2+} transport is supposed to equal total Ca^{2+} transport in apical direction). Note that his view has also been questioned repeatedly (Biddulph et al. 1959; Ringoet et al. 1968). But apart from these uncertainties, Ca^{2+} predominantly moves in the apoplast (interacting with fixed negative charges of the cell wall, see above). Local apoplastic Ca^{2+} accumulation allows to identify sites at which increased cellular water uptake by adjacent cells takes place. Ca^{2+} and water flow interact mutually, though, since Ca^{2+} can also exert feedback on water flow, e.g. by a regulation of aquaporins (Gilliham et al. 2011), or by its effect on stomatal aperture (Atkinson et al. 1992). An elegant model on this interaction was proposed by Gilliham et al. (2011). At elevated apoplastic Ca^{2+} levels, cytosolic Ca^{2+} will also increase with time, leading to a down-regulation of aquaporin activity; as a consequence, transcellular water flow is down-regulated and translocation of water will be restricted to the apoplast, contributing to a wash out of local apoplastic Ca^{2+} accumulation. The scenario can be extended and refined by taking Ca^{2+} secretion via the plasma membrane by Ca^{2+} ATPases as well as Ca^{2+} exchange between cytosol and vacuole into account. Physiological models with this degree of complexity will require quantitative modelling to identify strategies for experimental validation.

In his study on Ca^{2+} transport in maize, Engels (1999) came to the conclusion that root-to-shoot translocation correlates to some extent with transpiration; additionally, radial transport in the root is adjusted to shoot demand.

9.3 Potassium

Despite its abundance in the plant and its importance for various physiological processes, the link of K^+ transport to water flow (and vice versa) seems to be less "spectacular" than for NO_3^- and Ca^{2+}. K^+ transport from roots to the shoot is under tight control of shoot demand (Engels 1999), and excess K^+ in leaves is circulated back to the roots via the phloem (White 1997; Lüttge 2013). Radial translocation of K^+ in the root is strongly enhanced by volume flow, as stated previously (Wegner and Zimmermann 2009; Schurr and Schulze 1995, see also above), most likely due to K^+ accumulation in stelar cells. It was also shown previously that xylem K^+ is buffered at short-term changes in the external K^+ concentration (Wegner and Zimmermann 2002).

Rather than the presence of K^+, its deprivation seems to have a marked effect on plant water relations, though. At low K^+ availability in the soil solution, root hydraulic conductance and transpirational water flow are increased with respect to values at normal K^+ supply, and water use efficiency is reduced (Quintero et al. 1998; Fournier et al. 2005). Low K^+ prevents stomatal closure under mild drought stress conditions (Benlloch-González et al. 2008, 2010). These symptoms may be part of a mechanism to enhance K^+ retrieval from the soil by solvent drag. Only at a severe K^+ starvation, stomatal function is compromised by reduced availability of K^+ as an osmoticum in guard cells required to maintain stomatal conductance (Humble and Raschke 1971). As a consequence, stomata tend to close under these conditions (Hsiao and Lauchli 1986).

10 Conclusion

Far from being comprehensive, this overview of nutrient and water transport in plants was meant to provide the reader with some insight into the complexity of their interactions. Information on these phenomena is ever increasing, e.g. by the advent of new techniques like the ZIM probe. Moreover, interactions among ion fluxes were hardly considered here, but obvious constraints like electro-neutrality of transport processes in steady state imply tight coupling of nutrient fluxes.

It is clear from these considerations that plant nutrition needs a conversion to become a quantitative science in the near future, making use of the fast progress in computer modelling of complex systems that is currently taking place. The uprise of meteorology and climate sciences provides a good example and can be seen as an encouragement. However, these environmental sciences also demonstrate that a firm physico-chemical basis is mandatory for such an approach to be successful. It is an advantage of transport physiology over other disciplines of plant sciences that its subject is readily treated in a quantitative way in the form of fluxes, and that well-established concepts like the thermodynamics of irreversible processes are available to describe coupling of these fluxes in a comprehensive way. More efforts are required in the future to extend and adjust these concepts, e.g. to include non-linearities and regulatory processes in the quantitative treatment of transport processes.

Acknowledgements I would like to thank Prof U. Zimmermann, Henningsdorf, for many stimulating discussions and for critical reading of the manuscript. Thanks are also due to Prof. Zimmermann and S. Rüger for providing diagrams showing the ZIM probe and a hitherto unpublished model experiment.

References

Atkinson CJ, Ruiz LP, Mansfield TA (1992) Calcium in xylem sap and the regulation of its delivery to the shoot. J Exp Bot 43:1315–1324. doi:10.1093/jxb/43.10.1315

Balling A, Zimmermann U (1990) Comparative measurements of the xylem pressure of Nicotiana plants by means of the pressure bomb and pressure probe. Planta 182:325–338. doi:10.1007/BF02411382

Bell CW, Biddulph O (1963) Translocation of calcium. Exchange versus mass flow. Plant Physiol 38:610–614

Benlloch-González M, Arquero O, Fournier JM et al (2008) K + starvation inhibits water-stress-induced stomatal closure. J Plant Physiol 165:623–630. doi:10.1016/j.jplph.2007.05.010

Benlloch-González M, Fournier JM, Benlloch M (2010) K + deprivation induces xylem water and K + transport in sunflower: evidence for a co-ordinated control. J Exp Bot 61:157–164. doi:10.1093/jxb/erp288

Bentrup FW (1979) Reception and transduction of electrical and mechanical stimuli. Encyclopedia Plant Physiol 7:42–70

Biddulph O, Cory R, Biddulph S (1959) Translocation of calcium in the bean plant. Plant Physiol 34:512–519

Bramley H, Turner NC, Turner DW, Tyerman SD (2007) Comparison between gradient-dependent hydraulic conductivities of roots using the root pressure probe: the role of pressure propagations and implications for the relative roles of parallel radial pathways. Plant Cell Environ 30:861–874. doi:10.1111/j.1365-3040.2007.01678.x

Bramley H, Ehrenberger W, Zimmermann U et al (2013) Non-invasive pressure probes magnetically clamped to leaves to monitor the water status of wheat. Plant Soil 369:257–268. doi:10.1007/s11104-012-1568-x

Britto DT, Kronzucker HJ (2006) Futile cycling at the plasma membrane: a hallmark of low-affinity nutrient transport. Trends Plant Sci 11:529–534. doi:10.1016/j.tplants.2006.09.011

Britto DT, Kronzucker HJ (2013) Flux measurements of cations using radioactive tracers. Methods Mol Biol Clifton NJ 953:161–170. doi:10.1007/978-1-62703-152-3_10

Canny MJ (1990) What becomes of the transpiration stream? New Phytol 114:341–368

Ciancio V, Verhás J (1994) On the nonlinear generalizations of Onsager's reciprocal relations. J Non-Equil Thermody 19:184–194

Clarkson DT (1993) Roots and the delivery of solutes to the xylem. Philos Trans R Soc Lond B Biol Sci 341:5–17

Clarkson DT, Carvajal M, Henzler T et al (2000) Root hydraulic conductance: diurnal aquaporin expression and the effects of nutrient stress. J Exp Bot 51:61–70

Colmenero-Flores JM, Martínez G, Gamba G et al (2007) Identification and functional characterization of cation-chloride cotransporters in plants. Plant J Cell Mol Biol 50:278–292. doi:10.1111/j.1365-313X.2007.03048.x

Conroy J, Hocking P (1993) Nitrogen nutrition of C3 plants at elevated atmospheric CO2 concentrations. Physiol Plant 89:570–576. doi:10.1111/j.1399-3054.1993.tb05215.x

Cramer MD, Hawkins H-J, Verboom GA (2009) The importance of nutritional regulation of plant water flux. Oecologia 161:15–24. doi:10.1007/s00442-009-1364-3

Davenport RJ, Muñoz-Mayor A, Jha D et al (2007) The Na + transporter AtHKT1;1 controls retrieval of Na + from the xylem in Arabidopsis. Plant Cell Environ 30:497–507. doi:10.1111/j.1365-3040.2007.01637.x

de Geijn SCV, Petit CM (1979) Transport of divalent cations. Cation exchange capacity of intact xylem vessels. Plant Physiol 64:954–958. doi:10.1104/pp. 64.6.954

Deeken R, Geiger D, Fromm J et al (2002) Loss of the AKT2/3 potassium channel affects sugar loading into the phloem of Arabidopsis. Planta 216:334–344. doi:10.1007/s00425-002-0895-1

Diamond JM, Bossert WH (1967) Standing-gradient osmotic flow a mechanism for coupling of water and solute transport in epithelia. J Gen Physiol 50:2061–2083. doi:10.1085/jgp.50.8.2061

Engels C (1999) Regulation of xylem transport of calcium from roots to shoot of maize by growth-related demand. J Plant Nutr Soil Sci 162:287–294. doi:10.1002/(SICI)1522-2624(199906)162:3<287::AID-JPLN287>3.0.CO;2-N

Enns LC, Canny MJ, McCully ME (2000) An investigation of the role of solutes in the xylem sap and in the xylem parenchyma as the source of root pressure. Protoplasma 211:183–197. doi:10.1007/BF01304486

Fournier JM, Roldán ÁM, Sánchez C et al (2005) K + starvation increases water uptake in whole sunflower plants. Plant Sci 168:823–829. doi:10.1016/j.plantsci.2004.10.015

Fricke W (2004) Solute sorting in grass leaves. The transpiration stream. Planta 219:507–514. doi:10.1007/s00425-004-1262-1

Fritz M, Ehwald R (2011) Mannitol permeation and radial flow of water in maize roots. New Phytol 189:210–217. doi:10.1111/j.1469-8137.2010.03452.x

Gajdanowicz P, Michard E, Sandmann M et al (2011) Potassium (K+) gradients serve as a mobile energy source in plant vascular tissues. Proc Natl Acad Sci USA 108:864–869. doi:10.1073/pnas.1009777108

Gilliham M, Dayod M, Hocking BJ et al (2011) Calcium delivery and storage in plant leaves: exploring the link with water flow. J Exp Bot 62:2233–2250. doi:10.1093/jxb/err111

Gorska A, Ye Q, Holbrook NM, Zwieniecki MA (2008a) Nitrate control of root hydraulic properties in plants: translating local information to whole plant response. Plant Physiol 148:1159–1167. doi:10.1104/pp. 108.122499
Gorska A, Zwieniecka A, Holbrook NM, Zwieniecki MA (2008b) Nitrate induction of root hydraulic conductivity in maize is not correlated with aquaporin expression. Planta 228:989–998. doi:10.1007/s00425-008-0798-x
Guo S, Kaldenhoff R, Uehlein N et al (2007) Relationship between water and nitrogen uptake in nitrate- and ammonium-supplied Phaseolus vulgaris L. plants. J Plant Nutr Soil Sci 170:73–80. doi:10.1002/jpln.200625073
Hartt CE (1970) Effect of potassium deficiency upon translocation of 14C in detached blades of sugarcane. Plant Physiol 45:183–187. doi:10.1104/pp. 45.2.183
Hedrich R (2012) Ion channels in plants. Physiol Rev 92:1777–1811. doi:10.1152/physrev.00038.2011
Herbette S, Cochard H (2010) Calcium is a major determinant of xylem vulnerability to cavitation. Plant Physiol 153:1932–1939. doi:10.1104/pp. 110.155200
Homblé F, Véry AA (1992) Coupling of water and potassium ions in K channels of the tonoplast of Chara. Biophys J 63:996–999
Hsiao TC, Lauchli A (1986) Role of potassium in plant-water relations. Adv Plant Nutr 2:281–312
Humble GD, Raschke K (1971) Stomatal opening quantitatively related to potassium transport. Evidence from electron probe analysis. Plant Physiol 48:447–453. doi:10.1104/pp. 48.4.447
Jacoby B (1979) Sodium recirculation and loss from Phaseolus vulgaris L. Ann Bot 43:741–744
Jeschke WD, Pate JS (1991) Modelling of the partitioning, assimilation and storage of nitrate within root and shoot organs of castor bean (Ricinus communis L.). J Exp Bot 42:1091–1103. doi:10.1093/jxb/42.9.1091
Katou K, Taura T, Furumoto M (1987) A model for water transport in the stele of plant roots. Protoplasma 140:123–132. doi:10.1007/BF01273721
Knipfer T, Fricke W (2010) Root pressure and a solute reflection coefficient close to unity exclude a purely apoplastic pathway of radial water transport in barley (Hordeum vulgare). New Phytol 187:159–170. doi:10.1111/j.1469-8137.2010.03240.x
Knipfer T, Das D, Steudle E (2007) During measurements of root hydraulics with pressure probes, the contribution of unstirred layers is minimized in the pressure relaxation mode: comparison with pressure clamp and high-pressure flowmeter. Plant Cell Environ 30:845–860. doi:10.1111/j.1365-3040.2007.01670.x
Kong X-Q, Gao X-H, Sun W et al (2011) Cloning and functional characterization of a cation-chloride cotransporter gene OsCCC1. Plant Mol Biol 75:567–578. doi:10.1007/s11103-011-9744-6
Lang A (1983) Turgor-regulated translocation. Plant Cell Environ 6:683–689. doi:10.1111/1365-3040.ep11589312
Lee J, Holbrook NM, Zwieniecki MA (2012) Ion induced changes in the structure of bordered pit membranes. Front Plant Sci 3:55. doi:10.3389/fpls.2012.00055
Liu K, Li L, Luan S (2006) Intracellular K + sensing of SKOR, a Shaker-type K + channel from Arabidopsis. Plant J 46:260–268. doi:10.1111/j.1365-313X.2006.02689.x
Loo DD, Hirayama BA, Meinild AK et al (1999) Passive water and ion transport by cotransporters. J Physiol 518(Pt 1):195–202
Lüttge U (1975) Salt glands. In: Hall LJ, Baker DA (eds) Ion transport plant cells tissues. North Holland Publishing, Amsterdam, pp 335–376
Lüttge U (2013) Whole-plant physiology: synergistic emergence rather than modularity. In: Lüttge U, Beyschlag W, Francis D, Cushman J (eds) Progress in botany. Springer, Berlin, pp 165–190
Marschner H (1995) Mineral nutrition of higher plants, 2nd edn. Academic Press, London, p 889
Matimati I, Verboom GA, Cramer MD (2014) Nitrogen regulation of transpiration controls mass-flow acquisition of nutrients. J Exp Bot 65:159–168. doi:10.1093/jxb/ert367

Maurel C, Verdoucq L, Luu D-T, Santoni V (2008) Plant aquaporins: membrane channels with multiple integrated functions. Annu Rev Plant Biol 59:595–624. doi:10.1146/annurev.arplant.59.032607.092734

Metzner R, Schneider HU, Breuer U, Schroeder WH (2008) Imaging nutrient distributions in plant tissue using time-of-flight secondary ion mass spectrometry and scanning electron microscopy. Plant Physiol 147:1774–1787. doi:10.1104/pp. 107.109215

Metzner R, Schneider HU, Breuer U et al (2010a) Tracing cationic nutrients from xylem into stem tissue of French bean by stable isotope tracers and cryo-secondary ion mass spectrometry. Plant Physiol 152:1030–1043. doi:10.1104/pp. 109.143776

Metzner R, Thorpe MR, Breuer U et al (2010b) Contrasting dynamics of water and mineral nutrients in stems shown by stable isotope tracers and cryo-SIMS. Plant Cell Environ 33:1393–1407. doi:10.1111/j.1365-3040.2010.02157.x

Morillon R, Liénard D, Chrispeels MJ, Lassalles JP (2001) Rapid movements of plants organs require solute-water cotransporters or contractile proteins. Plant Physiol 127:720–723

Münch E (1930) Die Stoffbewegungen in der Pflanze. Fischer, Jena

Nardini A, Gascò A, Trifilò P et al (2007) Ion-mediated enhancement of xylem hydraulic conductivity is not always suppressed by the presence of Ca2+ in the sap. J Exp Bot 58:2609–2615. doi:10.1093/jxb/erm105

Nardini A, Salleo S, Jansen S (2011) More than just a vulnerable pipeline: xylem physiology in the light of ion-mediated regulation of plant water transport. J Exp Bot 62:4701–4718. doi:10.1093/jxb/err208

Nardini A, Dimasi F, Klepsch M, Jansen S (2012) Ion-mediated enhancement of xylem hydraulic conductivity in four Acer species: relationships with ecological and anatomical features. Tree Physiol 32:1434–1441. doi:10.1093/treephys/tps107

Newman IA (2001) Ion in roots: measurement of fluxes using ion-selective microelectrodes to characterize transporter function. Plant Cell Environ 24:1–14. doi:10.1046/j.1365-3040.2001.00661.x

Nobel PS (1991) Physicochemical and environmental plant physiology. Academic, San Diego, CA

Oertli JJ (1966) Active water transport in plants. Physiol Plant 19:809–817. doi:10.1111/j.1399-3054.1966.tb07067.x

Palta JP, Stadelmann EJ (1980) On simultaneous transport of water and solute through plant cell membranes: evidence for the absence of solvent drag effect and insensitivity of the reflection coefficient. Physiol Plant 50:83–90. doi:10.1111/j.1399-3054.1980.tb02689.x

Pate JS, Gunning BES (1972) Transfer cells. Annu Rev Plant Physiol 23:173–196. doi:10.1146/annurev.pp. 23.060172.001133

Pedersen O, Sand-Jensen K (1993) Water transport in submerged macrophytes. Aquat Bot 44:385–406. doi:10.1016/0304-3770(93)90079-C

Pedersen O, Sand-Jensen K (1997) Transpiration does not control growth and nutrient supply in the amphibious plant Mentha aquatica. Plant Cell Environ 20:117–123. doi:10.1046/j.1365-3040.1997.d01-12.x

Peuke AD (2010) Correlations in concentrations, xylem and phloem flows, and partitioning of elements and ions in intact plants. A summary and statistical re-evaluation of modelling experiments in Ricinus communis. J Exp Bot 61:635–655. doi:10.1093/jxb/erp352

Peuke AD, Rokitta M, Zimmermann U et al (2001) Simultaneous measurement of water flow velocity and solute transport in xylem and phloem of adult plants of Ricinus communis over a daily time course by nuclear magnetic resonance spectrometry. Plant Cell Environ 24:491–503. doi:10.1046/j.1365-3040.2001.00704.x

Pohl P, Saparov SM, Antonenko YN (1997) The effect of a transmembrane osmotic flux on the ion concentration distribution in the immediate membrane vicinity measured by microelectrodes. Biophys J 72:1711–1718

Quintero JM, Fournier JM, Ramos J, Benlloch M (1998) K + status and ABA affect both exudation rate and hydraulic conductivity in sunflower roots. Physiol Plant 102:279–284. doi:10.1034/j.1399-3054.1998.1020216.xJ

Radin JW, Boyer JS (1982) Control of leaf expansion by nitrogen nutrition in sunflower plants: role of hydraulic conductivity and turgor. Plant Physiol 69:771–775

Radin JW, Matthews MA (1989) Water transport properties of cortical cells in roots of nitrogen- and phosphorus-deficient cotton seedlings. Plant Physiol 89:264–268. doi:10.1104/pp. 89.1. 264, http://dx.doi.org/10.1104/pp.89.1.264

Renninger HJ, Schäfer KVR (2012) Comparison of tissue heat balance- and thermal dissipation-derived sap flow measurements in ring-porous oaks and a pine. Front Plant Sci 3:103. doi:10.3389/fpls.2012.00103

Ringoet A, Sauer G, Gielink AJ (1968) Phloem transport of calcium in oat leaves. Planta 80:15–20

Rokitta M, Peuke AD, Zimmermann U, Haase A (1999) Dynamic studies of phloem and xylem flow in fully differentiated plants by fast nuclear-magnetic-resonance microimaging. Protoplasma 209:126–131. doi:10.1007/BF01415708

Rüger S, Ehrenberger W, Arend M et al (2010) Comparative monitoring of temporal and spatial changes in tree water status using the non-invasive leaf patch clamp pressure probe and the pressure bomb. Agric Water Manag 98:283–290. doi:10.1016/j.agwat.2010.08.022

Schneider H, Zhu J, Zimmermann U (1997a) Xylem and cell turgor pressure probe measurements in intact roots of glycophytes: transpiration induces a change in the radial and cellular reflection coefficients. Plant Cell Environ 20:221–229. doi:10.1046/j.1365-3040.1997.d01-65.x

Schneider H, Wistuba N, Miller B, Geßner P, Thürmer F, Melcher P, Meinzer F, Zimmermann U (1997b) Diurnal variation in the radial reflection coefficient of intact maize root determined with the xylem pressure probe. J Exp Bot 48:2045–2053. doi:10.1093/jxb/48.12.2045

Schroeder JI, Delhaize E, Frommer WB et al (2013) Using membrane transporters to improve crops for sustainable food production. Nature 497:60–66. doi:10.1038/nature11909

Schulze-Till T, Kaufmann I, Sattelmacher B et al (2009) A 1H NMR study of water flow in Phaseolus vulgaris L. roots treated with nitrate or ammonium. Plant Soil 319:307–321. doi:10.1007/s11104-008-9871-2

Schurr U, Schulze E-D (1995) The concentration of xylem sap constituents in root exudate, and in sap from intact, transpiring castor bean plants (Ricinus communis L.). Plant Cell Environ 18:409–420. doi:10.1111/j.1365-3040.1995.tb00375.x

Sellin A, Ounapuu E, Karusion A (2010) Experimental evidence supporting the concept of light-mediated modulation of stem hydraulic conductance. Tree Physiol 30:1528–1532. doi:10.1093/treephys/tpq091

Shabala S, Pang J, Zhou M et al (2009) Electrical signalling and cytokinins mediate effects of light and root cutting on ion uptake in intact plants. Plant Cell Environ 32:194–207. doi:10.1111/j. 1365-3040.2008.01914.x

Shabala S, Shabala S, Cuin TA et al (2010) Xylem ionic relations and salinity tolerance in barley. Plant J Cell Mol Biol 61:839–853. doi:10.1111/j.1365-313X.2009.04110.x

Shabala S, Cuin TA, Shabala L, Newman I (2012) Quantifying kinetics of net ion fluxes from plant tissues by non-invasive microelectrode measuring MIFE technique. Methods Mol Biol Clifton NJ 913:119–134. doi:10.1007/978-1-61779-986-0_7

Shackel KA, Johnson RS, Medawar CK, Phene CJ (1992) Substantial errors in estimates of sap flow using the heat balance technique on woody stems under field conditions. J Am Soc Hortic Sci 117:351–356

Smith DM, Allen SJ (1996) Measurement of sap flow in plant stems. J Exp Bot 47:1833–1844

Steudle E (2000a) Water uptake by roots: effects of water deficit. J Exp Bot 51:1531–1542

Steudle E (2000b) Water uptake by plant roots: an integration of views. Plant Soil 226:45–56. doi:10.1023/A:1026439226716

Steudle E, Jeschke WD (1983) Water transport in barley roots: measurements of root pressure and hydraulic conductivity of roots in parallel with turgor and hydraulic conductivity of root cells. Planta 158:237–248. doi:10.1007/BF01075260

Steudle E, Peterson CA (1998) How does water get through roots? J Exp Bot 49:775–788. doi:10.1093/jxb/49.322.775

Tanner W, Beevers H (1990) Does transpiration have an essential function in long-distance ion transport in plants? Plant Cell Environ 13:745–750. doi:10.1111/j.1365-3040.1990.tb01089.x
Tanner W, Beevers H (2001) Transpiration, a prerequisite for long-distance transport of minerals in plants? Proc Natl Acad Sci USA 98:9443–9447. doi:10.1073/pnas.161279898
Tomos AD, Leigh RA (1999) The pressure probe: a versatile tool in plant-cell biology. Annu Rev Plant Biol 50:447–472. doi:10.1146/annurev.arplant.50.1.447
Trifilò P, Lo Gullo MA, Salleo S et al (2008) Xylem embolism alleviated by ion-mediated increase in hydraulic conductivity of functional xylem: insights from field measurements. Tree Physiol 28:1505–1512
Trifilò P, Nardini A, Raimondo F et al (2011) Ion-mediated compensation for drought-induced loss of xylem hydraulic conductivity in field-growing plants of Laurus nobilis. Funct Plant Biol 38:606–613
Uehlein N, Lovisolo C, Siefritz F, Kaldenhoff R (2003) The tobacco aquaporin NtAQP1 is a membrane CO2 pore with physiological functions. Nature 425:734–737. doi:10.1038/nature02027
Van Ieperen W (2007) Ion-mediated changes of xylem hydraulic resistance in planta: fact or fiction? Trends Plant Sci 12:137–142. doi:10.1016/j.tplants.2007.03.001
Van Ieperen W, van Meeteren U, van Gelder H (2000) Fluid ionic composition influences hydraulic conductance of xylem conduits. J Exp Bot 51:769–776
Volkov VS, Zholkevich VN (1993) Exudation parameters depending on spatial orientation of root segments of Zea mays L. Doklady Akad Nauk 332:526–528
Vreugdenhil D, Koot-Gronsveld EAM (1989) Measurements of pH, sucrose and potassium ions in the phloem sap of castor bean (Ricinus communis) plants. Physiol Plant 77:385–388. doi:10.1111/j.1399-3054.1989.tb05657.x
Wang Y-Y, Hsu P-K, Tsay Y-F (2012) Uptake, allocation and signaling of nitrate. Trends Plant Sci 17:458–467. doi:10.1016/j.tplants.2012.04.006
Wegner LH (2012) Using the multifunctional xylem probe for *in situ* studies of plant water and ion relations under saline conditions. In: Shabala SN, Cuinn TA (eds) Salt tolerance—methods and protocols, vol 913, Methods in molecular biology., pp 35–66. doi:10.1007/978-1-61779-986-0_3
Wegner LH (2014) Root pressure and beyond: energetically uphill water transport into xylem vessels? J Exp Bot 65:381–393. doi:10.1093/jxb/ert391
Wegner LH, De Boer AH (1997) Properties of two outward-rectifying channels in root xylem parenchyma cells suggest a role in K^+ homeostasis and long-distance signaling. Plant Physiol 115:1707–1719. doi:10.1104/pp. 115.4.1707, http://dx.doi.org/10.1104/pp.115.4.1707
Wegner LH, Raschke K (1994) Ion channels in the xylem parenchyma of barley roots (A Procedure to Isolate protoplasts from this tissue and a patch-clamp exploration of salt passageways into xylem vessels). Plant Physiol 105:799–813
Wegner LH, Zimmermann U (2002) On-line measurements of K + activity in the tensile water of the xylem conduit of higher plants. Plant J Cell Mol Biol 32:409–417
Wegner LH, Zimmermann U (2009) Hydraulic conductance and K + transport into the xylem depend on radial volume flow, rather than on xylem pressure, in roots of intact, transpiring maize seedlings. New Phytol 181:361–373. doi:10.1111/j.1469-8137.2008.02662.x
Wegner LH, De Boer AH, Raschke K (1994) Properties of the K + inward rectifier in the plasma membrane of xylem parenchyma cells from barley roots: effects of TEA+, Ca2+, Ba2+ and La3+. J Membr Biol 142:363–379
Westhoff M, Reuss R, Zimmermann D et al (2009) A non-invasive probe for online-monitoring of turgor pressure changes under field conditions. Plant Biol 11:701–712. doi:10.1111/j.1438-8677.2008.00170.x
White PJ (1997) Cation channels in the plasma membrane of rye roots. J Exp Bot 48 Spec No:499–514. doi:10.1093/jxb/48.Special_Issue.499
White PJ (2001) The pathways of calcium movement to the xylem. J Exp Bot 52:891–899. doi:10.1093/jexbot/52.358.891

Wilkinson S, Bacon MA, Davies WJ (2007) Nitrate signalling to stomata and growing leaves: interactions with soil drying, ABA, and xylem sap pH in maize. J Exp Bot 58:1705–1716. doi:10.1093/jxb/erm021

Wolterbeek HT (1987) Cation exchange in isolated xylem cell walls of tomato. I. Cd2+ and Rb + exchange in adsorption experiments. Plant Cell Environ 10:39–44. doi:10.1111/j.1365-3040.1987.tb02077.x

Ye Q, Holbrook M, Zwieniecki MA (2008) Cell-to-cell pathway dominates xylem-epidermis hydraulic connection in *Tradescantia fluminensis* (Vell. Conc.) leaves. Planta 227:1311–1319. doi:10.1007/s00425-008-0703-7

Zeuthen T (1996) Molecular mechanisms of water transport. RG Landes Compagny, Austin, TX

Zeuthen T (2010) Water-transporting proteins. J Membr Biol 234:57–73. doi:10.1007/s00232-009-9216-y

Zeuthen T, MacAulay N (2012) Transport of water against its concentration gradient: fact or fiction? Wiley Interdiscip Rev Membr Transp Signal 1:373–381. doi:10.1002/wmts.54

Zimmermann MH (1978) Hydraulic architecture of some diffuse-porous trees. Can J Bot 56:2286–2295. doi:10.1139/b78-274

Zimmermann U, Steudle E (1978) Physical aspects of water relations of plant cells. Adv Bot Res 6:45–117

Zimmermann U, Räde H, Steudle E (1969) Kontinuierliche Druckmessung in Pflanzenzellen. Naturwissenschaften 56:634. doi:10.1007/BF01185741

Zimmermann U, Schneider H, Wegner LH, Haase A (2004) Water ascent in tall trees: does evolution of land plants rely on a highly metastable state? New Phytol 162:575–615. doi:10.1111/j.1469-8137.2004.01083.x

Zimmermann D, Reuss R, Westhoff M et al (2008) A novel, non-invasive, online-monitoring, versatile and easy plant-based probe for measuring leaf water status. J Exp Bot 59:3157–3167. doi:10.1093/jxb/ern171

Zimmermann U, Bitter R, Marchiori PER et al (2013) A non-invasive plant-based probe for continuous monitoring of water stress in real time: a new tool for irrigation scheduling and deeper insight into drought and salinity stress physiology. Theor Exp Plant Physiol 25:2–11. doi:10.1590/S2197-00252013000100002

Zwieniecki MA, Melcher PJ, Michele Holbrook NM (2001) Hydrogel control of xylem hydraulic resistance in plants. Science 291:1059–1062. doi:10.1126/science.1057175

Active and Passive Electrical Signaling in Plants

Alexander G. Volkov and Vladislav S. Markin

Contents

Abstract Electrical signaling on long and short distances exists in plants. There are three major types of electrical signaling in plants and animals: action potentials, electrotonic potentials, and graded potentials. The action potential in plants can propagate over the entire length of the cell membrane and along the conductive

A.G. Volkov (✉)
Department of Chemistry and Biochemistry, Oakwood University, 7000 Adventist Blvd., Huntsville, AL 35896, USA
e-mail: agvolkov@yahoo.com

V.S. Markin
Department of Neurology, University of Texas, Southwestern Medical Center, Dallas, TX 75390-8813, USA
e-mail: MarkinA@swbell.net

U. Lüttge, W. Beyschlag (eds.), *Progress in Botany*, Progress in Botany 76,
DOI 10.1007/978-3-319-08807-5_6

bundles of tissue with constant amplitude, duration, and speed. Electrotonic potentials exponentially decrease with distance. An intermediate place takes so-called graded potentials that involve the process of electrical excitation but do not evolve into full-fledged action potentials. A graded potential is an electrical signal that corresponds to the size of the stimulus. Electrical signals can propagate along the plasma membrane on short distances in plasmodesmata, and on long distances in a phloem. In this chapter, we discuss electrical signaling in the Venus flytrap and *Mimosa pudica*.

1 Introduction

There are many different types of signaling in plants. Electrical signaling is a detectable physical quantity or impulse such as a voltage or electrical current by which information can be transmitted. The cells of many biological organs generate electrical potentials that can result in the flow of electric currents (Mohr and Schopfer 1995; Volkov et al. 1998; Volkov 2006a, b, 2012a, b). Electrical impulses, as a result of stimulation, can propagate to adjacent excitable cells. This propagation can be either active, representing an action potential, or passive, described as electrotonic potential. There are three major types of electrical signaling in plants and animals: action potentials, electrotonic potentials, and graded potentials. The action potential can propagate over the entire length of the cell membrane and along the conductive bundles of tissue with constant amplitude, duration, and speed (Beilby 2007; Eschrich 1989; Hedrich 2012; Shimmen 2006; Volkov 2000). Electrotonic potentials in plants exponentially decrease with distance (Volkov et al. 2013b). An intermediate place takes graded potentials that involve the process of electrical excitation but do not evolve into full-fledged action potential. A graded potential is a wave of electrical excitation that corresponds to the size of the stimulus. Electrical signals can propagate along the plasma membrane on short distances in plasmodesmata, and on long distances in a phloem.

Plants are continuously exposed to a wide variety of perturbations; these include variation of temperature, light, mechanical forces, gravity, air and soil pollution, drought, deficiency or surplus of nutrients, attacks by insects and pathogens, etc. Therefore, it is essential for all plants to have survival sensory mechanisms against such perturbations. Plants have evolved sophisticated systems to sense environmental stimuli for adaptation and signals from other cells for coordinated action. Consequently, plants generate intracellular and intercellular electrical signals in response to these environmental changes (Bertholon 1783; Bose 1907, 1913, 1918, 1926, 1928; Ritter 1811; Takamura 2006). Neurotransmitter-like compounds, such as acetylcholine, dopamine, histamine, noradrenaline, and serotonin, participate in the information processes in plants (Roshchina 2001). Wounding an *Arabidopsis* leaf results in the propagation of electrical signaling that stimulates the production of jasmonates, plant hormones that confer resistance against herbivores and pathogens, at undamaged sites some distance from the wound. The process is mediated by cation channels encoded by the *GLR* genes (Christmann and Grill 2013).

This field has both theoretical and practical significance. Discovering these mechanisms would greatly advance our knowledge of natural phytosensors, principles of their functioning and integration into general system of defense and attack. These systems play a very important role in the life of plants, but their nature is still very poorly understood. The cells, tissues, and organs of plants transmit electrochemical impulses over short and long distances. It is conceivable that action potentials are the mediators for intercellular and intracellular communication in response to environmental irritants. An action potential is a momentary change in electric potential on the surface of a cell that takes place when it is stimulated. Initially, plants respond to irritants at the site of stimulation; however, excitation waves can be conducted along the membranes throughout the entire plant. Bioelectrical impulses travel from the root to the stem and vice versa (Volkov et al. 2013a, b). Chemical treatment, intensity of the irritation, mechanical wounding, previous excitations, temperature, and other irritants influence the speed of propagation.

Plants can perceive mechanical stimuli. This process involves mechanosensitive channels that were found in all types of cells, from animal and plant cells to fungi and bacteria. The omnipresence of these channels underlines their important physiological function in the regulation of osmolarity, cell volume, and growth. These channels are ideal transducers of physiologically relevant mechanical forces. Mechanosensory ion channels in plants are activated by mechanical stress and then transduce this information into electrical signals. These channels are involved in the growth, development, and response to environmental stress in higher plants.

In terms of electrophysiology, plant responses can be considered in three stages: (1) stimulus perception, (2) signal transmission, and (3) induction of response. In *Chara* the first stage is due to the receptor potential, a transient depolarization with a critical threshold that triggers action potentials, which are responsible for stages (2) and (3). Receptor potentials are generated by mechanosensory ion channels. Action potentials involve a transient influx of Ca^{2+} to the cytoplasm, effluxes of K^+ and Cl^-, and a temporary decrease of turgor pressure. Like the action potential, a critical threshold depolarization triggers Ca^{2+} influx, opening of Ca^{2+}-sensitive Cl^- channels and K^+ channels; effluxes that last over an hour and result in turgor regulation. However, since higher plants are composed of complex tissues, detailed analysis of electrical phenomena is rather difficult, and so the mechanism for generating the receptor potential has not yet been established.

In this review, we shall consider both active and passive propagation of electrical signals in the Venus flytrap (*Dionaea muscipula* Ellis) and *Mimosa pudica* L.

2 Signaling in *Mimosa pudica*

The *Mimosa pudica* is one of the marvels of plant electrical, mechanical, and biochemical engineering. Among several other interesting features, it has the ability to change the shape of its leaves. Movements in this sensitive plant are associated with fast responses to environmental stimuli that appear to be regulated through electrical signal transduction. The change of shape can be caused by touch or by

Fig. 1 The structure of *Mimosa pudica*

burns. In this section, we analyze propagation of electrical signals and hydromechanical responses in *Mimosa pudica* induced by brief flaming. Propagation of impulses was studied along the plant shown in Fig. 1. The *Mimosa pudica* contains long slender branches, called petioles, which can fall due to mechanical, thermal, or electrical stimuli. The petioles contain small pinnules, arranged on the rachis or midrib of the pinna. The pinnules are the smallest leaflets while the entire leaf contains the petioles, pinnae, and pinnules. A pulvinus is a joint-like thickening at the base of a plant leaf or leaflet that facilitates thigmonastic movements (Fig. 1). Primary, secondary, and tertiary pulvini are responsible for the movement of the petiole, pinna, and leaflets, respectively (Shimmen 2006).

2.1 Propagation of Mechanical Response and Branching Points

Touching of a pinnule induces the closing of pinnules in the pinna in an upward fashion and propagation of this process to the branching point with speed of about 1 cm/s. Signal goes from the very end of the pinna, i.e., from the farthest pinnule touched in the experiment to the secondary pulvinus, causing in the process the pinnules in this pinna to ascend and to close. As soon as the "alert" wave reaches the secondary pulvinus, pinnules in the other three pinnae start to close beginning from

Fig. 2 Closing of *Mimosa pudica* pinnules and falling down of the petiole after stimulation of a pinnule by brief flaming

the secondary pulvinus to apices of the leaflet along midribs. The excitation does not propagate across the secondary pulvinus to a petiole. Therefore, this branching point together with pulvinus represents a barrier for propagating signal and a logical point. Block of the signal means that in this case the rest of the plant does not need to know about this local alert.

However, this branching point barrier can be overcome if a brief flaming involves a pinnule or a rachis (Fig. 2). Flaming induces a mechanical movement of pinnules, and excitation propagates through the secondary pulvinus to a petiole. Brief flaming of a pinnule and a rachis for 1 s induces closing of pinnules in the stimulated pinna, in three neighboring pinnae of the leaf, and hangs down petiole by petiole along the *Mimosa pudica* plant. In 3-month-old plants all pinnules close and the petioles hang down after brief flaming, but in large 5- to 6-month-old plants some leaves at the bottom of a plant do not move. The "alert" signal breaks through the branching point and secondary pulvinus when a brief flaming involves rachis. This damage is perceived as a severe threat to the plant and the signal is transferred to the whole plant or almost the whole plant. If nothing else occurs, the *Mimosa pudica* is able to relax within 20 min (Fig. 3). In this process, the shape of the pulvinus changes. Its abaxial side has higher volume and curvature when a petiole is in a relaxed state (Fig. 3, 0 s and 23 min). After stimulation reaches the pulvinus, the volume and curvature in the adaxial part of the pulvinus increase and a petiole hangs down (Fig. 3). A pulvinus changes its shape during the movement of a petiole. Figure 3 shows the kinetics of a single petiole bending, triggered by brief flaming stimulation. A petiole bending is synchronized with the increased volume of the adaxial part of a pulvinus and the decreased volume of the abaxial part of a pulvinus. Petiole bending can be caused by a rapid shrinking of the abaxial side of a pulvinus and synchronous increasing of the adaxial part of the pulvinus.

Fig. 3 Morphing structures of a pulvinus and a petiole movement of *Mimosa pudica* after thermal stimulation of pinnules and the top of a rachis by a flame

2.2 Velocity of Propagation

Movements in the *Mimosa pudica* are related to changes of turgor and "alert" wave's propagation (Bose 1918; Haberlandt 1914; Hooke 1667; Houwink 1935, 1938; Malone 1994). Malone (1994) monitored pinnules' thickness after brief flaming and found propagation of an "alert" wave with initial speed of 1.5 cm/s.

Propagation of mechanical responses of leaves along *Mimosa pudica* was recorded by video camera. The dots in Fig. 4 represent the speed of mechanical response of petioles falling after a brief flaming. Initial speed of mechanical response varies from 0.8 to 1.5 cm/s (Mean 1.011 cm/s, Median 1.000 cm/s, Std. Dev. 0.079 cm/s, Std. Err. 0.008 cm/s, $n = 99$) in different *Mimosa pudica* plants. Due to this reason, the normalized speed of "alert" wave propagation in Fig. 4 was presented. This speed is approximately constant in the beginning, between point of flame stimulation of pinna and secondary pulvinus, but then slowly decreases at each next pulvinus during propagation of an "alert" wave along the stem.

Fig. 4 Dependence of normalized speed V/V_{max} of "alert" wave propagation in *Mimosa pudica* induced by 1 s flame stimulation of pinnules and the top of a rachis on the distance. Experimental points are an average of 16 measurements

2.3 Electrical Nature of the "Alert" Wave

Brief flaming of a petiole induces propagation of electrical signals along the plant tissue as a transient change of voltage (Fromm and Lautner 2007; Kaiser and Grams 2006; Koziolek et al. 2003) or electrical current (Houwink 1935, 1938). Malone (1994) suggested that the "alert" wave in *Mimosa pudica* might be a hydraulic wave. However, we believe that the "alert" wave might have an electric nature. To prove this, we used very efficient chemical tools: ion channel blockers. We deposited a few 10 μL drops of inhibitors on all pinna of *Mimosa pudica* plant for 20 h before the experiments. We selected the anion channels blocker 9-anthracenecarboxylic acid (A-9-C), Ca^{2+}-penetrable ion channels blockers $ZnCl_2$ and $BaCl_2$, and also K^+ voltage-gated ion channel blocker tetraethylammonium chloride (TEACl). All of these blockers exponentially decrease the speed of an "alert" wave propagation along the plant induced by a brief flaming of pinnules and rachis (Fig. 5a–c) in comparison with a plant non-treated by inhibitors (Fig. 4). Speed of "alert" wave propagation along the *Mimosa pudica* was recorded by video camera.

Uncouplers carbonylcyanide-3-chlorophenylhydrazone (CCCP) and 2,4-dintrophenol (DNP) also exponentially decrease the speed of "alert" wave propagation along the stem of *Mimosa pudica* (Fig. 5d). Uncouplers, which are soluble in both water and lipid phases, permeate the lipid phase of a membrane and transfer protons across the membrane, thus eliminating the transmembrane proton concentration gradient. Uncouplers do not interact with voltage-gated ion channels, but they can indirectly induce opening or closing of voltage-gated ion channels caused by depolarization of membranes.

We have also studied effect of anesthetics on morphing process in this plant. If *Mimosa pudica* and a Petri dish with 5 mL of chloroform are covered by a 12-l glass jar for 10 min, anesthetic agent chloroform inhibits pinnule closing and leaflets movement after stimulation of a leaf by flaming. The effect of anesthetic is

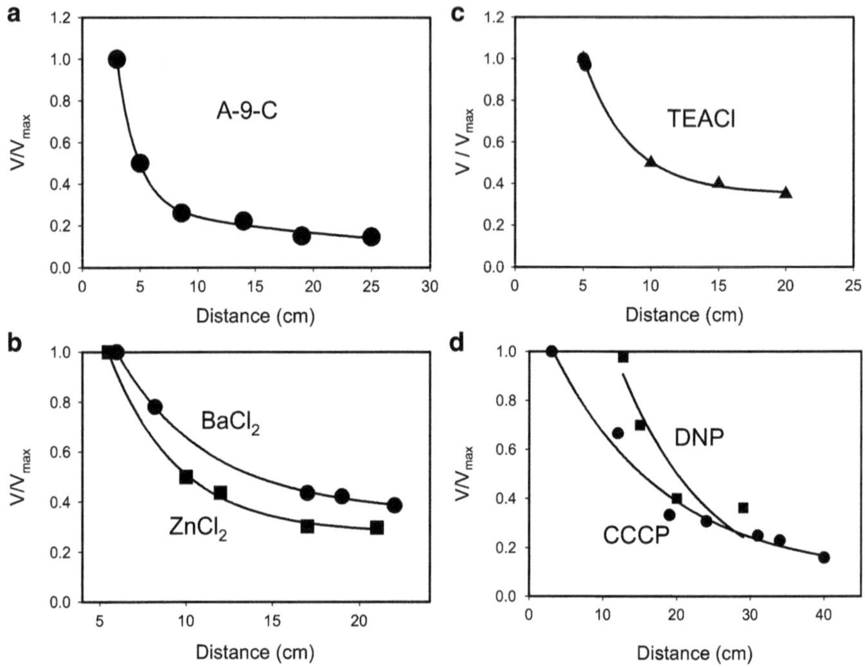

Fig. 5 Dependencies of a normalized speed V/V_{max} of mechanical response propagation in *Mimosa pudica* induced by 1 s flame stimulation of pinnules and the top of a rachis on the distance 20 h after deposition on all pinna of 10 μL drops of inhibitors: 5 mM 9-anthracenecarboxylic acid (A-9-C) (**a**); 5 mM $ZnCl_2$ or 5 mM $BaCl_2$ (**b**); 10 mM tetraethylammonium chloride (TEACl) (**c**); 10 μM carbonylcyanide-3-chlorophenylhydrazone (CCCP); or 0.5 mM 2,4-dinitrophenol (DNP) (**d**)

reversible: when the anesthetic is taken away, the leaves recover, and pinnules can close again. Effects of anesthetics on mechanical stimulation and responses in the *Mimosa pudica* were first described by Bose (1913, 1918). The disappearance of thigmonastic mobility of *Mimosa pudica*, caused by volatile or local anesthetic agents, was observed by Bernard (1878), Milne and Beamish (1999), Okazaki et al. (1993), Paes and De Luccia (2012), and Wallace (1931).

Variation of speed of the "alert" wave propagation in *Mimosa pudica* with distance x in Fig. 5 can be parameterized by the two-exponential function:

$$\frac{V}{V_{max}} = c + a \times \mathrm{Exp}(-b \times x). \tag{1}$$

Parameters a, b, and c, are different in each case; they are presented in Table 1. The most informative parameter in this function is b: it gives the rate of velocity decrease along the path of the wave. The inverse value $1/b$ gives the distance at which velocity decreases e-times. Interestingly enough this distance for most of these agents is only few centimeters, varying from 2 to 5 cm, and only for CCCP it

Table 1 Parameters of Eq. (1)

Inhibitor	*a*	*b* (cm^{-1})	*c*
9-Anthracenecarboxylic acid (A-9-C)	3.0846	0.4405	0.1732
$BaCl_2$	1.9932	0.1893	0.3595
$ZnCl_2$	2.8602	0.2500	0.2760
Tetraethylammonium chloride (TEACl)	2.7274	0.2870	0.3501
2,4-dinitrophenol (DNP)	21.6115	0.2768	0.3413
Carbonylcyanide-3-chlorophenylhydrazone (CCCP)	1.1291	0.0660	0.0872

rises to 15 cm. The most potent agent is A-9-C, which eliminates the wave at characteristic distance of 2 cm.

Ion channel inhibitors and uncouplers can regulate the plasma membrane potential. The sensitivity of the propagation of "alert" waves to the blockers of ionic channels shows that electrical signal propagation is certainly involved in the process.

To study the propagation of electrical signals along the *Mimosa pudica* plant, we used two sets of electrodes inserted in the stem; their recordings are presented in Fig. 6. Results show two-phase electrical responses that propagate in the stem between two pulvini with constant amplitude and approximately the same duration. The time delay between two traces shows that the speed of these electrical signals was 1 cm/s. Electrical signals in a stem of *Mimosa pudica* propagate in both basipetal and acropetal directions. Electrical signals in *Mimosa pudica* caused by a brief flaming are referred to as variation potentials or currents (Houwink 1935), because their amplitude decreases with distance of propagation along the stem. Houwink (1935) suggested that the propagation of variation electrical current is caused by a diffusion of an unknown chemical compound along the plant, "which is sucked in from the wound by the negative pressure in the vessels, and is transported by the transpiration stream." Fromm and Lautner (2007) found that two-phase electrical signals propagate in *Mimosa pudica* and their amplitude decreases with distance from the site of flaming. However, the authors did not study the amplitude dependence of the electrical signals on distance. A decreasing electrical response can be monotonous along the plant stem, or it can drop after passing through pulvini. To address this point, we analyzed amplitude of electrical signals transduction in pulvini and in a stem of *Mimosa pudica*.

If electrodes are located in the stem at two different sides from a pulvinus, they show that the drop in the amplitude of electrical signal is more pronounced after passing through the pulvinus (Fig. 7).

Electrical signals propagate from the point of a flaming application through the pinna, secondary pulvinus, petiole, pulvinus, and stem. This two-phase electrical signal propagates along the petiole with a speed of about 1 cm/s and duration of about 2.5 s; amplitude between peaks is about 0.05–0.06 V (Figs. 6 and 7).

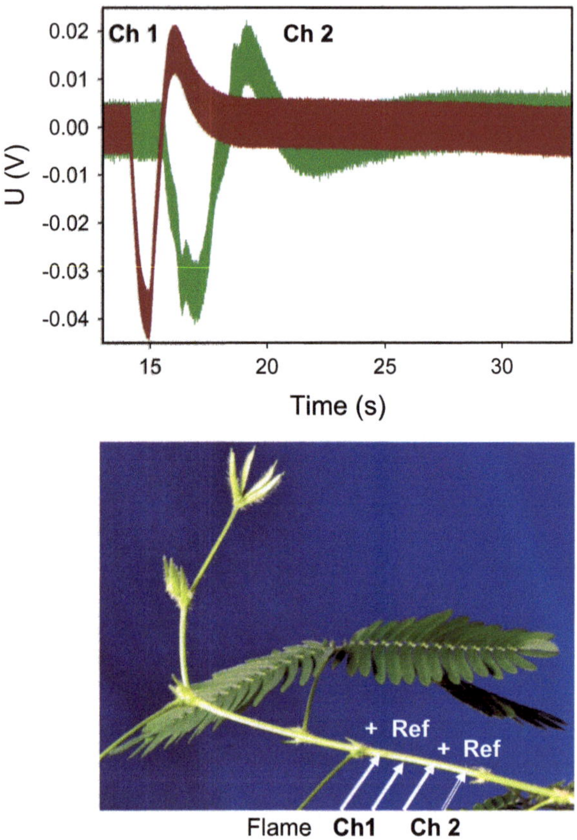

Fig. 6 Electrical signal propagation in a stem of *Mimosa pudica* between Ag/AgCl electrodes. Position of electrodes in the stem is shown on the inserted photo. Distance from 1 s flaming application was about 22 cm. U is voltage between electrodes in channels (Ch 1 and Ch 2)

2.4 *Distribution of Electrical Potentials in Different Parts of* Mimosa pudica

Our studies showed some peculiar properties of potential distribution in a running electrical wave. Figure 8 compares electrical response in a petiole along it and in the perpendicular direction. The idea of this experiment might look crazy, because nobody would expect to find different potentials in the same cross section of the petiole. We did not expect it either, and actually we did it after we observed something unexpected in the pulvinus. We found a two-phase signal in the longitude direction (channel 2) and no signal in the perpendicular direction—zero line for channel 1. Both sides of the petiole at the same cross section remained at the same potential when the signal propagated along the petiole.

When we came to the pulvinus (Fig. 9), we found that the electrical response along the pulvinus (in the x direction) was much shorter than in the petiole (Figs. 6, 7, and 8); it lasted for only 0.2–0.3 s. This response also had two-phase character with similar amplitude of 0.05–0.06 V between peaks.

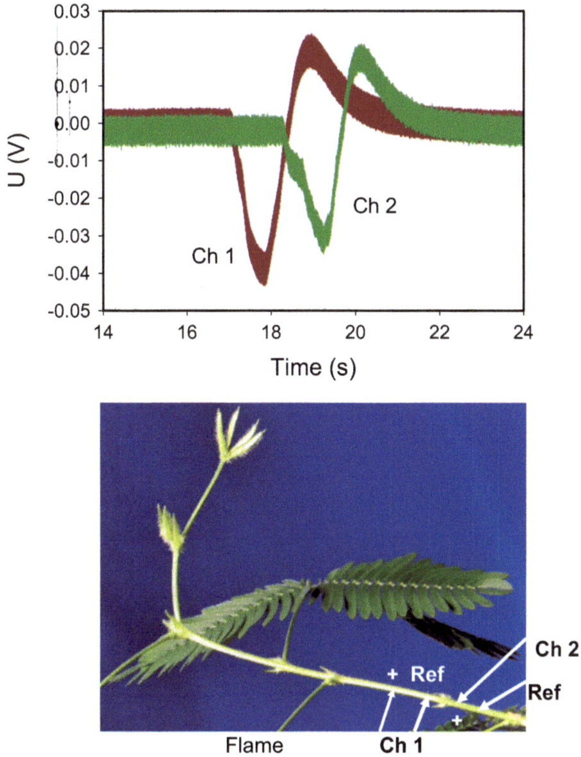

Fig. 7 Electrical signal propagation in a stem of *Mimosa pudica* between Ag/AgCl electrodes. Position of electrodes in the stem is shown in the inserted photo

In this experiment, presented in Fig. 9, the second set of electrodes was inserted across the pulvinus in the vertical direction along the *z*-axis (channel 2). Interestingly enough, it shows that the abaxial sides of the pulvinus acquire different potentials when the "alarm wave" arrives. In the beginning the adaxial side becomes negative, up to −0.04 V, and then positive, up to +0.02 V. This signal looks similar to action potential; this action potential might induce redistribution of pressure between two layers of flexor and extensor cells, changing shape of the pulvinus, and causing the down movement of the petiole. The transporters in the upper part are influenced by the incoming action potentials in a way different from those in the lower part (e.g., different voltage-controlled channels). We believe that the voltage difference generated across the pulvinus triggers the chain of events bringing the change of osmotic pressure in upper and lower parts of pulvinus and eventually to the movement of the petiole.

Duration of both signals is the same, but the amplitude of the electrical signal in the direction across the pulvinus is a little smaller (Fig. 9). To complete this experiment we checked if potential difference is generated in the horizontal direction along the *y*-axis. We did not register any electrical response in channel 3.

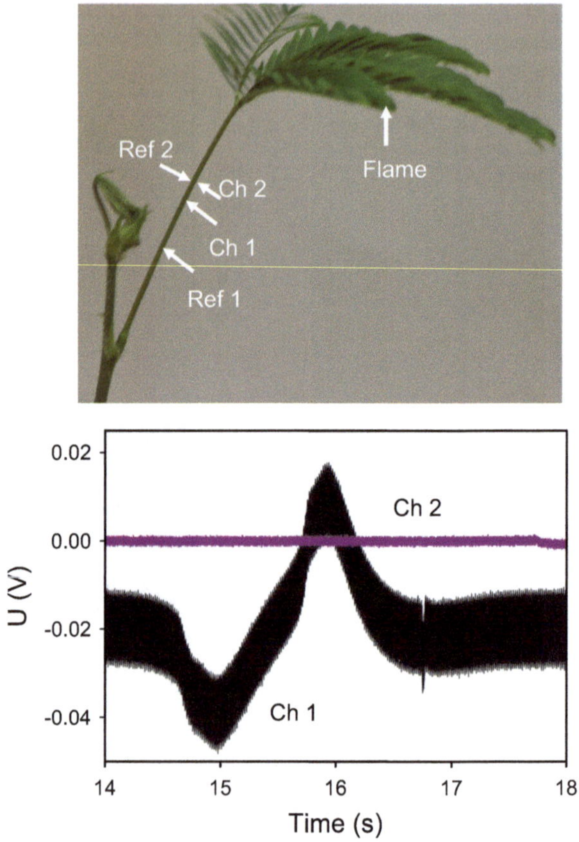

Fig. 8 Two-channel recording of electrical signal propagation in a petiole of *Mimosa pudica* between Ag/AgCl electrodes induced by 1 s flame stimulation of pinnules and the top of a rachis. Position of electrodes in a petiole and a pulvinus is shown in the inserted photo

Propagation of electrical signals through pulvini is bidirectional (Figs. 9 and 10). After a brief flaming, electrical signal propagates along a petiole through the pulvinus to the stem, in which it propagates in basipetal and acropetal directions (Fig. 10a). When the electrical signal reaches the next pulvinus, it propagates in the direction from the stem to a pinna inducing pinnules closing in the acropetal direction against the gravitational field.

When signal reaches the pulvinus, it produces electrical potential difference not only along a pulvinus, but also in the vertical direction along plasmodesmata. This causes a redistribution of water and ions in a pulvinus with reversible elastic change in the shape of the pulvinus (Fig. 10). There is an electrical coupling between vascular bundles and plasmodesmata in the pulvinus, which is responsible for the propagation of electrical signal in the direction perpendicular to vascular bundles in the pulvinus. Fleurat-Lessard and Bonnemain (1978) found large number of branched plasmodesmata in the pulvinus. Plasmodesmata are membrane-lined channels which are functionally comparable to animal gap junctions (Roberts 2005).

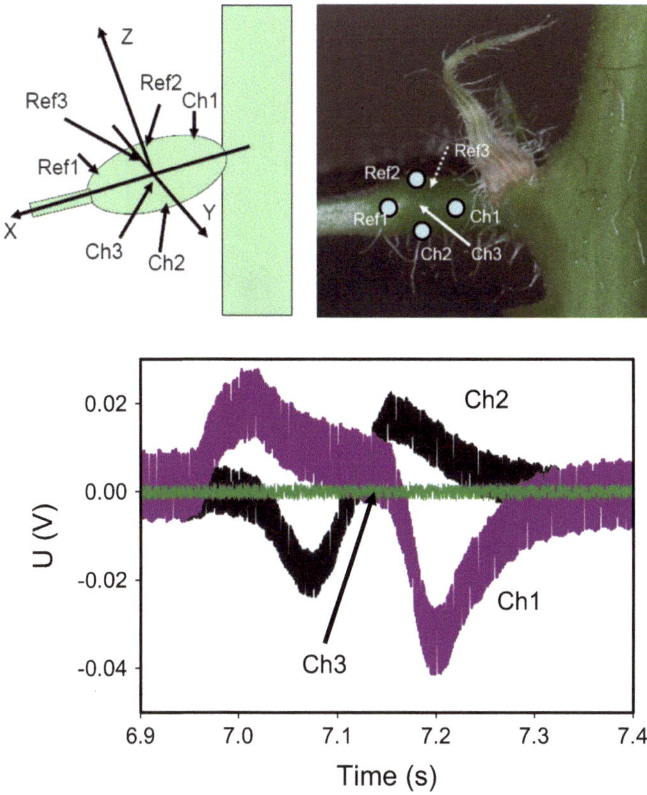

Fig. 9 Three-channel recording of electrical signal propagation in a pulvinus of *Mimosa pudica* between Ag/AgCl electrodes induced by 1 s flame stimulation of pinnules and the top of a rachis. Position of electrodes in the pulvinus is shown in the inserted photo

To measure the dependence of amplitude of electrical "alert" wave on distance, Ag/AgCl electrodes were inserted in the stem (stars) and at both sides of each pulvinus (closed circles) (Fig. 11). The amplitude of the electrical signal in a stem is constant between any two neighboring pulvini (Figs. 6 and 11), but decreases in the stem after each pulvinus (Figs. 7 and 11). This property of the pulvinus is similar to the property of electrical synapse in neuronal networks (Hormuzdi et al. 2004). There is a correlation between decreasing the amplitude of the electrical signal propagation through pulvini (Fig. 11) and speed of mechanical "alert" wave propagation along the plant (Fig. 4). The speed of electrical signal propagation along the *Mimosa pudica* plant is constant, but the speed of the mechanical "alert" wave propagation and amplitude of electrical signal decreases with distance after propagation through pulvini.

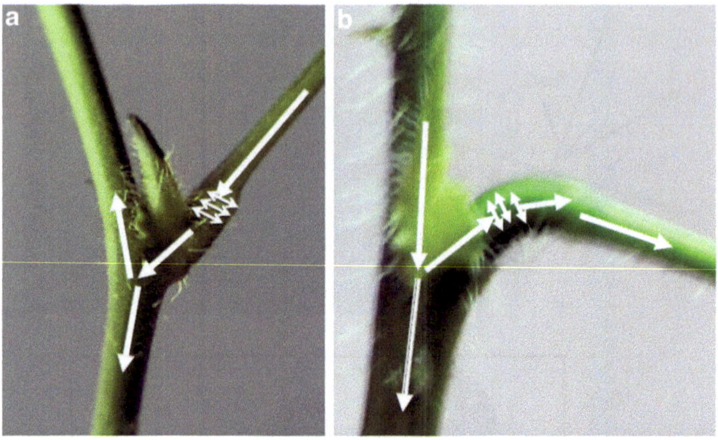

Fig. 10 A schematic diagram of propagation of electrical signals in *Mimosa pudica* induced by 1 s flame stimulation of pinnules and the top of a rachis from a petiole to a stem (**a**) and along a stem (**b**). *Long arrows* correspond to signal transduction along vascular bundles and *short arrows* show electrical signal propagation in a pulvinus through branched plasmodesmata

3 Mechanisms of Electrical Signaling

Mechanical movements in the Venus flytrap and *Mimosa pudica* are quite different though both have electrical background. Literature data on electrical signaling in *Mimosa pudica* has very significant discrepancies between different groups of researchers and even between different publications from the same group. According to literature, the amplitude of action potentials varies from 16 to 210 mV, and the duration of electrical signals varies from 1.2 to 2,000 s (Abe 1980, 1981; Abe and Oda 1976; Eschrich 1989; Fromm 1991; Fromm and Eschrich 1988; Fromm and Lautner 2007; Houwink 1935; Kaiser and Grams 2006; Koziolek et al. 2003; Oda and Abe 1972; Roblin 1979, 1982; Sibaoka 1962, 1966, 1969, 1991; Stoeckel and Takeda 1993; Umrath 1937).

Several factors which account for this large discrepancy include but are not limited to: (1) aliasing due to a low scanning rate of data acquisition systems without low-pass filters or with slow voltmeters (Shannon 1949); (2) slow ion-sensitive electrodes with membranes; and (3) high impedance of *Mimosa pudica* tissue which does not permit the use of fast oscilloscopes or high speed data acquisition systems with a low input impedance. While the effect of aliasing on the reproducibility of electrical signals measurements in plants was discussed in detail by Jovanov and Volkov (2012), electrical signaling in *Mimosa pudica* requires additional study. It is possible that various stimuli generate different electrical signals in the pulvinus, stem, and leaves of *Mimosa pudica*.

There are two possible mechanisms of effect of brief flaming of a pinnule: generation of action potentials, which can induce an electroosmotic wave or induction by a hydraulic wave electrical streaming potential in the plant. There are two possible pathways of signal transduction through a xylem or phloem.

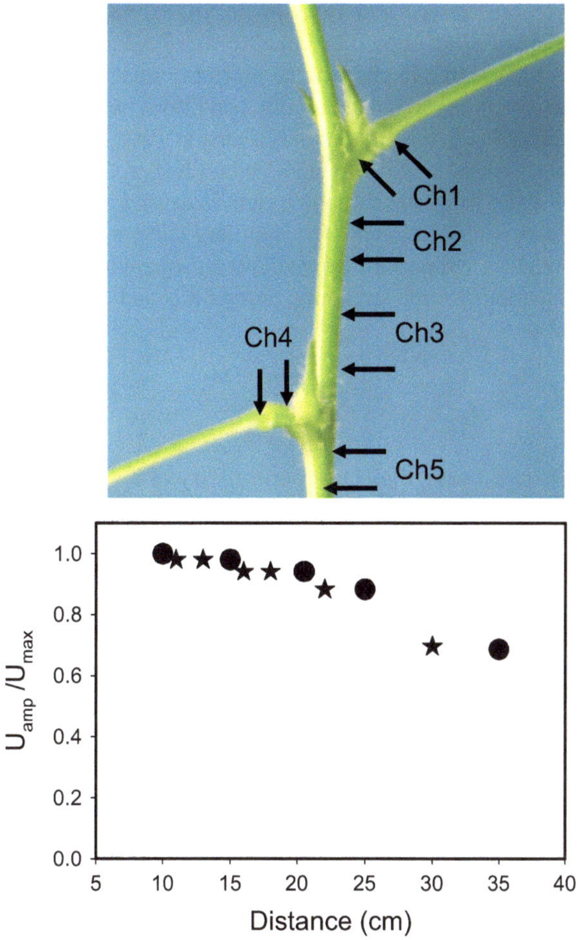

Fig. 11 Dependence of amplitude of electrical signals in pulvini (*closed circles*) and in a stem (*stars*) on distance from flame stimulation of pinnules and the top of a rachis. Ag/AgCl electrodes were inserted along pulvini and a stem

3.1 Hydraulic Wave and a Streaming Potential in a Xylem

The propagation of hydraulic waves in a xylem was described by Malone (1994) and various other authors. If hydrostatic pressure is much higher near a point of flaming than in a plant, it could induce a hydraulic wave in the direction of decreasing hydrostatic pressure, which can be described by the Poiseuille law:

$$J_v = -\frac{r^2}{8\eta}\frac{\partial P}{\partial l}, \tag{2}$$

where r is a radius of a capillary, η is the viscosity of the aqueous phase, l is the capillary length, and P is pressure. Since a radius of a xylem vessel is about 17 μm

(Fleurat-Lessard and Bonnemain 1978), $\eta = 10^{-3}$ Pa s, and J_v is 10^{-2} m/s, the gradient dP/dl should be equal to 0.277 MPa/m.

Electrokinetic phenomena play important roles in the transport of water through the xylem. Even in iso-osmotic conditions, water can flow from the soil through the xylem due to electro-osmosis (Fensom 1980; Fensom and Spanner 1969; Ksenzhek and Volkov 1998). Heyl (1933) and Keller (1930) found that electro-osmotic flow of water from soil through xylem is caused by root pressure. The movement of an electrolyte solution along a capillary generates a streaming potential and a streaming electrical current between the upstream and downstream ends of a liquid flow. Equations for the streaming potential U and hydraulic pressure P can be written as follows:

$$U = \frac{\varepsilon_0 \varepsilon}{\eta \kappa} P \zeta, \tag{3}$$

where ε_0 is the dielectric permittivity of a vacuum (8.85×10^{-12} C/Vm), ε is the dielectric permittivity of the aqueous phase, P stands for the pressure gradient causing the relative movement of the phases along the interface, η is the viscosity of the aqueous phase, κ is the conductivity of a aqueous phase, and ζ is the electrokinetic or zeta potential equal to the potential difference between the immobilized and mobile phases of the electric double layer which varies from zero to a few millivolts. Dainty (1963) estimated from Eq. (3) that pressure of 1 MPa is required for 10 mV streaming potential. It means that at least 6 MPa of pressure difference is required for propagation of 60 mV streaming potential in a xylem. Such pressure gradient will lead to much higher velocity of a hydraulic wave than 1 cm/s. Velocity of a hydraulic wave in a xylem should not depend on blockers of voltage-gated ion channels and anesthetics. Therefore, propagation of hydraulic wave with synchronous 60 mV streaming potential in a xylem is doubtful.

3.2 *Electrical Signal Propagation Can Induce Electroosmotic Flow*

The phloem is a possible pathway for transduction of electrical signals in the *Mimosa pudica*. Action potential in the phloem can induce electroosmotic flow with velocity v_{osm}:

$$v_{osm} = \frac{\varepsilon_0 \varepsilon \zeta E}{\eta}. \tag{4}$$

Amplitude of electrical signal is about 0.06 V, the length of signal is 2 s·1 cm/s = 0.02 m and $E = 3$ V/m. If $\varepsilon = 80$, $\zeta = -20$ mV, $\eta = 10^{-3}$ Pa s, one can find from Eq. (4) that $v_{osm} = 42$ nm/s. Tinz-Füchmeier and Gradmann (1990) found that speed of hydraulic flow in the *Mimosa pudica* induced by brief flaming is in the range of 10–100 nm/s. Their results confirm the primary role of electrical signaling in mechanical responses of *Mimosa pudica* after localized thermal stress.

3.3 Electrical or Hydraulic Conductance of Excitation?

There are many different opinions in the literature about how an electrical signal induces the streaming of a hydraulic wave or a hydraulic wave induces the streaming potential, which looks like an action or variation potential. Tinz-Füchmeier and Gradmann (1990) used a laser-interferometer and an electrometer and found the primary role of electrical signaling in rapid conductance of the "alert" wave with speed of 1 cm/s in *Mimosa pudica* induced by a flaming. Inhibition of the "alert" wave propagation in *Mimosa pudica* by blockers of voltage-gated ion channels (Fig. 5) shows that the generation and propagation of electrical signals is a primary effect responsible for turgor change and propagation of the "alert" wave. Chloroform, a volatile anesthetic agent, also blocks closing pinnules and propagation of the "alert" wave proving an electrical nature of signaling.

4 Action Potentials in the Lamina of the Venus Flytrap

Plant response to mechanical stimulation has long been known. Perhaps all plants can react in response to the mechanical stimuli, though only certain plants with rapid and highly noticeable touch-stimulus response have received much attention, for example, the trap closure of the Venus flytrap or pinnules closing in *Mimosa pudica*. Mechanosensation is a physiological process by which a distortion of the cell membrane is converted into an electrical and/or biochemical signal. Since mechanical forces exist in plants, it would have been essential for all living cells, including the earliest microorganisms on earth, to have a survival mechanism against these forces. For this reason, mechanosensation is considered to have evolved as one of the oldest sensory mechanisms in living organisms. The rapid closure of the Venus flytrap lamina in about 0.2 s is one of the fastest movements in the plant kingdom. When a prey touches the trigger hairs, these mechanosensors trigger a receptor potential (Benolken and Jacobson 1970; Buchen et al. 1983; Jacobson 1965), which generates an electrical action potential. There are eight steps between closing and opening of the trap (Fig. 12). Two stimuli generate two action potentials, which close the trap at room temperature in a fraction of a second. Propagation of action potentials and the trap closing can be blocked by uncouplers, inhibitors of

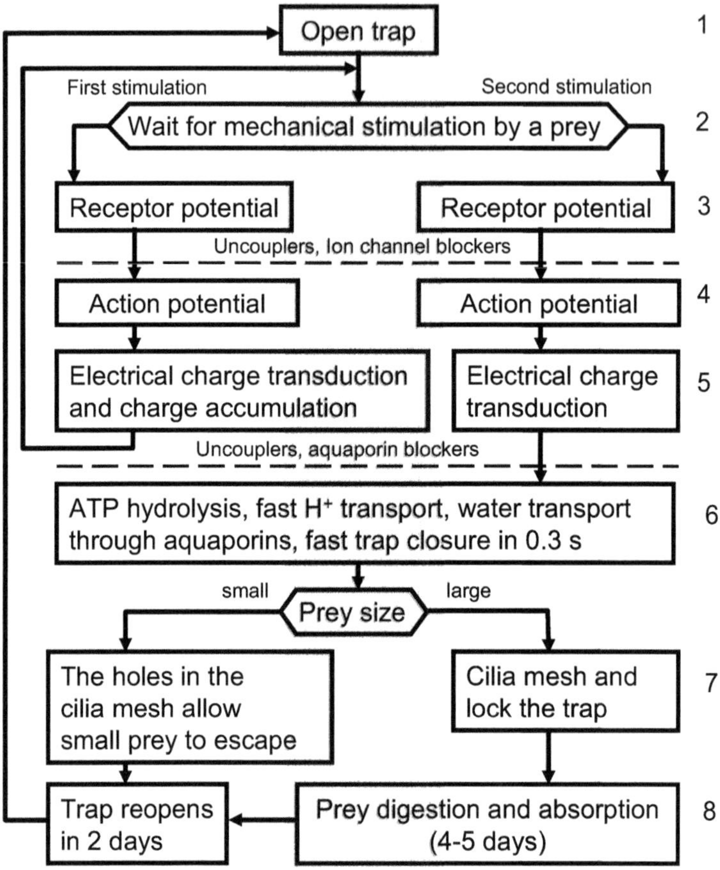

Fig. 12 The mechanism of trap closure (modified from Volkov et al. 2008b)

voltage-gated channels, and aquaporins (Volkov et al. 2008d). The trap closure has been investigated by the mechanical stimulation of the trigger hairs by a cotton thread (Fig. 12, step 1; Darwin 1880), by 7.35 μm laser with 50 μW power (Fig. 12, step 2; Eisen et al. 2013), by electrical stimulation between the lobes and midrib of the Venus flytrap (Fig. 12, step 4; Volkov et al. 2008b), and by various chemicals.

Electrical signaling and rapid closure of the carnivorous plant *Dionaea muscipula* Ellis (Venus flytrap) have been attracting the attention of researchers since nineteenth century (Burdon-Sanderson 1873; Darwin 1880; Hedrich 2012; Volkov et al. 2009a; Yang et al. 2010). In contrast to chemical signals such as hormones, electrical signals are able to transmit information rapidly over long distances. Electrical form of energy has no entropy content and 100 % of this energy can be used for work, synthesis of chemical compounds, and information transduction (Ksenzhek and Volkov 1998; Volkov et al. 1998). Biologically closed electrical circuits performing these functions operate over large distances in biological tissues (Nordestrom 1983; Volkov et al. 2009b, 2010a–e, 2011a). The activation of such circuits can lead to various physiological and biophysical responses. It is often convenient to represent the real electrical and electrochemical properties of biointerfaces with idealized equivalent electrical circuit models consisting of discrete electrical components. We investigated the biologically closed electrical circuits in the lamina of the Venus flytrap and proposed the equivalent electrical circuit.

The Venus flytrap can be closed by mechanical stimulation of two of the six trigger hairs inside the lamina of the Venus flytrap using a cotton thread or wooden stick. The Venus flytrap can also be closed by an electrical pulse with amplitude of 1.5 V between the midrib and the lobe of the lamina without mechanical stimulation. The closing was achieved by electrical stimulation with a positive electrode connected to the midrib and a negative electrode located in one of the lobes (Volkov et al. 2007, 2008b, 2009b; Eisen et al. 2013). It is noted that the inverted polarity pulse was not able to close the plant and that the closed trap would not open by electrical stimulus of either sign lasting up to 100 s.

The trap can also be closed by electrostimulation of the petiole, if one of the stimulating electrodes is located a few millimeters from the midrib or in a lobe (Volkov et al. 2013b). Electrical stimulation of the petiole induces electrical signals, which resemble action potentials in the trap between lobes and the midrib. The trap closes if the stimulating voltage is above the threshold level of 4.4 V with a positive poll near the midrib and a negative poll a few millimeters down the petiole. Since both electrical potential amplitude and electrical charge are important for the trap closing, we used low-frequency electrostimulation.

Possible pathways of action potential propagation include vascular bundles and plasmodesmata in the lamina of the Venus flytrap (Volkov et al. 2007). Markin et al. (2008) observed that the trap closure by electrical stimulus obeys the all-or-none law, which states that there is no reaction for under-threshold stimulus, and the speed of closing does not depend on stimulus strength above threshold.

Mechanical stimulation of a trigger hair in the trap induces action potential propagating between the trigger hairs in a lobe and the midrib with amplitude of 0.16–0.18 V (Escalante-Pérez et al. 2011; Volkov et al. 2007).

5 Passive Propagation of Electrical Signals

5.1 Electrotonic Potentials

Action potentials in the Venus flytrap do not penetrate to the petiole. However, we did find small electrical potentials in the petiole, which look similar to graded potentials or electrotonic potentials (Volkov et al. 2007, 2008a). To understand the

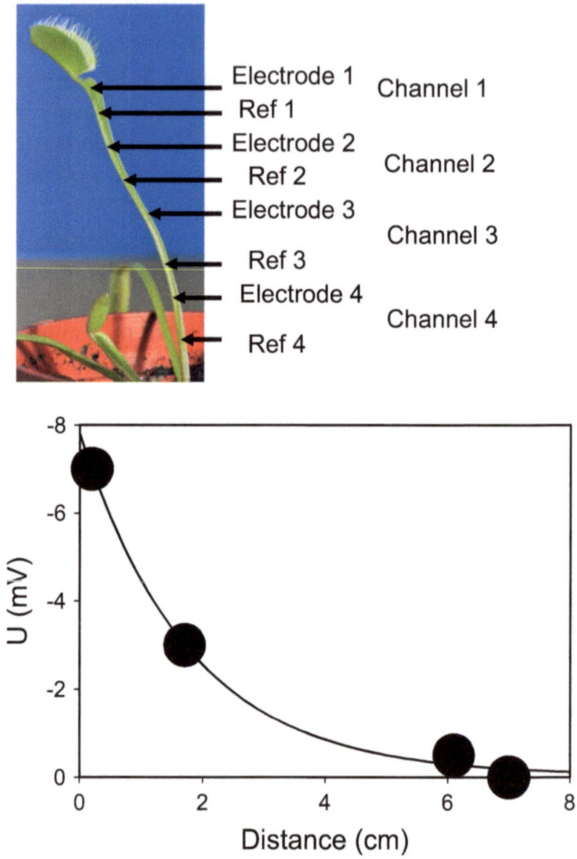

Fig. 13 Dependence of the electrical signal on distance from the midrib in the petiole induced by a gentle touch of two trigger hairs inside the lamina of the Venus flytrap by a cotton thread. The *solid line* was calculated according to Eq. (1). The frequency of scanning was 250,000 samples/s with a low-pass filter at 125,000 scans/s

nature of these electrical potentials in the petiole, we measured their dependence on the distance from the midrib (Fig. 13). The amplitude of these electrical potentials decreases exponentially with distance (Fig. 13) and can be described by the cable equation:

$$U = U_0 \times \text{Exp}(\text{Distance}/\lambda) \tag{5}$$

with parameters $U_0 = -7.7812$ mV and $\lambda = 1.77$ cm. Therefore, the constant of length, which indicates how far an electrical signal will spread in the petiole of the Venus flytrap, is $\lambda = 1.77$ cm. In small neurons, electrical potentials decreasing exponentially are referred to as electrotonic potentials (Jack et al. 1975).

5.2 Electrical Stimulation at Low Voltages

For electrostimulation of the Venus flytrap, we used the function generator selecting either square pulses or sinusoidal waves applied between the midrib and a lobe or to a petiole. Since the Venus flytrap can close the trap if electrical stimulation of 1.5 V and electrical charge of 14 μC are applied between a midrib and a lobe of the trap (Markin and Volkov 2012; Volkov et al. 2007, 2008a, b, 2011b), we applied voltages of 1 V or less from a function generator to the lamina and petiole.

Figure 14 shows results of electrostimulation of the Venus flytrap by square steps from the function generator at $U_0 = \pm 1$ V from the bottom of the petiole of the Venus flytrap. The electrical responses on electrostimulation are nonlinear with a duration time about 10 ms (Fig. 14b, c).

These electrical responses are not action potentials since their amplitude and polarity depend on the applied voltage. The amplitude of these nonlinear electrical responses exponentially decreases with distance (Fig. 15) as typical electrotonic potentials and can be described by Eq. (2) with the constant of length $\lambda = 1.32$ cm. Channel 3 in Fig. 14 is located in the trap and from Fig. 16 we can conclude that the difference in RC time constants in the petiole and in the lamina is about 1.5 times. As a result, the time dependence of electrical response in the trap crossed the time dependence in the petiole.

Figure 16 shows dependencies of electrotonic potential on time measured at two different channels in the petiole of the Venus flytrap during electrostimulation by the function generator. To estimate the speed of electrotonic potentials, we sent a single pulse and measured arriving time at each electrode pair. The speed of electrotonic potentials in the petiole was estimated by dividing the distance between channels by the time difference between two minimums in Fig. 16. The speed of electrotonic potential propagation is equal to 1 cm/2.5 ms = 4 m/s (Mean 3.91 m/s, Median 4.00 m/s, Std. Dev. 0.11 m/s, Std. Err. 0.03 m/s, $n = 14$).

Amplitude and polarity of electrical responses depend on the applied voltage (Fig. 17). One can notice that points in Fig. 17 represent two straight lines with slope of 0.178 V/V at the right side and slope of 0.067 V/V at the left. Transition between them occurs at $U_0 = -0.4$ V. Application of a sinusoidal electrical stimulation from the function generator to the lobe of the Venus flytrap generates linear electrical responses in the petiole in the form of graded potentials which are changes in polarization where the magnitude of the change varies with the strength of the stimulus (Fig. 18). Amplitude of this response linearly depends on applied voltage with a slope of 0.049 V/V (Fig. 19a). Dependence on distance initially is also linear, but then the curve quickly approaches zero (Fig. 19b). The output is entirely consistent with the tissue behaving as an RC circuit.

The response in Fig. 18 indicates that the time constant is short compared to the period of the input and the tissue of the Venus flytrap behaves as a differentiator with voltage determined by equation:

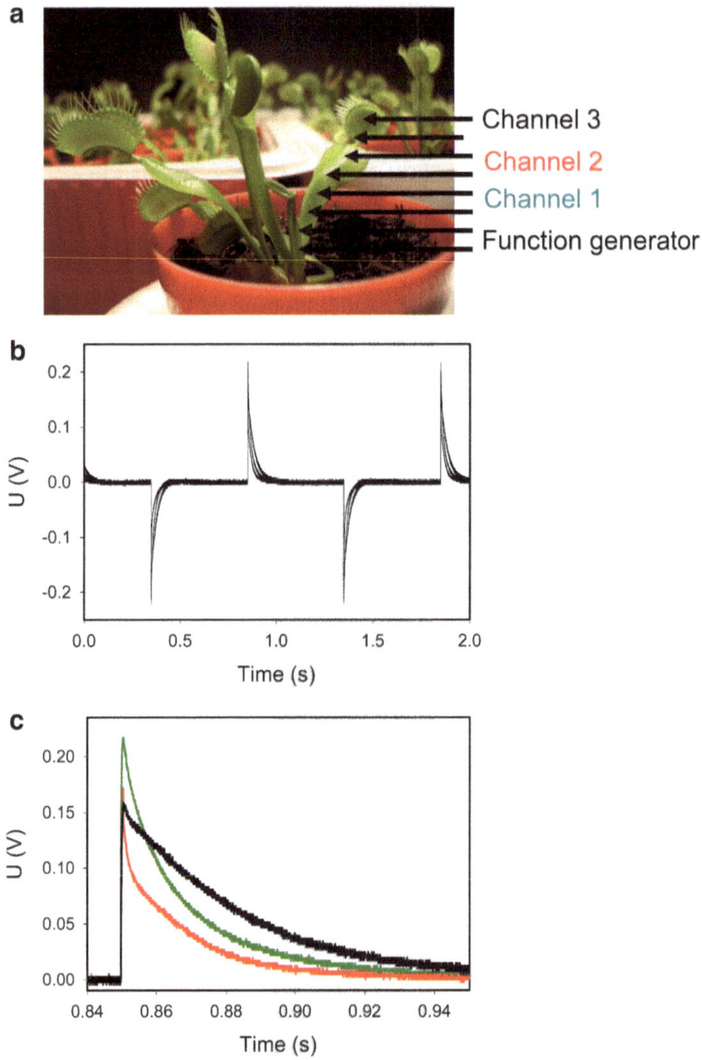

Fig. 14 Electrostimulation of the Venus flytrap by a square pulse from function generator at $U_0 = \pm 1.0$ V from the bottom of the petiole of the Venus flytrap. (**a**) Location of 8 Ag/AgCl electrodes in the Venus flytrap; (**b**, **c**) Electrical responses in channels 1, 2, and 3. The frequency of scanning was 250,000 samples/s with a low-pass filter at 125,000 scans/s. Frequency of electrostimulation was 1 Hz

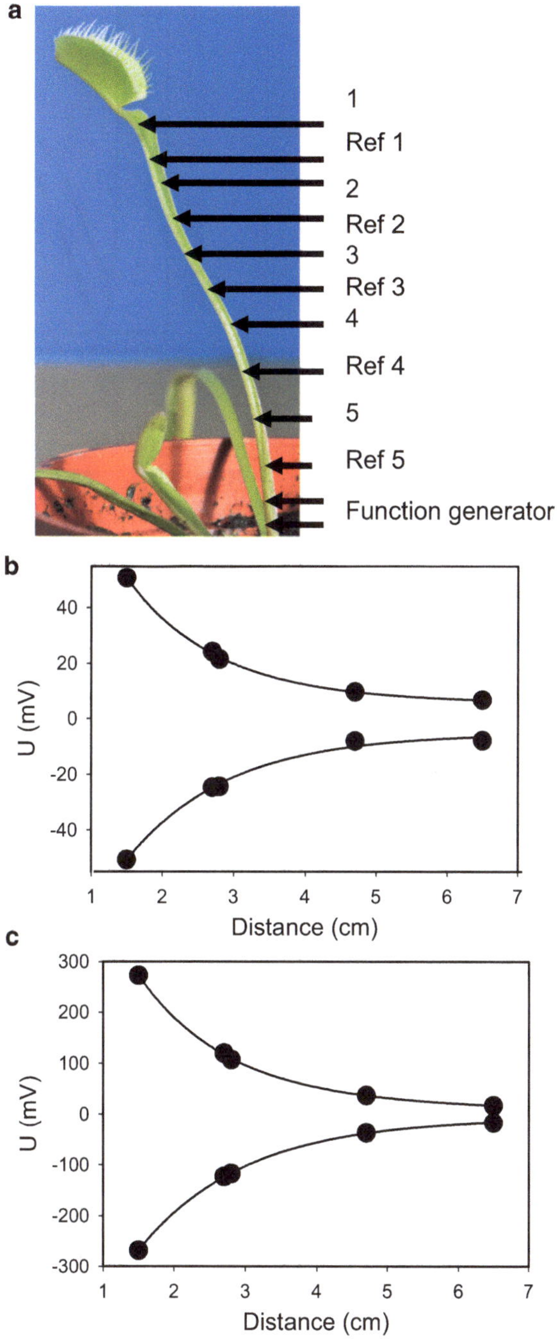

Fig. 15 (**a**) Location of Ag/AgCl electrodes in the Venus flytrap. Dependence of electrical signal in the petiole on distance from the electrodes induced by a function generator at (**b**) $U_0 = \pm 0.25$ V and (**c**) $U_0 = \pm 1.0$ V. The frequency of scanning was 250,000 samples/s with a low pass filter at 125,000 scans/s

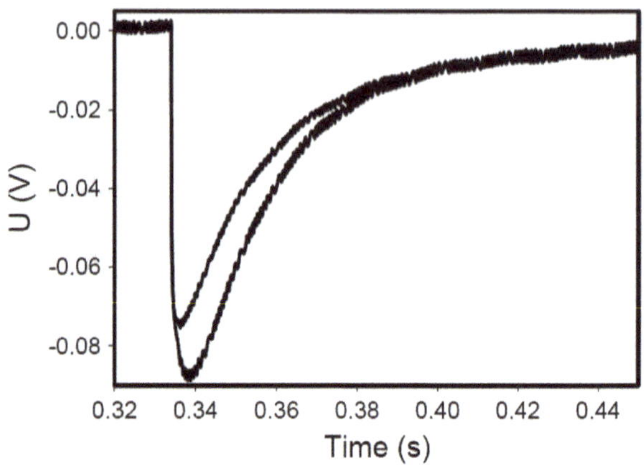

Fig. 16 Time dependence of electrical signal in the petiole induced by a square pulse from a function generator at $U_0 = \pm 0.5$ V. The frequency of scanning was 100,000 samples/s with a low-pass filter at 50,000 scans/s. Frequency of electrostimulation was 1 Hz

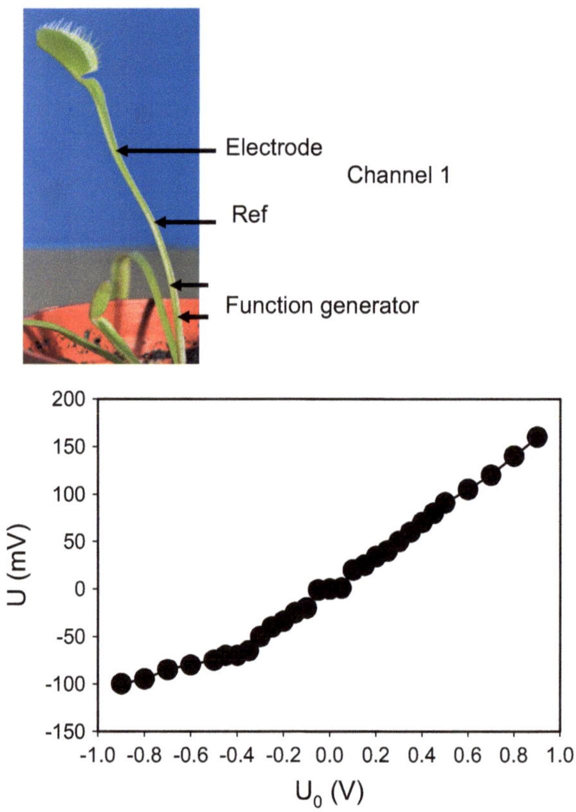

Fig. 17 Dependence of maximal amplitude U of electrical signals induced by a square pulse from the function generator on applied voltage U_0 between Ag/AgCl electrodes (channel 1) located in the Venus flytrap petiole at 2 cm distance from electrodes connected to the function generator. Frequency of electrostimulation was 1 Hz. Measurements were performed at 100,000 scans/s with a low-pass filter at 50,000 scans/s

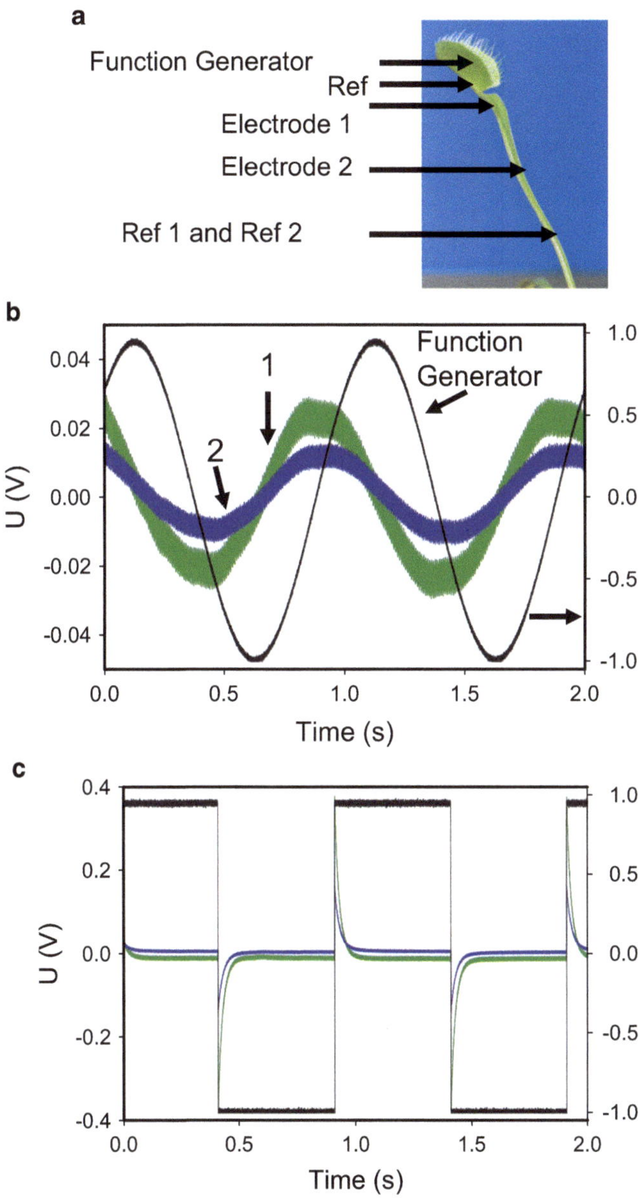

Fig. 18 (**a**) Location of Ag/AgCl electrodes in the Venus flytrap. Time dependence of electrical signal in the petiole induced by sinusoidal wave (**b**) and by square pulse (**c**) from function generator at U_0=±1.0 V. The frequency of scanning was 100,000 samples/s with a low-pass filter at 50,000 scans/s. Frequency of electrostimulation was 1 Hz

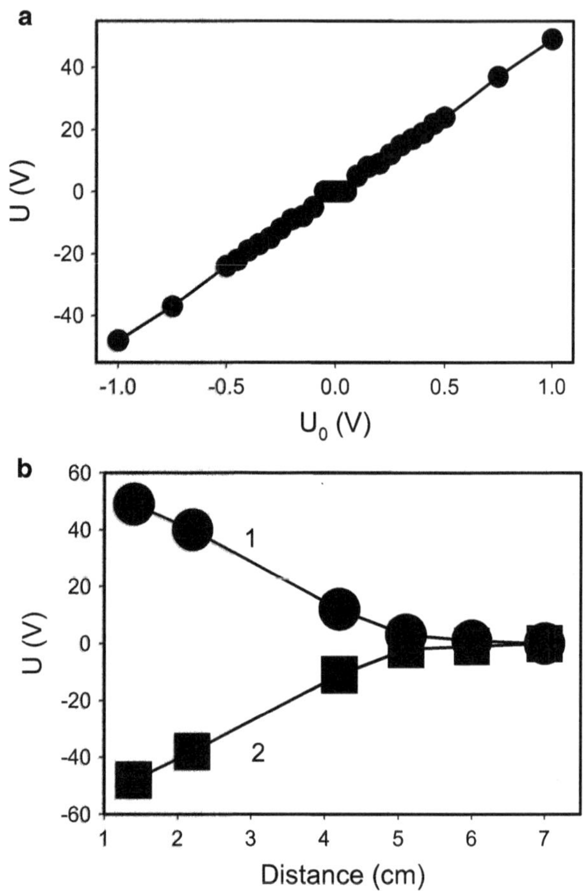

Fig. 19 Dependence of maximal amplitude U of electrical signals induced by a sinusoidal wave from the function generator connected to electrodes in the trap on applied voltage U_0 (**a**) and distance (**b**) between Ag/AgCl electrodes located in the Venus flytrap petiole. $U_0 = +1$ V (1) or -1 V (2). Frequency of electrostimulation was 1 Hz. Measurements were performed at 100,000 scans/s with a low-pass filter at 50,000 scans/s

$$U = \mathrm{RC}(\mathrm{d}U_0/\mathrm{d}t) - \mathrm{RC}(\mathrm{d}U/\mathrm{d}t). \tag{6}$$

This equation predicts the phase shift of the response in the case of electrostimulation by a sine wave shown in Fig. 18b. The analysis of the response to the sine wave shows the dependence of the amplitude and phase shift of the response on the frequency and amplitude of the signal. The square electrostimulating waves generate nonlinear responses as spikes (Fig. 18c), as predicted by Eq. (3) for a differentiator in the electrical network of the Venus flytrap.

Figure 20 shows nonlinear responses in the Venus flytrap induced by a square pulse electrostimulation from the function generator, which was applied to the lamina of the Venus flytrap. The dependence of the amplitude of response on the amplitude of stimulus is well described by Boltzmann function with appropriate parameters:

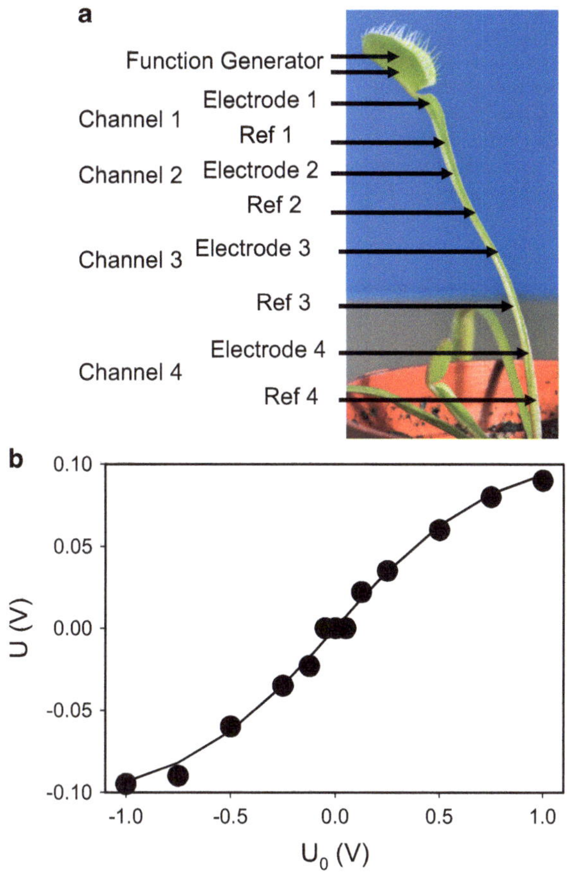

Fig. 20 Dependence of maximal amplitude U of electrical signals between Ag/AgCl electrodes located in the Venus flytrap petiole (channel 1) induced by a square pulse from the function generator connected to Ag/AgCl electrodes in the Venus flytrap lamina on applied voltage U_0. Frequency of electrostimulation was 1 Hz. Measurements were performed at 100,000 scans/s with a low-pass filter at 50,000 scans/s. *Solid line* was plotted according to Eq. (2)

$$U = 0.10747[1 - \exp(-2.6777U_0)] / [1 + \exp(-2.6777U_0)]. \quad (7)$$

In order to induce a nonlinear response, there must be an instantaneous increase or decrease in the stimulus voltage. Any stimulation that is not instantaneous, such as a sinusoidal or triangular function, results in electrotonic potentials. The duration of electrotonic potentials does not depend on the amplitude of the applied voltage (Figs. 14 and 16). Only one electrical signal, which resembles a spike, was generated in the lamina and in the petiole of the Venus flytrap during a single square pulse from the function generator. The amplitude of electrical response is proportional to the amplitude of the stimulus and does not obey the all-or-none rule. Therefore, it is not an action potential but rather corresponds to the propagating electrotonic potential. Amplitude of electrotonic potentials does not depend on frequency of electrostimulation (Fig. 21), if the applied square pulse stimulus has a frequency less than 10 scans/s and decreases at higher frequency. The reason is that if the frequency of the square wave is too fast compared to the RC constant, the plant tissue capacitors do not have time to charge or discharge.

Fig. 21 Dependence of electrical responses induced by a square pulse from function generator on frequency of electrostimulation ($U_0 = 0.5$ V, distance from function generator is 4 cm)

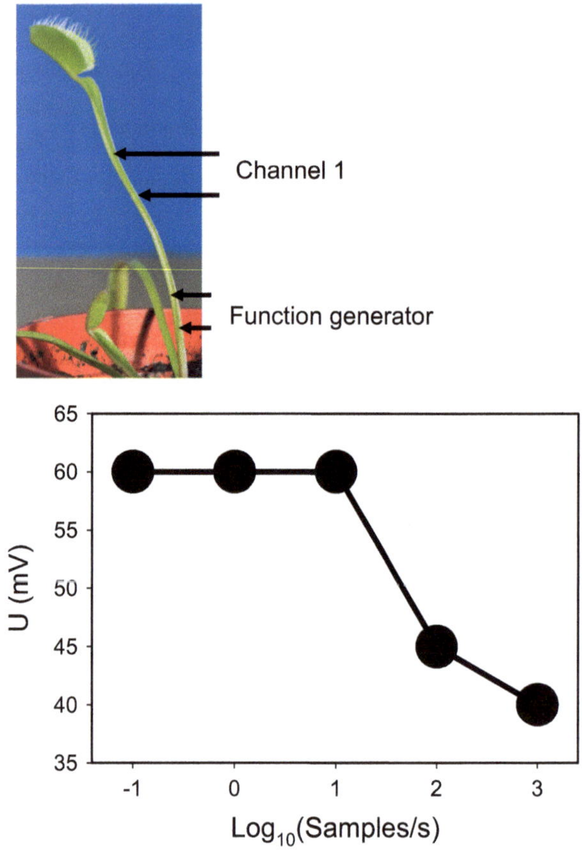

5.3 *Electrical Stimulation at High Voltages: From Electrotonic to Action Potential*

Electrotonic potentials can induce action potentials in the lamina of the Venus flytrap if the applied voltage of electrostimulation to the petiole exceeds the threshold value. Figure 22 shows electrical responses in the Venus flytrap between the midrib and the lobe when square electrical stimuli were applied to the petiole with a distance of 5 mm from the midrib. If the amplitude of the applied voltage was less than 4.4 V, only electrotonic potentials were generated in the trap with the same polarity as polarity of the stimulating voltage (Fig. 22a, b). Figure 22a shows electrical response in the trap when the amplitude of applied voltage was 1 V; Fig. 21b corresponds to the applied voltage of ±4 V. Figure 22c, d shows that if the applied voltage has an amplitude of 4.5 V, electrotonic potentials with an amplitude of 0.85 V and action potential with amplitude of 0.175 V propagate in the trap. Figure 22d shows in extended scale a portion of Fig. 22c. Polarity of electrotonic

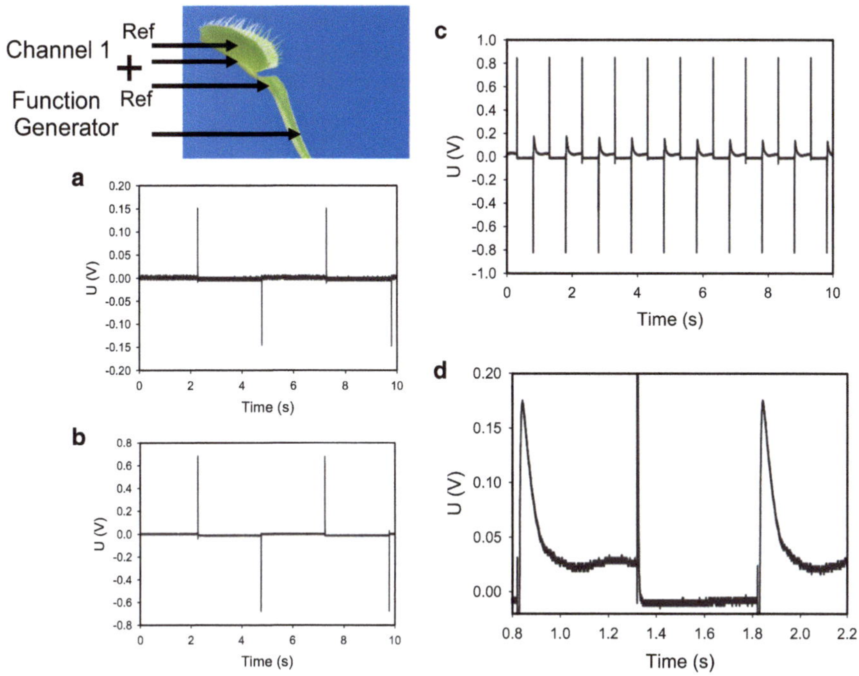

Fig. 22 Electrical signaling in the lamina of the Venus flytrap induced by electrostimulation of the petiole. (**a**) $U_0 = \pm 1$ V, 0.2 Hz; (**b**) $U_0 = \pm 4$ V, 0.2 Hz; (**c**, **d**) $U_0 = \pm 4.5$ V, 1 Hz (Volkov et al. 2013b)

potentials coincides with polarity of electrostimulating voltage (Fig. 21c), but negative electrotonic potential induce positive nonlinear responses in the trap, which are action potentials (Fig. 22c, d). Amplitude of these action potentials does not depend on applied voltage in the range from 4.4 to 6.0 V.

If the same electrical stimulation by a square wave with amplitude of 4.5 V is applied to the petiole of the Venus flytrap with an open trap near a midrib, the trap closes. If the amplitude is less than 4.5 V, the trap does not close. If the additional square pulse is applied to the petiole of the Venus flytrap with a closed trap, action potential between a midrib and the lobes generates, which then causes the tightening and constriction of the trap. Action potentials can be generated by applied voltage between 4.5 and 6.0 V. Higher voltages give complicated shapes of electrical responses and induce downward movement of the trap.

According to Fig. 22c, amplitude of electrical responses is 0.175 V and duration is about 0.2 s. Similar amplitudes of action potentials in the Venus flytrap were reported in the literature (Escalante-Pérez et al. 2011; Hodick and Sievers 1988; Trebacz and Sievers 1998; Volkov et al. 2007). DiPalma et al. (1966) reported similar duration of action potential (0.24 s). Duration and amplitude of an action potential in the Venus flytrap depend on Ca^{2+} concentration (Hodick and Sievers

1986, 1988) and in the presence of EDTA amplitude decreases and duration of an action potential increases (Krol et al. 2012). With increasing external Ca^{2+} concentration the amplitude of action potential grew while the duration shortened (Krol et al. 2006).

5.4 Interaction Between Active and Passive Impulses in the Venus Flytrap

Electrostimulation by a square pulse of the lamina or petiole of the Venus flytrap induces electrotonic potentials, which can propagate from the petiole to the lamina and from the lamina to the petiole. Electrostimulation of the petiole by a square pulse with amplitude of 4.4 V induces the propagation of electrotonic potential in the petiole and lamina and action potential in the trap, which induce the trap closing. Action potential can propagate from mechanosensitive trigger hairs in the lamina to the midrib, but does not penetrate to the petiole. Trebacz et al. (1996) studied action potentials in the trap induced by electrostimulation and found that the threshold is between 0.5 and 0.6 V. Krol et al. (2006) applied up to 4 V for electrostimulation of the trap for generation of action potentials in the Venus flytrap.

Action potentials that are induced by the touching of trigger hairs do not propagate from the lamina to the petiole (Volkov et al. 2007, 2008c). This is consistent with the findings of Pavlovič et al. (2010) who found that the irritation of trigger hairs and a subsequent generation of action potentials resulted in a decrease in the effective photochemical quantum yield of photosystem II and the rate of net photosynthesis in the trap, but not in the petiole of the Venus flytrap.

Electrotonic potentials are also well known in animal tissue such as in neurons, the heart, and muscles (Sampson and Henriquez 2005; Shepherd 1994). Amplitude of an electrotonic potential exponentially decreases with distance in both plants and animal tissues. These potentials play a rather important role. For example, the spread of a receptor potential is accomplished by means of electrotonic potentials (Shepherd 1994). Some small neurons have only electrotonic potentials; some neurons utilize electrotonic potentials to trigger the action potential. Electrotonic potentials can influence on duration of action potentials (Sampson and Henriquez 2005). In animal tissue, the study of both action potentials and electrotonic potentials is very advanced and can be done at the level of single excitable cell. This is almost impossible with plant tissue. The rare exceptions are *Chara* and *Nitella* (Beilby 2007; Shimmen 2006). In other cases, we deal with ensemble of cells and we are not sure about position of electrodes. Therefore, it is not surprising that researchers observe different shapes of electrical signals with single phase or two phases of electrical oscillations (DiPalma et al. 1961). Plant tissues are rather complex and do not make these studies easy. Even the application of the popular cable model to observed data is not always very justified. Nevertheless, the joint efforts of many laboratories have helped to better understand plant electrophysiology.

6 Conclusions

Plants are ideal adaptive structures with smart sensing capabilities. The knowledge gained from studying electrical, mechanical, and hydraulic processes in plants is key input for designing adaptive structures and intelligent materials. Understanding mechanisms in functioning of phytosensors and phytoactuators is very important not only for botany and plant physiology, but has important applications in bio-inspired engineering and technology (Lee et al. 2010; Shahinpoor 2011; Taya 2003; Volkov and Markin 2012).

Acknowledgments This review is based upon work supported in part by the National Science Foundation under Grant No. CBET-1064160 and in part by the U.S. Army Research Office under contract/grant number W911NF-11-1-0132.

References

Abe T (1980) The shortening and action potential of the cortex in the main pulvinus of *Mimosa pudica*. Bot Mag Tokyo 93:247–251
Abe T (1981) Chloride ion efflux during an action potential in the main pulvinus of *Mimosa pudica*. Bot Mag Tokyo 94:379–383
Abe T, Oda K (1976) Resting and action potentials of excitable cells in the main pulvinus of *Mimosa pudica*. Plant Cell Physiol 17:1343–1346
Beilby MJ (2007) Action potentials in charophytes. Int Rev Cytol 257:43–82
Benolken RM, Jacobson SL (1970) Response properties of a sensory hair excised from Venus's flytrap. J Gen Physiol 56:64–82
Bernard C (1878) Lectures on phenomena of life common to animals and plants. J.B. Balliere and Son, Paris
Bertholon M (1783) De l'electricite des vegetaux: ouvrage dans lequel on traite de l'electricite de l'atmosphere sur les plantes, de ses effets sur leconomie des vegetaux, de leurs vertus medico. P.F. Didot Jeune, Paris
Bose JC (1907) Comparative electro-physiology, a physico-physiological study. Longmans, Green & Co, London
Bose JC (1913) Researches on irritability of plants. Longmans, London
Bose JC (1918) Life movements in plants. B.R. Publishing Corp, Delhi
Bose JC (1926) The nervous mechanism of plants. Longmans, Green and Co., London
Bose JC (1928) The motor mechanism of plants. Longmans Green, London
Buchen B, Hensel D, Sievers A (1983) Polarity in mechanoreceptor cells of trigger hairs of *Dionaea muscipula*. Planta 158:458–468
Burdon-Sanderson J (1873) Note on the electrical phenomena which accompany stimulation of the leaf of *Dionaea muscipula*. Philos Proc R Soc Lond 21:495–496
Christmann A, Grill E (2013) Plant biology: electric defence. Nature 500:404–405
Dainty J (1963) Water relations of plant cells. Adv Bot Res 1:279–326
Darwin C (1880) The power of movements in plants. John Murra, London
DiPalma JR, Mohl R, Best W (1961) Action potential and contraction of *Dionaea muscipula* (Venus flytrap). Science 133:878–879
DiPalma JR, McMichael R, DiPalma M (1966) Touch receptor of Venus flytrap, Dionaea muscipula. Science 152:539–540

Eisen D, Janssen D, Chen X, Choa FS, Kostov D, Fan J (2013) Closing a Venus Flytrap with electrical and mid-IR photon stimulations. In: Proceedings of SPIE 8565, Photonic therapeutics and diagnostics IX, 85655I (March 8, 2013); doi:10.1117/12.2005351
Escalante-Pérez M, Krol E, Stange A, Geiger D, Al-Rasheid KAS, Hause B, Neher E, Hedrich RA (2011) Special pair of phytohormones controls excitability, slow closure, and external stomach formation in the Venus flytrap. Proc Natl Acad Sci U S A 108:15492–15497
Eschrich W (1989) Stofftransport in Bäumen. Sauerländer, Frankfurt
Fensom DS (1980) Problems arising from Munch-type pressure flow mechanism of sugar transport in phloem. Can J Bot 59:425–432
Fensom DS, Spanner DC (1969) Electro-osmotic and biopotential measurement on phloem strands of Nymhoides. Planta 88:321–331
Fleurat-Lessard P, Bonnemain J (1978) Structural and ultrastructural characteristics of vascular apparatus of the sensitive plant (*Mimosa pudica* L). Protoplasma 94:127–143
Fromm J (1991) Control of phloem unloading by action potentials in Mimosa. Physiol Plantarum 83:529–533
Fromm J, Eschrich W (1988) Transport processes in stimulated and non-stimulated leaves of *Mimosa pudica*. I. The movement of ^{14}C-labelled photoassimilates. Trees 2:7–17
Fromm J, Lautner S (2007) Electrical signals and their physiological significance in plants. Plant Cell Environ 30:249–257
Haberlandt G (1914) Physiological plant anatomy. Macmillan, London
Hedrich R (2012) Ion channels in plants. Physiol Rev 92:1777–1811
Heyl JG (1933) Der Einfluss von Aussenfaktoren auf das Bluten der Pflanzen. Planta 20:294–353
Hodick D, Sievers A (1986) The influence of Ca^{2+} on the action potential in mesophyll cells of *Dionaea muscipula* Ellis. Protoplasma 133:83–84
Hodick D, Sievers A (1988) The action potential of *Dionaea muscipula* Ellis. Planta 174:8–18
Hooke R (1667) Micrographia. The Royal Society, London
Houwink AL (1935) The conduction of excitation in *Mimosa pudica*. Recuril des Travaux Botaniques Neerlandais 32:51–91
Houwink AL (1938) The conduction of excitation in Clematis zeylanica and in *Mimosa pudica*. Annales du Jardin Botanique de Buitenzorg 48:10–16
Hormuzdi SG, Filippov MA, Mitropoulou G, Monyer H, Bruzzone R (2004) Electrical synapses: a dynamic signaling system that shapes the activity of neuronal networks. Biochim Biophys Acta 1662:113–137
Jack JJ, Noble D, Tsien RW (1975) Electric current flow in excitable cells. Clarendon, Oxford
Jacobson SL (1965) Receptor response in Venus's flytrap. J Gen Physiol 49:117–129
Jovanov E, Volkov AG (2012) Plant electrostimulation and data acquisition. In: Volkov AG (ed) Plant electrophysiology – methods and cell electrophysiology. Springer, Berlin, pp 45–67
Kaiser H, Grams TEE (2006) Rapid hydropassive opening and subsequent active stomatal closure follow heat-induced electrical signals in *Mimosa pudica*. J Exp Bot 57:2087–2092
Keller R (1930) Der elektrische Factor des Wassertransporte in Luhte der Vitalfarbung. Ergeb Physiologie 30:294–407
Koziolek C, Grams TEE, Schreiber U, Matyssek R, Fromm J (2003) Transient knockout of photosynthesis mediated by electrical signals. New Phytol 161:715–722
Krol E, Dziubinska H, Stolarz M, Trebacz K (2006) Effects of ion channel inhibitors on cold- and electrically-induced action potentials in *Dionaea muscipula*. Biol Plantarum 50:411–416
Krol E, Płachno BJ, Adamec L, Stolarz M, Dziubińska H, Trebacz K (2012) Quite a few reasons for calling carnivores 'the most wonderful plants in the world'. Ann Bot 109:47–64
Ksenzhek OS, Volkov AG (1998) Plant energetics. Academic Press, San Diego
Lee H, Xia C, Fang N (2010) First jump of microgel: actuation speed enhancement by elastic instability. Soft Matter 6:4342–4345
Malone M (1994) Wound-induced hydraulic signals and stimulus transmission in *Mimosa pudica* L. New Phytol 128:49–56

Markin VS, Volkov AG (2012) Morphing structures in the Venus flytrap. In: Volkov AG (ed) Plant electrophysiology – signaling and responses. Springer, Berlin, pp 1–31
Markin VS, Volkov AG, Jovanov E (2008) Active movements in plants: mechanism of trap closure by *Dionaea muscipula* Ellis. Plant Signal Behav 3:778–783
Milne A, Beamish T (1999) Inhalation and local anesthetics reduce tactile and thermal responses in *Mimosa pudica*. Can J Anaesth 46:287–289
Mohr H, Schopfer P (1995) Plant physiology. Springer, Berlin, pp 528–529
Nordestrom BEW (1983) Biologically closed electrical circuits. Clinical, experimental and theoretical evidence for an additional circulatory system. Nordic Medical Publications, Uppsala
Oda K, Abe T (1972) Action potential and rapid movement in the main pulvinus of *Mimosa pudica*. Bot Mag Tokyo 85:135–145
Okazaki N, Takai K, Sato T (1993) Immobilization of a sensitive plant, *Mimosa pudica* L., by volatile anesthetics. Mausi 42:1190–1193
Paes T, De Luccia B (2012) *Mimosa pudica*, *Dionaea muscipula* and anesthetics. Plant Signal Behav 7:1163–1167
Pavlovič A, Demko V, Hudak J (2010) Trap closure and prey retention in Venus flytrap (*Dionaea muscipula*) temporarily reduces photosynthesis and stimulates respiration. Ann Bot 105:37–44
Ritter JW (1811) Electrische Versuche an der *Mimosa pudica* L. In Parallel mit gleichen Versuchen an Fröschen. Denkschr Köningl Akad Wiss (München) 2:345–400
Roberts AG (2005) Plasmodesmal structure and development. In: Oparka KJ (ed) Plasmodesmata. Blackwell, Oxford, pp 1–32
Roblin G (1979) *Mimosa pudica*: a model for the study of the excitability in plants. Biol Rev 54:135–153
Roblin G (1982) Movements, bioelectrical events and proton excretion induced in the pulvini of *Mimosa pudica* L. by a period of darkness. Zeitschrift Pflanzenphysiologie 108:295–304
Roshchina VI (2001) Neurotransmitters in plant life. Science Publishers, Inc., Enfield, NH
Sampson KJ, Henriquez CS (2005) Electrotonic influences on action potential duration dispersion in small hearts: a simulation study. Am J Physiol Heart Circ Physiol 289:H350–H360
Shahinpoor M (2011) Biomimetic robotic Venus flytrap (*Dionaea muscipula Ellis*) made with ionic polymer metal composites. Bioinspir Biomim 6:046004-1-11. doi:10.1088/1748-3182/6/4/046004
Shannon CE (1949) Communication in the presence of noise. Proc Inst Radio Eng 37:10–21
Shepherd GM (1994) Neurobiology. Oxford University Press, New York
Shimmen T (2006) Electrophysiology in mechanosensing and wounding response. In: Volkov AG (ed) Plant electrophysiology. Theory and methods. Springer, Berlin, pp 319–339
Sibaoka T (1962) Excitable cells in Mimosa. Science 137:226–226
Sibaoka T (1966) Action potentials in plant organs. Symp Soc Exp Biol 20:49–74
Sibaoka T (1969) Physiology of rapid movements in higher plants. Ann Rev Plant Physiol 20:165–184
Sibaoka T (1991) Rapid plant movements triggered by action potentials. Bot Mag Tokyo 104:73–95
Stoeckel H, Takeda K (1993) Plasmalemmal, voltage-dependent ionic currents from excitable pulvinar motor cells of *Mimosa pudica*. J Membr Biol 131:179–192
Takamura T (2006) Electrochemical potential around the plant root in relation to metabolism and growth acceleration. In: Volkov AG (ed) Plant electrophysiology – theory and methods. Springer, Berlin, pp 341–374
Taya M (2003) Bio-inspired design of intelligent materials. Proc SPIE 5051:54–65
Tinz-Füchmeier A, Gradmann D (1990) Laser-Interferometric re-examination of rapid conductance of excitation in *Mimosa pudica*. J Exp Bot 41:15–19
Trebacz K, Bush MB, Hejnowicz Z, Sievers A (1996) Cyclopiazonic acid disturbs the regulation of cytosolic calcium when repetitive action potentials are evoked in *Dionaea* traps. Planta 198:623–626

Trebacz K, Sievers A (1998) Action potential evoked by light in traps of *Dionaea muscipula* Ellis. Plant Cell Physiol 39:369–372
Umrath K (1937) Der Erregungsvorgang bei höheren Pflancen. Ergebnis Biologie 14:1–42
Volkov AG (2000) Green plants: electrochemical interfaces. J Electroanal Chem 483:150–156
Volkov AG (2006a) Electrophysiology and phototropism. In: Balushka F, Manusco S, Volkman D (eds) Communication in plants. Springer, Berlin, pp 351–367
Volkov AG (ed) (2006b) Plant electrophysiology – theory and methods. Springer, Berlin
Volkov AG (ed) (2012a) Plant electrophysiology. Methods and cell electrophysiology. Springer, Berlin
Volkov AG (ed) (2012b) Plant electrophysiology. Signaling and responses. Springer, Berlin
Volkov AG, Markin VS (2012) Phytosensors and phytoactuators. In: Volkov AG (ed) Plant electrophysiology – signaling and responses. Springer, Berlin, pp 173–206
Volkov AG, Deamer DW, Tanelian DL, Markin VS (1998) Liquid interfaces in chemistry and biology. Wiley, New York
Volkov AG, Adesina T, Markin VS, Jovanov E (2007) Closing of Venus flytrap by electrical stimulation of motor cells. Plant Signal Behav 2:139–144
Volkov AG, Adesina T, Jovanov E (2008a) Charge induced closing of *Dionaea muscipula* Ellis trap. Bioelectrochemistry 74:16–21
Volkov AG, Adesina T, Markin VS, Jovanov E (2008b) Kinetics and mechanism of *Dionaea muscipula* trap closing. Plant Physiol 146:694–702
Volkov AG, Carrell H, Adesina T, Markin VS, Jovanov E (2008c) Plant electrical memory. Plant Signal Behav 3:490–492
Volkov AG, Coopwood KJ, Markin VS (2008d) Inhibition of the *Dionaea muscipula* Ellis trap closure by ion and water channels blockers and uncouplers. Plant Sci 175:642–649
Volkov AG, Carrell H, Baldwin A, Markin VS (2009a) Electrical memory in Venus flytrap. Bioelectrochemistry 75:142–147
Volkov AG, Carrell H, Markin VS (2009b) Biologically closed electrical circuits in Venus flytrap. Plant Physiol 149:1661–1667
Volkov AG, Foster JC, Ashby TA, Walker RK, Johnson JA, Markin VS (2010a) *Mimosa pudica*: electrical and mechanical stimulation of plant movements. Plant Cell Environ 33:163–173
Volkov AG, Foster JC, Baker KD, Markin VS (2010b) Mechanical and electrical anisotropy in *Mimosa pudica*. Plant Signal Behav 5:1211–1221
Volkov AG, Foster JC, Jovanov E, Markin VS (2010c) Anisotropy and nonlinear properties of electrochemical circuits in leaves of *Aloe vera* L. Bioelectrochemistry 81:4–9
Volkov AG, Foster JC, Markin VS (2010d) Signal transduction in *Mimosa pudica*: biologically closed electrical circuits. Plant Cell Environ 33:816–827
Volkov AG, Foster JC, Markin VS (2010e) Molecular electronics in pinnae of *Mimosa pudica*. Plant Signal Behav 5:826–831
Volkov AG, Baker K, Foster JC, Clemmens J, Jovanov E, Markin VS (2011a) Circadian variations in biologically closed electrochemical circuits in *Aloe vera* and *Mimosa pudica*. Bioelectrochemistry 81:39–45
Volkov AG, Pinnock MR, Lowe DC, Gay MS, Markin VS (2011b) Complete hunting cycle of *Dionaea muscipula*: consecutive steps and their electrical properties. J Plant Physiol 168:109–120
Volkov AG, O'Neal L, Volkova MI, Markin VS (2013a) Morphing structures and signal transduction in *Mimosa pudica* L. induced by localized thermal stress. J Plant Physiol 170:1317–1327
Volkov AG, Vilfranc CL, Murphy VA, Mitchell CM, Volkova MI, O'Neal L, Markin VS (2013b) Electrotonic and action potentials in the Venus flytrap. J Plant Physiol 170:838–846
Wallace RH (1931) Studies of sensitivity of *Mimosa pudica*. 1. The effect of certain animal anesthetics upon sleep movements. Ann J Bot 18:102–111
Yang RT, Scott CL, Zhang MJ, Xia LJ (2010) A mathematical model on the closing and opening mechanism for Venus flytrap. Plant Signal Behav 5:968–978

Adaptations of Chloroplastic Metabolism in Halophytic Plants

Ewa Niewiadomska and Monika Wiciarz

Contents

Abstract Halophytism is of complex nature and polyphyletic origin in plants. One of the common points of various strategies is a protection of the photosynthetic machinery for a long term. Here we summarize the recent data obtained with halophytic, as with glycophytic plants, concerning the protective strategies against salinity stress operating in chloroplasts composed of protecting both the light capturing machinery and the step of photosynthetic CO_2 assimilation. The involvement of reactive oxygen species (ROS) and redox tuning is also discussed.

E. Niewiadomska (✉)
The F. Górski Institute of Plant Physiology Polish Academy of Sciences, 30-239, Kraków, Poland
e-mail: e.niewiadomska@ifr-pan.krakow.pl

M. Wiciarz
Department of Plant Physiology and Biochemistry, Faculty of Biochemistry, Biophysics and Biotechnology, Jagiellonian University, Gronostajowa 7, 30-387, Kraków, Poland

U. Lüttge, W. Beyschlag (eds.), *Progress in Botany*, Progress in Botany 76,
DOI 10.1007/978-3-319-08807-5_7

1 Introduction

Increasing soil salinity is one of the major abiotic stress factors limiting yield of crop plants. As hypothesized by Tsugane et al. (1999), plants initially had a high salinity resistance, but it was lost during evolution. This idea has been illustrated by a recessive photoautotrophic salt tolerance1 (pst1) mutant of *Arabidopsis thaliana* (L.) Heynh. which appeared resistant to salt, but also to methyl viologen and high irradiance. Since these factors affect the chloroplastic metabolism directly one may suppose that resistance to salinity may be associated, at least in part, with mechanisms operating in this organelle.

However, resistance to salinity is a much more complex phenomenon. The overall amount of 2,171 salt-responsive protein identities has been dug out by an integrated analysis of available proteomics-based studies (Zhang et al. 2012). They represent 561 unique proteins identified from leaves, roots, shoots, seedlings, grains, hypocotyls, radicles, and panicles from 34 plant species. In a halophytic monocot *Puccinellia tenuiflora* Scribn. & Merr. more than 218 proteins were found (Yu et al. 2011). From a number of papers, a broad spectrum of salt stress responses may be categorized into the following groups: (1) control of ion uptake and compartmentation, (2) stress signaling, (3) scavenging of reactive oxygen species (ROS), (4) osmotic adjustment, (5) repair, (6) redirection of cell metabolism, and (7) acceleration of development (for review see Zhu 2001; Wang et al. 2003; Flowers and Colmer 2005; Parida and Das 2005; Tuteja 2007; Munns and Tester 2008).

All these strategies are not unique and several lines of evidences point out that halophytes are using similar metabolic pathways as glycophytes to cope with salinity, but in a more effective way and at different transcriptional control (Bohnert et al. 1995; Dassanayake et al. 2011). Hence, a salinity tolerance may be largely attributed to a different regulation of the same genes, as demonstrated by Kant et al. (2006). A detailed transcript profiling after salinity stress made by Gong et al. (2005) documented that in *A. thaliana* a bigger pool of salinity-induced transcripts was associated with stress defense in comparison to *Thellungiella halophila* O.E. Schulz, in which upregulation of transcripts linked with protein folding, posttranslational modification, and protein distribution were found. A similar tendency was noted in salinity-treated *Mesembryanthemum crystallinum* L. plants (Cushman et al. 2008b). Besides, a bigger pool of genes of unknown function appeared to be upregulated by salinity, and modulate salinity tolerance (Cushman et al. 2008b). The abovementioned data may indicate that halophytic species invest more energy in the changes made on existing proteins rather than in the de novo synthesis, what may be more profitable for the overall energy budget.

Halophytes are often so strictly adapted to saline environment that they grow better under salinity than without it. A positive effect of salinity on growth was noted for example in *Atriplex centralasiatica* Iljin (Qiu et al. 2003), *Sarcocornia fruticosa* (L.) A.J. Scott (Redondo-Gómez et al. 2006), *Sesuvium portulacastrum* (L.) L. (Rabhi et al. 2010), *M. crystallinum* (Haider et al. 2012), and with a wild halophytic rice *Porteresia coarctata* (Roxb.) Tateoka (Sengupta and Majumder 2009). On the other hand, one of the obvious effects of salinity is closure of stomata, which restricts entry of carbon dioxide into the leaves. This reaction is typical for both halophytic and glycophytic plants. In consequence of damped CO_2 assimilation, more ROS may be formed aside the photosynthetic electron transport (PET), which is harmful for the cells. Therefore it might be hypothesized that specific adaptations of chloroplastic metabolism belong to key elements of resistance to salinity stress. In order to verify that, we analyzed the recent state of research obtained with halophytic and glycophytic plants with an emphasis of photosynthetic metabolism.

2 Protection of Photosystem II

2.1 Photoinhibition

Studies made in algae and higher plants documented that salinity, similarly to the other stress factors (cold, moderate heat, oxidative stress, etc.), enhances the extent of photoinhibition of PSII. This process is determined by the balance between the rate of photodamage to PSII and the rate of its repair. Within this balance the most sensitive part seems to be the repair, in particular, transcription and translation of D1 protein (Allakhverdiev and Murata 2008; Takahashi and Murata 2008), and not a PSII photodamage itself.

According to the current view on photoinhibition, damage to PSII starts from its donor side due to the disruption of the manganese cluster of the oxygen evolving complex (OEC) upon absorption of light (Tyystjärvi 2008). As a result of blocked supply of electrons from water, a primary electron donor of PSII, P680 remains oxidized, which is harmful for the reaction center. In view of this, it is striking that in the extreme halophyte *Salicornia veneta* Pignatti & Lausi some OEC proteins are lacking (PsbQ) or are less abundant (PsbP) without any negative effect on PSII activity (Pagliano et al. 2009). It is suggested that these two polypeptides are probably not so important for electron flow through OEC, but play other functions, for example, during the assembly of PSII and retention of inorganic cofactors Ca^{2+} and Cl^- (Nelson and Yocum 2006). It is worth noting that increased concentration of Cl^- ions may exert some protection on OEC, since the lowered rate of O_2 evolution due to the removal of another polypeptide (PsbO) might be reconstituted simply by high concentrations of this ion (Seidler 1996). A different strategy is used by a halophytic rice *P. coarctata* and two mangrove species *Bruguiera gymnorrhiza* (L.) Lam. and *Tamarix hispida* Willd., in which salinity caused an induction of PsbO protein as a necessary step to stabilize PSII structure (Sengupta and Majumder 2009). A proteomic comparison of two cultivars of tomato differing in the salinity resistance revealed a higher salinity-dependent increase of PsbO and PsbP in the more resistant one (Chen et al. 2009).

On the other hand, photoinhibition, as long as reversible during the night, is a protective process which limits the extent of excitation of PSII antennae (Lüttge 2000). A reversible midday photoinhibition has been noted frequently in halophytic species (*A. centralasiatica*—Qiu et al. 2003, *Artemisia anethifolia* Weber ex Stechm.—Lu et al. 2003a, *S. fruticosa* and *Arthrocnemum macrostachyum* (Moric.) C. Koch—Redondo-Gómez et al. 2006 and 2010, and *M. crystallinum*—Schöttler et al. 2002; Barker et al. 2004; Broetto et al. 2007). However, the maximal PSII efficiency (F_v/F_m) seems to be more resistant to salinity than PSII efficiency in a light adapted state (*A. centralasiatica*—Qiu et al. 2003, *Suaeda salsa* Pall.—Lu et al. 2003b, *A. anethifolia*—Wen et al. 2005, *T. halophila*—M'rah et al. 2006; Stepien and Johnson 2009), and *M. crystallinum*—Broetto et al. 2007; Gawronska et al. 2013). This feature seems to be in common also for glycophytes, such as *Zea mays* L. and *A. thaliana* (Yang and Lu 2005; Stepien and Johnson 2009). Parameters describing electron transport in the proximity of PSII are sometimes even enhanced under salinity, as noted in the halophytic plant *S. fruticosa* (Redondo-Gómez et al. 2006). This implies that efficient protective strategies are induced and that further steps of PET are more sensitive to salinity than electron transfer at PSII.

2.2 Reduction of PSII Antenna Size

Among the powerful strategies to diminish the excitation pressure at PSII under salinity is a reduction of photosynthetic antennae (Parida and Das 2005). A reduction of the antennae size is represented by decreased chlorophyll content in *A. macrostachyum* (Redondo-Gómez et al. 2010), diminished amount of proteins belonging to the light harvesting complex of PSII (LHCII) in *S. portulacastrum* (Rabhi et al. 2010), and decreased amount of chlorophyll a/b binding proteins in *P. tenuiflora* (Yu et al. 2011). An illustration of that strategy might be also results of Wang et al. (1999) obtained with *Amaranthus tricolor* L. that leaf fragments with less chlorophyll and LHCII appeared more salinity resistant. A similar conclusion might be taken from an increased chl a/b ratio in salinity-resistant species compared with sensitive ones: *Hordeum maritimum* Stokes vs *H. vulgare* L. (Innocenti et al. 2009) and *T. halophila* vs *A. thaliana* (M'rah et al. 2006).

However, such a reaction is not always the case. The two halophytic species *Aster tripolium* L. and *S. portulacastrum* have a different response in this respect (Ramani et al. 2006). Also, in *S. salsa* and *Suaeda aegyptiaca* (Hasselq.) Zohary an induction of PSII-related proteins, i.e., chlorophyll a/b binding proteins, was noted after salinity (Yu et al. 2011). Besides, a halophytic C4 monocot *Aeluropus lagopoides* (L.) Thrin. ex Thwaites was turned to harvest more light under salt conditions due to the increase of chlorophyll a/b binding protein of LHCII type III under a mild NaCl treatment (Sobhanian et al. 2010).

2.3 *Xanthophyll Cycle*

Excess excitation energy at PSII could be dissipated harmlessly as heat with involvement of the xanthophyll cycle. This can be illustrated by increased non-photochemical quenching, NPQ (Niyogi 1999). This strategy is employed by some halophytic species, but not all. Increased NPQ has been documented in *A. centralasiatica* (Qiu et al. 2003) and *S. fruticosa* (Redondo-Gómez et al. 2006). Conversely, low or unchanged NPQ due to salinity was reported in *A. anethifolia* (Lu et al. 2003a), *T. halophila* (Stepien and Johnson 2009), *A. macrostachyum* (Redondo-Gómez et al. 2010), and *M. crystallinum* (Broetto et al. 2007; Niewiadomska et al. 2011; Gawronska et al. 2013). Variable protective strategies employed under salinity in the two halophytes *A. tripolium* and *S. portulacastrum* are also envisaged by a different extent of NPQ increase (Ramani et al. 2006). It may be hypothesized that NPQ is strongly engaged under salinity as long as PSII antenna are not reduced.

3 Protection at PSI and Changes in the Chloroplast Ultrastructure

PSI seems to be more stress resistant than PSII, with the exception of certain combinations of chilling temperatures and moderate light intensities (Scheller and Haldrup 2005). However, low CO_2 causes an acceptor side limitation of PSI which may increase ROS production in the proximity of PSI. Some protection against photoinhibition and ROS formation could be given by a cyclic electron transport (CET) around PSI (Golding and Johnson 2003; Joliot and Joliot 2006; Takahashi et al. 2009). Activation of CET under salinity has been documented in cyanobacteria [as reviewed in Bukhov and Carpenter (2004)], while data on higher plants are still missing.

With respect to PSI data on halophytes are scarce. A salinity-caused stimulation of transcripts for PSI reaction center was demonstrated in halophytic rice (Sengupta and Majumder 2009) and in *M. crystallinum* (Niewiadomska et al. 2011), suggesting an increased protection of PSI. Keiller et al. (1994) supplied an indication of the state 1–state 2 transition (stt) in *M. crystallinum*-CAM plants by stimulation of the slowly relaxing component of non-photochemical quenching. An increased acceptor side limitation of PSI due to salinity was also noted in this species together with a decreased Rubisco content (Niewiadomska et al. 2011). In contrast to that, an increased donor side limitation of PSI represents a typical effect of stress, as described repeatedly (Klughammer and Schreiber 1994; Johnson 2005). This phenomenon results from the acidification of the thylakoid lumen which, on the one hand, activates the xanthophyll cycle, while on the other hand, it inhibits cyt b_6f activity (Sacksteder and Kramer 2000; Johnson 2005). In agreement with this view, in salinity-treated *A. thaliana* increased NPQ was associated with a more oxidized P700 under a steady-state photosynthesis, while in salinity-treated *T. halophila* the value of NPQ stayed low and P700 stayed reduced (Stepien and Johnson 2009). Intriguingly, in *T. halophila* an indication of stt was found in controls, but not after salinity treatment, as shown by changes in a 77K chlorophyll emission spectra (Wiciarz et al. 2014). Because a reduced PQ pool promotes stt (Tikkanen and Aro 2012), a less reduced PQ pool might be expected after salinity treatment than in controls. This supposition is in agreement with the activation of chlororespiration in a salt-treated *T. halophila* plants demonstrated by Stepien and Johnson (2009).

The actual engagement of the two photosystems may be visualized by chloroplast ultrastructure, as shown repeatedly (Edwards et al. 2004; Darie et al. 2006; Chuartzman et al. 2008). However, under salinity changes in the osmotic pressure and ionic composition should also be considered as a modifying factor. As described by Parida and Das (2005), salt stress caused a notable disorganization of the thylakoid structure in halophytic *Bruguiera parviflora* Wight & Arn. ex W. Griffth. and *Eucalyptus microcorys* F. Muell. In *M. crystallinum* salinity diminished the number of thylakoids and content of starch, while it increased the number of plastoglobuli (Paramonova et al. 2004), similarly to *S. veneta* collected in natural saline conditions (Pagliano et al. 2009). However, it appears that in *M. crystallinum* thylakoid swelling is, at least partly, reversible during the night (Niewiadomska et al. 2011). Due to the similar picture of daytime swelling to that described for *A. thaliana* under state 1–state 2 transition (Chuartzman et al. 2008) it has been hypothesized that it may visualize stt (Niewiadomska et al. 2011). However, further study is necessary to verify whether this phenomenon is confined to CAM or to salinity in this species.

4 Changes in Carbon Metabolism

The rate of photosynthesis (P_N) is declining under salinity due to several limitations imposed by the stomata and other factors limiting the diffusion of CO_2 to and in mesophyll cells (Chaves et al. 2009). In this respect it is worth noting that some of the halophytic species are characterized by a lower P_N rate already under control conditions. Such a feature was reported for *H. maritimum* in comparison to *H. vulgare* (Innocenti et al. 2009). On the other hand, some of the halophytes show an increase in P_N after treatment with salinity. This reaction has been documented for: *S. salsa* (Lu et al. 2003b), *S. fruticosa* (Redondo-Gómez et al. 2006), and *A. macrostachyum* (Redondo-Gómez et al. 2010). As noted by Munns and Tester (2008), a paradox of increased P_N at decreased stomatal conductance might be explained by the changes in anatomy into the smaller and thicker leaves with a higher density of chloroplasts per leaf area.

The inhibition of several enzymes of the Calvin cycle under salt stress has been noted repeatedly. Salinity-induced changes of proteome of *S. salsa* leaves detected a significant decrease in several proteins involved in the Calvin cycle, such as glyceraldehyde-3-phosphate dehydrogenase, sedoheptulose-1,7-bisphosphatase, and Rubisco large subunit (Li et al. 2011). However, it does not diminish the rate of CO_2 assimilation. On the other hand, stimulation of photorespiration under salinity might be expected, by similarity to plants' response to drought, documented by Wingler et al. (1999). In agreement with that a transgenic rice overexpressing a photorespiratory enzyme glutamine synthetase from tobacco was characterized by enhanced photorespiratory flux and was more resistant to salinity (Hoshida et al. 2000). Similarly, in a halophytic rice *P. coarctata* salt-enhanced production of glutamine synthetase was found (Sengupta and Majumder 2009). Apart from being an important electron sink for electrons from PET, this pathway does also supply metabolites for synthesis of some compatible solutes and is a possible route for the dissipation of excess reducing power originated in chloroplasts. However, due to the sophisticated way of Rubisco activation (see Houtz and Portis (2003) for a review), several limitations of photorespiration have to be considered. Thus, photorespiration may represent an important strategy against photoinhibition as long as enzymatic and nonenzymatic machinery responsible for Rubisco activation is efficient.

4.1 Rubisco Activity and Activation

Several factors, associated with salinity, may affect the activity of Rubisco directly. Oxidative stress is capable to trigger degradation of Rubisco (Desimone et al. 1998). Moreover, low CO_2/high O_2 conditions cause a loss of Rubisco activity, named often as "Rubisco follover." Rubisco follover happens due to the increased production of misfire products which block the active sites of the enzyme, among them are 3-ketoarabinitol-1,5-bisphosphate, and D-glycero-2,3-pentodiulose-1,5-bisphosphate (Kim and Portis 2006; Parry et al. 2008). Therefore, the supporting enzyme—Rubisco activase (RCA)—is necessary to revitalize the Rubisco active sites. Due to this function, RCA has to be included in a still-growing pool of so-called stress enzymes.

Indeed, an increased amount of Rubisco activase has been reported in salinity-treated *Oryza sativa* L. (Parker et al. 2006), *P. coarctata* (Sengupta and Majumder 2009), and *S. salsa* (Li et al. 2011). In contrast, in *M. crystallinum* a decreased amount of soluble RCA was found after salinity treatment. Despite that, Rubisco appeared to be more highly activated in vitro (Davies and Griffiths 2012). Our recent study made on *T. halophila* revealed that a majority of RCA is attached to the thylakoid membranes, in contrast to *A. thaliana* where the soluble form is predominant (Wiciarz et al. 2014). This is associated by a higher Rubisco activation state in the former species. Eichelmann et al. (2009) suggested that a temporary attachment of RCA to the thylakoid membranes (close to cyt b_6f) is necessary for the restoration of its active form in close correlation with PSI activity. A further light into this phenomenon was put by Chen et al. (2010), who demonstrated that attachment of RCA is correlated with a low pH gradient across the thylakoid membrane, hence with low ATP production. In view of this, it is likely that some halophytes may demand enhanced Rubisco activation, achieved by a closer contact of RCA with PSI, as a compensation for a low ATP supply.

4.2 *CO_2 Concentrating Mechanisms*

A drastic decline of photosynthesis at limited CO_2 supply may be, at least partly, overcome by metabolic strategies which concentrate CO_2 around the Rubisco active sites. The two known CO_2 Concentrating Mechanisms (CCMs) operating in higher plants are crassulacean acid metabolism (CAM) and C_4 metabolism. Several CAM species are to be found in the salinas; however, their resistance depends mainly on the avoidance of stress, i.e., ability to perform photosynthesis behind closed stomata and scavenging respiratory CO_2 (as reviewed by Lüttge 2004). A comparison of the halophytic C_3 plant *S. salsa* with a drought-resistant CAM plant *Kalanchoë daigremontiana* Raym.-Hamet & H. Perrier demonstrated that in halophytic species efficient mechanisms are activated by salinity, among them tonoplast H^+-ATPase and H^+-PPase, which are more critical than CAM decisive for salinity resistance than CAM itself (Kefu et al. 2003). In the facultative halophyte and C_3-CAM intermediate plant *M. crystallinum* metabolism switch takes place in response to salinity or drought (Winter and Holtum 2007). Nevertheless, a characterization of a null-CAM mutant of *M. crystallinum*, defective in the production of starch, revealed that the change of photosynthetic metabolism into CAM is not an absolute demand for the high salinity resistance of this species (Cushman et al. 2008a; Haider et al. 2012).

Some halophytic species perform the C_4 type of photosynthesis. In the C_4 halophyte *Atriplex halimus* L. subjected to NaCl and osmotic stress (treatment with PEG) beneficial effects of salinity were found (Martinez et al. 2005). NaCl induced a more efficient osmotic adjustment, probably through the increase in total soluble sugars, proline and glycine betaine (GB). A study made on another C_4 halophyte *Spartina densiflora* Brongn. revealed a positive effect of elevated CO_2 in the atmosphere on plant growth under salinity (Mateos-Naranjo et al. 2010). This was associated with improved water relations, greater leaf area, and enhanced PEPC carboxylation capacity. A comparison of C_4 (maize) and C_3 (wheat) utilizing monocots revealed that maize is better protected against salinity due to the more efficient antioxidant scavenging system (Stepien and Klobus 2005).

5 Redox Regulation

Decreased CO_2 assimilation, caused by salinity, elicits accumulation of NADPH and deficit of $NADP^+$ (Takahashi and Murata 2008). Such a situation leads to the over-reduction of PET carriers and intensive generation of ROS. Hence, recycling of NADPH under salt stress, with help of several dehydrogenases: ferredoxin-NADP reductase (FNR), glucose-6-phosphate dehydrogenase (G6PDH), isocitrate dehydrogenase (ICDH), and malic enzyme (ME), may contribute to the tolerance of salinity (Valderrama et al. 2006).

In *T. halophila* a higher accumulation of the stromal thioredoxin CDSP32 due to salinity was documented, in comparison to *A. thaliana* (M'rah et al. 2007). This may suggest the increased reduction of the chloroplast stroma or a stronger activation of defense mechanisms in the halophytic species. A part of reducing power may be exported to the cytosol or mitochondria by the so-called malate valve with engagement of chloroplastic and cytosolic malate dehydrogenases (Scheibe et al. 2005). A chloroplastic NADP-MDH is activated in the light by a reduced thioredoxin, which links its activity with the increased redox state of the stroma (Voss et al. 2008). Data obtained on *M. crystallinum* suggest that activation of the chloroplastic NADP-MDH may be under control of salinity, since its activation was found under salinity and not after high irradiance (Gawronska et al. 2013).

It is well established that the redox state at the plastoquinone pool modulates numerous stress responses in plants (Pfannschmidt 2003; Pfannschmidt et al. 2009). Most typically, stress situations increase an excitation pressure at PSII which leads to the enormous reduction of the PQ pool. In consequence, not fully defined redox signals arise, which upregulate the expression of genes associated with cell defense. Intriguingly, in the halophytic *M. crystallinum* plant kind of the opposite was found, namely, transcripts for *FeSOD* and *CuZnSOD* were not induced by conditions reducing the PQ pool (high light, DBMIB), but by conditions oxidizing it (DCMU), as demonstrated by Ślesak et al. (2003). This may indicate that salinity resistance is associated with the specific modulation of the redox state of PQ pool. It may also be suggested that the PQ pool is actually more reduced in control conditions than after stress treatment. In support for this view, a comparison of the glycophytic *H. vulgare* with the more resistant *H. maritimum* revealed a higher PQ reduction in the latter one in control and under salinity than in its glycophytic counterpart *H. vulgare* (Innocenti et al. 2009).

One possibility to oxidize the PQ pool under stress is the chlororespiratory pathway. It diminishes PQ pool reduction due to the directing of electrons to oxygen by a plastidic terminal oxidase (PTOX, IMMUTANS). Therefore, this enzyme is considered as a safety valve for PET under stress (Rumeau et al. 2007). However, its action is not fully understood and overexpression of PTOX did not bring about an enhanced resistance toward photooxidative stress in *Arabidopsis* and tobacco (Joët et al. 2002; Rosso et al. 2006; Heyno et al. 2009). Nonetheless, several highly resistant species possess a high constitutive activity of this enzyme, for example, the high mountain plant *Ranunculus glacialis* L. (Streb et al. 2005), and the stress-resistant *Brassica fruticulosa* Cirillo (Díaz et al. 2007). Stepien and Johnson (2009) reported on the increased activity and amount of PTOX in *T. halophila* after salt treatment. An increased transcripts for IMMUTANS were also found in the salinity-treated *M. crystallinum* (Cushman et al. 2008b). Further work is necessary to define the precise function of this enzyme in prevention or response to stress.

6 Cross-Tolerance

When a plant encounters a stress situation of low or moderate extent, it often becomes more resistant to another unfavorable environmental factor (so-called "cross-tolerance"). This phenomenon is an important element of plants' adaptation to changing environment where they are subjected simultaneously or in sequence to different stress stimuli. The halophytic *A. anethifolia* after treatment with NaCl became more tolerant to high light (Lu et al. 2003a) and to high temperature (Wen et al. 2005). Similarly, after salinity treatment of *A. centralasiatica* and *M. crystallinum* an increased resistance to photoinhibition was noted (Qiu et al. 2003; Gawrońska et al. 2013), whereas salinity-treated *A. halimus* revealed an increased resistance to drought (Martinez et al. 2005).

Studies made on *S. salsa* and *A. anethifolia* revealed that salinity treatment increases the thermostability of the PSII reaction center, LHCII and OEC (Lu et al. 2003b; Wen et al. 2005). This phenomenon arises mainly from increased abundance of heat shock proteins and of late embryogenesis abundant (LEA) proteins, as well as from the accumulation of GB, which have a stabilizing function on the multimeric pigment–protein photosynthetic complexes and on Rubisco holoenzyme (Sakamoto and Murata 2002; Wang et al. 2003; Chen et al. 2009). In support of that, numerous studies documented alleviation of stress by the application of GB or its overproduction in transgenic plants (reviewed in Sakamoto and Murata 2002). Giri (2011) summarized the recent achievements obtained in the field of stress tolerance due to the introduction of GB biosynthetic genes into plants.

Another aspect of a multifactor stress tolerance might be that halophytic species are "from definition" more resistant to the other stresses such as drought and cold. An example is *T. halophila* which is also highly resistant to cold and freezing (Griffith et al. 2007; Amtmann 2009). Stress responses to salinity and drought, although having a lot of similarities, have also specific traits. Ma et al. (2006) used an extensive transcript profiling to make a distinction between genes responding specifically to salinity and to the other factors of abiotic and biotic stress, respectively. Similarities and differences in plant responses to the salinity and water stress have also been summarized by Chaves et al. (2009).

7 Antioxidants and Role of ROS

A huge body of data links resistance to salinity with upregulation of the enzymatic and nonenzymatic antioxidants (Ashraf and Harris 2004; Parida and Das 2005, and references therein). In chloroplasts upregulation of thylakoid-bound and stromal superoxide dismutase (SOD) and ascorbate peroxidase (APX), as well as soluble glutathione peroxidase (GPX), glutathione reductase (GR), and enzymes of glutathione synthesis, is a common response to salinity (Miszalski et al. 1998; Mittova et al. 2003; Cai-Hong et al. 2005; Qiu-Fang et al. 2005). In agreement with that transgenic tobacco plants overexpressing *Arabidopsis* CuZnSOD in the chloroplast stroma revealed enhanced tolerance to NaCl (Badawi et al. 2004). However, in *T. halophila* antioxidants were not significantly induced under salinity, in contrast to *A. thaliana*, as shown by M'rah et al. (2006). This may suggest that more efficient strategies are most likely those which prevent a high ROS formation.

In the phenomenon of stress tolerance ROS have been recognized as important players (Mittler et al. 2011). Gomez et al. (2004) provided evidence that a H_2O_2 signaling originating from PET may activate a stromal APX in salinity-treated pea. Verslues et al. (2007) found an indication that H_2O_2 is a point of cross talk between the salinity-specific signaling (involving SOS2) and ROS signaling. Another illustration of such a protective involvement of ROS is a beneficial effect of H_2O_2 pretreatment on the salinity resistance of maize plants (Azevedo Neto et al. 2005). In view of this it may not be so surprising that *Arabidopsis* mutants deficient in the H_2O_2 scavenging enzymes (cytosolic and thylakoid ascorbate peroxidase) appeared more resistant to salinity stress (Miller et al. 2007).

In *T. halophila*, in comparison to *A. thaliana*, elevated levels of several compatible solutes and transcripts associated with their synthesis, as well as with synthesis of abscisic acid (ABA) and ABA-responsive genes, have been documented already in control conditions (Hasegawa et al. 2000; Inan et al. 2004; Taji et al. 2004; Gong et al. 2005; Kant et al. 2006). Our recent data point to a much higher generation of H_2O_2 by PET in *T. halophila* in comparison to *A. thaliana* (Wiciarz et al. 2014). Similarly, in leaves of *S. salsa* the level of H_2O_2 was found to be lower after salinity than in control conditions (Cai-Hong et al. 2005). It is therefore tempting to speculate that a high H_2O_2 production by PET under steady-state photosynthesis is amongst the pre-adaptive features for a high stress resistance.

8 Concluding Remarks

Saline soils make a particularly harsh environment for plants, because of combined action of osmotic and ionic stresses, which may be further intensified by high irradiance and high temperature. Therefore, it is a tempting task to reveal powerful strategies of stress resistance demonstrated by halophytic plants. The genetic and molecular potential of halophytes has been intensively explored for the last few decades. This enabled to recognize several stress-associated proteins and pathways, as well as whole plant's metabolic strategies. With respect to chloroplastic metabolism, halophytic plants developed very effective mechanisms to avoid the excessive reduction of the PQ pool and to diminish ROS formation under stress. However, several aspects still need to be elucidated, among them: role of chlororespiration, involvement of CET, mechanism of Rubisco activation secured by activase, and role of H_2O_2 in a stress preparedness of plants.

Acknowledgments Authors are thankful to anonymous reviewers for their comments on the text, and to the Polish National Science Center project 2011/03/B/NZ9/01619 for a financial support.

References

Allakhverdiev SI, Murata N (2008) Salt stress inhibits photosystems II and I in cyanobacteria. Photosynth Res 98:529–539

Amtmann A (2009) Learning from evolution: *Thellungiella* generates new knowledge on essential and critical components of abiotic stress tolerance in plants. Mol Plant 2:3–12

Ashraf M, Harris PJC (2004) Potential biochemical indicators of salinity tolerance in plants. Plant Sci 166:3–16

Azevedo Neto AD, Prisco JT, Eneas-Filho J, Medeiros JV, Gomes-Filho E (2005) Hydrogen peroxide pre-treatment induces salt-stress acclimation in maize plants. J Plant Physiol 162:1114–1122

Badawi GH, Yamauchi Y, Shimada E, Sasaki R, Kawano N, Tanaka K, Tanaka K (2004) Enhanced tolerance to salt stress and water deficit by overexpressing superoxide dismutase in tobacco (*Nicotiana tabacum*) chloroplasts. Plant Sci 166:919–928

Barker DH, Marszalek J, Zimpfer JF, Adams WW III (2004) Changes in photosynthetic pigment composition and absorbed energy allocation during salt stress and CAM induction in *Mesembryanthemum crystallinum*. Funct Plant Biol 31:781–787

Bohnert HJ, Nelson DE, Jensen RG (1995) Adaptations to environmental stresses. Plant Cell 7:1099–1111

Broetto F, Duarte HM, Lüttge U (2007) Responses of chlorophyll fluorescence parameters of the facultative halophyte and C_3–CAM intermediate species *Mesembryanthemum crystallinum* to salinity and high irradiance stress. J Plant Physiol 164:904–912

Bukhov N, Carpenter R (2004) Alternative photosystem I-driven electron transport routes: mechanisms and functions. Photosynth Res 82:17–33

Cai-Hong P, Su-Jun Z, Zhi-Zhong G, Bao-Shan W (2005) NaCl treatment markedly enhances H_2O_2-scavenging system in leaves of halophyte *Suaeda salsa*. Physiol Plant 125:490–499

Chaves MM, Flexas J, Pinheiro C (2009) Photosynthesis under drought and salt stress: regulation mechanisms from whole plant to cell. Ann Bot 103:551–560

Chen S, Gollop N, Heuer B (2009) Proteomic analysis of salt-stressed tomato (*Solanum lycopersicum*) seedlings: effect of genotype and exogenous application of glycinebetaine. J Exp Bot 60:2005–2019

Chen J, Wang P, Mi HL, Chen GY, Xu DG (2010) Reversible association of ribulose-1,5-bisphosphate carboxylase/oxygenase activase with the thylakoid membrane depends upon the ATP level and pH in rice without heat stress. J Exp Bot 11:2939–2950

Chuartzman SG, Nevo R, Shimoni E, Charuvi D, Kiss V, Ohad I, Brumfeld V, Reich Z (2008) Thylakoid membrane remodeling during state transition in Arabidopsis. Plant Cell 20:1029–1039

Cushman JC, Agarie S, Albion RL, Elliot SM, Taybi T, Borland AM (2008a) Isolation and characterization of mutants of common ice plant deficient in Crassulacean acid metabolism. Plant Physiol 147:228–238

Cushman JC, Tillett RL, Wood JA, Branco JM, Schlauch KA (2008b) Large-scale mRNA expression profiling in the common ice plant, *Mesembryanthemum crystallinum*, performing C_3 photosynthesis and crassulacean acid metabolism (CAM). J Exp Bot 59:1875–1894

Darie CC, De Pascalis L, Mutschler B, Haehnel W (2006) Studies of the Ndh complex and photosystem II from mesophyll and bundle sheath chloroplasts of the C_4-type plant *Zea mays*. J Plant Physiol 163:800–808

Dassanayake M, Oh D-H, Hong H, Bohnert HJ, Cheeseman JM (2011) Transcription strength and halophytic lifestyle. Trends Plant Sci 16:1–3

Davies BN, Griffiths H (2012) Competing carboxylases: circadian and metabolic regulation of Rubisco in C_3 and CAM *Mesembryanthemum crystallinum* L. Plant Cell Environ 35:1211–1220

Desimone M, Wagner E, Johanningmeier U (1998) Degradation of active-oxygen-modifed ribulose-1,5-bisphosphate carboxylase/oxygenase by chloroplastic proteases requires ATP-hydrolysis. Planta 205:459–466

Díaz M, De Haro V, Munoz R, Quiles MJ (2007) Chlororespiration is involved in the adaptation of *Brassica* plants to heat and high light intensity. Plant Cell Environ 30:1578–1585

Edwards GE, Franceschi VR, Vosniesenskaya EV (2004) Single-cell C4 photosynthesis versus the dual-cell (Kranz) paradigm. Annu Rev Plant Biol 55:173–196

Eichelmann H, Talts E, Oja V, Padu E, Laisk A (2009) Rubisco in planta kcat is regulated in balance with photosynthetic electron transport. J Exp Bot 60:4077–4088

Flowers TJ, Colmer TD (2005) Salinity tolerance in halophytes. New Phytol 179:945–963

Gawronska K, Romanowska E, Miszalski Z, Niewiadomska E (2013) Limitation of C_3–CAM shift in the common ice plant under high irradiance. J Plant Physiol 170:129–135

Giri J (2011) Glycinebetaine and abiotic stress tolerance in plants. Plant Signal Behav 6:1746–1751

Golding AJ, Johnson GN (2003) Down-regulation of linear and activation of cyclic electron transport during drought. Planta 218:107–114

Gómez JM, Jiménez A, Olmos E, Sevilla F (2004) Location and effects of long-term NaCl stress on superoxide dismutase and ascorbate peroxidase isoenzymes of pea (Pisum sativum cv. Puget) chloroplasts. J Exp Bot 55: 119–130.

Gong Q, Li P, Ma S, Rupassara SI, Bohnert HJ (2005) Salinity stress adaptation competence in the extremophile *Thellungiella halophila* in comparison with its relative *Arabidopsis thaliana*. Plant J 44:826–839

Griffith M, Timonin M, Wong ACE, Gray GR, Akhter SR, Saldanha M, Rogers MA, Weretilnyk EA, Moffatt B (2007) *Thellungiella*: an *Arabidopsis*-related model plant adapted to cold temperatures. Plant Cell Environ 30:529–538

Haider MS, Barnes JD, Cushman JC, Borland AM (2012) A CAM- and starch-deficient mutant of the facultative CAM species *Mesembryanthemum crystallinum* reconciles sink demands by repartitioning carbon during acclimation to salinity. J Exp Bot 63:1985–1996

Hasegawa PM, Bressan RA, Zhu JK, Bohnert HJ (2000) Plant cellular and molecular responses to high salinity. Annu Rev Plant Physiol Plant Mol Biol 51:463–499

Heyno E, Gross CM, Laureau C, Culcasi M, Pietri S, Krieger-Liszkay A (2009) Plastid alternative oxidase (PTOX) promotes oxidative stress when overexpressed in tobacco. J Biol Chem 284:31174–31180

Hoshida H, Tanaka Y, Hibino T, Hayashi Y, Tanaka A, Takabe T, Takabe T (2000) Enhanced tolerance to salt stress in transgenic rice that overexpresses chloroplast glutamine synthetase. Plant Mol Biol 43:103–111

Houtz RL, Portis AR (2003) The life of ribulose 1,5-bisphosphatecarboxylase/oxygenase – posttranslational facts and mysteries. Arch Biochem Biophys 414:150–158

Inan G, Zhang Q, Li P, Wang Z, Cao Z, Zhang H, Zhang C, Quist TM, Goodwin SM, Zhu J, Shi H, Damsz B, Charbaji T, Gong Q, Ma S, Fredricksen M, Galbraith DW, Jenks MA, Rhodes D, Hasegawa PM, Bohnert HJ, Joly RJ, Bressan RA, Zhu JK (2004) Salt Cress. A halophyte and cryophyte *Arabidopsis* relative model system and its applicability to molecular genetic analyses of growth and development of extremophiles. Plant Physiol 135:1718–1737

Innocenti ED, Hafsi Ch, Guidi L, Navari-Izzo F (2009) The effect of salinity on photosynthetic activity in potassium-deficient barley species. J Plant Physiol 166:1968–1981

Johnson GN (2005) Cyclic electron transport in C3 plants: fact or artefact? J Exp Bot 56:407–416

Joliot P, Joliot A (2006) Cyclic electron flow in C_3 plants. Biochim Biophys Acta 1757:362–368

Joët T, Genty B, Josse EM, Kuntz M, Cournac L, Peltier G (2002) Involvement of a plastid terminal oxidase in plastoquinone oxidation as evidenced by expression of the *Arabidopsis thaliana* enzyme in tobacco. J Biol Chem 277:31623–31630

Kant S, Kant P, Raveh E, Barak S (2006) Evidence that differential gene expression between the halophyte, *Thellungiella halophila*, and *Arabidopsis thaliana* is responsible for higher levels of the compatible osmolyte proline and tight control of Na^+ uptake in *T. halophila*. Plant Cell Environ 29:1220–1234

Kefu Z, Hai F, San Z, Jie S (2003) Study on the salt and drought tolerance of *Suaeda salsa* and *Kalanchoë daigremontiana* under iso-osmotic salt and water stress. Plant Sci 165:837–844

Keiller DR, Slocombe SP, Cockburn W (1994) Analysis of chlorophyll a fluorescence in C_3 and CAM forms of *Mesembryanthemum crystallinum*. J Exp Bot 45:325–334

Kim K, Portis AR Jr (2006) Kinetic analysis of the slow inactivation of Rubisco during catalysis: effects of temperature, O_2 and Mg^{++}. Photosynth Res 87:195–204

Klughammer C, Schreiber U (1994) An improved method, using saturating light pulses, for the determination of photosystem I quantum yield via $P700^+$-absorbance changes at 830 nm. Planta 192:261–268

Li W, Zhang Ch LQ, Wen X, Lu C (2011) The combined effect of salt stress and heat shock on proteome profiling in *Suaeda salsa*. J Plant Physiol 168:1743–1752

Lu C, Qiu N, Lu Q (2003a) Photoinhibition and the xanthophyll cycle are not enhanced in the salt-acclimated halophyte *Artimisia anethifolia*. Physiol Plant 118:532–537

Lu C, Qiu N, Wang B, Zhang J (2003b) Salinity treatment shows no effects on photosystem II photochemistry, but increases the resistance of photosystem II to heat stress in halophyte *Suaeda salsa*. J Exp Bot 54:851–860

Lüttge U (2000) Light-stress and crassulacean acid metabolism. Phyton 40:65–82

Lüttge U (2004) Ecophysiology of crassulacean acid metabolism (CAM). Ann Bot 93:629–652

M'rah S, Ouerghi Z, Berthomieu C, Havaux M, Jungas C, Hajji M, Grignon C, Lachaâl M (2006) Effects of NaCl on the growth, ion accumulation and photosynthetic parameters of *Thellungiella halophila*. J Plant Physiol 163:1022–1103

M'rah S, Ouerghi Z, Eymery F, Rey P, Hajji M, Grignon C, Lachaâl M (2007) Efficiency of biochemical protection against toxic effects of accumulated salt differentiates *Thellungiella halophila* from *Arabidopsis thaliana*. J Plant Physiol 164:375–384

Ma S, Gong Q, Bohnert HJ (2006) Dissecting salt stress pathways. J Exp Bot 57:1097–1107

Martinez J-P, Kinet J-M, Bajji M, Lutts S (2005) NaCl alleviates polyethylene glycol-induced water stress in the halophyte species *Atriplex halimus* L. J Exp Bot 56:2421–2431

Mateos-Naranjo E, Redondo-Gómez S, Álvarez R, Cambrollé J, Gandullo J, Figueroa ME (2010) Synergic effect of salinity and CO_2 enrichment on growth and photosynthetic responses of the invasive cordgrass *Spartina densiflora*. J Exp Bot 61:1643–1654

Miller G, Suzuki N, Rizhsky L, Hegie A, Koussevitzky S, Mittler R (2007) Double mutants deficient in cytosolic and thylakoid ascorbate peroxidase reveal a complex mode of interaction between reactive oxygen species, plant development, and a response to abiotic stress. Plant Physiol 144:1777–1785

Miszalski Z, Ślesak I, Niewiadomska E, Baczek-Kwinta R, Lüttge U, Ratajczak R (1998) Subcellular localization and stress responses of superoxide dismutase isoforms from leaves in the C_3-CAM intermediate halophyte *Mesembryanthemum crystallinum* L. Plant Cell Environ 21:169–179

Mittler R, Vanderauwera S, Suzuki N, Miller G, Tognetti VB, Vandepoele K, Gollery M, Shulaev V, Van Breusegem F (2011) ROS signaling: the new wave? Trends Plant Sci 16:300–309

Mittova V, Tal M, Volokita M, Guy M (2003) Up-regulation of the leaf mitochondrial and peroxisomal antioxidative systems in response to salt-induced oxidative stress in the wild salt-tomato species *Lycopersicon pennellii*. Plant Cell Environ 26:845–856

Munns R, Tester M (2008) Mechanisms of salinity tolerance. Annu Rev Plant Biol 59:651–681

Nelson N, Yocum CF (2006) Structure and function of photosystems I and II. Annu Rev Plant Biol 57:521–565
Niewiadomska E, Bilger W, Gruca M, Mulisch M, Miszalski Z, Krupinska K (2011) CAM-related changes in chloroplastic metabolism of *Mesembryanthemum crystallinum* L. Planta 233:275–285
Niyogi KK (1999) Photoprotection revisited: genetic and molecular approaches. Annu Rev Plant Physiol Plant Mol Biol 50:333–359
Pagliano C, La Rocca N, Andreucci F, Deák Z, Vass I, Rascio N, Barbato R (2009) The extreme halophyte *Salicornia veneta* is depleted of the extrinsic PsbQ and PsbP proteins of the oxygen-evolving complex without loss of functional activity. Ann Bot 103:505–515
Paramonova NV, Shevyakova NI, Kuznetsov VV (2004) Ultrastructure of chloroplasts and their storage inclusions in the primary leaves of *Mesembryanthemum crystallinum* affected by putrescine and NaCl. Russ J Plant Physiol 51:86–96
Parida AK, Das AB (2005) Salt tolerance and salinity effects on plants: a review. Ecotoxicol Environ Saf 60:324–349
Parker R, Flowers TJ, Moore AL, Harpham NVJ (2006) An accurate and reproducible method for proteome profiling of the effects of salt stress in the rice leaf lamina. J Exp Bot 57:1109–1118
Parry MAJ, Keys AJ, Madgwick PJ, Carmo-Silva AE, Andralojc PJ (2008) Rubisco regulation: a role for inhibitors. J Exp Bot 59:1569–1580
Pfannschmidt T (2003) Chloroplast redox signals: how photosynthesis controls its own genes. Trends Plant Sci 8:33–41
Pfannschmidt T, Bräutigam K, Wagner R, Dietzel L, Schröter Y, Steiner S, Nykytenko A (2009) Potential regulation of gene expression in photosynthetic cells by redox and energy state: approaches towards better understanding. Ann Bot 103:599–607
Qiu N, Lu Q, Lu C (2003) Photosynthesis, photosystem II efficiency and the xanthophyll cycle in the salt-adapted halophyte *Atriplex centralasiatica*. New Phytol 159:479–486
Qiu-Fang Z, Yuan-Yuan L, Cai-Hong P, Cong-Ming L, Bao-Shan W (2005) NaCl enhances thylakoid-bound SOD activity in the leaves of C3 halophyte *Suaeda salsa* L. Plant Sci 168:423–430
Rabhi M, Giuntini D, Castagna A, Remorini D, Baldan B, Smaoui A, Abdelly C, Ranieri A (2010) *Sesuvium portulacastrum* maintains adequate gas exchange, pigment composition, and thylakoid proteins under moderate and high salinity. J Plant Physiol 167:1336–1341
Ramani B, Reeck T, Debez A, Stelzer R, Huchzermeyer B, Schmidt A, Papenbrock J (2006) *Aster tripolium* L. and *Sesuvium portulacastrum* L.: two halophytes, two strategies to survive in saline habitats. Plant Physiol Biochem 44:395–408
Redondo-Gómez S, Wharmby C, Castillo JM, Mateos-Naranjo E, Luque CJ, de Cires A, Luque T, Davy AJ, Figueroa ME (2006) Growth and photosynthetic responses to salinity in an extreme halophyte, *Sarcocornia fruticosa*. Physiol Plant 128:116–124
Redondo-Gómez S, Mateos-Naranjo E, Figueroa ME, Davy AJ (2010) Salt stimulation of growth and photosynthesis in an extreme halophyte, *Arthrocnemum macrostachyum*. Plant Biol 12:79–87
Rosso D, Ivanov AG, Fu A, Geisler-Lee J, Hendrickson L, Geisler SG, Krol M, Hurry V, Rodermel SR, Maxwell DP, Hüner NPA (2006) IMMUTANS does not act as a stress-induced safety valve in the protection of the photosynthetic apparatus of *Arabidopsis* during steady-state photosynthesis. Plant Physiol 142:574–585
Rumeau D, Peltier G, Cournac G (2007) Chlororespiration and cyclic electron flow around PSI during photosynthesis and plant stress response. Plant Cell Environ 30:1041–1051
Sacksteder CA, Kramer DM (2000) Dark-interval relaxation kinetics (DIRK) of absorbance changes as a quantitative probe of steady state electron transfer. Photosynth Res 66:145–158
Sakamoto A, Murata N (2002) The role of glycine betaine in the protection of plants from stress: clues from transgenic plants. Plant Cell Environ 25:163–171
Scheibe R, Backhausen JE, Emmerlich V, Holtgrefe S (2005) Strategies to maintain redox homeostasis during photosynthesis under changing conditions. J Exp Bot 56:1481–1489

Scheller HV, Haldrup A (2005) Photoinhibition of photosystem I. Planta 221:5–8
Schöttler MA, Kirchhoff H, Siebke K, Weis E (2002) Metabolic control of photosynthetic electron transport in crassulacean acid metabolism-induced *Mesembryanthemum crystallinum*. Funct Plant Biol 29:697–705
Seidler A (1996) The extrinsic polypeptides of Photosystem II. Biochim Biophys Acta 1277:35–60
Sengupta S, Majumder AL (2009) Insight into the salt tolerance factors of a wild halophytic rice, *Porteresia coarctata*: a physiological and proteomic approach. Planta 229:911–929
Ślesak I, Karpinska B, Surówka E, Miszalski Z, Karpinski S (2003) Redox changes in the chloroplast and hydrogen peroxide are essential for regulation of C_3–CAM transition and photooxidative stress responses in the facultative CAM plant *Mesembryanthemum crystallinum* L. Plant Cell Physiol 44:573–581
Sobhanian H, Motamed N, Jazii FR, Nakamura T, Komatsu S (2010) Salt stress induced differential proteome and metabolome response in the shoots of *Aeluropus lagopoides*, a halophyte C4 monocot plant. J Proteome Res 9:2882–2897
Stepien P, Johnson GN (2009) Contrasting responses of photosynthesis to salt stress in the glycophyte *Arabidopsis* and the halophyte *Thellungiella*: role of the plastid terminal oxidase as an alternative electron sink. Plant Physiol 149:1154–1165
Stepien P, Klobus G (2005) Antioxidant defense in the leaves of C3 and C4 plants under salinity stress. Physiol Plant 125:31–40
Streb P, Josse E-M, Gallouët E, Baptist F, Kuntz M, Cornic G (2005) Evidence for alternative electron sinks to photosynthetic carbon assimilation in the high mountain plant species *Ranunculus glacialis*. Plant Cell Environ 28:1123–1135
Taji T, Seki M, Satou M, Sakurai T, Kobayashi M, Ishiyama K, Narusaka Y, Narusaka M, Zhu J-K, Shinozaki K (2004) Comparative genomics in salt tolerance between Arabidopsis and Arabidopsis-related halophyte salt cress using Arabidopsis microarray. Plant Physiol 135:1697–1709
Takahashi S, Murata N (2008) How do environmental stresses accelerate photoinhibition? Trends Plant Sci 13:178–182
Takahashi S, Milward SE, Fan D-Y, Chow WS, Badger MR (2009) How does cyclic electron transport flow alleviate photoinhibition in Arabidopsis. Plant Physiol 149:1560–1567
Tikkanen M, Aro EM (2012) Thylakoid protein phosphorylation in dynamic regulation of photosystem II in higher plants. Biochim Biophys Acta 1817:232–238
Tsugane K, Kobayashi K, Niwa Y, Ohba Y, Wada K, Kobayashia H (1999) A recessive Arabidopsis mutant that grows photoautotrophically under salt stress shows enhanced active oxygen detoxification. Plant Cell 11:1195–1206
Tuteja N (2007) Mechanisms of high salinity tolerance in plants. Methods Enzymol 428:419–438
Tyystjärvi E (2008) Photoinhibition of photosystem II and photodamage of the oxygen evolving manganese cluster. Coord Chem Rev 252:361–376
Valderrama R, Corpas FJ, Carreras A, Gómez-Rodríguez MV, Chaki M, Pedrajas JR, Fernández-Ocaña A, Del Río LA, Barroso JB (2006) The dehydrogenase-mediated recycling of NADPH is a key antioxidant system against salt-induced oxidative stress in olive plants. Plant Cell Environ 29:1449–1459
Verslues PE, Batelli G, Grillo S, Agius F, Kim YS, Zhu J, Agarwal M, Katiyar-Agarwal S, Zhu J-K (2007) Interaction of SOS2 with nucleoside diphosphate kinase 2 and catalases reveals a point of connection between salt stress and H_2O_2 signaling in *Arabidopsis thaliana*. Mol Cell Biol 27:7771–7780
Voss I, Koelmann M, Wojtera J, Holtgrefe S, Kitzmann C, Backhausen JE, Scheibe R (2008) Knockout of major leaf ferredoxin reveals new redox-regulatory adaptations in *Arabidopsis thaliana*. Physiol Plant 133:584–598
Wang Y, Meng Y-L, Ishikawa H, Hibino T, Tanaka Y, Nii N, Takabe T (1999) Photosynthetic adaptation to salt stress in three-color leaves of a C4 plant *Amaranthus tricolor*. Plant Cell Physiol 40:668–674

Wang W, Vinocur B, Altman A (2003) Plant responses to drought, salinity and extreme temperatures: towards genetic engineering for stress tolerance. Planta 218:1–14

Wen X, Qiu N, Lu Q, Lu C (2005) Enhanced thermotolerance of photosystem II in salt-adapted plants of the halophyte *Artemisia anethifolia*. Planta 220:486–497

Wiciarz M, Gubernator B, Kruk J, Niewiadomska E. (2014) Enhanced chloroplastic generation of H_2O_2 in stress-resistant Thellungiella salsuginea in comparison to Arabidopsis thaliana. Physiol Plant (2014). doi:10.1111/ppl.12248

Wingler A, Quick WP, Bungard RA, Bailey KJ, Lea PJ, Leegood RC (1999) The role of photorespiration during drought stress: an analysis utilizing barley mutants with reduced activities of photorespiratory enzymes. Plant Cell Environ 22:361–373

Winter K, Holtum JAM (2007) Environment or development? Lifetime net CO_2 exchange and control of the expression of crassulacean acid metabolism in *Mesembryanthemum crystallinum*. Plant Physiol 143:98–107

Yang X, Lu C (2005) Photosynthesis is improved by exogenous glycine betaine in salt-stressed maize plants. Physiol Plant 124:343–352

Yu J, Chen S, Zhao Q, Wang T, Yang C, Diaz C, Sun G, Dai S (2011) Physiological and proteomic analysis of salinity tolerance in *Puccinellia tenuiflora*. J Proteome Res 10:3852–3870

Zhang H, Han B, Wang T, Chen S, Li H, Zhang Y, Dai S (2012) Mechanisms of plant salt response: insights from proteomics. J Proteome Res 11:49–67

Zhu J-K (2001) Plant salt tolerance. Trends Plant Sci 6:66–71

CAM-Like Traits in C_3 Plants: Biochemistry and Stomatal Behavior

Paulo Tamaso Mioto, Maria Aurineide Rodrigues, Alejandra Matiz, and Helenice Mercier

Contents

Abstract Although it is generally accepted that Crassulacean Acid Metabolism (CAM) originated from C_3 ancestors through a co-option process, this is rarely discussed in terms of specific characteristics and putative mechanisms behind this event. Here we discuss the available data concerning the biochemical and stomatal traits that are present in C_3 plants and could have been enrolled in the CAM cycle. In summary, the biochemical machinery of CAM seems to have originated from a potential stress-driven recruitment of key non-photosynthetic enzymes of the C_3 background which have entrained circadian rhythm. CAM stomatal behavior could be either a direct consequence of an upregulation of the biochemical machinery or it might require additional changes in the signaling/perception pathways controlling stomatal aperture. Considering that CAM has multiple origins, it is likely that each plant group developed it through different combinations of biochemical/stomatal changes, resulting in various degrees of plasticity of this photosynthetic pathway.

P.T. Mioto • M.A. Rodrigues • A. Matiz • H. Mercier (✉)
Department of Botany, Institute of Biosciences, University of São Paulo, CEP 05508-090 São Paulo, SP, Brazil
e-mail: hmercier@usp.br

U. Lüttge, W. Beyschlag (eds.), *Progress in Botany*, Progress in Botany 76,
DOI 10.1007/978-3-319-08807-5_8

1 Introduction

It has been over 70 years since the term Crassulacean Acid Metabolism (CAM) was used for the first time to indicate the nocturnal acidification observed in species of the genus *Kalanchoë* (Thomas and Beevers 1949; Ranson and Thomas 1960). Ever since, scientists have tried to unravel the mechanisms behind this phenomenon and how it could have appeared along evolution. The emergence of CAM plants multiple times in different taxa and habitats suggests that CAM might have originated by a co-option process in which ancient metabolic pathways were reorganized to generate new functions through modifications in some already-existing key proteins involved in numerous non-photosynthetic processes of C_3 plants (Silvera et al. 2010; Aubry et al. 2011; West-Eberhard et al. 2011; Berry et al. 2013). Apparently, the recruitment of these biochemical elements into the CAM pathway depended on significant increases in the expression of genes involved in both production and transport of C_4-organic acids, as well as alteration in their diel rhythm, coupled with an inversion of stomatal aperture pattern (Taybi et al. 2004; Silvera et al. 2010; Borland et al. 2014). However, a discussion of how elements from C_3 plants could be recruited into CAM is still uncommon. In this chapter we intend to share some insights into the mechanisms that may have led to CAM behavior.

2 General Features of Typical CAM Behavior

The basic C_3 pathway serves as the primary mechanism for the photosynthetic carbon fixation employed by most terrestrial plant species. This mode of CO_2 assimilation, also known as Calvin–Benson cycle, operates with the central participation of RuBP carboxylase-oxygenase (Rubisco, EC 4.1.1.39) as the sole carboxylating enzyme in C_3 plants. Concurrently, in plants performing typical CAM, Rubisco re-fixates the carbon that was previously assimilated by phosphoenolpyruvate carboxylase (PEPC, EC 4.1.1.31) during the nocturnal phase of the CAM cycle (Berry et al. 2013). The atmospheric carbon fixed by PEPC at night is stored in the form of malic acid in the vacuole, also known as Phase I. Then, on the following day, while the stomata remain closed, the malic acid stored during the night is decarboxylated, allowing the CO_2 generated to be photosynthetically reduced in the chloroplasts via Calvin cycle (Phase III), concentrating CO_2 around Rubisco in this phase (Cushman and Bohnert 1999; Dodd et al. 2002; Lüttge 2002, 2004; Keeley and Rundel 2003; Crayn et al. 2004; Silvera et al. 2010; Matiz et al. 2013). Phases II and IV are transitional states between Phases I and III. Phase II occurs at early light period through open stomata, when Rubisco is becoming active while PEPC is being inactivated and CO_2 fixation can happen via both enzymes. In the transition from light period to dark period, the Phase IV occurs, which is characterized by the reopening of the stomata when the storage of organic acids is already exhausted, allowing atmospheric CO_2 assimilation via Rubisco (Osmond 1978; Dodd et al. 2003; Lüttge 2008; Kluge 2008).

3 Driving Forces of CAM Evolution

It has been proposed that atmospheric CO_2/O_2 ratio reduction in the early Miocene allowed the uprising of CO_2-concentrating mechanisms, such as CAM and C_4 photosynthesis (Ehleringer et al. 1991; Ehleringer and Monson 1993; Raven and Spicer 1996; Winter and Smith 1996; Edwards and Ogburn 2012). Decreasing atmospheric CO_2 had an important impact for terrestrial plants, not only favoring photorespiration by the increasing oxygenase activity of Rubisco (Edwards and Ogburn 2012) but also increasing transpirational cost per unit of carbon fixed (Raven and Spicer 1996; Brodribb and Feild 2010). Thus, by opening stomata during the night and closing them during most of the day, CAM plants achieve a higher water use efficiency than C_3 plants, providing a selective advantage in dry environments (Ehleringer and Monson 1993; Drennan and Nobel 2000; Keeley and Rundel 2003; Winter et al. 2008; Borland et al. 2009, 2014). In fact, terrestrial CAM plants are commonly found in habitats with low water availability, such as deserts and the canopy of tropical forests. Therefore, water deficit and low CO_2 concentrations were important selective driving forces for the emergence of CAM photosynthesis (Keeley and Rundel 2003).

Although the biochemistry of CAM is frequently coupled with nocturnal stomatal opening, sometimes they seem to be independent. *Isoëtes* species, for example, commonly grow in aquatic environments with depleted CO_2 and bicarbonate during the day, due to high photosynthetic activity of the other organisms. As a result, CAM allows the uptake of inorganic carbon (both from water and respiration) at night by storing it as organic acid (Keeley 1985; Ting 1985; Keeley and Rundel 2003). In these taxa the stomata are generally absent or nonfunctional, but all the biochemical machinery of CAM is active. More evidence showing an uncoupling between CAM biochemistry and nocturnal stomatal opening was found in some pseudobulbs and roots of orchids, which are capable of expressing the biochemical reactions of the CAM pathway, but are unable to express typical CAM due to a lack of stomata (Rodrigues et al. 2013). In other words, the biochemistry of CAM may happen despite the absence of stomata or in the case of plants performing CAM cycling and idling, without nocturnal stomatal aperture. In the following sections, it will be discussed how CAM biochemistry could have been selected independently of the nocturnal stomatal opening. Besides, some evidence will be presented about the occurrence of nocturnal stomatal opening without the expression of CAM biochemical machinery. Finally, we are going to address some insights into the potential interactions between the biochemical and stomatal modules operating in the CAM cycle, and their impact on controlling the plasticity of this photosynthetic pathway.

4 Key Biochemical Candidates for Adaptive Selection of CAM-Related Features

Although our current knowledge regarding the evolutionary progression of specific genes selected for CAM expression is somewhat limited, studies of the PEPC gene family have indicated that similar changes to those described for the C_4 pathway might have occurred during the recruitment of non-photosynthetic enzymes from C_3 background into CAM (Silvera et al. 2010). Some important candidates for the evolution of C_4 and CAM photosynthesis appear to involve genes that encode the key enzymes for the carboxylation and decarboxylation processes, such as PEPC, PEPC kinase, malate dehydrogenase (MDH, EC 1.1.1.37), pyruvate orthophosphate dikinase (PPDK, EC 2.7.9.1), and NADH or NADPH-dependent malic enzyme (NADP-ME, EC 1.1.1.40, or NAD-ME, EC 1.1.1.39) (Doubnerová and Ryslavá 2011; Berry et al. 2013).

PEPC activity is thought to be a major factor in limiting the magnitude of the CAM pathway (Taybi et al. 2004); therefore, the essential characteristics of this enzyme should be considered in the context of potential mechanisms involved in determining the evolution and expression plasticity of CAM photosynthesis. In fact, PEPC is a tightly regulated enzyme that is present in the cytosol of all vascular plants and is also broadly distributed in green algae and bacteria. This enzyme represents a crucial regulatory point at a key branch of plant metabolism that confers a highly flexible aspect for synchronizing the carbon partitioning with changing environmental conditions (O'Leary et al. 2011; Shane et al. 2013). The diverse PEPC functions include the regulation of malate production/homeostasis during stomatal conductance modulation, environmental stress responses, and N_2-fixing nodule development in legume roots, among others (Nimmo 2000; Aubry et al. 2011). Furthermore, plant-type PEPCs are particularly relevant for supplying carbon skeletons to the tricarboxylic acid (TCA) cycle, which allows the anaplerotic replenishment of the TCA intermediates redirected for biosynthesis and ammonium assimilation (Gennidakis et al. 2007; Masumoto et al. 2010; Aubry et al. 2011; O'Leary et al. 2011; Shane et al. 2013). O'Leary et al. (2011) considered that, although certainly valid, such a traditional view of the non-photosynthetic PEPC participating only in the replenishment of TCA intermediates oversimplifies the broader contribution of this enzyme to plant metabolism.

In CAM and C_4 plants, PEPC catalyzes the first and pivotal step in CO_2 assimilation which involves the irreversible β-carboxylation of phosphoenolpyruvate (PEP) to yield oxaloacetate (OAA) and inorganic phosphate (Nimmo 2000; Gennidakis et al. 2007). All plant-type PEPCs are regulated by a complex set of posttranslational mechanisms that control their day/night activities, which includes allosteric effectors, phosphorylation, monoubiquitination, and other protein–protein interactions (O'Leary et al. 2011; Shane et al. 2013). Since plant-type PEPCs are allosteric enzymes inhibited by malate and activated by glucose-6-phosphate (Glc-6-P), phosphorylation represents one essential activator of PEPC activity by simultaneously reducing PEPC sensitivity to malate inhibition while enhancing

Glc-6-P activation. In CAM species, therefore, such a posttranslational modification allows this enzyme to overcome feedback inactivation by the end product of nighttime CO_2 fixation (e.g., malate), enabling the abundant nocturnal accumulation of C_4-organic acids required for the proper operation of the CAM cycle (Nimmo 2000; Taybi et al. 2004; Gennidakis et al. 2007; Kluge 2008; Aubry et al. 2011; Berry et al. 2013; Shane et al. 2013).

The PEPC phosphorylation is catalyzed by the presence of a specific Ca^{2+}-independent PEPC protein kinase termed PEPC kinase. In C_3 and C_4 plants, PEPC kinase seems to be activated exclusively by light (Gousset-Dupont et al. 2005; Shenton et al. 2006), while in CAM plants this is a night-specific enzyme whose transcription is mostly under the influence of an internal circadian rhythm (Hartwell et al. 1999, 2002; Nimmo 2000, 2003). Although it is possible that there is a direct connection between the circadian oscillator and the expression of PEPC kinase in CAM plants, through a potential transcription factor directly associated with the endogenous circadian clock, compelling evidence in favor of such a link is still elusive (Nimmo 2000). Another hypothesis concerning the connection between the regulation of PEPC kinase and the circadian clock during the CAM cycle suggests that the circadian rhythm of PEPC kinase expression may be a consequence of fluctuations in the primary metabolism related to the cellular distribution/levels of malate. This hypothesis is based on results showing that the abundance of PEPC kinase transcripts was inversely correlated with cytoplasmic malate concentrations, thus indicating that malate levels could negatively affect PEPC kinase expression and/or its mRNA stability (Borland et al. 1999; Nimmo 2000; Borland and Taybi 2004; Cushman et al. 2008). All together, these evidence indicate that PEPC kinase modulation (by gene expression and/or enzyme activity) might represent one of the strongest candidates required for both the establishment and the maintenance of the core CAM machinery, due to its influence on PEPC expression.

4.1 The Synchronous Modulation of Non-photosynthetic PEPC, MDH, PPDK, and ME by Abiotic Constraints: A "Precondition" for CAM Cycle?

Environmental challenges such as drought, unfavorable temperatures, salinity, and other harsh conditions can considerably hamper the photosynthesis in most plants due to consequences of stress-induced impairment of the photosystems, which, therefore, limit the CO_2 reduction process and can generate oxidative stress (Ashraf and Harris 2013). However, these challenging conditions might have contributed to the adaptive recruitment of specific non-photosynthetic enzymes from the C_3 background into photosynthetic functions in C_4 and CAM plants (Silvera et al. 2010; Berry et al. 2013; Cowling 2013). The selective recruitment of non-photosynthetic genes to a photosynthetic role generally involves modifications in their default C_3-expression patterns that cause greatly enhanced transcription levels, thus leading to the accumulation of their respective proteins in leaves (Hibberd and Covshoff 2010; Langdale 2011; Williams et al. 2012; Berry et al. 2013). Böcher and Kluge (1978) have already suggested that a pathway of carbon flow similar to CAM could be established in some C_3 plants. In fact, it is generally accepted that CAM evolved through increased expression of C_3 genes involved in both production and transport of organic acids (Taybi et al. 2004).

Some essential components for the CO_2-concentrating process during CAM cycle, such as representatives of the families PEPC, MDH, PPDK, and ME, frequently show increased expression and/or activities in virtually all plants under various types of abiotic constraints (Gonzalez et al. 2003; Aubry et al. 2011; Doubnerová and Ryslavá 2011; Langdale 2011; Cowling 2013). As illustrated in Fig. 1, it is suggested that these enzymes under adverse conditions can form an alternative cycle, which may confer adaptive metabolic adjustments for C_3 plants exposed to challenging environments (Doubnerová and Ryslavá 2011). The coupled activities of PEPC and cytosolic MDH can generate organic acids (such as malate) with important implications in the cellular redox balance (Sriram et al. 2007). Furthermore, the oxidation of malate to pyruvate by ME results in both NAD(P)H production and carbon supply at the involved cellular compartment, which could contribute to a redistribution of the reducing power among different compartments of the cell. Additionally, the combined activities of PEPC, MDH, and ME can form an alternative metabolic flux which provides the ability to respire OAA generated from PEP, instead of relying only on the reaction catalyzed by the cytosolic pyruvate kinase (EC 2.7.1.40) to generate pyruvate. Finally, PPDK activity can regenerate PEP which can be used as substrate for the PEPC reaction. Therefore, this potential alternative cycle formed by PEPC, MDH, ME, and PPDK (Fig. 1) provides and/or redistributes CO_2 and NAD(P)H that can be used by the TCA cycle, antioxidant system, and amino acid metabolism (Doubnerová and

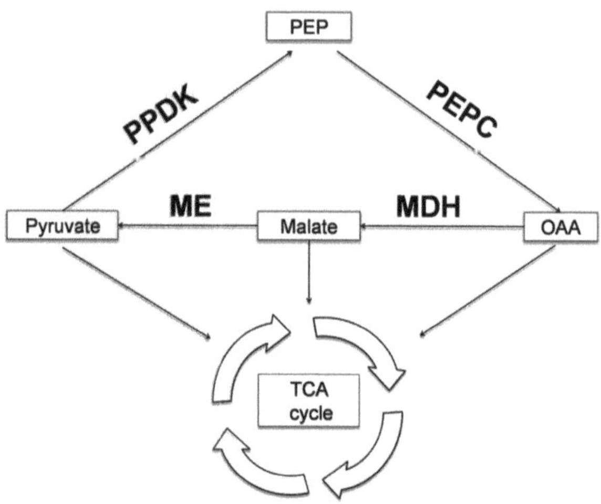

Fig. 1 Scheme of a hypothetical cycle formed by the major key enzymes of CAM in C_3 plants under adverse conditions. PEPC carboxylates PEP, yielding OAA, which undergoes reduction by MDH into malate. Malate is decarboxylated by ME into pyruvate, which, in turn, is converted to PEP by PPDK, closing the cycle. OAA, malate, and pyruvate can be also used to replenish the intermediates of the TCA cycle

Ryslavá 2011; O'Leary et al. 2011; Rodrigues et al. 2014). It is tempting to hypothesize that the recruitment of these metabolic elements used by C_3 plants as a potential strategy to couple with unfavorable conditions, together with the selection of a circadian control of these reactions, might represent important steps for the origin of CAM.

Undoubtedly, a better understanding of the non-photosynthetic roles of these proteins in C_3 species would be useful in predicting the metabolic alterations in a C_3 tissue when components of the CAM pathway are artificially introduced. This is especially relevant when considering that CAM can be interpreted as the most flexible and adaptive photosynthetic pathway and that it has been suggested that economically and ecologically important CAM species should be exploited to support sustainable production in the future(Borland et al. 2011; Cowling 2013). Furthermore, some exciting prospects have been recently envisioned by the scientific community concerning the development of bioenergy feedstocks and food crops engineered with a functional CAM system into C_3 crops (Borland et al. 2014).

5 The Establishment of CAM Stomatal Behavior Could Happen Independently of Biochemistry?

Curiously, stomatal opening during the night does not seem to be exclusive of CAM, as it has already been reported in C_3 and C_4 plants. However, when C_3 and C_4 plants open stomata during nighttime, there is no CO_2 assimilation (Caird et al. 2007), indicating that nighttime stomatal opening in these cases seems to be independent of the enzymatic machinery required for CAM. A recent review pointed out that the possible factors controlling C_3 and C_4 nocturnal stomatal opening may include microclimatic conditions both in soil and in leaves, species-specific variations, and plant and/or leaf age (Zeppel et al. 2013). In the same review, the authors speculate on possible advantages of nocturnal stomatal opening without CO_2 fixation, including embolism removal and nutrient transport (Zeppel et al. 2013). It was also suggested that root temperature may influence nocturnal stomatal conductance in *Vitis vinifera* (Rogiers and Clarke 2013). Interestingly, it was recently discovered that there are specialized stomata in leaves of *Nelumbo nucifera* that open during the night (besides the "normal" ones that open during daytime) and this opening seems to be mainly regulated by darkness (Matthews and Seymour 2013). Undoubtedly, the phenomenon of nocturnal stomatal opening in C_3 and C_4 plants deserves more attention in order to determine its exact consequences for the plant metabolism. Stomatal closure during the day can happen in C_3 plants mainly in response to environmental factors, as will be discussed below.

5.1 Is Stomatal Control in CAM Similar to Its C_3 Counterpart?

As a general assumption, stomata can respond to several environmental factors, such as light, CO_2, drought, pathogens/elicitors, and also endogenous factors, such as circadian rhythm (Klüsener et al. 2002; Chen et al. 2012). Since under drought or pathogen attack both C_3 and CAM plants close their stomata (resulting in CAM idling when CAM biochemistry is present), the differences in stomatal behavior between them are likely to depend on signaling by light, CO_2, or the endogenous clock. For this reason, we will focus on how stomata respond to these factors and the possible changes that may have occurred to yield CAM.

The control of stomata by light, especially in blue wavelength, is already well established for C_3 plants. For instance, AtMYB60 and AtMYB61 are *A. thaliana* transcription factors involved with stomatal control that appear to be regulated by photoreceptors such as cryptochrome and phototropins (Chen et al. 2012). While AtMYB60 is a positive regulator of stomatal aperture and accumulates in the light, AtMYB61 appears to have the opposite function of closing the stomata and accumulates during the dark period (Cominelli et al. 2005; Liang et al. 2005). Additionally, it has been recently shown in *A. thaliana* that the transcription factor ELF3 (*EARLY FLOWERING 3*) is negatively involved in stomatal aperture, while FT (*FLOWERING LOCUS T*) is positively linked to stomatal control (Kinoshita et al. 2011). The same authors suggest that the transcription factor FT either controls an intermediary component in blue-light signaling pathway that mediates stomatal opening or it is the component itself. Interestingly, both ELF3 and FT are also strongly regulated by the circadian clock (Hicks et al. 1996; Covington et al. 2001; Liu et al. 2001; Onai et al. 2004; Hubbard and Webb 2011). In fact, *elf3-201* mutants showed continued open stomata in continuous light with a 50-fold increase in FT expression, while *ft-1* mutants showed continued closed stomata in the same conditions (Kinoshita et al. 2011). Therefore, at least in the C_3 plant *A. thaliana*, light and circadian clock appear to work together to promote the opening of stomata during the day.

CAM plants, however, would require a dampening of stomatal response to light, possibly relying more on circadian rhythms and/or CO_2 levels instead, in order to close stomata during the day. In fact, it was observed that in both *Mesembryanthemum crystallinum* and *Portulacaria afra*, the induction of CAM suppresses the stomatal opening in response to blue light (Lee and Assmann 1992; Tallman et al. 1997). The mechanisms of how this dampening occurs, however, are still unknown. In CAM-induced *M. crystallinum*, an *ELF3* ortholog shows a pattern of expression very similar to that of its C_3 counterpart, accumulating its transcripts during the evening. Therefore, the possible differences between light-regulated stomatal control after induction of CAM do not change expression of ELF3 and, possibly, FT (Boxall et al. 2005). These results indicate that although the central clock remains the same, the output for stomatal aperture in CAM plants may be somehow different from that of *A. thaliana* and other C_3 plants. Mechanisms of stomatal opening during the night in CAM plants are not known in detail, but it seems there is a strong circadian component since even in continuous light condition stomata of CAM plants continue to open during the subjective night (Wilkins 1984; Lüttge and Beck 1992; Wyka and Lüttge 2003).

5.2 *CO_2 Sensing: The Interaction Between Biochemistry and Stomatal Control*

It has been known that stomata can respond to intercellular CO_2 concentration, but the mechanism underlying this observation still remains largely unknown. It is discussed, for example, whether the guard cells can perceive internal CO_2 directly or whether this gas is perceived by the mesophyll cells (Flexas et al. 2008; Mott et al. 2008; Araújo et al. 2011). In *A. thaliana*, the kinase HT1 (HIGH LEAF TEMPERATURE 1, EC 2.7.11.1) seems to be one of the few components that promotes stomatal aperture and is highly influenced by CO_2 concentrations (Hashimoto et al. 2006). It was also proposed that two carbonic anhydrases (βCA1 and βCA4, EC 4.2.1.1) somehow appear to sense high CO_2 concentrations and promote stomatal closure by inhibiting HT1 activity, indicating that the sensing of CO_2 depends on HCO_3^- generation (Kim et al. 2010). It is important to note that changes in these components affect only CO_2-induced stomatal closure, while stomatal closure in response to the phytohormone abscisic acid (ABA) and blue light remains largely unaffected.

More recently, Merilo et al. (2013) found that OST1 (OPEN STOMATA 1, EC 2.7.11.1), responsible for phosphorylation of SLAC1 (SLOW ANION CHANNEL-ASSOCIATED 1), appears to be essential in CO_2-mediated stomatal closure. SLAC1 was demonstrated to activate Ca^{2+}-dependent slow anion channels and promote stomatal closure (Vahisalu et al. 2008). It is also suggested that there are possibly several points of interaction between the signaling pathways of CO_2, darkness, ozone, drought, and ABA during stomatal closure, including OST1 (Merilo et al. 2013).

Until now, biochemical pathways of stomatal control in charge of sensing intracellular CO_2 concentration were not investigated in plants expressing CAM. Perhaps the expression patterns of HT1 and OST1 orthologs could provide some insight into how CO_2 mediates stomatal behavior in CAM plants; furthermore, since CO_2 could be perceived as HCO_3^-, carbonic anhydrases may also be an important target for research. If there are no changes in these components, then it is probable that malate decarboxylation could generate a sufficiently high internal CO_2 concentration during daytime to induce stomatal closure, while CO_2 assimilation by PEPC during the night may lead to CO_2 concentrations low enough to cause stomata to open during this period (Lüttge 2002; Kluge 2008). Alternatively, stomata of CAM plants could increase their sensibility to CO_2 to follow the organic acid fluctuations.

In a very interesting group of experiments, plants of the CAM species *Kalanchoë daigremontiana* were kept in N_2 during one night, which resulted in a severe reduction in nocturnal malate accumulation during Phase I (Borland and Griffiths 1997; Borland et al. 1999). The results showed that on the following day CO_2 assimilation values were higher and lasted longer during Phase II. The authors suggested that this effect could be due to higher PEPC activity as a consequence of lower malate content (as malate inhibits PEPC) and activation by PEPC kinase. However, these plants still showed a Phase III and not much difference was detected in Phase IV. Further work in the same species under continuous light showed that the circadian rhythm of CO_2 uptake and stomatal conductance was not heavily affected by nocturnal malate depletion (Wyka et al. 2004).

Von Caemmerer and Griffiths (2009) tested stomatal CO_2 responses in both *K. daigremontiana* and *K. pinnata* by manipulating CO_2 availability during different moments in the CAM cycle and also by depleting intracellular malate accumulated during the night. Interestingly, they found that stomata did not open during phase III, even when combining a lowering of internal CO_2 (reduction in malate accumulation in the previous night) and atmospheric CO_2. They suggest that there must be a signal other than CO_2 that causes stomata to close during phase III. The developmental changes in expression of CAM in *Peperomia scandens*, a plant capable of going from CAM cycling to typical CAM, showed that the stomatal behavior changed regardless of alterations in the amount of organic acids accumulated during the night (Holthe et al. 1987), suggesting that in this species it was not an upregulation of biochemical machinery that caused the changes in stomatal behavior.

Recently, Owen and Griffiths (2013) developed a model to predict CAM behavior based on *K. daigremontiana*, showing that metabolic control may be a major factor in determining the CAM phases. It was also shown that, at least theoretically, it is possible to extinguish phase III with a severe downregulation of malate decarboxylation. Although this model was built mainly over stomatal control by metabolic factors, this leads to the hypothesis that a simple upregulation of CAM biochemistry could generate CO_2 variations high enough to result in CAM stomatal behavior. Accordingly, Kluge (1968) already demonstrated that phase III is shortened under high light due to more rapid consumption of nocturnally stored malate, resulting in earlier stomatal opening for phase IV than in low light.

Gathering all these observations, it is still not clear whether the stomatal behavior of CAM plants could simply be a consequence of the biochemical machinery (generation of CO_2 variations large enough to supplant other stimuli) or whether it would require changes in other control mechanisms (abolishment of opening in response to light, inversion of circadian rhythms, increased sensitivity to CO_2, etc.). More likely, both factors contribute differently in each species, conferring different degrees of plasticity. A biochemistry-driven stomatal control could probably result in a more rapid and plastic expression of CAM, allowing a species to be capable of going from C_3 to CAM and back in response mainly to the environment. Examples of this plasticity are rare so far, as it was only confirmed that species such as *Calandrinia polyandra*, *Clusia pratensis*, and *Clusia minor* are capable of such event (Lüttge 2008; Winter et al. 2008; Winter and Holtum 2014). On the other hand, species such as *M. crystallinum* are not capable of returning to a

C_3 state once CAM has been established (Winter and Holtum 2007; Winter et al. 2008), perhaps due to permanent changes in stomatal control. Undoubtedly, even irreversible CAM plants show some degree of metabolic control over stomatal aperture that confers some plasticity regarding the strength of CAM.

6 Conclusions and Perspectives

The discussion presented in this review, although still speculative to some extent, raises some interesting questions that deserve further attention in future research. It is still not known whether the stomata of CAM plants function, in terms of perception and response to signals, are similar to those of C_3 plants. We believe that permanent changes in stomatal behavior would lead to a less plastic CAM. The understanding of circadian clock elements and their functions is definitely vital for the comprehension of how crucial enzymes such as PEPC, MDH, ME, and PPDK started to show diverse patterns of activity along the day/night cycle. A key point seems to rest on understanding the upstream controllers of PEPC kinase expression and activity.

A very interesting subject of study is the so-called C_3-CAM facultative plants. Winter et al. (2008) demonstrated that the switch from C_3 to CAM can occur in response to the environment as well as to ontogeny, in a degree that varies with the plant species: some species are heavily affected by the environment, while others rely mainly on ontogeny, with numerous behaviors between these extremes. These massive changes in metabolism could answer some questions as to how CAM stomatal behavior is achieved: is it simply through upregulation of the biochemical machinery or through changes in perception of signals related to stomatal control? Does the biochemical machinery consist of specific isoenzymes for CAM or does it originate from the same isoenzymes present in the C_3 mode?

The answers to those questions would certainly lead to important targets to work on engineering CAM into C_3 crops, allowing these plants to grow in semiarid habitats and, therefore, increase agricultural production (Borland et al. 2014).

Acknowledgements The authors would like to thank Dr. Luciano Freschi for valuable discussions on the topic of this review.

References

Araújo WL, Fernie AR, Nunes-Nesi A (2011) Control of stomatal aperture, a renaissance of the old guard. Plant Signal Behav 6:1305–1311

Ashraf M, Harris PJC (2013) Photosynthesis under stressful environments: an overview. Photosynthetica 51:163–190

Aubry S, Brown NJ, Hibberd JM (2011) The role of proteins in C_3 plants prior to their recruitment into the C_4 pathway. J Exp Bot 62:3049–3059

Berry JO, Yerramsetty P, Zielinski AM, Mure CM (2013) Photosynthetic gene expression in higher plants. Photosynth Res 117:91–120
Böcher M, Kluge M (1978) The C_4-pathway of C-fixation in Spinacea olearacea II. Pulse chase experiments with suspended leaf slices. Zeitschriftfür Pflanzenphysiologie 86:405–421
Borland AM, Griffiths H (1997) A comparative study on the regulation of C_3 and C_4 carboxylation processes in the constitutive crassulacean acid metabolism (CAM) plant *Kalanchoe daigremontiana* and the C_3-CAM intermediate *Clusia minor*. Planta 201:368–378
Borland A, Taybi T (2004) Synchronization of metabolic processes in plants with crassulacean acid metabolism. J Exp Bot 55:1255–1265
Borland AM, Hartwell J, Jenkins GI, Wilkins MB, Nimmo HG (1999) Metabolite control overrides circadian regulation of phosphoenolpyruvate carboxylase kinase and CO_2 fixation in crassulacean acid metabolism. Plant Physiol 121:889–896
Borland AM, Griffiths H, Hartwell J, Smith JAC (2009) Exploiting the potential of plants with crassulacean acid metabolism for bioenergy production on marginal lands. J Exp Bot 60:2879–2896
Borland AM, Zambrano VAB, Ceusters J, Shorrock K (2011) The photosynthetic plasticity of crassulacean acid metabolism: an evolutionary innovation for sustainable productivity in a changing world. New Phytol 191:619–633
Borland AM, Hartwell J, Weston DJ, Schlauch KA, Tschaplinski TJ, Tuskan GA, Yang X, Cushman JC (2014) Engineering crassulacean acid metabolism to improve water-use efficiency. Trends Plant Sci 19:327–338
Boxall SF, Foster JM, Bohnert HJ, Cushman JC, Nimmo HG, Hartwell J (2005) Conservation and divergence of circadian clock operation in a stress-inducible crassulacean acid metabolism species reveals clock compensation against stress. Plant Physiol 137:969–982
Brodribb TJ, Feild TS (2010) Leaf hydraulic evolution led a surge in leaf photosynthetic capacity during early angiosperm diversification. Ecol Lett 13:175–183
Caird MA, Richards JH, Donovan LA (2007) Nighttime stomatal conductance and transpiration in C_3 and C_4 plants. Plant Physiol 143:4–10
Chen C, Xiao Y-G, Li X, Ni M (2012) Light-regulated stomatal aperture in Arabidopsis. Mol Plant 5:566–572
Cominelli E, Galbiati M, Vavasseur A, Conti L, Sala T, Vuylsteke M, Leonhardt N, Dellaporta SL, Tonelli C (2005) A guard-cell-specific MYB transcription factor regulates stomatal movements and plant drought tolerance. Curr Biol 15:1196–1200
Covington MF, Panda S, Liu XL, Strayer CA, Wagner DR, Kay SA (2001) ELF3 modulates resetting of the circadian clock in Arabidopsis. Plant Cell 13:1305–1315
Cowling SA (2013) Did early land plants use carbon concentrating mechanisms? Trends Plant Sci 18:120–124
Crayn MC, Winter K, Smith JAC (2004) Multiple origins of Crassulacean acid metabolism and the epiphytic habit in the Neotropical family Bromeliaceae. Proc Natl Acad Sci U S A 101:3703–3708
Cushman JC, Bohnert HJ (1999) Crassulacean acid metabolism: molecular genetics. Annu Rev Plant Physiol Plant Mol Biol 50:305–332
Cushman JC, Tillett RL, Wood JA, Branco JM, Schlauch KA (2008) Large-scale mRNA expression profiling in the common ice plant, *Mesembryanthemum crystallinum*, performing C_3 photosynthesis and Crassulacean acid metabolism (CAM). J Exp Bot 59:1875–1894
Dodd AN, Borland AM, Haslam RP, Griffiths H, Maxwell K (2002) Crassulacean acid metabolism: plastic, fantastic. J Exp Bot 53:569–580
Dodd AN, Griffiths H, Taybi T, Cushman JC, Borland AM (2003) Integrating diel starch metabolism with the circadian and environmental regulation of crassulacean acid metabolism in *Mesembryanthemum crystallinum*. Planta 216:789–797
Doubnerová V, Ryslavá H (2011) What can enzymes of C_4 photosynthesis do for C_3 plants under stress? Plant Sci 180:575–583
Drennan PM, Nobel PS (2000) Responses of CAM species to increase atmospheric CO_2 concentrations. Plant Cell Environ 23:767–781

Edwards EJ, Ogburn RM (2012) Angiosperm responses to a low-CO_2 world: CAM and C_4 photosynthesis as parallel evolutionary trajectories. Int J Plant Sci 173:724–733
Ehleringer JR, Monson JK (1993) Evolutionary and ecological aspects of photosynthetic pathway variation. Annu Rev Ecol Syst 24:411–439
Ehleringer JR, Sage RF, Flanagan LB, Pearcy RW (1991) Climate change and the evolution of C_4 photosynthesis. Trends Ecol Evol 6:95–99
Flexas J, Ribas-Carbó M, Diaz-Espejo A, Galmés J, Medrano H (2008) Mesophyll conductance to CO_2: current knowledge and future prospects. Plant Cell Environ 31:602–621
Gennidakis S, Rao S, Greenham K, Winter K, Uhrig RG, O'Leary B, Snedden WA, Lu C, Plaxton WC (2007) Bacterial- and plant-type phosphoenolpyruvate carboxylase polypeptides interact in the hetero-oligomeric Class-2 PEPC complex of developing castor oil seeds. Plant J 52:839–849
Gonzalez MC, Sanchez R, Cejudo FJ (2003) Abiotic stresses affecting water balance induce phosphoenolpyruvate carboxylase expression in roots of wheat seedlings. Planta 216:985–992
Gousset-Dupont A, Lebouteiller B, Monreal J, Echevarria C, Pierre JN, Hodges M, Vidal J (2005) Metabolite and post-translational control of phosphoenolpyruvate carboxylase from leaves and mesophyll cell protoplasts of Arabidopsis thaliana. Plant Sci 169:1096–1101
Hartwell J, Nimmo G, Wilkins M, Jenkins G, Nimmo H (1999) Phosphoenolpyruvate carboxylase kinase is a novel protein kinase regulated at the level of gene expression. Plant J 20:333–342
Hartwell J, Nimmo GA, Wilkins MB, Jenkins GI, Nimmo HG (2002) Probing the circadian control of phosphoenolpyruvate carboxylase kinase expression in *Kalanchoë fedtschenkoi*. Funct Plant Biol 29:663–668
Hashimoto M, Negi J, Young J, Israelsson M, Schroeder JI, Iba K (2006) Arabidopsis HT1 kinase controls stomatal movements in response to CO_2. Nat Cell Biol 8:391–397
Hibberd JM, Covshoff S (2010) The regulation of gene expression required for C_4 photosynthesis. Annu Rev Plant Biol 61:181–207
Hicks KA, Millar AJ, Carré IA, Somers DE, Straume M, Meeks-Wagner DR, Kay SA (1996) Conditional circadian dysfunction of the Arabidopsis early-flowering 3 mutant. Science 274:790–792
Holthe PA, Sternberg LSL, Ting IP (1987) Developmental control of CAM in *Peperomia scandens*. Plant Physiol 84:743–747
Hubbard KE, Webb AAR (2011) Circadian rhythms: FLOWERING LOCUS T extends opening hours. Curr Biol 21:636–638
Keeley JE (1985) The role of CAM in the carbon economy of the submerged-aquatic *Isoetes howellii*. Verhandlungen des Internationalen Verein Limnologie 22:2909–2911
Keeley JE, Rundel PW (2003) Evolution of CAM and C_4 carbon-concentrating mechanisms. Int J Plant Sci 164:55–77
Kim T-H, Böhmer M, Hu H, Nishimura N, Schroeder JI (2010) Guard cell signal transduction network: advances in understanding abscisic acid, CO_2, and Ca^{2+} Signaling. Annu Rev Plant Biol 61:561–591
KinoshitaT ON, HayashiY MS, Nakamura S, Soda M, Kato Y, Ohnishi M, Nakano T, Inoue S, Shimazaki K (2011) FLOWERING LOCUS T regulates stomatal opening. Curr Biol 21:1232–1238
Kluge M (1968) Untersuchungenüber den Gaswechsel von *Bryophyllum* während der Lichtperiode. Planta 80:359–377
Kluge M (2008) Ecophysiology: migrations between different levels of scaling. Prog Bot 69:5–34
Klüsener B, Young JJ, Murata Y, Allen GJ, Mori IC, Hugouvieux V, Schroeder JI (2002) Convergence of calcium signaling pathways of pathogenic elicitors and abscisic acid in *Arabidopsis* guard cells. Plant Physiol 130:2152–2163
Langdale JA (2011) C_4 cycles: past, present, and future research on C_4 photosynthesis. Plant Cell 23:3879–3892
Lee DM, Assmann SM (1992) Stomatal responses to light in the facultative crassulacean acid metabolism species, *Portulacaria afra*. Physiol Plant 85:35–42

Liang YK, Dubos C, Dodd IC, Holroyd GH, Hetherington AM, Campbell MM (2005) AtMYB61, an R2R3-MYB transcription factor controlling stomatal aperture in *Arabidopsis thaliana*. Curr Biol 15:1201–1206

Liu XL, Covington MF, Fankhauser C, Chory J, Wagner DR (2001) ELF3 encodes a circadian clock–regulated nuclear protein that functions in an Arabidopsis PHYB signal transduction pathway. Plant Cell 13:1293–1304

Lüttge U (2002) CO_2-concentrating: consequences in crassulacean acid metabolism. J Exp Bot 53:2131–2142

Lüttge U (2004) Ecophysiology of crassulacean acid metabolism (CAM). Ann Bot 93:629–652

Lüttge U (2008) Clusia: Holy Grail and enigma. J Exp Bot 59:1503–1514

Lüttge U, Beck F (1992) Endogenous rhythms and chaos in crassulacean acid metabolism. Planta 188:28–38

Masumoto C, Miyazawa SI, Ohkawa H, Fukuda T, Taniguchi Y, Murayama S, KusanoM SK, Fukayama H, Miyao M (2010) Phosphoenolpyruvate carboxylase intrinsically located in the chloroplast of rice plays a crucial role in ammonium assimilation. Proc Natl Acad Sci U S A 107:5226–5231

Matiz A, Mioto PT, Mayorga AY, Freschi L, Mercier H (2013) CAM photosynthesis in bromeliads and agaves: what can we learn from these plants? In: Dubinsky Z (ed) Photosynthesis. Intech, Rijeka, Croatia, pp 91–134

Matthews PGD, Seymour R (2013) Stomata actively regulate internal aeration of the sacred lotus *Nelumbo nucifera*. Plant Cell Environ 37:402–413

Merilo E, Laanemets K, Hu H, Xue S, Jakobson L, Tulva I, González-Guzman M, Rodruiguez PL, Schroeder JI, Brosché M, Kollist H (2013) PYR/RCAR receptors contribute to ozone-, reduced air humidity-, darkness-, and CO_2-induced stomatal regulation. Plant Physiol 162:1652–1668

Mott KA, Sibbernsen ED, Shope JC (2008) The role of the mesophyll in stomatal responses to light and CO_2. Plant Cell Environ 31:1299–1306

Nimmo HG (2000) The regulation of phosphoenolpyruvate carboxylase in CAM plants. Trends Plant Sci 5:75–80

Nimmo HG (2003) Control of the phosphorylation of phosphoenolpyruvate carboxylase in higher plants. Arch Biochem Biophys 414:189–196

O'Leary B, Park J, Plaxton WC (2011) The remarkable diversity of plant PEPC (phosphoenolpyruvate carboxylase): recent insights into the physiological functions and post-translational controls of non-photosynthetic PEPCs. Biochem J 436:15–34

Onai K, Okamoto K, Nishimoto H, Morioka C, Hirano M, Kami-ike N, Ishiura M (2004) Large-scale screening of Arabidopsis circadian clock mutants by a high-throughput real-time bioluminescence monitoring system. Plant J 40:1–11

Osmond CB (1978) Crassulacean acid metabolism: a curiosity in context. Annu Rev Plant Physiol 29:379–414

Owen NA, Griffiths H (2013) A system dynamics model integrating physiology and biochemical regulation predicts extent of crassulacean acid metabolism (CAM) phases. New Phytol 200:1116–1131

Ranson SL, Thomas M (1960) Crassulacean acid metabolism. Annu Rev Plant Physiol 11:81–110

Raven JA, Spicer RA (1996) The evolution of crassulacean acid metabolism. In: Winter K, Smith JAC (eds) Crassulacean acid metabolism. Biochemistry, ecophysiology and evolution. Springer, Berlin, pp 360–385

Rodrigues MA, Matiz A, Cruz AB, Matsumura AT, Takahashi CA, Hamachi L, Félix LM, Pereira PN, Latansio-Aidar SP, Aidar MPM, Demarco D, Freschi L, Mercier H, Kerbauy GB (2013) Spatial patterns of photosynthesis in thin- and thick-leaved epiphytic orchids: unravelling C_3-CAM plasticity in an organ-compartmented way. Ann Bot 112:17–29

Rodrigues MA, Freschi L, Pereira PN, Mercier H (2014) Interactions between nutrients and crassulacean acid metabolism. Prog Bot 75:167–186

Rogiers SY, Clarke SJ (2013) Nocturnal and daytime stomatal conductance respond to root-zone temperature in 'Shiraz' grapevines. Ann Bot 111:433–444

Shane MW, Fedosejevs ET, Plaxton WC (2013) Reciprocal control of anaplerotic phosphoenolpyruvate carboxylase by in vivo monoubiquitination and phosphorylation in developing proteoid roots of phosphate-deficient. Plant Physiol 161:1634–1644
Shenton M, Fontaine V, Hartwell J, Marsh JT, Jenkins GI, Nimmo HG (2006) Distinct patterns of control and expression amongst members of the PEP carboxylase kinase gene family in C_4 plants. Plant J 48:45–53
Silvera K, Neubig KM, Whitten M, Williams NH, Winter K, Cushman JC (2010) Evolution along the crassulacean acid metabolism continuum. Funct Plant Biol 37:995–1010
Sriram G, Fulton DB, Shanks JV (2007) Flux quantification in central carbon metabolism of *Catharanthus roseus* hairy roots by C-13 labeling and comprehensive bondomer balancing. Phytochemistry 68:2243–2257
Tallman G, Zhu J, Mawson BT, Amodeo G, Nouhi Z, Levy K, Zeiger E (1997) Induction of CAM in *Mesembryanthemum crystallinum* abolishes the stomatal response to blue light and light-dependent zeaxanthin formation in guard cell chloroplasts. Plant Cell Physiol 38:236–242
Taybi T, Nimmo HG, Borland AM (2004) Expression of phosphoenolpyruvate carboxylase and phosphoenolpyruvate carboxylase kinase genes. Implications for genotypic capacity and phenotypic plasticity in the expression of Crassulacean acid metabolism. Plant Physiol 135:587–598
Thomas M, Beevers H (1949) Physiological studies on acid metabolism in green plants. II. Evidence of CO_2 fixation in *Bryophyllum* and the study of diurnal variation of acidity in this genus. New Phytol 48:421–447
Ting IP (1985) Crassulacean acid metabolism. Annu Rev Plant Physiol 36:595–622
Vahisalu T, Kollist H, Wang YF, Nishimura N, Chan WY, Valerio G, Lamminmäki A, Brosché M, Moldau H, Desikan R, Schroeder JI, Kangasjärvi J (2008) SLAC1 is required for plant guard cell S-type anion channel function in stomatal signaling. Nature 452:487–491
Von Caemmerer S, Griffiths H (2009) Stomatal responses to CO_2 during a dielcrassulacean acid metabolism cycle in *Kalanchoë daigremontiana* and *Kalanchoë pinnata*. Plant Cell Environ 32:567–576
West-Eberhard MJ, Smith JAC, Winter K (2011) Photosynthesis, reorganized. Science 332:311–312
Wilkins MB (1984) A rapid circadian rhythm of carbon-dioxide metabolism in *Bryophyllum fedtschenkoi*. Planta 161:381–384
Williams BP, Aubry S, Hicks JM (2012) Molecular evolution of genes recruited into C_4 photosynthesis. Trends Plant Sci 17:213–220
Winter K, Holtum JAM (2007) Environment or development? Lifetime net CO_2 exchange and control of the expression of Crassulacean acid metabolism in *Mesembryanthemum crystallinum*. Plant Physiol 143:98–107
Winter K, Holtum JAM (2014) Facultative crassulacean acid metabolism (CAM) plants: powerful tools for unravelling the functional elements of CAM photosynthesis. J Exp Bot. doi:10.1093/jxb/eru063
Winter K, Smith JAC (1996) Crassulacean acid metabolism. Current status and perspectives. In: Winter K, Smith JAC (eds) Crassulacean acid metabolism. Biochemistry, ecophysiology and evolution. Springer, Berlin, pp 389–426
Winter K, Garcia M, Holtum JAM (2008) On the nature of facultative and constitutive CAM: environmental and developmental control of CAM expression during early growth of *Clusia*, *Kalanchoë*, and *Opuntia*. J Exp Bot 59:1829–1840
Wyka TP, Lüttge U (2003) Contribution of C_3 carboxylation to the circadian rhythm of carbon dioxide uptake in a Crassulacean acid metabolism plant *Kalanchoë daigremontiana*. J Exp Bot 54:1471–1479
Wyka TP, Bohn A, Duarte HM, Kaiser F, Lüttge U (2004) Perturbations of malate accumulation and the endogenous rhythms of gas exchange in the crassulacean acid metabolism plant *Kalanchoë daigremontiana*: testing the tonoplast-as-oscillator model. Planta 219:705–713
Zeppel M, Logan B, Lewis JD, Phillips N, Tissue D (2013) Why lose water at night? Disentangling the mystery of nocturnal sap flow, transpiration and stomatal conductance – when, where, who? Acta Hortic 991:307–312

Stability as a Phenomenon Emergent from Plasticity–Complexity–Diversity in Eco-physiology

Gustavo M. Souza and Ulrich Lüttge

Contents

Abstract The network of *plasticity*, *diversity*, *complexity*, and *stability* is drawn as a quadruped-scheme. Plasticity is on the top of the scheme and stability is in the center. Plasticity is discussed in some detail. Examples are given of intraspecific plasticity, especially of photosynthesis. Plasticity allows escape from the dilemma of growth or defense of the growth differentiation balance theory (GDB). Analysis by principal component analysis (PCA) of multi-variant traits and their integration explain plastic emergence of phenotypes. Via the phenotypes plasticity can both

G.M. Souza
Laboratório de Cognição Ecológica em Plantas – ECOLAB, Rod. Raposo Tavares km 572, Presidente Prudente, SP, Brasil
e-mail: gumaia.gms@gmail.com

U. Lüttge (✉)
Department of Biology, Technical University of Darmstadt, Schnittspahnstraße 3 – 5, 64287, Darmstadt, Germany
e-mail: luettge@bio.tu-darmstadt.de

U. Lüttge, W. Beyschlag (eds.), *Progress in Botany*, Progress in Botany 76,
DOI 10.1007/978-3-319-08807-5_9

impede or support diversity by speciation. Diversity, or as we say in the realm of life "biodiversity," governs spatiotemporal dynamics of competition/facilitation equilibria in stress gradients (stress gradient hypothesis, SGH). Diversity is the basis of complexity. Both, biodiversity and complexity, are challenging and stabilizing in terms of sustaining ecosystems. Via the different connections in the network of the quadruped plasticity, diversity and complexity affect stability. Understanding of the quadruped network offers an outlook on potential applications in environmental management and agro-forest ecology.

1 Introduction

Biological systems are, by definition, open systems thermodynamically in non-equilibrium (Nicolis and Prigogine 1977; Schneider and Kay 1994). They represent multi-level organization in spatiotemporal terms (Novikoff 1945; Ravasz et al. 2002). System organization (considered herein as the integration between the structure and functionality throughout the system's persistence) depends, therefore, on continuous flow of energy driving exchanges of matter and information with the surrounding environment (Schneider and Kay 1994). The physical environment (abiotic and biotic) of a living system is naturally variable. The amplitude and frequency of such variation is contextual, yet always present at some level. Thus, biological systems must constantly deal with a variation of the available resources (energy, water, nutrients) that eventually may limit or threaten their organization and survival (Souza et al. 2005; Wagner 2005).

Therefore, one of the fundamental aspects of a biological system is the maintenance of its stability. Considering stability in this essay we realize, as said above, that all living systems, such as individual organisms, populations, and ecosystems, are open systems through which a continuous flow of material and energy occurs. Hence, the living systems are not stable in a static sense of thermodynamics. However, the open systems arrive at dynamic equilibria, also called pseudo-steady states, which may prove to show stability within certain time windows. This is immediately evident for example when we realize organisms including plants as unitary self-organized living systems with internal integrity and extend this view to higher scalar levels of biological organization (Lüttge 2012b). Thus, here we use stability in the context of General Systems Theory and System Dynamic Theory. This accounts for both homeostatic capacity of the system (steady-state; maintenance of the internal structural and functional integrity of the system) and its flexibility (resilience upon organizational disturbances, the return to initial condition after perturbation). With this understanding of stability we adopt the concept of Lyapunov stability (Souza et al. 2004; Solé and Bascompte 2006) referring to the dynamic equilibrium of the system, i.e., its pseudo-steady state and NOT its thermodynamic steady state. This is appropriate use in ecology, i.e., stability conferring robustness to the whole system as a prerequisite of the system's persistence.

One problem that emerges from the complex nature of biological systems is to determine the characteristic settings and mechanisms of the system's internal integrity. Under a systemic approach (Lüttge 2012b) the determination of the internal integrity depends directly on the spatiotemporal observation scale used to define the object/phenomenon under investigation, and as a consequence, also sets up what is the environment of the system. Thus, if individuals are the objects of study, the internal integrity arises from the cellular constitution and related physiological and metabolic processes. Conversely, if the system under study is a population or even a community, the internal integrity arises from individuals/species and their respective dynamics (for instance, temporal changes in gene frequencies, fluctuations in the number of individuals).

The ability to face the complexity of the environmental changes is particularly essential for sessile organisms such as plants. Because of the limited capacity of locomotion, plants must deal with all sorts of environmental variation in their surroundings. As a result, their stability requires dynamic elements that confer some degree of organizational flexibility. The set of possible changes in response to external stimuli is what we call phenotypic plasticity (DeWitt and Scheiner 2004). Therefore, there is a close correlation between the stability of a system and the plasticity of phenotypic responses (which are naturally limited by the genetic constitution of organisms). In essence phenotypic plasticity gives the system the ability to expand its capacities of physiological acclimation (in the case of individuals) or genetic adaptation (in the case of populations/species). The plasticity, particularly the modulation/control of plasticity, is related to patterns of organization of biological systems which are characterized by complex networks (Hütt and Lüttge 2005; Souza et al. 2009). Indeed, biological systems are essentially complex adaptive systems.

Complex systems are formed by elements, in general of many different natures, which have cause-effect based relationships to each other, among which necessarily some ones must be nonlinear. Such nonlinear relationships, often found in natural systems, commonly follow positive feedbacks, which amplify signals, or negative feedbacks, which attenuate signals, giving the system the ability to self-regulation (Camazine et al. 2001; Mitchell 2009). The set of relations between the elements of a complex system is often topologically structured as a network with small-world characteristics. In addition, depending on the size and hierarchic organization of networks, distribution and disposition of elements within relationships may also be scale-invariant. Small-world networks typically display small structural distances on average between elements. Scale-free networks are small-world networks in which the distribution of nodes (elements) with a number of relationships (links) follows a power law. In practical terms, networks that have a very small number of elements are densely connected (hubs—elements with many relations with other elements) and most elements have few connections (Barabási 2002). Such topological network structure shows higher robustness than random networks. Herein, robustness is considered as the capacity of a network to stay stable despite of random removing of nodes. For instance, in ecological food webs showing scale-free topology the removal of the most connected species (hubs) causes more disturbance than does the random removal of species (Proulx et al. 2005). Moreover, networks with power-law distribution have high information transmission efficiency through the entire system, for instance, allowing the spread of diseases with low transmission rates (May and Lloyd 2001). This supports the establishment of nonlocal interactions between different parts of the system, allowing the possibility of synchronization and collective behavior (emergent behavior). Such features, in the case of living beings, enable new phenotypic states, creating morphotypes or physiotypes (Hütt and Lüttge 2005).

In different contexts, such systems have the ability to shape themselves through changes in the pattern of connection of the network elements, even in different scales (in the case of hierarchical systems), providing self-regulation to the system (Souza et al. 2009; Lucas et al. 2011; Bertolli et al. 2013, 2014). Such systemic changes occur due to flow of information in the system, allowing local signals from the system to reorganize certain structures and/or processes, or to reorganize it as a whole (nonlocal actions). The reorganization processes of the system can be incorporated into its informational network that, with repeated reinforcement of similar experiences, can create recurring behaviors (habits) in the establishment and maintenance of new connections (Barabási and Oltvai 2004; Hütt and Lüttge 2005; Csermely 2006).

Conversely, diversity may reach a local maximum. This is related to the extent to which individuals and populations become more stable by forming new habits. Thus, new processes of speciation may become reduced due to the maintenance of the stability of individuals, thereby stabilizing current populations. This, naturally, limits the ecological complexity, taking into account both biological diversity and the space-time dynamics of the populations. The stability of a system depends positively on the complexity of the system up to a certain limit, because at the same time that diversity gives plasticity to the system, redundancy in some essential elements of the system (e.g., copies of regulatory genes or greater number of individuals of a key-species in a community) provides greater system robustness (Edelman and Gally 2001; Wagner 2005).

Therefore, despite of a plethora of stability definitions and uses (Grimm and Wissel 1997), herein stability is defined not as a property of the system itself, but rather as a concept that involves three fundamental properties woven in complexity: (1) Homeostasis, the system's resistance to change, keeping self-regulation of the system and being reflected by resilience (Cannon 1932); (2) resilience, the ability of the system to return to its initial state after a disturbance (McCann 2000); (3) persistence, the maintenance of the identity of the system as a whole (Grimm and Wissel 1997), i.e., representing robustness as reflecting some kind of stability. Hence, systemic features are kept invariant, for example, genetic constitution of an individual, or the typical physiognomy of a vegetation. The three properties can be related as follows: under changing environmental conditions, the first reaction of the system (if the changes are actually perceived) is to maintain its homeostasis by a sort of mechanisms counteracting the external influence, conferring resistance capacity to the system. However, if the homeostatic mechanisms are overloaded, homeostasis can be lost. Depending on the level of injury, after cessation of the external disturbance, the system can return to previous state (manifestation of resilience capacity), can assume a new state of organization (reorganization of the system, reaching a new homeostatic state) or, if injury was too harmful, the system

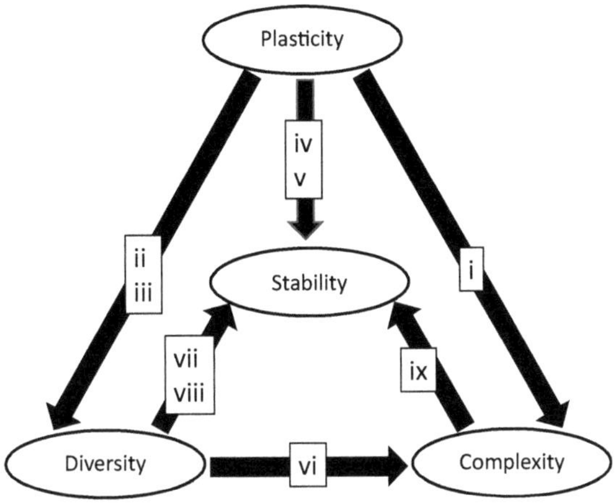

Fig. 1 Quadruped-scheme of plasticity, diversity, and complexity and their interactions creating stability. Complexity is closely related to diversity (vi). Both support stability (vii, viii, ix). Plasticity can do this directly (iv, v), via affecting diversity (ii, iii) or via its inherent qualities of complexity (i). (The text refers to the roman numbers where these relations are elaborated)

may collapse. In this context, homeostatic maintenance and resilience capacity support the persistence of the system.

In Fig. 1, we suggest a scheme (the quadruped-scheme) showing the relationships among plasticity, diversity, and complexity creating stability. Thus, in the following sections we shall give a number of examples of intraspecific plasticity and show how it can facilitate choices between options of plant performance in dilemmas in resource allocation imposed by stress situations. We shall see how plasticity can support emergence of phenotypes positively or negatively affecting speciation and diversity. In this way understanding plasticity will allow to assess the roles of diversity and complexity for stability.

2 Plasticity

2.1 Intraspecific Plasticity

Intraspecific plasticity is given and interspecific diversity becomes visible when species and often even individual plants express different phenotypes under the influence of environmental cues. These may be morphological variations of morphotypes and functional variations of performance or physiotypes (Bradshaw 1965). Functional plasticity is a very wide field in plant performance. As sessile organisms which cannot move around in the search of appropriate external conditions, plants must be particularly well equipped with options for acclimation to varying conditions at their growth sites. This is given by multifunctional regulatory capacity. It is based on network structures with multiple branching points for feedback and feed-forward regulations. Such networks provide high degrees of flexibility for reversible expression of physiotypes (Hütt and Lüttge 2005; Lucas et al. 2011). Plant life is full of examples. We chose here (1) the network of energy metabolism driving ion transport at membranes, (2) the modes of photosynthesis, where morphotypic and physiotypic plasticity interact, (3) the growth-defense dilemma, where plasticity provides pathways of escape.

2.1.1 Metabolic Energy Networks for Energization of Membrane Transport of Ions

It is just a historical recollection here, when we are recalling intensive studies in the 1970s investigating whether there were specific reactions of energy metabolism driving specific processes of membrane transport of ions. However, this is a very illustrative example of functional physiological plasticity given by network dynamics in plants. The starting point was the hypothesis by Lundegårdh that mitochondrial electron transport would drive anion uptake by plant cells (Lundegårdh and Burström 1933, 1935; Lundegårdh 1950, 1955). Similarly it was proposed that photosynthetic electron transport would directly power anion uptake in green algal cells (MacRobbie 1965; Raven 1967). This elicited a large number of studies where various techniques were used to separate specific mechanisms of energy metabolism in order to see if per se they could drive ion transport. These included use of various metabolic inhibitors of electron transport and phosphorylation, anaerobiosis, variation of wavelengths distinguishing activation of the two photosystems, photosynthesis mutants, variegated leaves, and greening etiolated leaves. It turned out that any given energy source if isolated in the appropriate way can drive ion transport. As long as any meshes in the network were maintaining cellular energy state with energy charge and redox poise, this could sustain energy-dependent ion transport at membranes (Lüttge and Higinbotham 1979).

2.1.2 Plastic Expression of Modes of Photosynthesis

Although for the stoichiometry of plant life sensu Körner (Körner 2013; see also Lüttge 2012a), we must not neglect the integration of the nutrient cycle with the carbon cycle, photosynthesis remains the primary event in the production of new biomass. In the physiological ecology of photosynthesis, we find a pronounced plasticity of plants in response to factors such as light, water and nutrients, and carbon. We also note that intermediateness can occur between the three major modes of photosynthesis, i.e., C_3, C_4, and Crassulacean Acid Metabolism (CAM). There are C_3/CAM, C_3/C_4, and C_4/CAM intermediate species.

Sun and Shade Forms

In the ecophysiology of photosynthesis we can distinguish sun plants and shade plants (Boardman 1977; Larcher 2001; Lichtenthaler et al. 2007; Lichtenthaler 2014), where the adaptation to and the demand of high and low irradiance, respectively, is genetically constitutive. In the context of plasticity of photosynthesis, it is important to mention, however, that sun and shade phenotypes can also be expressed by the same genotype. A well-known example of this intraspecific plasticity is that of the leaves of European beech, *Fagus sylvatica* L. Exposed leaves at the surface of the crown are sun type and shaded leaves in the inner part of the crown are shade type leaves. The sun leaves are thicker and have multilayer-palisade parenchyma. Sun leaves have higher leaf area-related respiration and maximum photosynthetic rates (but not so when based on dry mass), higher light compensation points, higher light saturation points, and lower quantum yield of photosynthesis than shade leaves. Compared to sun plants shade plants have lower chlorophyll a/b ratios, lower rates of electron flow along the redox chain in the thylakoids related to chlorophyll, less soluble protein (especially less enzyme protein of ribulose-bis-phosphate carboxylase/oxygenase, RubisCO) in relation to chlorophyll, larger photosystem II/photosystem I ratios, larger chloroplasts, and more grana formation (summarized in Lüttge 2008a). With this acclimation to light regimes, C_3-photosynthesis shows plasticity.

Another well-studied example of sun and shade forms is the Crassulacean acid metabolism (CAM) plant *Bromelia humilis* Jacq. growing in tropical deciduous dry forests and on sand plains where the tank-forming leaf rosettes which they develop on the same ramets are dark green when shaded under shrubs and trees, pale green in semi-shade, and lemon-yellow in open sun exposure (Lee et al. 1989; Fetene et al. 1990; Lüttge 2008a). These morphotypic differences are correlated with physiotypic differences especially in the expression of CAM-phases and the internal recycling of CO_2 with CAM-idling as explained in the following subsection.

Phases of Crassulacean Acid Metabolism

Crassulacean acid metabolism (CAM) is a mode of photosynthesis which in itself is an outstanding example of plasticity. This is due to the most flexible expression of the four phases of CAM as they were first defined by Osmond and Ziegler (Osmond and Ziegler 1975; Osmond 1978). Phase I is nocturnal uptake and fixation of CO_2 via phospho*enol*pyruvate carboxylase (PEPC) and accumulation of the resulting malic acid in the vacuoles. This is the phase responsible for the water saving capacity of CAM as stomata are opened in the dark period with low evaporative demand of the atmosphere and therefore low transpiratory loss of water vapor. Phase II is a transition phase in the early morning when PEPC is down-regulated and RubisCO is up-regulated. Phase III is daytime remobilization of the organic acid from the vacuoles and decarboxylation and refixation of the CO_2 via RubisCO behind closed stomata preventing transpiration when water potential of the atmosphere is low. Phase IV in the afternoon occurs when the nocturnally accumulated organic acid is consumed and stomata open for CO_2 uptake and direct C_3-photosynthesis via RubisCO. Expression of these phases (reviewed in Lüttge 2004, 2006, 2007b) is modulated by the availability of water. Under scarcity of water they are gradually suppressed which begins with reduction and elimination of phases II and IV followed by a delay of and reduction of the amplitude of phase I in the dark period as water stress becomes more severe. This is observed in the laboratory (Smith and Lüttge 1985) and also with the yellow morphotypes of *B. humilis* in the field (Lee et al. 1989). The climax is stomatal closure all of the time during day and night when respiratory CO_2 is refixed in the dark period and recycled in the light period, a mode of CAM called CAM-idling. It allows the plants to overcome extended periods of dry conditions. Another way of using the phases is stomatal closure in the dark period and recycling of respiratory CO_2 accompanied by stomatal opening during the light period when both external CO_2 and CO_2 coming from the organic acid stored during recycling of CO_2 serve C_3-photosynthesis via RubisCO, a mode of CAM called CAM-cycling. Among the species of the genus *Kalanchoë* in Madagascar, obligate C_3 and CAM species are localized to the wetter and drier regions of the island, respectively, while C_3-CAM cycling species are more common under varying degrees of drought stress (Kluge et al. 2001).

The extraordinary flexible expression of the CAM-phases makes CAM plants particularly fit for tropical environments where environmental stress mostly is not due to a single given factor but to multiple factors including water, irradiance, temperature, CO_2, and minerals (nutrients and salinity) (Lüttge 2004, 2010a; Herrera 2009; Rodrigues et al. 2014). These environmental parameters form a network (Figure 1 in Lüttge 2004) which regulates the multifactorial stress reactions of CAM plants.

C_3/CAM Intermediate Plants

The plasticity of CAM is topped by the C_3/CAM intermediate plants which can alternatively express CAM with all its phases and pure genuine C_3-photosynthesis, respectively. Quite a number of C_3/CAM intermediate genera and species have been described in the literature among them—to name just a few—*Codonanthe crassifolia* (Guralnick et al. 1986), *Guzmania monostachia* (Maxwell 2002), *Kalanchoë blossfeldiana* (Brulfert et al. 1973), *Portulacaria afra* (Huerta and Ting 1988), *Sedum telephium* (Lee and Griffiths 1987). Two of them need special mentioning: One is the annual plant *Mesembryanthemum crystallinum*, where CAM is induced by salinity (Winter and von Willert 1972), and which has now developed to a widely studied model species of plant stress biology. Induction of CAM is associated with the development of more succulent leaves. CAM induction is only partially reversible. The other one is the tropical tree genus *Clusia* comprising both obligate C_3 and CAM species and also many C_3/CAM intermediate species of which *Clusia minor* L. is the most well-studied one (monographs: Lüttge 2006, 2007a, b, 2008b). This plant in terms of its photosynthetic modes is the most flexible we know of. Switches between C_3 and CAM are rapidly reversible. Two leaves of the same node in the same plant can perform either mode if artificially kept at different atmospheric water potentials (Schmitt et al. 1988). *C. minor* is an example for plasticity providing broad amplitude of niche occupation (Lüttge 2007c; Sect. 2.3.2).

C_3/C_4 Intermediate Plants

The C_4-mode of photosynthesis ecophysiologically adapts plants to high irradiance and low water availability. In the typical case the green leaf tissue is differentiated in a peripheral mesophyll tissue and an internal bundle-sheath tissue surrounding central bundle sheaths. In the mesophyll cells CO_2 is fixed via PEPC, malate is transported to the bundle-sheath cells where it is decarboxylated, and the CO_2 gained is refixed via RubisCO (Hatch and Osmond 1976; Hatch 1987; Sage 2004). This is a CO_2 concentrating mechanism as it results in a considerable increase of CO_2 partial pressures at RubisCO in the bundle sheaths, i.e., about sixfold as compared with atmospheric CO_2. The major ecophysiolgical advantages are (1) good CO_2 acquisition, even if stomata are partially closed, accompanied by reduced transpiratory loss of water (probably being the crucial point in C_4 evolution), because PEPC has an about 60-fold higher affinity to CO_2 than RubisCO, (2) a lowered CO_2 compensation point of photosynthesis, and (3) a reduction or even elimination of photorespiration due to the high CO_2 levels at RubisCO, and therefore higher productivity as observed in water non-limited agricultural C_4 plants. There are a number of structural and biochemical variations and modifications which we do not need to detail here for the purpose of drawing attention to C_3/C_4 intermediate species and for demonstration of the separation of C_3 and C_4 species in principle component analysis of multi-variant traits (Sect. 2.2).

The formation of C_3/C_4 intermediateness, however, is different from C_3/CAM intermediateness. The latter is dynamic short-term phenotypic plasticity on the basis of given genotypes in relation to the dominant interest of the present essay. C_3/C_4 intermediateness is genetically constitutive anchorage of photosynthetic characteristics intermediate between the two modes of photosynthesis. In other words, it is more long-term evolutionary plasticity. In fact it is discussed as a possible path on the way to the evolution of the full C_4-syndrome (Peisker 1986; Moore et al. 1987; Rawsthorne 1992; Westhoff and Gowik 2010; Ueno 2011). It is also considered, however, to offer avenues for molecular engineering for creating novel phenotypes with reduced photorespiration and higher productivity (Sect. 5).

C_4/CAM Intermediate Plants

C_4/CAM plasticity has been observed to occur in some succulent C_4 dicotyledons which are capable of diurnal fluctuations of organic acids. Dark-respiratory CO_2 is trapped in bundle sheaths by PEPC and the water storage tissue in the succulent leaves may also participate in the fixation of internally released CO_2 (Ku et al. 1981). This may also be a form of CAM-cycling occurring in C_4-plants like in C_3-plants. Examples of C_4/CAM intermediate species are *Peperomia camptotricha* (Nishio and Ting 1993), *Portulaca oleracea* (Koch and Kennedy 1980, 1982; Mazen 1996), and *Portulaca grandiflora* (Koch and Kennedy 1980, 1982; Ku et al. 1981; Kraybill and Martin 1996; Guralnick and Jackson 2001; Guralnick et al. 2002).

2.1.3 Plasticity and the Escape from the Growth-Defense Dilemma

Competition and defense are contrasting strategies under abiotic and biotic environmental stress. Competition requires vigorous growth. Defense under biotic stress requires structural and in plants predominantly chemical means. Both compete in resource allocation. This dilemma is addressed by the growth-differentiation balance (GDB) theory, which originally was posed as a hypothesis (Loomis 1953). However, recent reviewing of it acquired evidence that it needs to be regarded as a theory at immature stage (Matyssek et al. 2012b).

The GDB theory assumes a trade-off with a branching point between the requirements of resources for growth and defense, respectively (Loomis 1953; Herms and Mattson 1992; Matyssek et al. 2002, 2005). With affluent resources and high primary production of biomass by photosynthesis (gross primary productivity, GPP), growth as an irreversible increase in biomass is stimulated. This is the strategy for remaining competitive in continuing acquisition of resources. Competition for carbon, for water and nutrients (Schenk 2006; Novoplansky 2009; Hodge 2009, 2010; Cahill 2013), for light (Küppers 1989; Grams and Lüttge 2011; Cahill 2013), and even just for space as a resource (Grams and Lüttge 2011; Grams et al. 2012; Grams 2013) supports growth, and vice versa, growth allows competition. An evident example is competition for light, where growth and with it size per se is the basis of successful competing via shading the competitors (Cahill 2013). With scarcity of resources, on the contrary, at the cost of growth resources are used for defense mechanisms in order to keep the status of the resources already incorporated. Defense mechanisms are particularly important under biotic stress for repelling pathogens, predators, and herbivores. (The term differentiation in GDB addresses the resource-investment in structural and chemical modifications serving defense.) For plants especially chemical defense mechanisms are characteristic including compounds such as phenolics, phenyl-propane derivatives, terpenoids, or tannins (Oßwald et al. 2012; Kolosova and Bohlmann 2012). One has often termed these compounds secondary metabolites. However, this is somewhat misleading and the dilemma of resource allocation sensu GDB is more intimate. For example phenyl-propane derivatives are needed for both lignin synthesis essential for growth and the synthesis of pathogen-repellent chemicals (phytoalexins). By producing phenyl-propanes the shikimic acid pathway, which is unique to plants, is a key switch between growth and defense (Gayler and Priesack 2013). Proteins may also have antimicrobial defense functions in addition to structural and functional growth functions. Hence, it is considered preferable to speak of growth and defense-related metabolism instead of primary and secondary metabolites, respectively (Matyssek et al. 2012b, c; Gayler and Priesack 2013).

Two recent books present a wealth of data and theoretical interpretations addressing the question as to whether GDB with the dilemma of the strict choice between two contrasting alternatives is valid under all circumstances or there is escape from the dilemma (Matyssek et al. 2012a, 2013). It is evident that defense metabolism diverts energy and material from growth metabolism and that growth priority reduces energy and material available for defense. Looking at it in this way, we see the plant at a branching point of resource allocation towards either of two alternative linear outcomes. However, the situation changes completely if we introduce nonlinear network dynamics (Hütt and Lüttge 2002, 2005) where links or pathways (edges) between processes or functional mechanisms and pools of metabolites (knots) provide multi-variant branching points. Only superficially it appears that competition may reduce diversity by outcompeting compatriots. Competition-plasticity relations are nonlinear spatiotemporal systems as can be nicely documented in mathematical-model simulations (Gayler and Priesack 2013). Multifunctional regulatory capacity confers plasticity, and the degree of plasticity is correlated with the degree of complexity of the networks. We cannot talk about plasticity without talking about complexity [(i) in Fig. 1]. The two books mentioned above provide examples. The dilemma is broken for instance where growth can serve both resource acquisition and defense. One example is localized growth with sacrificing redundant and easily replaceable organs. In this way surplus GPP serves defense by the plant giving up organs whose growth is limited by mechanisms other than GPP. Another example is secondary metabolism driving both growth and defense, e.g., by lignification (Matyssek et al. 2012b, c). In conclusion, the GDB theory needs modification (Matyssek et al. 2012c) as we realize that plasticity and complexity modulate growth-defense relations and in this way provide stability to the existence of plant species and individuals (Fig. 1).

2.2 *Analysis of Multi-variant Traits and Their Integration for Plastic Emergence of Phenotypes*

The responses of plants to environmental complexity are the sum of all of their modular responses to local conditions plus all of the interaction effects resulting from the integration of individual modules (de Kroon et al. 2005). Indeed, the integration of modules allows for the emergent properties of biological systems to be revealed (Lüttge 2012b).Therefore, there is no single scale that represents the whole-plant plasticity (Vítolo et al. 2012; Bertolli et al. 2013, 2014).

An approach to assess the role of multi-variant traits for plastic emergence of phenotypes is to measure complementary sets of data relevant in ecophysiological performance and to submit them to principal component analysis (PCA). The main purpose of PCA is to condense the information from a large number of original variables into a smaller set of new compound dimensions with a minimal loss of information (McGarigal et al. 2000). Such a procedure allows to verify the grouping of the different plant responses to environmental cues, taking into account the entire set of measured physiological parameters, and further to simulate how the plant responses were grouped when different sets of data are analyzed separately according to different scales of observation (Vítolo et al. 2012).

In a study comparing the plasticity between two species with different photosynthetic metabolism (C_3 and C_4) in response to water deficit and high temperature, a whole set of parameters at different scales (from biochemistry to whole-plant biomass) was evaluated in the same population. The results were virtually different when single scales or sub-sets of parameters were analyzed separately (Vítolo et al. 2012; Bertolli et al. 2013, 2014). It was expected that the C_3 species would be more plastic than the C_4 species (Sage and McKown 2006); however, the multivariant analyses showed that the level of plasticity was scale-dependent. In particular, just the whole set of parameters was representative of the plant as an entire organism, indicating the non-reductive character of plant plasticity (Vítolo et al. 2012; Bertolli et al. 2013, 2014).

2.3 *Plasticity Impeding or Supporting Diversity by Speciation*

Phenotypic plasticity is an outcome of genetic diversity (Booy et al. 2000). Plasticity "may buffer or promote various evolutionary processes" (Novoplansky 2002). That genotype and phenotype are intimately related is the reason for why there is controversial debate of whether plasticity enhances or impedes diversity. By selection of new genotypes from diverse phenotypes it may support development of species diversity. Conversely, plasticity may hinder speciation by protecting genotypes from environmental pressures, and hence, from speciation [(ii) and (iii) in Fig. 1; Grime et al. 1986; West-Eberhard 1986, 1989, 2003; Solbrig 1994; Lüttge 1995a,b, 2000; Gehrig et al. 2001].

2.3.1 Hindrance of Speciation Due to Flexibility of Stress Responses and Therefore Lower Species Diversity [(ii) in Fig. 1]

Plasticity with flexibility of species can stabilize the persistence of individuals and individual species [(iv) in Fig. 1]. An example may be the shift from C_3-photosynthesis to CAM in *M. crystallinum* elicited by salinity (Sect. 2.1.2). The annual plant begins its life cycle with C_3-photosynthesis. For stabilization it is noteworthy that in principle the plant can complete its annual life cycle without shifting to CAM. There is no age-related induction of CAM. However, stress of salinity and drought induces CAM (Winter and Holtum 2005, 2007). This is only partially reversible upon stress removal as plants age (Ratajczak et al. 1994), and age also makes the plant more sensitive to stress-dependent CAM induction (Winter and Holtum 2005, 2007). Under feedback from the phenotype, the genotype switches the plant from the C_3-physiotype to the CAM-physiotype and stabilizes completion of the annual cycle of the plant (Lüttge 2005).

We may consider this type of metabolic C_3-CAM plasticity as an example of plasticity with the more focused meaning of "developmental plasticity" which is normally irreversible (Novoplansky 2002). The case of salinity-induced C_3/CAM-shift in *M. crystallinum* is also an excellent example of the documentation of links between the environmental influence (salinity) and the genotypic capacity in the molecular expression of the CAM-phenotype (Cushman and Bohnert 1999, 2002; Borland and Taybi 2004; Taybi et al. 2004).

2.3.2 Support of Speciation and Therefore Increase of Diversity [(iii) in Fig. 1]

Dwelling on the plasticity of CAM we reach completely different dimensions when we consider the versatile realization of the various CAM-phases and CAM-modes (CAM idling, CAM cycling) (Sect. 2.1.2) in the perennial woody plants of *Clusia*, especially *C. minor*, as compared to the simple C_3-CAM switch in the annual herbaceous *M. crystallinum* (Sect. 2.3.1). This is plasticity in the broader sense, i.e., not purely developmental but broadly adaptive to environmental conditions. We just need to recall that the two opposite leaves of the same node of *C. minor* in experiments can perform simultaneously C_3-photosynthesis and CAM, respectively, depending on atmospheric moisture in leaf chambers (Sect. 2.1.2). It has never been focused on in the C_3-CAM literature, but it is intriguing that we find the more specific developmental plasticity of *M. crystallinum* and the broad adaptive plasticity of *C. minor*, respectively, in the same metabolic frame of C_3/CAM.

The versatile flexibility equips *C. minor* with a particularly large ecophysiological niche amplitude so that it can occupy a niche width extended from semi-shaded sites of tropical dry forests into fully sun-exposed savannas (Herzog et al. 1999; Lüttge 1999). Possibly the broad niche amplitudes are an important basis of the high degree of evolutionary changes and speciation and therefore high diversity currently apparent in the large neo-tropical tree genus *Clusia* (Lüttge 2005, 2007b; Gustafsson et al. 2007; Vaasen et al. 2007).

Plasticity has its intrinsic costs for plants. Not always and not necessarily plasticity is adaptive. Prima facie expansion of niche width decreases species diversity (Novoplansky 2002). However, ecological amplitudes may also separate populations with reduced sets of genotypes which are specially adapted to particular sites. Such new genetically stable populations are called ecotypes (Kinzel 1982; Turesson 1992). Subsequently to separation of ecotypes, segregation leads to speciation in the path of the wanderings of populations in the space of genotypes (Schuster 1998). Environmental cues are the input to the network built up of genotypes, phenotypes, and ecotypes (Figure 1 in Lüttge 2005), and the relationship between genotype and phenotype only becomes meaningful in environmental context (Schlichting and Smith 2002). As an output of the network type interactions the development of phenotypes from genotypes proves to be the real origin of complexity (Schuster 1998). Conversely, network modulation can promote system stabilization across different environmental conditions (Bertolli et al. 2013, 2014). Under environmental constrains, the network organization of plants, at different scales, can change the level of network connectance (Amzallag 2001; Souza et al. 2005) providing physiological adjustments under different external conditions. In a study with tropical tree species of different forest-successional groups (pioneer and late secondary) grown under contrasting environments (gap and understory), regardless of the successional group, all species showed similar network changes under the same forest conditions. Under gap conditions, the potentially most potential stressful environment, the plants increased the network connectance, which allows fine-tuning adjustments (Souza et al. 2005), supporting plant stability (Souza et al. 2009).

In the Darwinian gradual evolution new species originate from the transformation of populations, i.e., selection replaces existing species by new species and so plasticity can destabilize the existence of individuals and individual species [(ii) in Fig. 1]. Thus, there is no change in diversity. Nevertheless, when plasticity creates different ecotypes via which new species are selected this also results in an increase in diversity [(iii) in Fig. 1]. Diversity always increases in Eldredgian/Gouldian punctual evolution because not the entire existing population and species is merging in the new species (Darwinian) but new species branch off from sub-populations of the existing species and speciation is due to splitting of lineages (Gould 2002). When, as noted above, the development of phenotypes from genotypes proves to be the actual origin of complexity, it is not astonishing that epigenetically based phenotypic variation can accelerate evolutionary speciation (Zhang et al. 2013). Stress-dependent epigenetic changes, e.g., due to methylation/demethylation of epigenetically plastic DNA are inheritable and can be an origin for the evolution of new species [Saze 2008; Verhoeven et al. 2010; for more references in relation to epigenetic memory see Thellier and Lüttge (2013)].

3 Biodiversity: Interspecific Diversity, Species Diversity (Also Implying Complexity)

3.1 Diversity Is the Basis of Complexity

Diversity in general, and specifically biodiversity in the realm of life, is implicating complexity. Interspecific diversity is given with different species as the players. Floristic diversity considers the number of different species occurring in a given area. We can distinguish α-diversity at the level of communities, ß-diversity at the level of ecosystems, and γ-diversity at the level of landscapes comprising several ecosystems (Whitmore 1990; Lüttge 2008a). In addition ecophysiological aspects are introduced in the assessment of diversity of species so that we arrive at functional interspecific diversity, which may also be termed as functional biodiversity. When the various species building up the biodiversity communicate in network pattern-type interactions, it becomes evident that the larger the richness of species, i.e., the higher the interspecific diversity, the higher is the complexity of systems [(vi) in Fig. 1].

3.2 Biodiversity Is Stabilizing (Eco-) Systems [(vii) in Fig. 1]

Although in late successional stages ecosystems may be persistent while diversity declines, in principle diversity is correlated positively with ecosystem stability, especially when species diversity is including functional diversity (McCann 2000; Tilman et al. 2006). The stabilizing function of diversity has been highlighted by Körner (2012): "The diversity of organisms .... is an intrinsic feature of life. .... Uniformity is fatal. There is no future when variation fades."

"The more species, the more different traits, the more functional niches in an ecosystem are explored/occupied, the better are resources utilized and converted into biomass" (Körner 2012). Modeling suggests that diversity of trees in forests increases productivity and that even a single environmental factor, light in this case, can be the decisive cue (Morin et al. 2011). Productivity of grasslands is related to species richness and functional group richness (Hector et al. 1999). Examples given by Körner (2012) are the resources of irradiance and nutrients in grassland ecosystems (meadows). Light transmission and with it light use is better in diverse grasslands (Spehn et al. 2000). Nutrient supply in ecosystems can determine biodiversity of plants. Conversely, biodiversity can determine nutrient availability and acquisition by different species (Richards et al. 2010). Below ground exploration of nutrients, e.g., nitrate, is more exhaustive, i.e., the nutrient efficiency of the entire system is higher, under larger diversity of plant species (Niklaus et al. 2001). In a Malaysian rainforest species diversity of trees and vines was found to be highest on soils with medium combined phosphorus and potassium concentrations and lower at lower and higher P + K indices of the soil, respectively (Tilman 1982). Medium and non-extreme conditions prove to be stabilizing.

A multiplicity of species can cause the so-called portfolio effect named after the stock exchange when a broadly scattered portfolio in which different stocks oscillate non-synchronously prevents extreme losses (or gains). The average performance of an ecosystem remains more stable with the contribution of diverse species with different levels of individual performances (Cottingham et al. 2001). Vegetation and arthropod diversity are correlated and control insect pests (Andow 1991; Altieri 1999). Illustrative examples also come from pollination biology, e.g., from broad studies of the coexistence of plant and pollinator species in meadows. In a reciprocal way a higher diversity of pollinator species contributes to increased pollination success of plants and a higher diversity of flowering plants is a resource for sustaining the species diversity of consumers. Flower diversity and flower-visitor diversity are positively correlated, and this functionally gives a good example of portfolio effects (Fründ et al. 2010; Blüthgen and Klein 2011; Weiner et al. 2011).

In addition to diversity also the existence of significant redundancy (Edelman and Gally 2001) within the network buffers the primary pathways or mechanisms within biological systems against external perturbations. Systems with sufficient redundancy provide robustness in performance even when the system suffers an external disturbance, e.g., via transmission across alternate pathways, providing overall stability to the ensemble system (Amzallag 2001; Edelman and Gally 2001). Indeed, complexity is, somehow, between homogeneity (regularity) and randomness (Mitchell 2009). Theoretically, there is a critical threshold between increasing complexity and gain of stability (Gardner and Ashby 1970). It seems reasonable to consider that in a hypothetical system without any redundancy (e.g., each species with only one individual) the loss of any component can be potentially disruptive, even more so if the component was a hub in the system network.

3.3 *Biodiversity and Relations of Competition and Facilitation*

A prominent example of interrelations between species in ecosystems is mycorrhizae (Teste et al. 2009). However, beyond such mutualism biotic interactions comprise competition and facilitation. Facilitation is given when neighboring species support each other indirectly as a kind of integrated outcome from interactions. Facilitation can reach as far as particular species exerting nurse effects so that other species can get established in their vicinity (Da Silva et al. 1995; Feyera et al. 2002; Grams and Lüttge 2011; Grams 2013). In this way new niches are created and new systems are established where diversity leads to ecological stabilization.

The role of plant–plant interactions, in particular based on the self- and non self-reference capacity, has been recognized as an important factor to establish ecological organizational patterns of distribution and diversity (Novoplansky 2009). Many are the evidences of the abilities of plants to discriminate their neighbors, developing in different ways due to the recognition of individuals of the same species or different species, showing a range of behaviors varying according to the genetic relationships (Kelly 1996; Dudley and File 2007), physiological responses (Gruntman and Novoplansky 2004), and ecotypes (Mahall and Callaway 1996). In the soil, despite the availability of resources to be a major factor determining the behavior of the roots (van Vuuren et al. 1996; Hodge et al. 1998), there are many evidences that the presence of neighboring plants significantly influences the response of roots. Plants are capable of self/non-self reference, with further growth in the presence of foreign plants and reduction of root elongation in the presence of roots of the same plant or plants with same genotype (Caldwell et al. 1996; Falik et al. 2006). The roots are capable of distinguishing "self" (roots of the same plant) of "non-self" (roots of other plants), even when plants are genetically identical, and even without physical contact between the roots (Falik et al. 2003, 2006). Root–root recognition has been evaluated in studies that compared the performance of plants which grew in groups of siblings and non-siblings individuals. The recognition mechanisms in *Miscanthus sinensis*, for example, were entirely genetic: root growth was strongly enhanced by contact with roots belonging to a different genotype, but was significantly inhibited when in contact with roots belonging to the same genotype (de Kroon et al. 2003). The mechanisms involved in self/non-self reference are still unclear, but possibilities include chemical communication through root exudates and release of volatile molecules and enzymes that act on the cell surface. The electrical signaling has also been considered as a possible signaling mechanism (Schenk et al. 1999; Falik et al. 2003).

Extirpation-resistant species among a large diversity of species present in a system can functionally compensate for the loss of species under stress and deterioration (Davies et al. 2012). However, this is not always given. A very interesting feature of biodiversity is that the power of facilitation can be hidden within the diversity and only comes out under sudden extreme stress (Körner 2012). At high diversity among the many species there may be hidden species without any obvious important functions for the ecosystem. However, such species may have capacities to protect the system under suddenly occurring extreme events, such as *Festuca valesiaca* stabilizing gully erosion of grasslands or species-diverse forests under extreme environmental pressures (Scherer-Lorenzen et al. 2005; Caprez et al. 2012; Körner 2012; Huck et al. 2013). Systems fail when because of low diversity such species do not happen to be there. These examples also underline that stability is not a static property of systems as they show that diversity supports stability under dynamic change as a dynamic equilibrium (Hillebrand and Fitter 2013).

Pseudo-steady states in system behavior are not only provided by facilitation but also by competition. With time facilitation can shift to competition (Callaway and Walker 1997). In networks competition and facilitation may interchange and interact; they do not act in isolation (Lin et al. 2012; Callaway 2013). They occur simultaneously within the same community and are difficult to separate from each other (Callaway and Walker 1997; Callaway 1998, 2013; Lin et al. 2012; del Río et al. 2014). The balance between them is the topic of the stress-gradient hypothesis (SGH). Facilitation dominates in harsh and stressful environments with abiotic as well as biotic stress, while competition rules in fertile benign environments under affluence (Bertness and Callaway 1994; Callaway and Walker 1997; Dangles et al. 2013). This spatial view of SGH has been recently extended by introducing the temporal aspect. For mixed-species forests of Central Europe at the stand level (Pretzsch 2013; del Río et al. 2014) show that there are temporal shifts between facilitation and competition, e.g., between different years being "bad" or "good" for productivity. A similar picture is inherent in a meta-analysis of the dynamic tree-grass systems of global savannas (Dohn et al. 2013). Moreover, the SGH requires refinement when different types of facilitation are considered. These reach from completely symmetric, where all plants receive the same degree of benefit, to completely asymmetric, where only the beneficiary plants receive benefit and no advantage is given to the benefactor, with intermediate modes in between (Lin et al. 2012). The nonlinear spatiotemporal stress-gradient effects are based on plasticity and provide a fine illustration of the interactions shown in Fig. 1.

Working with microcosms Grime et al. (1987) showed that high richness of species and stabilization of diversity occurred within a window of stress conditions. At high stress only a few well adapted specialists survive. At low stress only a few very robust and competitive species gain dominance. As we noted above there is stabilization under medium conditions of mineral nutrition, and stabilizing diversity is lost under extreme conditions of low or high resource availability. Only medium stress allows the unfolding of variability and its stabilizing effects. This has led us earlier to suggest that the rules of deterministic chaos may govern species diversity and stability of ecosystems (Lüttge 2008a). At low resource densities, r, systems in deterministic chaos are monotonous, and at high r they become chaotic, while at medium r there appear stable oscillations between different states (Schuster 1995). Here we also may recall Lyapunov stability (see Introduction). The time it takes a system to move from small differences in the initial conditions into chaos is related to an exponent named Lyapunov exponent, λ ($e^{\lambda t}$). This time is equal to $1/\lambda$. In the region of the system with stable oscillations, λ is always positive whereas in the chaotic region it is mostly negative (Schuster 1995).

Ecologically the best examples are tropical rain forests with their extraordinary plant biodiversity and their characteristic multifactorial stress of medium intensity (Lüttge 2004, 2008a; Sect. 2.1.2). These forests are characterized by an oscillating spatiotemporal mosaic pattern of series of states of successions (Remmert 1985, 1991; Whitmore 1990). Oscillatory systems can be regulated by nonstructured completely irregular so-called white noise. When peaks of oscillations remain below a threshold only above which rhythmicity would be overtly expressed, this can be modulated by noise. Nothing happens when the noise is weak. Conversely noise overrides the oscillations when it is too strong. However, at medium intensity noise may just lift the peaks of oscillations above threshold. In this way noise creates the output of overt rhythmicity, or in other words, noise creates order and stability [(iv) and (v) in Fig. 1]. This is called stochastic coherence or stochastic resonance (Hütt and Lüttge 2002, 2007). In this way noise can have stabilizing effects via diversity and complexity.

3.4 *Biodiversity Challenging Ecosystems [(viii) in Fig. 1]*

Biodiversity may also challenge ecosystems. For example disturbance and clearings, e.g., building roads across ecosystems, will provide new and different conditions for plants and therefore increase γ-diversity. Another process of at least temporarily increasing species diversity is the mostly anthropogenic introduction of invasive species or neophytes (Elton 1958; Kowarik 2010). Such species often have high phenotypic plasticity (Williams et al. 1995; Willis et al. 2000; Alpert and Simms 2002). Their arrival first increases species diversity when they coexist with native species, and they also can create new niches. They may have adverse effects on ecosystems not only via competition with native species. Hybridization may affect the gene pool of native species. New species can emerge. Eventually invasive species threaten biodiversity. Existing systems can be destabilized (Klotz 2013).

4 Complexity

The previous sections repeatedly have referred to complexity. Complexity

- is based on biodiversity [(ii) > (iii) > (vi) in Fig. 1],
- is modulated by plasticity [(i) in Fig. 1],
- stabilizes systems [(ix) in Fig. 1].

This is due to the fact that complexity is an intrinsic property of interactions within networks, such as that of Fig. 1 itself and any others. There are alternatives via different pathways of connections of edges and knots in networks. There is no one-sidedness of either positive or negative interactions and in the effects on stability. It is a basic feature of the organization of networks that they always comprise positive and negative feedbacks (Hütt and Lüttge 2005; Souza et al. 2009).

5 Outlook on Practical Implications: Stability and Sustainability

The theoretical considerations presented in this essay on how plasticity, diversity, and complexity support stability have enormous practical implications because stability of systems, in the sense discussed here, is a prerequisite of sustainability. Menace and hope are currently deeply attached to sustainability. Failure and success, respectively, of sustainability determine the fate of mankind on Earth. Hence, the mechanisms supporting stability must be considered and studied in relation to agriculture, forestry, agro-forestry, and other human activities.

Plasticity, diversity, and complexity, all of them are based on integration of modules at different scalar levels which leads to emergence of phenotypes with new functional properties as compared to the properties of the modules they incorporate (Lüttge 2012a, b). The new approach of applying principle component analysis (PCA) to multi-variant functional traits, i.e., the modules, should be applied more broadly. This then will prove the great power of functional PCA for understanding emergent integrated systems at scalar levels of ecology (Sect. 2.2; Vítolo et al. 2012; Bertolli et al. 2013, 2014). It unravels many practical outlooks for applications on environmental management and agro-forest ecology.

The interrelations of plasticity, diversity, and complexity in the establishment of stability (Fig. 1) are setting the scene for projects of conservation (Lüttge 2010b). Plasticity is the basic property conferring flexibility to species, which recently have been named "stem-species" serving the emergent establishment of new ecosystems on deteriorated sites (Lüttge et al. 2012; Scarano and Garbin 2013).

Plasticity of modes of photosynthesis, especially as seen in the C_3/C_4 intermediate species (Sect. 2.1.2), has been considered to open avenues for molecular engineering of crop plants for the higher productivity attained in the C_4-syndrom. The C_3/C_4-intermediate phenotypes as they occur especially in the genus *Flaveria* (Westhoff and Gowik 2010) may help to identify master genes for engineering (see Lüttge 2013).

PCA of multi-variant functional traits should be applied to agro- and forest ecosystems and their combination in agro-forestry. In currently fashionable so-called bio-agriculture, we can distinguish (1) biodynamic agriculture, which is an anthroposophic occultism (Treue 2002); (2) organic farming, which is highly expanding worldwide and acquiring markets for its products but has lower productivity and causes a variety of ecological problems; and, by great contrast to the two former ways, (3) agro-ecology as a serious interdisciplinary approach applying ecological principles to agriculture (Wikipedia 2011, see Lüttge 2013).With the challenge of feeding increasing human populations on Earth functional PCA shall be important for developing approaches of sustainable agro-ecology.

The discussion of ecological agriculture (agro-ecology) traditionally focuses on ecophysiolgical autecology, i.e., ecophysiology at the level of individual (crop-) plants or species. This comprises ecological evaluations of relations of the individual species to resources, such as water, nutrients, CO_2, light, and others. Genotype improvements and engineering also concern the autecological level. Considering the quadruped (Fig. 1) we can ask if one could extend ecological agriculture to ecophysiological synecology, i.e., ecophysiology at the community level (Lüttge and Scarano 2004, 2007; Lüttge 2005). It would involve the terms plasticity and diversity of the quadruped and evaluate the aim of stability (i.e., "sustainability" in eco-agriculture). It would address the question of the role of plasticity in challenges related to the growth differentiation balance theory (GDB) and the stress gradient hypothesis (SGH) as discussed in this essay. It would need to assess the spatiotemporal dynamics of interacting competition and facilitation in species diverse agro-ecosystems.

A major question would be the relation to biodiversity. For a start we might learn from intercropping or cocultivation, e.g., of cereal and legume species (Bedoussac and Justes 2010), or mixed stands of forest trees (Scherrer et al. 2011; Pretzsch 2013; Pretzsch et al. 2013; del Río et al. 2014). Often these only have a "diversity" of two species (Richards et al. 2010). We also may learn much from agro-forestry. Agro-forestry conserves biodiversity in various ways, e.g., by providing habitat structures, by enhancing soil fertility, by reducing erosion, by improving water quality, and by other ecosystem services (McNeely 2004; Jose 2009). Can agro-ecosystems be further diversified? Agro-ecosystems with higher crop diversity are of great interest (Altieri 1999), because species diversity can increase ecosystem services and productivity (Balvanera et al. 2006; Tilman et al. 2006). Can ecological agriculture manage agricultural landscapes with land use mosaics to host increased biodiversity where crop diversity and diversity of adjacent "natural" vegetation interactively add complexity (Altieri 1999)? Stabilizing effects of complexity could foster sustainability of agro-ecosystems (Fig. 1) (Altieri 1999; Scherr and McNeely 2008). Such expectations are *qualitatively* underlined by many observations, such as nutrient cycling, effects on micro-climate including hydrology, pest control, etc. (Altieri 1999). However, is all that feasible in view of the challenge to increase production for feeding increasing humankind (Lüttge 2013)? Ecological agriculture might bring about some immediate or short-term reduction of productivity. However, there could be a profitable long-term return by minimization of degradation and support of agro-ecosystems' stability. Can the stabilizing effects of diversity and complexity in ecological agriculture maintain or increase agricultural output meeting an increased demand for agricultural products (Scherr and McNeely 2008)? *Quantitative* assessments are urgently required.

Acknowledgment We thank Rainer Matyssek, Weihenstephan, Freising, for valuable comments on the manuscript.

References

Alpert P, Simms EL (2002) The relative advantages of plasticity and fixity in different environments: when is it good for a plant to adjust? Evol Ecol 16:285–297
Altieri M (1999) The ecological role of biodiversity in agroecosystems. Agric Ecosyst Environ 74:19–31
Amzallag GN (2001) Data analysis in plant physiology: are we missing the reality? Plant Cell Environ 24:881–890
Andow DA (1991) Vegetational diversity and arthropod population response. Ann Rev Entomol 36:561–586
Balvanera P, Pfisterer AB, Buchmann N, He J-S, Nakashizuka T, Raffaelli D, Schmid B (2006) Quantifying the evidence for biodiversity effects on ecosystem functioning and services. Ecol Lett 9:1146–1156
Barabási A-L (2002) Linked: the new science of networks. Perseus, London
Barabási A-L, Oltvai ZN (2004) Network biology: understanding the cell's functional organization. Nat Rev Genetics 5:101–113
Bedoussac L, Justes E (2010) The efficiency of a durum wheat-winter pea intercrop to improve yield and wheat grain protein concentration depends on N availability during early growth. Plant Soil 330:19–35
Bertness MD, Callaway R (1994) Positive interactions in communities. Trends Ecol Evol 9:191–193
Bertolli SC, Vitolo HF, Souza GM (2013) Network connectance analysis as a tool to understand homeostasis of plants under environmental changes. Plants 2:473–488
Bertolli SC, Mazzafera P, Souza GM (2014) Why is it so difficult to identify a single indicator of water stress in plants? A proposal for a multivariate analysis to assess emergent properties. Plant Biol 16:578–585
Blüthgen N, Klein A-M (2011) Functional complementarity and specialization: the role of biodiversity in plant-pollinator interactions. Basic Appl Ecol 12:282–291

Boardman N (1977) Comparative photosynthesis of sun and shade plants. Annu Rev Plant Physiol 28:355–377
Booy G, Hendriks RJJ, Smulders MJM, van Groenendael JM, Vosman B (2000) Genetic diversity and the survival of populations. Plant Biol 2:379–395
Borland AM, Taybi T (2004) Synchronization of metabolic processes in plants with crassulacean acid metabolism. J Exp Bot 55:1255–1265
Bradshaw AD (1965) Evolutionary significance of phenotypic plasticity in plants. Adv Genet 13:115–155
Brulfert J, Guerrier D, Queiroz O (1973) Photoperiodism and enzyme activity: balance between inhibition and induction of the crassulacean acid metabolism. Plant Physiol 51:220–222
Cahill JF (2013) Plant competition: can understanding trait-behavior linkages offer a new perspective on very old questions? Nova Acta Leopoldina 114(391):115–125
Caldwell MM, Manwaring JH, Durham SL (1996) Species interactions at the level of fine roots in the field: influence of soil nutrient heterogeneity and plant size. Oecologia 106:440–447
Callaway RM (1998) Competition and facilitation on elevation gradients in subalpine forests of the northern Rocky Mountains, USA. Oikos 82:561–573
Callaway RM (2013) Facilitation, competition and the organization of plant communities. Nova Acta Leopoldina 114(391):147–157
Callaway RM, Walker LR (1997) Competition and facilitation: a synthetic approach to interactions in plant communities. Ecology 78:1958–1965
Camazine S, Deneubourg J-L, Franks NR, Sneyd J, Theraulaz G, Bonabeau E (2001) Self-organization in biological systems. Princeton University Press, Princeton, NJ
Cannon WB (1932) The wisdom of the body. Norton, New York, NY
Caprez R, Spehn E, Nakhutrishvili G, Körner C (2012) Drought at erosion edges selects for a "hidden" keystonae species. Plant Ecol Divers. doi:10.1080/17550874.2011.600343
Cottingham KL, Brown BL, Lennon JT (2001) Biodiversity may regulate the temporal variability of ecological systems. Ecol Lett 4:72–85
Csermely P (2006) Weak links: stabilizers of complex systems from proteins to social networks. Springer, Berlin
Cushman JC, Bohnert HJ (1999) Crassulacean acid metabolism: molecular genetics. Annu Rev Plant Physiol Plant Mol Biol 50:305–332
Cushman JC, Bohnert HJ (2002) Induction of crassulacean acid metabolism by salinity – molecular aspects. In: Läuchli A, Lüttge U (eds) Salinity: environment – plants – molecules. Kluwer, Dordrecht, pp 362–393
Da Silva MC, Scarano FR, De Souza CF (1995) Regeneration of an Atlantic forest formation in the understory of a *Eucalyptus grandis* plantation in south-eastern Brazil. J Trop Ecol 11:147–152
Dangles O, Herrera M, Anthelme F (2013) Experimental support of the stress-gradient hypothesis in herbivore-herbivore interactions. New Phytol 197:405–408
Davies TW, Jenkins SR, Kingham R, Hawkins SJ, Hiddink JG (2012) Extirpation-resistant species do not always compensate for the decline in ecosystem processes associated with biodiversity loss. J Ecol 1000:1475–1481
de Kroon H, Mommer L, Nishiwaki A (2003) Root competition: towards a mechanistic understanding. In: de Kroon H, Visser EJW (eds) Root ecology. Springer, Berlin
de Kroon H, Huber H, Stuefer JF, van Groenendael JM (2005) A modular concept of phenotypic plasticity in plants. New Phytol 166:73–82
del Río M, Schütze G, Pretzsch H (2014) Temporal variation of competition and facilitation in mixed species forests in Central Europe. Plant Biol 16:166–176
DeWitt TJ, Scheiner SM (2004) Phenotypic plasticity: functional and conceptual approaches. Oxford University Press, New York, NY
Dohn J, Dembélé F, Karembé M, Moustakas A, Amévor KA, Hanan N (2013) Tree effects of grass growth in savannas: competition, facilitation and the stress-gradient hypothesis. J Ecol 101:202–209
Dudley SA, File AL (2007) Kin recognition in an annual plant. Biol Lett 3:435–438

Edelman GM, Gally JA (2001) Degeneracy and complexity in biological systems. Proc Natl Acad Sci USA 98:13763–13768

Elton CS (1958) The ecology of invasions by animals and plants. The University of Chicago Press, Chicago

Falik O, de Kroon H, Novoplansky A (2003) Self/non-self discrimination in roots. J Ecol 91: 525–531

Falik O, de Kroon H, Novoplansky A (2006) Physiologically mediated self/non-self root discrimination in: *Trifolium repens* has mixed effects on plant performance. Plant Signal Behav 1:116–121

Fetene M, Lee HSJ, Lüttge U (1990) Photosynthetic acclimation in a terrestrial CAM bromeliad, *Bromelia humilis* Jacq. New Phytol 114:399–406

Feyera S, Beck E, Lüttge U (2002) Exotic trees as nurse-trees for the regeneration of natural tropical forests. Trees 16:245–249

Fründ J, Linsenmair KE, Blüthgen N (2010) Pollinator diversity and specialization in relation to flower diversity. Oikos 119:1581–1590

Gardner MR, Ashby R (1970) Connectance of large dynamic (cybernetic) systems: critical values for stability. Nature 228:784

Gayler S, Priesack E (2013) Carbohydrate allocation to growth and defense-related metabolism – a modeling approach at the whole-plant level. Nova Acta Leopoldina 114(391):99–114

Gehrig H, Gaussmann O, Marx H, Schwarzott D, Kluge M (2001) Molecular phylogeny of the genus *Kalanchoë* (Crassulaceae) inferred from the nucleotide sequences of the IST-1 and IST-2 regions. Plant Sci 160:827–835

Gould JG (2002) The structure of evolutionary theory. Harvard University Press, Cambridge, MA

Grams TEE (2013) A space-related perspective on plant-plant competition. Nova Acta Leopoldina 114(391):127–134

Grams TEE, Lüttge U (2011) Space as a resource. Prog Bot 72:349–370

Grams TEE, Daigo MJ, Winkler JB, Gayler S, Matyssek R (2012) Growth and space use in competitive interactions between juvenile trees. In: Matyssek R, Schnyder H, Oßwald W, Ernst D, Munch JC, Pretzsch H (eds) Growth and defence in plants. Resource allocation at multiple scales. Springer, Heidelberg, pp 273–286

Grime JP, Crick JC, Rincon JE (1986) The ecological significance of plasticity. In: Jennings DH, Trewavas AJ (eds) Plasticity in plants. Cambridge University Press, Cambridge, pp 5–29

Grime JP, Mackey JML, Hillier SH, Read DJ (1987) Floristic diversity in a model system using experimental microcosms. Nature 328:420–422

Grimm V, Wissel C (1997) Babel, or the ecological stability discussions: an inventory and analysis of terminology and a guide for avoiding confusion. Oecologia 109:323–334

Gruntman M, Novoplansky A (2004) Physiologically mediated self/non-self discrimination in roots. Proc Natl Acad Sci USA 101:3863–3867

Guralnick LJ, Jackson MD (2001) The occurrence and phylogenetics of Crassulacean acid metabolism in the Portulacaceae. Int J Plant Sci 162:257–262

Guralnick LJ, Ting IP, Lord EM (1986) Crassulacean acid metabolism in the Gesneriaceae. Am J Bot 53:336–345

Guralnick LJ, Edwards G, Ku MSB, Hockema B, Franceschi VR (2002) Photosynthetic and anatomical characteristics in the C_4-crassulacean acid metabolism-cycling plant, *Portulca grandiflora*. Funct Plant Biol 29:763–773

Gustafsson MHG, Winter K, Bittrich V (2007) Diversity, phylogeny and classification of *Clusia*. In: Lüttge U (ed) *Clusia*. A woody neotropical genus of remarkable plasticity and diversity, vol 194, Ecological studies. Springer, Berlin, pp 95–116

Hatch MD (1987) C_4 photosynthesis: a unique blend of modified biochemistry, anatomy and ultrastructure. Biochim Biophys Acta 895:81–106

Hatch MD, Osmond CB (1976) Compartmentation and transport in C_4 photosynthesis. In: Stocking CR, Heber U (eds) Transport in plants III. Intracellular interactions and transport processes, vol 3, Encyclopedia of plant physiology NS. Springer, Berlin, pp 144–184

Hector A, Schmid B, Beierkuhnlein C, Caldeira MC, Diemer M, Dimitrakopoulus PG, Finn JA, Freitas H, Giller PS, Good J, Harris R, Högberg P, Huss-Danell K, Joshi J, Jumpponen A, Körner C, Leadley PW, Loreau M, Minns A, Mulder CPH, O'Donovan G, Otway SJ, Pereira JS, Prinz A, Read DJ, Scherer-Lorenzen M, Schulze E-D, Siamantziouras A-SD, Spehn EM, Terry AC, Troumbis AY, Woodward FI, Yachi S, Lawton JH (1999) Plant diversity and productivity experiments in European grasslands. Science 286:1123–1127

Herms DA, Mattson WJ (1992) The dilemma of plants: to grow or defend. Q Rev Biol 67:283–335

Herrera A (2009) Crassulacean acid metabolism and fitness under water deficit stress: if not for carbon gain, what is facultative CAM good for? Ann Bot 103:645–653

Herzog B, Hoffmann S, Hartung W, Lüttge U (1999) Comparison of photosynthetic responses of the sympatric tropical C_3-species *Clusia multiflora* H.B.K. and the C_3-CAM intermediate species *Clusia minor* L. to irradiance and drought stress in a phytotron. Plant Biol 1:460–470

Hillebrand H, Fitter A (2013) Biodiversität und Stabilität. Neue Erkenntnisse zu einem ökologischen Paradigma. In: Beck E (ed) Die Vielfalt des Lebens. Wiley-VCH, Weinheim, pp 111–181

Hodge A (2009) Root decisions. Plant Cell Environ 32:628–640

Hodge A (2010) Roots: the acquisition of water and nutrients from the heterogeneous soil environment. Prog Bot 71:307–337

Hodge A, Stewart J, Robinson D, Griffiths BS, Fitter AH (1998) Root proliferation, soil fauna and plant nitrogen capture from nutrient-rich patches in soil. New Phytol 139:479–494

Huck C, Körner C, Hitbrunner E (2013) Plant species dominance shifts across erosion edge-meadow grassland soils. Oecologia 171:693–703

Huerta AJ, Ting IP (1988) Effects of various levels of CO_2 on the induction of crassulacean acid metabolism in *Portulacaria afra* (L.) Jacq. Plant Physiol 88:183–188

Hütt M-T, Lüttge U (2002) Nonlinear dynamics as a tool for data analysis and modeling in plant physiology. Plant Biol 4:281–297

Hütt M-T, Lüttge U (2005) Network dynamics in plant biology: current progress in historical perspective. Prog Bot 66:277–310

Hütt M-T, Lüttge U (2007) Noise-induced phenomena and complex rhythms: theoretical considerations, modeling and experimental evidence. In: Manusco S, Shabala S (eds) Rhythms in plants: phenomenology, mechanisms, and adaptive significance. Springer, Berlin, pp 313–339

Jose S (2009) Agroforestry for ecosystem services and environmental benefits: an overview. Agroforestry Syst 76:1–10

Kelly JK (1996) Kin selection in the annual plant *Impatiens capensis*. Am Naturalist 147:899–918

Kinzel H (1982) Pflanzenökologie und Mineralstoffwechsel. Ulmer, Stuttgart

Klotz S (2013) Globaler Austausch, eine neue Herausforderung: Biologische Invasionen – Gefahr im Verzug? In: Beck E (ed) Die Vielfalt des Lebens. Wiley-VCH, Weinheim, pp 191–201

Kluge M, Razanoelisoa B, Brulfert J (2001) Implications of genotypic diversity and phenotypic plasticity in the ecophysiological success of CAM plants, examined by studies on the vegetation of Madagascar. Plant Biol 3:214–222

Koch K, Kennedy RA (1980) Characteristics of Crassulacean acid metabolism in the succulent C_4 dicot, *Portulaca oleracea* L. Plant Physiol 65:193–197

Koch K, Kennedy RA (1982) Crassulacean acid metabolism in the succulent C_4 dicot, *Portulaca oleracea* L. under natural environmental conditions. Plant Physiol 69:757–761

Kolosova N, Bohlmann J (2012) Conifer defense against insects and fungal pathogens. In: Matyssek R, Schnyder H, Oßwald W, Ernst D, Munch JC, Pretzsch H (eds) Growth and defence in plants. Resource allocation at multiple scales, vol 220, Ecological studies. Springer, Heidelberg, pp 85–109

Körner C (2012) Biological diversity – the essence of life and ecosystem functioning. Nova Acta Leopoldina 116(394):147–159

Körner C (2013) Growth controls photosynthesis – mostly. Nova Acta Leopoldina 114(391):273–283

Kowarik I (2010) Biologische Invasionen, Neophyten und Neozoen in Mitteleuropa, 2nd edn. Ulmer, Stuttgart

Kraybill AA, Martin CE (1996) Crassulacean acid metabolism in three species of the genus *Portulaca*. Int J Plant Sci 157:103–109

Ku S-B, Shieh YJ, Reger BJ, Black CC (1981) Photosynthetic characteristics of *Portulaca grandiflora* a succulent C_4 dicot. Cellular compartmentation of enzymes and acid metabolism. Plant Physiol 68:1073–1080

Küppers M (1989) Ecological significance of above-ground architectural patterns in woody plants – a question of cost-benefit relationships. Trends Ecol Evol 4:375–379

Larcher W (2001) Ökophysiologie der Pflanzen, 6th edn. Ulmer, Stuttgart

Lee HSJ, Griffiths H (1987) Induction and repression of CAM in *Sedum telephium* L. in response to photoperiod and water stress. J Exp Bot 38:834–841

Lee HSJ, Lüttge U, Medina E, Smith JAC, Cram WJ, Diaz M, Griffiths H, Popp M, Schäfer C, Stimmel K-H, Thonke B (1989) Ecophysiology of xerophytic and halophytic vegetation of a coastal alluvial plain in northern Venezuela. III. *Bromelia humilis* Jacq., a terrestrial CAM bromeliad. New Phytol 111:253–271

Lichtenthaler HK (2014) Fifty-five years of research on photosynthesis, chloroplasts and stress physiology of plants: 1958–2013. In: Lüttge U, Beyschlag W (eds) Progress in botany: genetics – physiology – systematics – ecology, vol 76. Springer, Heidelberg

Lichtenthaler HK, Ac A, Marek MV, Kalina J, Urban O (2007) Differences in pigment composition, photosynthetic rates and chlorophyll fluorescence images of sun and shade leaves of four tree species. Plant Physiol Biochem 45:577–588

Lin Y, Berger U, Grimm V, Ji Q-R (2012) Differences between symmetric and asymmetric facilitation matter: exploring the interplay between modes of positive and negative plant interactions. J Ecol 100:1482–1491

Loomis WE (1953) Growth and differentiation – an introduction and summary. In: Loomis WE (ed) Growth and differentiation in plants. Iowa State College Press, Ames, IA, pp 1–17

Lucas M, Laplaze L, Bennett MJ (2011) Plant systems biology: network matters. Plant Cell Environ 34:535–553

Lundegårdh H (1950) The translocation of salts and water through wheat roots. Plant Physiol 2:103–151

Lundegårdh H (1955) Mechanisms of absorption, transport, accumulation, and secretion of ions. Ann Rev Plant Physiol 6:1–24

Lundegårdh H, Burström H (1933) Untersuchungen über die Salzaufnahme der Pflanzen. III. Quantitative Beziehungen zwischen Atmung und Anionenaufnahme. Biochem Z 261:235–251

Lundeghårdh H, Burström H (1935) Untersuchungen über die Atmungsvorgänge in Pflanzenwurzeln. Biochem Z 277:223–249

Lüttge U (1995a) Ecophysiological basis of the diversity of tropical plants: the example of the genus *Clusia*. In: Heinen HD, San José JJ, Caballero-Arias H (eds) Nature and human ecology in the neotropics, vol 5, Scientia Guaianae, pp 23–36

Lüttge U (1995b) *Clusia*: Ein Modellfall der ökophysiologischen Plastizität in einer tropischen Gattung. In: Rundgespräche der Kommission für Ökologie der BAdW. 10: 173–186. Bayerische Tropenforschung einst und jetzt. Dr. Pfeil, München

Lüttge U (1999) One morphotype, three physiotypes: sympatric species of *Clusia* with obligate C_3-photosynthesis, obligate CAM and C_3-CAM intermediate behaviour. Plant Biol 1:138–148

Lüttge U (2000) Photosynthese-Physiotypen unter gleichen Morphotypen, Species und bei Klonen: Kann ökophysiologische Plastizität zur Entstehung von Diversität beitragen? Ber Reinhold Tüxen Ges 12:319–334

Lüttge U (2004) Ecophysiology of crassulacean acid metabolism (CAM). Ann Bot 93:629–652

Lüttge U (2005) Genotypes – phenotypes – ecotypes: relations to crassulacean acid metabolism. Nova Acta Leopoldina 92(342):177–193

Lüttge U (2006) Photosynthetic flexibility and ecophysiological plasticity: questions and lessons from *Clusia*, the only CAM tree, in the tropics. New Phytol 171:7–25

Lüttge U (2007a) *Clusia*: a woody neotropical genus of remarkable plasticity and diversity, vol 194, Ecological studies. Springer, Berlin
Lüttge U (2007b) Photosynthesis. In: Lüttge U (ed) *Clusia*. A woody neotropical genus of remarkable plasticity and diversity, vol 194, Ecological studies. Springer, Berlin, pp 135–186
Lüttge U (2007c) Physiological ecology. In: Lüttge U (ed) *Clusia*. A woody neotropical genus of remarkable plasticity and diversity, vol 194, Ecological studies. Springer, Berlin, pp 187–234
Lüttge U (2008a) Physiological ecology of tropical plants, 2nd edn. Springer, Heidelberg
Lüttge U (2008b) *Clusia*: holy grail and enigma. J Exp Bot 59:1503–1514
Lüttge U (2010a) Ability of crassulacean acid metabolism plants to overcome interacting stresses in tropical environments. AoB Plants 2010. doi:10.1093/aobpla/plq005
Lüttge U (2010b) Plasticity and conservation. Natureza Conservãço 8:120–126
Lüttge U (2012a) Whole-plant physiology: synergistic emergence rather than modularity. Prog Bot 74:165–190
Lüttge U (2012b) Modularity and emergence: biology's challenge in understanding life. Plant Biol 14:865–871
Lüttge U (2013) The planet Earth: can it feed nine billion people. Nova Acta Leopoldina 114 (391):345–364
Lüttge U, Higinbotham N (1979) Transport in plants. Springer, New York, NY
Lüttge U, Scarano FR (2004) Ecophysiology. Rev Bras Bot 27:1–10
Lüttge U, Scarano FR (2007) Synecological comparisons sustained by ecophysiological finger-printing of intrinsic photosynthetic capacity of plants as assessed by measurements of light response curves. Braz J Bot 30:355–364
Lüttge U, Garbin ML, Scarano FR (2012) Evo-devo-eco and ecological stem species: potential repair systems in the planetary biosphere crisis. Prog Bot 74:191–212
MacRobbie EAC (1965) The nature of the coupling between light energy and active ion transport in *Nitella translucens*. Biochim Biophys Acta 94:64–73
Mahall BE, Callaway RM (1996) Effects of regional origin and genotype on intraspecific root communication in the desert shrub *Ambrosia dumosa* (Asteraceae). Am J Bot 83:83–98
Matyssek R, Schnyder H, Elstner E-F, Munch J-C, Pretzsch H, Sandermann H (2002) Growth and parasite defence in plants: the balance between resource sequestration and retention. Plant Biol 4:133–136
Matyssek R, Agerer R, Ernst D, Munch J-C, Oßwald W, Pretzsch H, Priesack E, Schnyder H, Treutter D (2005) The plant's capacity in regulating resource demand. Plant Biol 7:560–580
Matyssek R, Gayler S, zu Castell W, Oßwald W, Ernst D, Pretzsch H, Schnyder H, Munch JC (2012a) Predictability of plant resource allocation: new theory needed? In: Matyssek R, Schnyder H, Oßwald W, Ernst D, Munch JC, Pretzsch H (eds) Growth and defence in plants. Resource allocation at multiple scales, vol 220, Ecological studies. Springer, Heidelberg, pp 433–449
Matyssek R, Koricheva J, Schnyder H, Ernst D, Munch JC, Oßwald W, Pretzsch H (2012b) The balance between resource sequestration and retention: a challenge in plant science. In: Matyssek R, Schnyder H, Oßwald W, Ernst D, Munch JC, Pretzsch H (eds) Growth and defence in plants. Resource allocation at multiple scales, vol 220, Ecological studies. Springer, Heidelberg, pp 3–24
Matyssek R, Schnyder H, Oßwald W, Ernst D, Munch JC, Pretzsch H (eds) (2012c) Growth and defence in plants. Resource allocation at multiple scales, vol 220, Ecological studies. Springer, Heidelberg
Matyssek R, Lüttge U, Rennenberg H (eds) (2013) The alternatives growth and defense: resource allocation at multiple scales in plants. Nova Acta Leopoldina 114(391):1–369
Maxwell K (2002) Resistance is useful: diurnal patterns of photosynthesis in C_3 and crassulacean acid metabolism epiphytic bromeliads. Funct Plant Biol 29:679–687
May RM, Lloyd AL (2001) Infection dynamics on scale-free networks. Phys Rev E 64(6):066112

Mazen AMA (1996) Changes in levels of phosphoenolpyruvte carboxylase with induction of crassulacean acid metabolism (CAM)-like behavior in the C_4 plant *Portulaca oleracea*. Physiol Plant 98:111–116

McCann KS (2000) The diversity – stability debate. Nature 405:228–233

McGarigal K, Cushman S, Stafford SG (2000) Multivariate statistics for wildlife ecology research. Springer, New York, NY

McNeely JA (2004) Nature vs. nurture: managing relationships between forests, agroforestry and wild biodiversity. Agroforestry Syst 61:155–166

Mitchell E (2009) Complexity: a guided tour. Oxford University Press, New York, NY

Moore BD, Franceschi VR, Cheng S-H, Wu J, Ku MSB (1987) Photosynthetic characteristics of the C_3-C_4 intermediate *Parthenium hysterophorus*. Plant Physiol 85:984–989

Morin X, Fahse L, Scherer-Lorenzen M, Bugmann H (2011) Tree species richness promotes productivity in temperate forests through strong complementarity between forests. Ecol Lett 14:1211–1219

Nicolis G, Prigogine I (1977) Self-organization in nonequilibrium systems. Wiley, New York, NY

Niklaus PA, Kandeler E, Leadley PW, Schmid B, Tscherko D, Körner C (2001) A link between plant diversity, elevated CO_2 and soil nitrate. Oecologia 127:540–548

Nishio JN, Ting IP (1993) Photosynthetic characteristics of the palisade mesophyll and spongy mesophyll in the CAM/C_4 intermediate plant *Peperomia camptotricha*. Bot Acta 106:120–125

Novikoff AB (1945) The concept of integrative levels and biology. Science 101:209–215

Novoplansky A (2002) Developmental plasticity in plants: implications of non-cognitive behavior. Evol Ecol 16:177–188

Novoplansky A (2009) Picking battles wisely: plant behaviour under competition. Plant Cell Environ 32:726–741

Osmond CB (1978) Crassulacean acid metabolism: a curiosity in context. Ann Rev Plant Physiol 29:379–414

Osmond CB, Ziegler H (1975) Schwere Pflanzen und leichte Pflanzen: Stabile Isotope im Photosynthesestoffwechsel und in der biochemischen Ökologie. Naturw Rundsch 28:323–328

Oßwald W, Fleischmann F, Treutter D (2012) Host-parasite interactions and trade-offs between growth- and defence-related metabolism under changing environments. In: Matyssek R, Schnyder H, Oßwald W, Ernst D, Munch JC, Pretzsch H (eds) Growth and defence in plants. Resource allocation at multiple scales, vol 220, Ecological studies. Springer, Heidelberg, pp 53–83

Peisker M (1986) Models of carbon metabolism in C_3-C_4 intermediate plants as applied to the evolution of C_4 photosynthesis. Plant Cell Environ 9:627–635

Pretzsch H (2013) Facilitation and competition in mixed-species forests analyzed along an ecological gradient. Nova Acta Leopoldina 114(391):159–174

Pretzsch H, Schütze G, Uhl E (2013) Resistance of European tree species to drought stress in mixed *versus* pure forests: evidence of stress release by inter-specific facilitation. Plant Biol 15:483–495

Proulx SR, Promislow DEL, Phillips PC (2005) Network thinking in ecology and evolution. Trends Ecol Evol 20:345–353

Ratajczak R, Richter J, Lüttge U (1994) Adaptation of the tonoplast V-type H^+-ATPase of *Mesembryanthemum crystallinum* to salt stress, C_3-CAM transition and plant age. Plant Cell Environ 17:1101–1112

Ravasz E, Somera AL, Mongru DA, Oltvai ZN, Barabási AL (2002) Hierarchical organization of modularity in metabolic networks. Science 297:1551–1555

Raven JA (1967) Light stimulation of active transport in *Hydrodictyon africanum*. J Gen Physiol 50:1627–1640

Rawsthorne S (1992) C_3-C_4 intermediate photosynthesis: linking physiology to gene expression. Plant J 2:267–274

Remmert H (1985) Was geschieht im Klimax-Stadium? Ökologisches Gleichgewicht durch Mosaik aus desynchronen Zyklen. Naturw 72:505–512

Remmert H (1991) The mosaic cycle of ecosystems, vol 85, Ecological studies. Springer, Berlin
Richards AE, Forrester DI, Bauhus J, Scherer-Lorenzen M (2010) The influence of mixed tree plantations on the nutrition of individual species: a review. Tree Physiol 30:1192–1208
Rodrigues MA, Freschi L, Pereira PN, Mercier H (2014) Interactions between nutrients and crassulacean acid metabolism. Prog Bot 75:167–186
Sage RF (2004) Evolution of C_4 photosynthesis. New Phytol 161:341–370
Sage RF, McKown AD (2006) Is C4 photosynthesis less phenotypically plastic than C3 photosynthesis? J Exp Bot 57:303–317
Saze H (2008) Epigenetic memory transmission through mitosis and meiosis in plants. Semin Cell Dev Biol 19:527–536
Scarano FR, Garbin ML (2013) Stem species: plant species that function as regenerating cells of Gaia. Nova Acta Leopoldina 114(391):317–324
Schenk HJ (2006) Root competition beyond resource depletion. J Ecol 94:725–739
Schenk HJ, Callaway RM, Mahall BE (1999) Spatial root segregation: are plants territorial? Adv Ecol Res 28:145–180
Scherer-Lorenzen M, Körner C, Schulze ED (eds) (2005) Forest diversity and function. Temperate and boreal systems, vol 176, Ecological studies. Springer, Berlin
Scherr SJ, McNeely JA (2008) Biodiversity conservation and agricultural sustainability: towards a new paradigm of 'ecoagriculture' landscapes. Phil Trans Roy Soc B Biol Sci 363:477–494
Scherrer D, Bader MKF, Körner C (2011) Drought-sensitivity ranking of deciduous tree species based on thermal imaging of forest canopies. Agric For Meteorol 151:1632–1640
Schlichting CD, Smith H (2002) Phenotypic plasticity: linking molecular mechanisms with evolutionary outcomes. Evol Ecol 16:189–211
Schmitt A, Lee HSJ, Lüttge U (1988) The response of the C_3-CAM tree, *Clusia rosea*, to light and water stress. J Exp Bot 39:1581–1590
Schneider ED, Kay JJ (1994) Life as a manifestation of the second law of thermodynamics. Mathematical Computation Modeling 19:25–48
Schuster HG (1995) Deterministic chaos. VCH, Weinheim
Schuster P (1998) Evolution in molekularer Auflösung. Ber Abh Berlin-Brandenb Akad Wiss 6:187–215
Smith JAC, Lüttge U (1985) Day-night changes in leaf water relations associated with the rhythm of crassulacean acid metabolism in *Kalanchoë daigremontiana*. Planta 163:272–282
Solbrig OT (1994) Plant traits and adaptive strategies: their role in ecosystem function. In: Schulze E-D, Mooney HA (eds) Biodiversity and ecosystem function, vol 99, Ecological studies. Springer, Berlin, pp 97–116
Solé RV, Bascompte J (2006) Self-organization in complex ecosystems. Princeton University Press, Princeton, NJ
Souza GM, de Oliveira RF, Cardoso VJM (2004) Temporal dynamics of stomatal conductance of plants under water deficit: can homeostasis be improved by more complex dynamics. Arquivos de Biologia e Tecnologia, Curitiba 47:423–431
Souza GM, Pincus SM, Monteiro JAF (2005) The complexity-stability hypothesis in plant gas exchange under water deficit. Braz J Plant Physiol 17:363–373
Souza GM, Ribeiro RV, Prado CHBS, Damineli DSC, Sato AM, Oliveira MS (2009) Using network connectance and autonomy analyses to uncover patterns of photosynthetic responses in tropical woody species. Ecological Complexity 6:15–26
Spehn EM, Joshi J, Schmid B, Diemer M, Körner C (2000) Above-ground resource use increases with plant species richness in experimental grassland ecosystems. Funct Ecol 14:326–337
Taybi T, Nimmo HG, Borland AM (2004) Expression of phospho*enol*pyruvate carboxylase and phospho*enol*pyruvate carboxylase kinase genes. Implications for genotypic capacity and phenotypic plasticity in the expression of crassulacean acid metabolism. Plant Physiol 135:587–598

Teste FP, Simard SW, Durall DM, Guy RD, Jones MD, Schoonmaker AL (2009) Access to mycorrhizal networks and roots of trees: importance of seedling survival and resource transfer. Ecology 90:2808–2822
Thellier M, Lüttge U (2013) Plant memory: a tentative model. Plant Biol 15:1–12
Tilman D (1982) Resource competition and community structure. Princeton University Press, Princeton, NJ
Tilman D, Reich PB, Knops JMH (2006) Biodiversity and ecosystem stability in a decade-long grassland experiment. Nature 441:629–632
Treue P (2002) Blut und Bohnen. Der Paradigmenwechsel im Künast-Ministerium ersetzt Wissenschaft durch Okkultismus. Frankfurter Allgemeine Zeitung, 13–03–2002, Nr 61, p 12
Turesson G (1992) The genotypic response of the plant species to the habitat. Hereditas 3:211–350
Ueno O (2011) Structural and biochemical characterization of the C_3-C_4 intermediate *Brassica gravinae* and relatives, with particular reference to cellular distribution of Rubisco. J Exp Bot 62:5347–5355
Vaasen A, Scarano FR, Hampp R (2007) Population biology of different *Clusia* species in the state of Rio de Janeiro. In: Lüttge U (ed) *Clusia*. A woody neotropical genus of remarkable plasticity and diversity, vol 194, Ecological studies. Springer, Berlin, pp 117–127
van Vuuren MMI, Robinson D, Griffiths BS (1996) Nutrient inflow and root proliferation during the exploitation of a temporally and spatially discrete source of nitrogen in soil. Plant Soil 178:185–192
Verhoeven KJF, Jansen JJ, van Dijk PJ, Biere A (2010) Stress-induced DNA methylation changes and their heritability in asexual dandelions. New Phytol 185:1108–1118
Vítolo HF, Souza GM, Silveira J (2012) Cross-scale multivariate analysis of physiological responses to high temperature in two tropical crops with C3 and C4 metabolism. Environ Exp Bot 80:54–62
Wagner A (2005) Robustness and evolvability in living systems. Princeton University Press, Princeton, NJ
Weiner CN, Wemer M, Linsenmair KE, Blüthgen N (2011) Land use intensity in grasslands: changes in biodiversity, species composition and specialisation in flower visitor networks. Basic Appl Ecol 12:292–299
West-Eberhard MJ (1986) Alternative adaptations, speciation, and phylogeny (a review). Proc Natl Acad Sci USA 83:1388–1392
West-Eberhard MJ (1989) Phenotypic plasticity and origins of diversity. Ann Rev Ecol Syst 20:249–278
West-Eberhard MJ (2003) Developmental plasticity and evolution. Oxford University Press, Oxford
Westhoff P, Gowik U (2010) Evolution of C_4 photosynthesis – looking for the master switch. Plant Physiol 154:598–601
Whitmore TC (1990) An introduction to tropical rainforests. Oxford University Press, Oxford
Wikipedia (2011) http://en.wikipedia.org/wiki/Agroecology
Williams DG, Mack RN, Black RA (1995) Ecophysiology of introduced *Pennisetum setaceum* on Hawaii – the role of phenotypic plasticity. Ecology 76:1569–1580
Willis AJ, Memmott J, Forrester RI (2000) Is there evidence for the post-invasion evolution of increased size among plant species. Ecol Lett 3:275–283
Winter K, Holtum JAM (2005) The effects of salinity, crassulacean acid metabolism and plant age on the carbon isotope composition of *Mesembryanthemum crystallinum* L., a halophytic C_3-CAM species. Planta 222:201–209
Winter K, Holtum JAM (2007) Environment or development? Lifetime net CO_2 exchange and control of the expression of crassulacean acid metabolism in *Mesembryanthemum crystallinum*. Plant Physiol 143:98–107
Winter K, von Willert DJ (1972) NaCl-induzierter Crassulaceensäurestoffwechsel bei *Mesembryanthemum crystallinum*. Z Pflanzenphysiol 67:166–170
Zhang Y-Y, Fischer M, Colot V, Bosdorf O (2013) Epigenetic variation creates potential for evolution of plant phenotypic plasticity. New Phytol 197:314–322

The Proposed Anti-herbivory Roles of White Leaf Variegation

Simcha Lev-Yadun

Contents

Abstract It has been suggested that white variegation, the outcome of various developmental, genetic, and physiological processes, may defend leaves and other plant organs from herbivory by several proposed mechanisms: camouflage, aposematism (including Müllerian and Batesian mimicry), mimicry of insect damage and fungal attacks, dazzle effects that make it hard for large herbivores to decide where to bite the leaves and for insects to land on them, and by visual repellence of insects from landing as well as by unknown mechanisms. Very few cases of these suggested leaf defenses by variegation have been examined in depth. Some such studied cases were indeed found to actually operate as defense from

S. Lev-Yadun (✉)
Department of Biology and Environment, Faculty of Natural Sciences, University of Haifa–Oranim, Tivon 36006, Israel
e-mail: levyadun@research.haifa.ac.il

U. Lüttge, W. Beyschlag (eds.), *Progress in Botany*, Progress in Botany 76,
DOI 10.1007/978-3-319-08807-5_10

herbivory either in nature or in experiments, suggesting the potential defensive function of others. However, the specific operating defensive mechanism by white variegation was not always identified or even proposed, even when variegation was found to be associated with reduced herbivory. Studying white variegation has a significant advantage over studying other types of plant defensive coloration because even bi-chromatic vision is sufficient to see these patterns. Moreover, white variegation is probably visible under most types of natural light conditions, including strong moonlight. While in this essay I wish to stimulate an effort for a broader and deeper understanding of the defensive roles of white variegation, the possible simultaneous physiological roles of white leaf variegation that will not be reviewed here should not be ignored.

1 Introduction

Defensive plant coloration (camouflage, aposematism, mimicry, undermining herbivorous insect camouflage, exploiting animals' perceptual biases, various types of signaling) has received very limited attention (e.g., Wiens 1978; Rothschild 1986; Givnish 1990; Lev-Yadun 2001, 2006a, 2009a; Lev-Yadun et al. 2004a; Archetti et al. 2009; Schaefer and Ruxton 2011) compared to defensive animal coloration (e.g., Cott 1940; Kettlewell 1973; Edmunds 1974; Majerus 1998; Ruxton et al. 2004; Caro 2005). Harper (1977) in his seminal book about plant demography wrote concerning his view of the potential defensive role of leaf variegation: "botanists were reluctant to accept things that are commonplace for zoologists and often seem reluctant to see the animal as a powerful selective force in plant evolution except in the curiously acceptable realm of adaptation to pollination! It may be that much of the fantastic variation in leaf form, variegation, dissection, and marking that is known in the plant kingdom is accounted for by the selective advantage to the plant of associating nonpalatability with a visual symbol." This, however, should be changed, and this essay on the specific issue of white variegation is a continuation of the recent efforts to bridge this gap, reviewed in Lev-Yadun (2006a, 2009a), Lev-Yadun and Gould (2007, 2009), Archetti et al. (2009) and Schaefer and Ruxton (2009, 2011). I focus on the potential defensive functions of white leaf variegation, a well known morphological phenomenon in thousands of plant species and thousands of additional white variegated mutants of usually non-variegated species that have been identified by horticulturists and cloned as ornamentals. Although outside the scope of this essay, the possible simultaneous physiological roles of white leaf variegation must always be kept in mind.

Since some of the proposed defensive functions of white variegated leaves involve mimicry, I cite the two major types (out of a longer list) of plant defensive mimicry, which certainly do not cover all possible mimicry types used by plants, an issue also outside the scope of this review. Müllerian mimicry is a phenomenon in which two or more species with effective defenses share a similar appearance or signaling, thus also sharing and by this reducing the cost of associative learning by their enemies. Batesian mimicry is a phenomenon in which members of a palatable species gain protection from predation by resembling an unpalatable or defended species (Cott 1940; Wickler 1968; Ruxton et al. 2004). There are, however, intermediate types between these two extremes (e.g., Rowland et al. 2010). Most of our knowledge about defensive mimicry has emerged from animal studies, and while even the better-studied animal mimicry systems, like butterfly aposematism and its Müllerian and Batesian mimicry, are still not fully understood (e.g., Forbes 2009), the level of understanding of the operation of defensive mimicry in plants and its ecology and evolution is much lower (e.g., Hinton 1973; Wiens 1978; Lev-Yadun 2009a, b; Schaefer and Ruxton 2009, 2011). The evolution of mimicry requires a model and a mimic. The model should be another species or a group of species, or their actions (e.g., release of chemicals or physical damage to other organisms) (Wickler 1968). Organisms may also mimic a biological or nonbiological substrate on which they grow as a camouflage against enemies or to hide from potential prey (Cott 1940; Wickler 1968; Ruxton et al. 2004). Masquerade (close resemblance of inedible and often inanimate objects) (Endler 1981; Allen and Cooper 1985; Skelhorn et al. 2010a, b) is a related visual defense but not by crypsis. This character may sometimes initially evolve not for defense but for physiological reasons like the common light plant coloration that reflects excess light in coastal or desert environments, resulting (probably as a secondary gain) in actual substrate mimicry that may potentially also reduce herbivory by camouflage or by just not being green (Lev-Yadun 2006b; Reeves 2011). Such multifunctional situations make it difficult to evaluate the relative role of the mimicry in various morphologies and types of coloration, but may explain the advantages during early stages of the evolution of such cases (Schaefer and Ruxton 2009). Another significant theoretical problem that was posited recently is the difficulty in distinguishing between cases in which plants exploit perceptual biases of animals that may by convergent evolution arrive at a morphology or coloration that just looks like mimicry but is not a true mimicry (Schaefer and Ruxton 2009). For the convenience of discussion and because of the very small number of experiments with defensive plant mimicry, I will refer only to mimicry, because of our current limited ability to distinguish between the situations of exploitation of perceptual biases of herbivores by plants, and true mimicry.

2 White Variegation in Plants: At Least Nine Different Types of Formation

Visually distinct white variegation in leaves or in other organs occurs as the outcome of one of the nine different currently known possible mechanisms (structural, biochemical, and genetic), which may sometimes partly overlap. In the first, the epidermis is not attached to the green photosynthetic parenchyma and the air

Fig. 1 A variegated spiny leaf of *Silybum marianum* from Israel. The very conspicuous *white markings* are the outcome of sub-epidermal air spaces

spaces thus formed reflect light, resulting in white coloration (Hara 1957; Scarchuk and Lent 1965; Tsukaya et al. 2004; Fig. 1). In this case the white color does not result in loss of photosynthetic ability, and the air spaces may even have various physiological functions (e.g., Konoplyova et al. 2008). The second and related type of white variegation is formed by the formation of a single loose layer of short palisade cells instead of a compact layer or two of long palisade cells in the green sectors of variegated leaves (La Rocca et al. 2011). The third type of variegation is formed by regulated lack of expression of chlorophyll in certain cell groups (Lev-Yadun et al. 2004b; Lee 2007). This pattern is usually developmentally regulated, and in order to distinguish between this type and white variegation resulting from air spaces, anatomical examination should be conducted. The fourth type of white variegation results from a genetic chimera, where two cell genotypes, one expressing chlorophyll and one not, are found in growth apices, and the incidence of directions and numbers of cell divisions in the apices results in variegation patterns that may greatly fluctuate from leaf to leaf, or even between various parts of the same leaf (Tilney-Bassett 1986; Poethig 1987; Evenari 1989). The fifth type of white variegation results from mutations in which an unbalanced redox state in the photosynthetic electron transport chain predisposes the chloroplasts to photooxidation, which bleaches them, resulting in variegation (Rosso et al. 2009). The sixth type of white variegation results from dense groups of short white spines (glochids in cacti's areoles) or trichomes in other taxa. The seventh type of white variegation results from carotenoid cleavage and is common in flowers (Ohmiya et al. 2006). The eighth type of white variegation is the outcome of the lack of transcript expression or activity of enzymes of anthocyanin synthesis (e.g., Kotepong et al. 2011). The ninth type is actually composed of a collection of various biochemical modifications that alter chloroplast development, structure, and biochemistry and may overlap types number 3, 4, and 5 (e.g., Aluru et al. 2006; Yu et al. 2007), many of which could probably be determined as specific variegation types if the mutations that give rise to them were known, but in the vast majority of variegated genotypes they have not been studied in detail.

3 The Anti-herbivory Role

In general, it is important not to overlook the role of vision in finding a host plant by insects (e.g., Rausher 1978; Prokopy and Owens 1983; Reeves 2011). It has been proposed that white plant variegation has several defensive (anti-herbivory) functions (Cole 1970; Cahn and Harper 1976a, b; Harper 1977; Wiens 1978; Niemelä et al. 1984; Smith 1986; Givnish 1990; Allen and Knill 1991; Brown et al. 1991; Lev-Yadun 2001, 2003a, 2006a, 2009a, b, c, 2013, 2014; Cole and Cole 2005; Lee 2007; Campitelli et al. 2008; Wilkinson and Sherratt 2008; Soltau et al. 2009; Skelhorn et al. 2010a; Yamazaki 2010; Zheng et al. 2010; Schaefer and Ruxton 2011) (Table 1) and these hypotheses, along with several new hypotheses proposed here will be described and discussed below.

3.1 Evidence (Evidence of Defense by Leaf Variegation, Functioning by Unknown Mechanisms)

Several studies showed that white leaf variegation is associated with lower herbivory, but without proposing a defensive mechanism. Cahn and Harper (1976a) showed that the frequency of variegated leaves in *Trifolium repens* L. decreased with increasing grass length and concluded that this was related to grazing. Later, Cahn and Harper (1976b) showed in a field experiment with rumen-fistulated sheep (which enabled sampling of what had just been grazed) that non-variegated leaves of *T. repens* were clearly preferred over variegated ones. Shifriss (1981) found that *Cucurbita* L. plants with silvery leaves are better protected from insect-transmitted virus infections. Similarly, Campitelli et al. (2008) showed that leaf variegation is associated with reduced herbivore damage in *Hydrophyllum virginianum* L., although the actual defensive mechanism is not known. Zheng et al. (2010) induced white leaf variegation by disruption of carotenoid biosynthesis in transgenes and found that *Pieris rapae* L. butterflies have an innate ability to visually discriminate between green and variegated plants, preferring the non-variegated type. Moreover, their caterpillars grew less when fed on variegated plants compared to those fed on the green ones.

3.2 The Global Perspective

If white leaf variegation has several defensive roles, then it should be more widespread in ecosystems with historically strong grazing or insect herbivory pressure than in areas with a lower level of herbivory, a question that must be studied with a global perspective. It is well known that defense by sharp appendages (thorns, spines, and prickles) is more common in hot arid regions (Grubb 1992;

Table 1 Various previous hypotheses for defense from herbivory *via* white variegation

Proposed defensive mechanism	Comments	References
Camouflage	*Lithops* sp., South African deserts	Cole (1970); Cole and Cole (2005); Wilkinson and Sherratt (2008); Schaefer and Ruxton (2011)
	Understory herbs in New England forests	Givnish (1990); Allen and Knill (1991)
	Unripe fleshy fruits	Lev-Yadun (2013)
Masquerade	*Lithops* sp., South African deserts	Skelhorn et al. (2010a)
	Smilax aspera in understory	Lev-Yadun (2009b)
Mimicry of tunneling	Vine species *Byttneria aculeata*	Smith (1986)[a]
	General note	Brown et al. (1991)
	Silybum marianum	Lev-Yadun (2003a, 2006a, 2009a, b)
	Various species	Lee (2007)
	Caladium steudneriifolium (Including experiments)	Soltau et al. (2009)[b]
	Theoretical discussion	Schaefer and Ruxton (2011)
	Various species	Yamazaki (2010)
Mimicry of bird droppings		Yamazaki (2010)
Aposematism	Leaf variegation in many species	Harper (1977); Wiens (1978)
	Many species belonging to cacti, *Euphorbia*, *Aloe*, wild plants in Israel	Lev-Yadun (2001, 2003a, 2006a, 2009a, b, c)
	Unripe fleshy fruits	Lev-Yadun (2013)
	Müllerian and Batesian mimicry	Lev-Yadun (2003a, 2006a, 2009a, b, c)
Deterring insect landing	*Silybum marianum*	Lev-Yadun (2003a, 2006a, 2009a)
Dazzle effect	*Silybum marianum*	Lev-Yadun (2003a, 2006a, 2009a)
	Various plants	Lev-Yadun (2014)
Diverting herbivores away from green tissues[a]	*Acer pseudoplatanus*	Niemelä et al. (1984)

(continued)

Table 1 (continued)

Proposed defensive mechanism	Comments	References
Unexplained defense	*Trifolium repens*, a field experiment with rumen-fistulated sheep	Cahn and Harper (1976a, b)[a]
	General comment	Harper (1977)
	Hydrophyllum virginianum reduced herbivore damage in variegated morph	Campitelli et al. (2008)[a]
	Variegation in transgenic plants	Zheng et al. (2010)[a]
	Broad theoretical discussions	Schaefer and Ruxton (2011)
Mimicry of insect eggs	Unripe fleshy fruits	Lev-Yadun (2013)

[a]Found to operate
[b]Found to operate and the type of defensive mechanism examined

Lev-Yadun 2001; Ronel et al. 2010; Ronel and Lev-Yadun 2012). This trend is in accordance with the general pattern of stronger anti-herbivory defense found in slow-growing plants in habitats with limited resources (McKey et al. 1978; Coley et al. 1985; Endara and Coley 2011). Within the milkweeds (*Asclepias* spp. L.) there is a clear geographical gradient in defense by cardenolides, with species from lower latitudes better defended and with higher inducible and greater diversity and stronger toxicity (Rasmann and Agrawal 2011), a pattern also true for aposematic coloration in animals (Schemske et al. 2009). While it is impossible for a single person to examine in depth the flora of the entire world, I conducted field surveys to compare the heavily grazed arid Near East with the floras of several temperate and even boreal countries (Canada, Estonia, Finland, Sweden, Norway, and Russia), each of which has different assemblages of local plant and animal taxa, and all with a much lower current and historical grazing impact than in the hot and arid Near East. The field work I conducted in the temperate and boreal regions clearly indicated a dramatically lower level of spininess in these ecosystems, along with much lower levels of genuine white leaf variegation. Thus, the geographical distribution of white leaf variegation supports (but does not prove) the various hypotheses of the probable common anti-herbivory functions of white leaf variegation discussed above.

3.3 Operating Under Stress in Nature Limits the Need for Perfect Defensive Mechanisms

Herbivores, which are usually not well defended against fierce carnivores at close range, operate under various risks (see Brown 1999; Preisser 2009; Sheriff et al. 2009; Hawlena and Schmitz 2010; Embar et al. 2011). Thus, part of the attention of herbivorous animals is always drawn away from their prey, the plants, and this helps the defending plants that have evolved characters that add various visual difficulties that prolong the time needed for decision making, thus exploiting for defense the herbivore's limited attention (Jones et al. 2006). Staying for a long time in one place to graze, browse, or collect seeds increases the risks of predation. If an herbivore has to spend a longer time in searching or decision making because of a plant's defensive coloration, it may increase the risk of herbivore predation, sometimes making it safer to skip the problematic plants altogether. Accumulating evidence shows that the risk of predation is so significant that it influences herbivore activity in ways that may even change vegetation structure (e.g., Pfister et al. 1990; Terlouw et al. 1998; Brown and Kotler 2004; Ripple and Beschta 2004). I propose that various types of plant coloration, including white variegation, may operate in this way because they result in certain cases in a longer foraging and decision-making time by the herbivores compared to plants without these types of visual defenses.

3.4 Possible Mechanisms

3.4.1 Aposematism

Mechanical (Thorns, Spines, Prickles)

Wiens (1978), in the first review ever to discuss many aspects of defensive plant mimicry, proposed in brief that leaf variegation might be aposematic and that it should be examined for aposematic effects and for possible mimicry by non-protected associated plants. Lev-Yadun (2001) proposed that in many spiny taxa (e.g., over 1,000 species of cacti, *Agave* L., *Aloe* L., and *Euphorbia* L.) white variegation and other white markings along with various other color patterns contribute to the putative aposematic signals of these plants. Lev-Yadun (2003a) proposed that not only the conspicuous white variegation of the very spiny annual species of open Mediterranean habitats, *Silybum marianum* (L.) Gaertner, a member of the Asteraceae, but also two related spiny and white variegated species (*Notobasis syriaca* (L.) Cass. and *Scolymus maculatus* L.), which resemble "green zebras" with their alternating white and dark bands, are visually aposematic. In *Silybum marianum* the widths of typical variegation bands correlate highly with the length of the longest spines at leaf margins and the number of spines along leaf circumference. The zebra-like white variegation has thus been proposed to be a special case of aposematic coloration, although other defensive and physiological functions of this variegation were also proposed to operate simultaneously (Lev-Yadun 2003a, 2006a, 2009a, b, c, 2011, 2014). Interestingly, many spiny animals, including both invertebrates and vertebrates of both terrestrial and aquatic habitats, have colorful or contrasting black and white spines (Cott 1940; Inbar and Lev-Yadun 2005; Caro 2009).

The Near East, the cradle of Old World agriculture (e.g., Zohary et al. 2012; Lev-Yadun et al. 2000), which included large-scale herding, has witnessed a very long history of intense grazing that selected for various anti-herbivory defenses (Zohary 1962, 1973, 1983; Ronel et al. 2009, 2010; Ronel and Lev-Yadun 2009, 2012). The preagricultural grazing impact during the Pleistocene by large mammals that thrived in the region lasted more than two million years during the Pleistocene (e.g., Tchernov 1979; Davis 1987; Bar-Oz 2004; Steiner 2005), probably also for millions of years already in the Tertiary. The long history of intensive grazing in the Near East has selected for plants that are better protected and resilient and thus suffer less from grazing, and induced an increase in their proportion practically everywhere (Zohary 1962, 1983; Noy-Meir et al. 1989; Perevolotsky and Seligman 1998; Nassar and Lev-Yadun 2009; Ronel et al. 2009, 2010; Ronel and Lev-Yadun 2012). Out of more than 20 spiny species with white-variegated leaves in the flora of Israel (Lev-Yadun 2009b), the most conspicuous white-variegated leaves are found in the three very spiny annual plants (*Silybum marianum*, *Notobasis syriaca*, and *Scolymus maculatus*). These three species are among the most common tall annuals that survive very strong mammalian grazing pressures and are especially common along unpaved field roads, where large numbers of cattle, sheep, and goats pass by, and their dry plant skeletons remain intact during the summer in rangelands when most other nonwoody plants are grazed to the roots. Their abundance under strong grazing pressure is excellent evidence of their very strong defensive character, and they serve as a natural experiment for the probable defensive role of their spines combined with conspicuousness (Lev-Yadun 2003a, 2006a, 2009a, b, c, 2011, 2014; Ronel and Lev-Yadun 2012). Of course, additional defensive mechanisms may be involved, as it has been found that the very spiny *S. marianum* produces pyrazine, which probably functions as olfactory aposematism (Rothschild and Moore 1987) and because the variegation may mimic leaf infestation or be involved in other types of defense (Lev-Yadun 2003a, 2009b, 2014).

The defensive potential of white variegation was further studied in spiny, thorny, prickly, and white variegated plants in relation to their potential aposematism, and the existence of Müllerian and Batesian mimicry rings of such plants has been proposed (Lev-Yadun 2003a, 2009a, b, c). Lev-Yadun (2009b) compiled the information on geographical distribution of 21 wild, spiny plant species in the flora of Israel that have white-variegated leaves. The general overlapping geographical distribution and the overlapping of specific habitats of many of these species and also that of the relevant large wild and domesticated mammalian herbivores that have fed on them for millennia indicate that they probably form Müllerian mimicry rings. Moreover, in the members of the genus *Launaea* Cass. (Asteraceae) growing in Israel and elsewhere in the Near East, there are several species that are both white variegated and spiny or thorny (a defended Müllerian mimicry ring), and four non-thorny but variegated plants (a Batesian mimicry ring). The latter may mimic both other thorny and spiny *Launaea* species as well as spiny variegated species belonging to other taxa with an overlapping distribution.

Symmetry has been proposed to increase the efficiency of visual aposematic displays in animals (Forsmann and Merilaita 1999; Forsman and Herrström 2004), and Lev-Yadun (2011) suggested that it may also be true for many aposematic spiny or poisonous plants, including those with white variegation, for instance both the spiny leaf rosettes and flowering heads of *S. marianum*, *N. syriaca*, and *S. maculatus*.

Micro-mechanical and Biochemical (Raphids, Toxins, Bacteria)

While the potential aposematic coloration of spiny plants has received significant recent attention (Lev-Yadun 2001, 2003a, b, 2006a, 2009a, b, c, 2011; Midgley et al. 2001; Lev-Yadun and Ne'eman 2004, 2006; Midgley 2004; Rubino and McCarthy 2004; Ruxton et al. 2004; Speed and Ruxton 2005; Halpern et al. 2007a, b, 2011; Lev-Yadun and Halpern 2008; Fadzly et al. 2009; Lev-Yadun and Gould 2009; Lev-Yadun et al. 2009; Ronel et al. 2009, 2010; Burns 2010; Fadzly and Burns 2010; Schaefer and Ruxton 2011; Ronel and Lev-Yadun 2012), the potential aposematism of poisonous plants forming internal spines (raphids—microscopic needles made of calcium oxalate) has received much less attention (Lev-Yadun and Halpern 2008; Lev-Yadun 2009a; Halpern et al. 2011). Thousands of plant species belonging to many families, including many white and otherwise variegated species, produce raphids with or without associated toxins (Franceschi and Horner 1980). Raphids are always elongated, needle-shaped, and have two sharp pointed ends. Raphids are usually formed in specific parenchymatic cells that differ from their neighboring cells and are called idioblasts. The raphids are formed in idioblasts in large numbers and are packed compactly, aligned parallel to each other, but spread when the tissue is wounded (Fahn 1990). Because of their small size, raphids can internally wound the mouth and digestive system not only of large vertebrate herbivores but also of insects and other small herbivores that manage to avoid thorns, spines, and prickles by passing between them.

The physical wounding of the herbivores by the sharp-ended raphids is not the whole story. Studies conducted with a scanning electron microscope have revealed that in many cases, the raphids may be barbed or may have deep grooves running along them. The grooves serve as channels through which plant toxins are introduced into the tissues of the herbivores (Sakai et al. 1972; Franceschi and Horner 1980) in a way resembling the action of hollow fangs of venomous snakes. Moreover, Lev-Yadun and Halpern (2008) and Halpern et al. (2011) proposed that pathogenic bacteria and fungi found on the plant surfaces and within the herbivore's digestive system may also enter the herbivore's body through these microscopic wounds and either operate in parallel to plant toxins or independently. These potentially severe and even lethal consequences of spines and raphids of white variegated leaves strengthen the honesty of the proposed aposematic signals of white variegation (Lev-Yadun and Halpern 2008; Lev-Yadun 2009a; Halpern et al. 2011).

Here I use the case of the poisonous American genus *Dieffenbachia* Schott to demonstrate the probably broad, but overlooked phenomenon of association and mutual defensive function of raphids and white (as well as nonwhite) leaf variegation. Leaves of wild members of the genus *Dieffenbachia* growing in the forests of Central and South America are typically variegated in white (Lee 2007). The plants are well defended from herbivores by the large amounts of sharp grooved raphids that, in addition to wounding, administer various plant toxins into the wounded tissues of the herbivores (or humans) that eat them (Arditti and Rodriguez 1982; Evans 1987; Gardner 1994; Bradbury and Nixon 1998). The combination of visual conspicuousness with unpalatability in the genus *Dieffenbachia* points to the probable aposematic effect of their leaf variegation. Interestingly, studies on field behavior of herbivores such as Dorcas gazelles (*Gazelle dorcas* L.) have shown that they specifically avoid plant tissues containing raphids (Ward et al. 1997; Salts and Ward 2000).

I propose that a broad comparative taxonomic study should be conducted to examine whether there is a nonrandom association of white- or colorful variegated leaves with the formation of raphids and possibly also with toxins. Comparing variegated and non-variegated species of various genera, or genera within families, or of various genotypes of species found in areas that significantly differ in herbivory pressure, is required in order to test the generality of this hypothesis.

Mixed Defensive Strategies in Plants with White Variegated Leaves

Because visual anti-herbivory defenses in plants have not been studied much, it is no surprise that there are only meager data or even untested hypotheses on mixed defensive strategies related to plants with white variegated leaves. The best described case seems to be that of the variegated Mary's thistle (*Silybum marianum*). Rothschild and Moore (1987) proposed that *S. marianum* uses olfactory aposematism via pyrazine. Later, Lev-Yadun (2003a, 2009b, 2011) proposed that the very conspicuous white variegation of this very spiny annual species of open habitats is a special case of aposematic coloration. Since spines in other species sharing its habitat harbor pathogenic bacteria that can defend plants from herbivores (Halpern et al. 2007a, b, 2011), there is no reason to assume that the spines of *S. marianum* are free of such bacteria. This adds to the significance of its physically based spine aposematism (see Halpern et al. 2007a, 2011). It is thus likely that in this species two types of aposematism (visual and olfactory) operate simultaneously, possibly towards different herbivores and simultaneously with other types of defense (Lev-Yadun 2003a, 2009a). Moreover, *S. marianum*, being a ruderal plant, sometimes accumulates large amounts of nitrate (like many other plant species), to a level making it toxic to large mammalian herbivores, killing cattle, sheep, and horses that consume these plants (Kendrick et al. 1955), thus adding to the other types of aposematism in this species, a potential chemically based one. In addition to the above, it has been proposed that *S. marianum* is also visually defended by tunneling mimicry, by dazzle effects, and by reducing the tendency of insect to land on its variegated leaves (Lev-Yadun 2003a, 2014). Thus, the fact that both olfactory and visual (possibly not only in the visible light but also in the U.V.) aposematism, or other defense mechanisms, both known and as yet unknown, may be involved in the defense of white variegated leaves and other variegated plant parts should always be kept in mind.

Toxic White Variegated Leaves

Since some of the plants with leaves variegated in white (or with other colors) (e.g., *Cyclamen* spp. L. and certain *Trifolium* spp. L. and *Medicago* spp. L.) may be poisonous, because *Cyclamen* species are known to be rich in alkaloids and saponins (Hornell 1941; Reznicek et al. 1989), and because there are cyanogenic types in *Trifolium* spp. and *Medicago* spp. (Crawford-Sidebotham 1972; Dirzo and Harper 1982), visual aposematism of such variegated leaves should be considered (see Lev-Yadun 2009a, 2013). Müllerian and Batesian mimicry rings are expected to exist in such cases.

3.4.2 Mimicry

Insect Tunneling Damage Mimicry

Insect damage mimicry, especially of tunneling, was the first specific ecological hypothesis that tried to explain the evolution and ecology of white leaf variegation. The first to discuss in detail white leaf variegation in the context of possible aposematism was Smith (1986), and while he rejected the aposematic hypothesis for the species (*Byttneria aculeata* Jacq.) he studied, he gave a clear and detailed formulation of the aposematic hypothesis for poisonous plants: "The benefits to the plant of chemical defense against herbivores would be greater if herbivores avoided such plants altogether, rather than testing leaves for palatability, and so causing some damage. A distinct leaf color pattern linked with chemical defense might function in this way. Polymorphism for leaf color should then coincide with polymorphisms for chemical defense. Müllerian and Batesian mimicry could result in evolution of similar patterns of variegation, with or without associated toxicity, among other species which have herbivore species in common with the model species." Mimicry of tunneling insect damage was also one of the several types of proposed defensive functions of white leaf variegation of *Silybum marianum* (Lev-Yadun 2003a), making it look as if it were already infested to prevent further insect attacks. Originally, this was the first potential defensive function of the variegation of this species that I thought about, but later, aposematism took priority because the three annual plant species of the Near Eastern flora that express this morphology to the highest extent (*Silybum marianum*, *Notobasis syriaca*, and *Scolymus maculatus*) are all very spiny (Lev-Yadun 2003a, 2006a, 2009a, b, c, 2011). Lee (2007) also proposed that certain types of white variegation may serve as a defense by mimicking leaf infestation. Soltau et al. (2009) experimentally manipulated the visual appearance of *Caladium steudneriifolium* Engl. (Araceae), an understory plant from the *Podocarpus* National Park in South East Ecuador plants by painting artificial white variegation on non-variegated green leaves. The leaves of *C. steudneriifolium* are either plain green or patterned with whitish variegation. In nature, about a third of the leaves are variegated and both morphs are frequently attacked by mining moth caterpillars. The variegated zones of the leaves strongly resemble recent mining damage and was hypothesized to mimic recent mining attacks. Infestation was found to be 4–12 times higher in plain green leaves than for variegated ones. They studied the level of herbivore damage and showed that painting artificial white variegation on plain green leaves resulted in reduced attacks from 7.88 % to 0.41 %, leading them to propose that the variegation is probably the mimicry of mining damage to deter ovipositing moths (Soltau et al. 2009). Yamazaki (2010) also proposed that tunneling damage in the form of white variegation may deter herbivores.

There is solid evidence that plants infested by insects are avoided by various other insects because of several risks: cannibalism or interspecific predation, competition, induction of host defensive mechanisms that reduce its palatability and increase its toxicity, and in the case of leaves, flowers, fruits, and young branches, also the risk of organ shed (habitat destruction) (e.g., Addicott 1982; Lev-Yadun and Gould 2007; Karban 2007; Yamazaki 2010; Schaefer and Ruxton 2011). The common phenomenon that wounded leaves signal via odor to attract predators and parasitoids (e.g., Kessler and Baldwin 2001; De Moraes et al. 2001; Kappers et al. 2005) should also be considered. For instance, Finch and Jones (1989) reported that large colonies of the cabbage aphid *Brevicoryne brassicae* L. and the peach aphid *Myzus presicae* Sulzer deter ovipositing by the root fly *Delia radicum* L. Inbar et al. (1999) demonstrated that homopterans (whiteflies) not only alter adult cabbage looper (*Trichoplusia ni* Hübner) host selection but also actually reduce the feeding efficiency of their offspring. Thus, variegation that mimics herbivore damage might serve as mimicry of an already infected leaf and deter female insects from laying eggs (see Smith 1986; Soltau et al. 2009).

Bird Droppings Mimicry

One of the known masquerade types in animals is the mimicry of bird droppings (Wickler 1968; Skelhorn et al. 2010c). Yamazaki (2010) proposed that some tunneling damage in leaves may reduce herbivory because it mimics bird droppings. I propose that various types of leaf variegation should be compared visually (including their U.V. spectra) to actual bird droppings and that the possibility of bird dropping mimicry by variegated plants should be examined in much more detail. Because bird droppings may contain both parasites and pathogenic bacteria, and because white variegation may look like bird droppings, this may be a basis for some of the defense provided by white variegation.

Fig. 2 A variegated leaf of *Acer pseudoplatanus* from Turku (Finland)

3.4.3 Mimicry of Fungal Attacks

In many cases white leaf variegation visually mimics fungal attacks to such an extent that only a close and careful examination allows us to distinguish between them. Fungal attack mimicry by white variegation has the potential of being ecologically important. Lev-Yadun (2006b) described and discussed the potential mimicry of fungal infestation by whitish-colored leaves. I found many leaves were mottled by white fungal hyphae in the flora of the Near East; in Central, Northern, and Eastern Europe; as well as in North America. Since white leaf variegation (Fig. 2) looks in many cases like fungal colonies on leaves, it is logical to propose that it may mimic fungal attacks. The question is why should herbivores refrain from consuming plant tissues infested with fungi? The answer seems to be clear: many fungi produce toxic chemicals that in many cases are known to defend their hosts from herbivory. Fungal endophyte-mediated alkaloids provide the basis for the acquired chemical defense against herbivory (Porter 1994; Justus et al. 1997; Saikkonen et al. 1998; Lev-Yadun and Halpern 2007) and recent field data support the aposematic function of ergot mutualism in *Festuca rubra* L. plants in grazed areas in Finland (Wäli et al. 2013). Moreover, leaves attacked by fungi may have lower palatability for various other reasons, since the attacks induce various defensive mechanisms and because some of their nutritive resources have already been exploited by the fungi. There are very good indications that plant parts that may be infested by fungi are rejected by herbivorous animals. Frugivores, for instance, regularly avoid eating damaged fruits, especially large ones (Janzen 1977; Herrera 1982; Manzur and Courtney 1984; Borowicz 1988; Buchholz and Levey 1990). Lev-Yadun (2006b) therefore proposed that certain white plant surfaces may mimic fungal-infested plants and that this character may reduce the tendency of herbivores to consume such plants.

3.4.4 Camouflage

Plant camouflage has received only marginal, almost anecdotal treatment. After many years of neglect, this aspect of visual plant defense has recently been both discussed and examined (Lev-Yadun 2006a, b; Fadzly et al. 2009; Klooster et al. 2009; Burns 2010; Fadzly and Burns 2010; Schaefer and Ruxton 2011; Lev-Yadun and Ne'eman 2013; La Rocca et al. 2014), but there is a clear need to conduct many more studies on defense from herbivory by camouflage in plants, like the many studies done with animals (e.g., Cott 1940; Edmunds 1974; Merilaita 1998; Ruxton et al. 2004; Caro 2005; Cuthill et al. 2005; Schaefer and Stobbe 2006; Stevens et al. 2006).

Cole (1970) and later Cole and Cole (2005) clearly described how difficult it is to identify in the field the camouflaged plants of various *Lithops* N. E. Br. species, which are variegated in various shades including white, among the gravel that covers the soil in the arid South African regions where they grow. *Lithops* species are found in open arid regions, where the light is strong and evenly distributed. Schaefer and Ruxton (2011) noted that even this apparent case of putative plant camouflage was not examined experimentally. However, a different hypothesis, about the defensive role via camouflage, has been proposed for various types of leaf variegation, including white, for herbs that occupy the forest understory where sun flecks are common. Givnish (1990) proposed that camouflage from color-blind vertebrate herbivores is the major selective agent for their commonly variegated leaves. Givnish's hypothesis was positively discussed by Allen and Knill (1991), but as with many aspects of plant camouflage it was not pursued further. Blanco and Martén-Rodríguez (2007) also proposed that colorful variegation in understory palm leaves may be a type of disruptive coloration, camouflaging them from large color-blind herbivores. A similar hypothesis concerning a possible role in camouflage (and in other defensive and physiological functions) of white variegation in green, unripe fleshy fruits was recently proposed (Lev-Yadun 2013).

At first glance, Givnish's (1990) hypothesis, i.e., that white variegation serves as camouflage, may seem to oppose that of the aposematic or dazzle hypotheses of leaf variegation (e.g., Lev-Yadun 2003a, 2009b, 2014), but since Givnish studied understory species growing in a habitat characterized by sun flecks, and aposematism of spiny plants or dazzle effects by white variegation was proposed for plants growing in open and well-illuminated areas (Lev-Yadun 2001, 2003a, 2009a, b, c, 2014), there is no contradiction. Thus, white leaf mottling (Fig. 3) can probably act both as camouflage in forest undergrowth, and as a conspicuous aposematic or dazzle coloration, in open, well-illuminated areas.

3.4.5 Masquerade

Masquerade (close resemblance to common inedible and inanimate objects) has been proposed to operate in *Lithops* species (Skelhorn et al. 2010a). Since this

Fig. 3 Variegated leaves that make it hard to identify their shape, in a collection of variegated species in the greenhouses in Edmonton, Canada

character was commonly considered as camouflage or mimicry (Skelhorn et al. 2010a), it is premature to discuss it in depth, but further research of this strategy in plants in general, and in variegated ones in particular, is certainly needed. Wiens (1978) mentioned several types of plant camouflage that according to their description fit the classification of masquerade even though they were not classified as such since the theoretical interest in masquerade as defense for animals and plants is very recent.

3.4.6 Visual Deterrence of Insect Landing

The best known (although still very far from both a good understanding and from a general acceptance of the hypothesis) case for which variegation (alternating dark and light bands) has been proposed to defend an organism is that of the zebra. Ortolani (1999) reviewed many old references concerning the putative functions of zebra stripes and stated that there is a considerable disagreement on the function of zebra stripes. The only hypothesis (out of several) for the defensive function of zebra stripes that was tested experimentally was that of reducing tsetse fly landing as the potential selective force for the evolution of zebra stripes (Waage 1981; Brady and Shereni 1988; Doku and Brady 1989; Gibson 1992). Ruxton (2002), who specifically reviewed the hypotheses concerning the potential defensive role of zebra stripes, concluded that the tsetse fly landing as the selective force for the evolution of zebra stripes was not proved, and Caro (2009) was strongly negative towards this possible defensive adaptation in zebras. The recent detailed experimental study (Egri et al. 2012) of the role of zebra stripes in horsefly (tabanids) repellence seems to give very strong indications that such patterns indeed function in insect repellence. This, however, does not negate other simultaneous functions of such coloration. Concerning the zebra-like coloration of *Silybum marianum* and other very spiny Mediterranean plant species, Lev-Yadun (2003a) proposed that reducing insect landing on the leaves in general may be one of several reasons for the evolution of this type of variegation. If indeed deterrence of insect landing operates when variegation in the form of light and dark bands exists, and this can be convincingly demonstrated, I propose it to be a case of exploitation of perceptual biases of herbivores *sensu* Schaefer and Ruxton (2009).

Although I think that the aposematic hypothesis and some physiological gains (which are under study in *Silybum marianum*) are the most significant explanations for zebra-like white plant variegation, there is no theoretical reason for not considering insect deterrence as a potential partial benefit of this unusual morphology.

3.4.7 Potential Defense by Dazzle Effects of White Leaf Variegation

Zebra-like white leaf variegation was first just proposed (with no broad and deep theoretical discussion) to serve as defensive dazzle coloration (*sensu* Wilkinson 1969) for the white-variegated Mediterranean annuals *Silybum marianum*, *Notobasis syriaca*, and *Scolymus maculatus* (Lev-Yadun 2003a, 2006a, 2009a, b). Recently, the theoretical basis was discussed in detail and it was proposed that this hypothesis can be extended to various species belonging to the genera *Haworthia* Duval, *Gasteria* Duval, and *Sansevieria* Thunb., which also have species with zebra-like white leaf variegation, and to any other plant taxa with such coloration (Lev-Yadun 2014). Such plants actually have classic dazzle coloration very similar to what was applied to naval vessels [compare photographs of the French cruiser Gloire in Williams (2001) or on the Internet, and other photographs in Wilkinson (1969), and Stanley (1998), with those in Lev-Yadun (2003a)]. The swaying of the leaves in the wind and the relative movements of plants in respect to flying insects or following the head movements of large herbivores approaching the plants increase the potential to form the desired dazzle effects on the visual nerve systems of the herbivores, which will make it difficult for them to land on or bite the leaf because of problems in locating its actual position in space. As with many other types of proposed defensive leaf variegation, dazzle effects by plants have not been studied experimentally.

Theoretically, coloration that produces an illusion of a different leaf shape may also cause identification problems for insects that search for specific leaf types or leaf developmental stages, as proposed for insects' search images for other leaf morphologies (e.g., Rausher 1978; Prokopy and Owens 1983; Mackay and Jones 1989; Brown et al. 1991; Reeves 2011).

3.4.8 White Sectors in Variegated Leaves as a Herbivore Diversion from Green Sectors

Niemelä et al. (1984) found selective herbivory on mosaic leaves of white variegated *Acer pseudoplatanus* L. They showed that in such variegated leaves, insect herbivores favored white areas over mixed and green areas. In *A. pseudoplatanus* this preference correlated well with the chemical properties of white areas that contained more nutrients and less defensive phenolic compounds. This chemical evidence was sufficient to explain the increased herbivory in the white sectors of *A. pseudoplatanus* leaves. However, I rephrase the explanation, taking the plant's point of view, that while the white sectors of the leaves suffer higher herbivory, the more productive green parts suffer less, and thus the white islands in this case practically serve as traps for the insects. Lüttge (1997) proposed this hypothesis concerning young colorful leaves that are common in tropical forests, which according to his hypothesis may attract herbivores and divert them from the more costly and productive older and green ones. In any case, a broad comparative chemical analysis of defensive substances in white versus green leaf parts of many variegated species is certainly needed in order to see if, how frequently and to what extent they differ in their defensive and nutritive potentials, but this has not been done.

3.4.9 White Leaf Variegation: Camouflage Versus Aposematism

In plants growing in open, well-illuminated habitats the white markings are conspicuous, especially where the marks are large and patterned like zebras, i.e., in the three common Mediterranean spiny annual rosettes of the Asteraceae (*Silybum marianum*, *Notobasis syriaca*, and *Scolymus maculatus*) that were proposed to be aposematic (Lev-Yadun 2003a, 2009b). This conclusion seems at first glance to oppose that of Givnish (1990), i.e., that white variegation (as well as of other variegation colors) serves as camouflage, but since Givnish studied understory species growing in a habitat characterized by dark areas interrupted by sun flecks, there is no contradiction. Thus, white mottling can act as camouflage in undergrowth in the forest and probably also within the canopy (see Lev-Yadun 2013) and as a conspicuous aposematic coloration in open, well-illuminated areas. Similar different habitat- and herbivore-specific functions may exist in *Cyclamen* spp. leaves. The white variegated and toxic leaves of *Cyclamen* spp. may serve as camouflage in the understory *sensu* Givnish (1990), or as dazzling under any type of illumination (e.g., Lev-Yadun 2014), and also as aposematic coloration, especially in well-illuminated habitats. The common *Cyclamen* species found in Israel (*C. persicum* Miller) grows in both the understory and in open areas. The camouflage function of variegated *Cyclamen* leaves may also operate in open habitats since many frogs and toads, including the two local toads *Bufo viridis* Laurenti and *Pelobates syriacus* Boettger, use color patterns resembling those of *Cyclamen* spp. leaves as camouflage (e.g., for frog camouflage see Osorio and Srinivasan 1991).

4 Problems and Questions

4.1 Evaluating Risk: The Problematic and Even Erroneous Common View of "No Damage or No Attack Equals No Risk"

A theoretical issue related to the functionality of defensive plant coloration and the operation of herbivores under stress that has been raised recently is the question of risk evaluation in ecological and evolutionary plant/herbivore studies (e.g., Lev-Yadun 2006b, 2009a; Lev-Yadun and Gould 2007, 2009). In order to both change the common erroneous view and enhance theoretical and experimental studies I repeat it. This is not just a mere theoretical issue; rather, the understanding of this principle may significantly influence the planning and interpreting of both experiments and actual herbivory data collected in nature. There are inherent theoretical difficulties in evaluation of defense by experiments. A good defense may operate so well towards certain herbivores that there are almost no attacks on the defended organism and thus, the experimental noise may be bigger than the signals. Many scientists find it difficult to accept that "no damage does not automatically indicate that there is no risk." They usually say, no attacks equals no or reduced risk. However, in many cases lower attack levels indicate just the opposite: that the defense is strong and well known to potential enemies. A case of no attack in spite of a high risk was demonstrated by various summer green plants growing near Bedouin settlements in the Negev Desert (Israel) and elsewhere in the deserts of the Near East. Several common alkaloid-rich poisonous or thorny plants form green islands in the dry summer when all surrounding plants in this desert have turned yellow or grey and in many cases have been grazed down to their roots over large areas. Even under such extreme grazing pressure those green plants are ignored by the large herds of sheep, goats, donkeys, and camels that pass them daily (Lev-Yadun and Ne'eman 2004). If each individual animal of these large flocks tasted a single leaf once a day, these green plants would disappear in a short time. The absence of attacks on these summer green desert plants when no other vegetal food is available is a clear indication of their very good defensive and repelling qualities rather than of a low level of risk. Such facts bear on the interpretation of experimental results. A well-defended plant may have so few attacks that the statistical analysis of experiments may be problematic. Therefore, understanding the principle of a low level of attacks indicating good defense rather than low risk is critical for studies of individual species (e.g., Soltau et al. 2009), or for comparative studies involving many taxa (e.g., Archetti 2000; Hamilton and Brown 2001; Lev-Yadun 2001). The classic study by Soltau et al. (2009) is an excellent demonstration of the issue of the problematic statistics of studying well-defended organisms. Schaefer and Ruxton (2011:165–166) were absolutely correct from the current classic statistical point of view when they were cautious with the field data presented in Soltau et al. (2009). However, the reason for a low level of attacks because of various defenses was not considered in depth by Schaefer and

Ruxton (2011). This common approach among "statistically correct" ecologists may allow for accepting results only when less-defended organisms are studied, and the role of a strong defense may thus be overlooked in many cases. For instance, there are fewer lion attacks on elephants, hippopotamus, and rhinoceros than on gazelles and zebras, and even with many years of field observations the statistics of lion attacks on these well-defended herbivorous animals will be extremely low compared to those on the much less-defended herbivores. I think that the statistically problematic, but in my understanding, real and very important results of Soltau et al. (2009) fall into the same category of the low level of lion attacks on elephants, hippopotamus, and rhinoceros. There is a strong need to develop (or borrow from other academic disciplines) such tests and make them a standard for testing cases of low frequency events against a huge background noise in ecology. Otherwise only large effects will be agreed on and published, while smaller effects, which are probably much more common and of considerable ecological and evolutionary importance, will not be recognized (see Martínez-Abraín 2008; Gotelli and Ulrich 2012; Mudge 2013). Alternatively, since statistics is not proving anything in any case, and cannot handle various other questions, the actual numbers may be sufficient in such cases. Very small differences in survival rate may over many generations result in strong evolutionary advantages (see von Helversen et al. 2013).

4.2 Testing the Hypotheses: A Very Complicated Issue

An important issue for hypothesis validity in general is the possibility of testing it. Intuitively it sometimes seems simple, but in reality it is very complicated to test defensive hypotheses in plants and there are contradicting theoretical considerations that I discuss below.

If white variegation has an aposematic role, then it should deter herbivores with previous foraging experience, but not naïve ones. However, some young and naïve herbivores learn from their parents to avoid certain plants (Landau et al. 1999). The problem is that since aposematism is a very old and common phenomenon, there are no genetically naïve animals towards aposematism, but only inexperienced (usually young) individuals. When several defense mechanisms operate simultaneously it is very hard to distinguish between their relative contributions. For instance, the possibility that thorny, spiny, and prickly plants simultaneously use visual and olfactory aposematism and that such plants are also poisonous was never studied systematically (see Lev-Yadun 2009a), and the same is true for the various pathogenic bacteria harboring in spines (Halpern et al. 2007a, 2011) that contribute an additional level of defense. La Rocca et al. (2014) found that in leaves of the early flowering understory *Erythronium dens-canis* L. that are variegated temporary in both brown and silvery flecks, the brown mottling probably functions only as camouflage and the persistent silvery flecks function also in attraction of pollinators later in the season. Such cases as well as those of many unripe white-mottled green fleshy fruits (Lev-Yadun 2013) demonstrate the functional and therefore ecologically and evolutionary complex situation. Moreover, when non-defensive gains due to white variegation occur, it may be difficult to distinguish them from direct defenses because better resource acquisition may allow a larger allocation to various types of defense and to growth and reproduction. All these mentioned factors and probably others complicate the analyses and evaluation of whatever results one gets (see Grubb 1992). All these complicated issues have not yet been studied in depth in plants to give a realistic and balanced view.

4.3 Open Questions

Because of the historical reluctance of botanists to consider defensive plant coloration (Harper 1977), what is known about it lags behind what we know about defensive animal coloration by a century if not more. I list several issues that should be studied concerning white plant variegation in order to better understand it: (1) The anatomical/developmental aspects of variegation. (2) Documenting variegation in all floras in general and according to the geographic and ecological distribution in particular. (3) The phylogenetic aspect of white leaf variegation. (4) Correlation of plant life history parameters with the defensive characters. (5) The genetic aspects. (6) The physiological aspects: is there a physiological component that operates simultaneously with anti-herbivory, what is its relative importance and does it fluctuate with time, with developmental stage and with environmental changes? (7) Is there an epigenetic component in the expression of variegation? (8) What types of defense does variegation provide (camouflage, aposematism, dazzle, mimicry, etc.)? (9) If there is more than one type of defense, what is the relative importance of each mechanism and are the relationships between them constant or do they fluctuate with time, age, and growth conditions? (10) Is variegation expressed more or less following herbivore attacks? (11) Do herbivores learn to avoid such plants? (12) Is there a genetic component in herbivore avoidance of white plant variegation? (13) If the variegation actually defends plants from herbivory, to what extent? (14) Mutants should be studied. (15) Are there fluctuations in the selection towards variegation with time according to biotic and abiotic environmental changes?

Advancing these questions and probably others that will emerge while studying the above questions, to the level of understanding of defensive coloration in animals, will probably take the whole twenty-first century if not more. For instance, Cott's (1940) hypotheses about defensive animal coloration are still being studied more than 70 years later.

5 Conclusions

White variegation of various developmental types has been proposed to defend leaves from herbivory by several mechanisms: camouflage, aposematism when it is associated with spines (including Müllerian and Batesian mimicry), toxins and raphids, mimicry of insect damage, mimicry of fungal attacks, mimicry of bird droppings, dazzle effects, and also by causing visual repellence of insects from landing. Very few cases of proposed leaf defense by white variegation have been studied in depth to date and found to actually operate as defense from herbivory in nature or in experiments. Moreover, the type of operating defensive mechanism was not always proposed or identified even when white variegation was found to be actually associated with reduced herbivory. Studying white variegation has a significant advantage over studying other types of defensive coloration in plants because even bichromatic vision, so common in mammalian herbivores, is sufficient to see these patterns. A broader and deeper effort to understand the defensive roles of white variegation and other types of defensive plant coloration is certainly needed. While studying defensive plant coloration, the possible simultaneous physiological roles of these types of coloration should not be ignored.

Acknowledgments I thank Moshe Flaishman, Amots Dafni, Shahal Abbo, Ron Sederoff, Gidi Ne'eman, Moshe Inbar, Gadi Katzir, Ido Izhaki, Michal Ronel, Malka Halpern, Kevin Gould, Pilli Urbas, Martin Schaeffer, Marco Archetti, Pekka Niemelä, Jarmo Holopainen, Tamar Keasar, Aki Sinkkonen, and Kazuo Yamazaki for stimulating discussions and for field work or trips concerning defensive plant coloration. I thank Professor Lüttge and an anonymous reviewer for their valuable comments and suggestions.

References

Addicott FT (1982) Abscission. University of California Press, Berkeley, CA
Allen JA, Cooper JM (1985) Crypsis and masquerade. J Biol Educ 19:268–270
Allen JA, Knill R (1991) Do grazers leave mottled leaves in the shade? Trends Ecol Evol 6:109–110
Aluru MR, Yu F, Fu A, Roderme S (2006) *Arabidopsis* variegation mutants: new insights into chloroplast biogenesis. J Exp Bot 57:1871–1881
Archetti M (2000) The origin of autumn colours by coevolution. J Theor Biol 205:625–630
Archetti M, Döring TF, Hagen SB, Hughes NM, Leather SR, Lee DW, Lev-Yadun S, Manetas Y, Ougham HJ, Schaberg PG, Thomas H (2009) Unravelling the evolution of autumn colours: an interdisciplinary approach. Trends Ecol Evol 24:166–173
Arditti J, Rodriguez E (1982) *Dieffenbachia*: uses, abuses and toxic constituents: a review. J Ethnopharmacol 5:293–302
Bar-Oz G (2004) Epipalaeolithic subsistence strategies in the Levant: a zooarchaeological perspective, The American School of Prehistoric Research (ASPR) monograph series. Brill Academic Publishers, Boston, MA
Blanco MA, Martén-Rodríguez S (2007) The stained-glass palm, *Geonoma epetiolata*. Palms 51:139–146

Borowicz VA (1988) Do vertebrates reject decaying fruit? An experimental test with *Cornus amomum* fruits. Oikos 53:74–78

Bradbury JH, Nixon RW (1998) The acridity of raphides from the edible aroids. J Sci Food Agric 76:608–616

Brady J, Shereni W (1988) Landing responses of the tsetse fly *Glossina morsitans morsitans* Westwood and the stable fly *Stomoxys calcitrans* (L.) (Diptera: Glossinidae & Muscidae) to black-and-white patterns: a laboratory study. Bull Entomol Res 78:301–311

Brown JS (1999) Vigilance, patch use and habitat selection: foraging under predation risk. Evol Ecol Res 1:49–71

Brown JS, Kotler BP (2004) Hazardous duty pay and the foraging cost of predation. Ecol Lett 7:999–1014

Brown VK, Lawton JH, Grubb PJ (1991) Herbivory and the evolution of leaf size and shape. Phil Trans R Soc Lond B 333:265–272

Buchholz R, Levey DJ (1990) The evolutionary triad of microbes, fruits, and seed dispersers: an experiment in fruit choice by cedar waxwings, *Bombycilla cedrorum*. Oikos 59:200–204

Burns KC (2010) Is crypsis a common defensive strategy in plants? Speculation on signal deception in the New Zealand flora. Plant Signal Behav 5:9–13

Cahn MG, Harper JL (1976a) The biology of the leaf mark polymorphism in *Trifolium repens* L. 1. Distribution of phenotypes at a local scale. Heredity 37:309–325

Cahn MG, Harper JL (1976b) The biology of the leaf mark polymorphism in *Trifolium repens* L. 2. Evidence for the selection of leaf marks by rumen fistulated sheep. Heredity 37:327–333

Campitelli BE, Steglik I, Stinchcombe JR (2008) Leaf variegation is associated with reduced herbivore damage in *Hydrophyllum virginianum*. Botany 86:306–313

Caro T (2005) Antipredator defenses in birds and mammals. University of Chicago Press, Chicago, IL

Caro T (2009) Contrasting coloration in terrestrial mammals. Phil Trans R Soc B 364:537–548

Cole DT (1970) Lithops in habitat. In: Sprechman DL (ed) Lithops. Fairleigh Dickinson University Press, Cranbury, NJ, pp 21–32

Cole DT, Cole NA (2005) Lithops – flowering stones. Cactus & Co., Venegono, Italy

Coley PD, Bryant JP, Chapin FS III (1985) Resource availability and plant antiherbivore defense. Science 230:895–899

Cott HB (1940) Adaptive coloration in animals. Methuen & Co., London

Crawford-Sidebotham TJ (1972) The role of slugs and snails in the maintenance of the cyanogenesis polymorphisms of *Lotus corniculatus* and *Trifolium repens*. Heredity 28:405–411

Cuthill IC, Stevens M, Sheppard J, Maddocks T, Párraga CA, Troscianko TS (2005) Disruptive coloration and background pattern matching. Nature 434:72–74

Davis SJM (1987) The archaeology of animals. Yale University Press, New Haven, CT

De Moraes CM, Mescher MC, Tumlinson JH (2001) Caterpillar-induced nocturnal plant volatiles repel conspecific females. Nature 410:577–580

Dirzo R, Harper JL (1982) Experimental studies on slug-plant interactions IV. The performance of cyanogenic and acyanogenic morphs of *Trifolium repens* in the field. J Ecol 70:119–138

Doku C, Brady J (1989) Landing site preferences of *Glossina morsitans morsitans* Westwood (Diptera: Glossinidae) in the laboratory: avoidance of horizontal features? Bull Entomol Res 79:521–528

Edmunds M (1974) Defence in animals. A survey of anti-predator defences. Longman, Harlow

Egri Á, Blahó M, Kriska G, Farkas R, Gyurkovszky M, Åkesson S, Horváth G (2012) Polarotactic tabanids find striped patterns with brightness and/or polarization modulation least attractive: an advantage of zebra stripes. J Exp Biol 215:736–745

Embar K, Kotler BP, Mukherjee S (2011) Risk management in optimal foragers: the effect of sightlines and predator type on patch use, time allocation, and vigilance in gerbils. Oikos 120:1657–1666

Endara MJ, Coley PD (2011) The resource availability hypothesis revisited: a meta-analysis. Funct Ecol 25:389–398

Endler JA (1981) An overview of the relationships between mimicry and crypsis. Biol J Linn Soc 16:25–31
Evans CRH (1987) Oral ulceration after contact with the houseplant *Dieffenbachia*. Br Dental J 162:467–468
Evenari M (1989) The history of research on white-green variegated plants. Bot Rev 55:106–139
Fadzly N, Burns KC (2010) Hiding from the ghost of herbivory past: evidence for crypsis in an insular tree species. Int J Plant Sci 171:828–833
Fadzly N, Jack C, Schaefer HM, Burns KC (2009) Ontogenetic colour changes in an insular tree species: signalling to extinct browsing birds? New Phytol 184:495–501
Fahn A (1990) Plant anatomy, 4th edn. Pergamon, Oxford
Finch S, Jones TH (1989) An analysis of the deterrent effect of Aphids on cabbage root fly (*Delia radicum*) egg-laying. Ecol Entomol 14:387–391
Forbes P (2009) Dazzled and deceived: mimicry and camouflage. Yale University Press, New Haven, CT
Forsman A, Herrström J (2004) Asymmetry in size, shape, and color impairs the protective value of conspicuous color patterns. Behav Ecol 15:141–147
Forsmann A, Merilaita S (1999) Fearful symmetry: pattern size and asymmetry affects aposematic signal efficacy. Evol Ecol 13:131–140
Franceschi VR, Horner HT Jr (1980) Calcium oxalate crystals in plants. Bot Rev 46:361–427
Gardner DG (1994) Injury to the oral mucous membranes caused by the common houseplant, diffenbachia. A review. Oral Surg Oral Med Oral Pathol 78:631–633
Gibson G (1992) Do tsetse flies 'see' zebras? A field study of the visual response of tsetse to striped targets. Physiol Entomol 17:141–147
Givnish TJ (1990) Leaf mottling: relation to growth form and leaf phenology and possible role as camouflage. Funct Ecol 4:463–474
Gotelli NJ, Ulrich W (2012) Statistical challenges in null model analysis. Oikos 121:171–180
Grubb PJ (1992) A positive distrust in simplicity – lessons from plant defences and from competition among plants and among animals. J Ecol 80:585–610
Halpern M, Raats D, Lev-Yadun S (2007a) Plant biological warfare: thorns inject pathogenic bacteria into herbivores. Environ Microbiol 9:584–592
Halpern M, Raats D, Lev-Yadun S (2007b) The potential anti-herbivory role of microorganisms on plant thorns. Plant Signal Behav 2:503–504
Halpern M, Waissler A, Dror A, Lev-Yadun S (2011) Biological warfare of the spiny plant: introducing pathogenic microorganisms into herbivore's tissues. Adv Appl Microbiol 74:97–116
Hamilton WD, Brown SP (2001) Autumn tree colours as a handicap signal. Proc R Soc Lond B 268:1489–1493
Hara N (1957) Study of the variegated leaves, with special reference to those caused by air spaces. Jap J Bot 16:86–101
Harper JL (1977) Population biology of plants. Academic, London
Hawlena D, Schmitz OJ (2010) Herbivore physiological response to predation risk and implications for ecosystem nutrient dynamics. Proc Natl Acad Sci U S A 107:15503–15507
Herrera CM (1982) Defense of ripe fruit from pests: its significance in relation to plant-disperser interactions. Am Nat 120:218–241
Hinton HE (1973) Natural deception. In: Gregory RL, Gombrich EH (eds) Illusion in nature and art. Duckworth, London, pp 97–159
Hornell J (1941) Fishing-poisons. Man 41:126–128
Inbar M, Lev-Yadun S (2005) Conspicuous and aposematic spines in the animal kingdom. Naturwiss 92:170–172
Inbar M, Doostdar H, Mayer RT (1999) Effects of sessile Whitefly nymphs (Homoptera: Aleyrodidae) on leaf-chewing larvae (Lepidoptera: Noctuidae). Environ Entomol 28:353–357
Janzen DH (1977) Why fruits rot, seeds mold, and meat spoils. Am Nat 111:691–713

Jones KA, Krebs JR, Whittingham MJ (2006) Interaction between seed crypsis and habitat structure influence patch choice in a granivorous bird, the chaffinch *Fringilla coelebs*. J Avian Biol 37:413–418

Justus M, Witte L, Hartmann T (1997) Levels and tissue distribution of loline alkaloids in endophyte-infected *Festuca pratensis*. Phytochem 44:51–57

Kappers IF, Aharoni A, van Herpen TWJM, Luckerhoff LLP, Dicke M, Bouwmeester HJ (2005) Genetic engineering of terpenoid metabolism attracts bodyguards to Arabidopsis. Science 309:2070–2072

Karban R (2007) Deciduous leaf drop reduces insect herbivory. Oecologia 153:81–88

Kendrick JW, Tucker J, Peoples SA (1955) Nitrate poisoning in cattle due to ingestion of variegated thistle, *Silybum marianum*. J Am Vet Med Assoc 126:53–56

Kessler A, Baldwin IT (2001) Defensive function of herbivore induced plant volatile emissions in nature. Science 291:2141–2144

Kettlewell B (1973) The evolution of melanism. Clarendon, Oxford

Klooster MR, Clark D, Culley TA (2009) Cryptic bracts facilitate herbivore avoidance in the mycoheterotrophic plant *Monotropsis odorata* (Ericaceae). Am J Bot 96:2197–2205

Konoplyova A, Petropoulou Y, Yiotis C, Psaras GK, Manetas Y (2008) The fine structure and photosynthetic cost of structural leaf variegation. Flora 203:653–662

Kotepong P, Ketsa S, van Doorn WG (2011) A white mutant of Malay apple fruit (*Syzygium malaccense*) lacks transcript expression and activity for the last enzyme of anthocyanin synthesis, and the normal expression of a MYB transcription factor. Funct Plant Biol 38:75–86

La Rocca N, Rascio N, Pupillo P (2011) Variegation in *Arum italicum* leaves. A structural-functional study. Plant Physiol Biochem 49:1392–1398

La Rocca N, Pupillo P, Puppi G, Rascio N (2014) *Erythronium dens-canis* L. (Liliaceae): an unusual case of change of leaf mottling. Plant Physiol Biochem 74:108–117

Landau SY, Ben-Moshe E, Shlosberg A, Bellaiche M, Perevolotsky A (1999) Conditioned aversion to minimize Ferula communis intake by orphaned lambs. J Range Manag 52:436–439

Lee D (2007) Nature's palette. The science of plant color. University Chicago Press, Chicago, IL

Lev-Yadun S (2001) Aposematic (warning) coloration associated with thorns in higher plants. J Theor Biol 210:385–388

Lev-Yadun S (2003a) Why do some thorny plants resemble green zebras? J Theor Biol 244:483–489

Lev-Yadun S (2003b) Weapon (thorn) automimicry and mimicry of aposematic colorful thorns in plants. J Theor Biol 244:183–188

Lev-Yadun S (2006a) Defensive coloration in plants: a review of current ideas about anti-herbivore coloration strategies. In: Teixeira da Silva JA (ed) Floriculture, ornamental and plant biotechnology: advances and topical issues, vol IV. Global Science Books, London, pp 292–299

Lev-Yadun S (2006b) Defensive functions of white coloration in coastal and dune plants. Isr J Plant Sci 54:317–325

Lev-Yadun S (2009a) Aposematic (warning) coloration in plants. In: Baluska F (ed) Plant-environment interactions, vol II, From sensory plant biology to active behavior. Springer, Berlin, pp 167–202

Lev-Yadun S (2009b) Müllerian and Batesian mimicry rings of white-variegated aposematic spiny and thorny plants: a hypothesis. Isr J Plant Sci 57:107–116

Lev-Yadun S (2009c) Müllerian mimicry in aposematic spiny plants. Plant Signal Behav 4:482–483

Lev-Yadun S (2011) Fearful symmetry in aposematic plants. Plant Signal Behav 6:1739–1740

Lev-Yadun S (2013) Theoretical and functional complexity of white variegation of unripe fleshy fruits. Plant Signal Behav 8:e25851

Lev-Yadun S (2014) Potential defence from herbivory by dazzle effects and trickery coloration of leaf variegation. Biol J Linn Soc 111:692–697

Lev-Yadun S, Gould KS (2007) What do red and yellow autumn leaves signal? Bot Rev 73:279–289
Lev-Yadun S, Gould KS (2009) Role of anthocyanins in plant defense. In: Gould KS, Davies KM, Winefield C (eds) Life's colorful solutions: the biosynthesis, functions, and applications of anthocyanins. Springer, Berlin, pp 21–48
Lev-Yadun S, Halpern M (2007) Ergot (*Claviceps purpurea*) - an aposematic fungus. Symbiosis J 43:105–108
Lev-Yadun S, Halpern M (2008) External and internal spines in plants insert pathogenic micro-organisms into herbivore's tissues for defense. In: Van Dijk T (ed) Microbial ecology research trends. Nova Scientific, New York, NY, pp 155–168
Lev-Yadun S, Ne'eman G (2004) When may green plants be aposematic? Biol J Linn Soc 81:413–416
Lev-Yadun S, Ne'eman G (2006) Color changes in old aposematic thorns, spines, and prickles. Isr J Plant Sci 54:327–333
Lev-Yadun S, Ne'eman G (2013) Bimodal colour pattern of individual *Pinus halepensis* Mill. seeds – an adaptation for crypsis. Biol J Linn Soc 109:271–278
Lev-Yadun S, Gopher A, Abbo S (2000) The cradle of agriculture. Science 288:1602–1603
Lev-Yadun S, Dafni A, Flaishman MA, Inbar M, Izhaki I, Katzir G, Ne'eman G (2004a) Plant coloration undermines herbivorous insect camouflage. Bioessays 26:1126–1130
Lev-Yadun S, Flaishman MA, Atzmon N (2004b) Nonchimeric variegated mutation in *Cupressus sempervirens* L. Int J Plant Sci 165:257–261
Lev-Yadun S, Ne'eman G, Izhaki I (2009) Unripe red fruits may be aposematic. Plant Signal Behav 4:836–841
Lüttge U (1997) Physiological ecology of tropical plants. Springer, Berlin
Mackay DA, Jones RE (1989) Leaf shape and the host-finding behaviour of two ovipositing monophagous butterfly species. Ecol Entomol 14:423–431
Majerus MEN (1998) Melanism. Evolution in action. Oxford University Press, Oxford
Manzur MI, Courtney SP (1984) Influence of insect damage in fruits of hawthorn on bird foraging and seed dispersal. Oikos 43:265–270
Martínez-Abraín A (2008) Statistical significance and biological relevance: a call for a more cautious interpretation of results in ecology. Acta Oecol 34:9–11
McKey D, Waterman PG, Mbi CN, Gartlan JS, Struhsaker TT (1978) Phenolic content of vegetation in two African rain forests: ecological implications. Science 202:61–64
Merilaita S (1998) Crypsis through disruptive coloration in an isopod. Proc R Soc Lond B 265:1059–1064
Midgley JJ (2004) Why are spines of African *Acacia* species white? Afr J Range Forage Sci 21:211–212
Midgley JJ, Botha MA, Balfour D (2001) Patterns of thorn length, density, type and colour in African Acacias. Afr J Range Forage Sci 18:59–61
Mudge JF (2013) Explicit consideration of critical effect sizes and costs of errors can improve decision-making in plant science. New Phytol 199:876–878
Nassar O, Lev-Yadun S (2009) How prickly is a prickly pear? Isr J Plant Sci 57:117–124
Niemelä P, Tuomi J, Siren S (1984) Selective herbivory on mosaic leaves of variegated *Acer pseudoplatanus*. Experientia 40:1433–1434
Noy-Meir I, Gutman M, Kaplan Y (1989) Responses of Mediterranean grassland plants to grazing and protection. J Ecol 77:290–310
Ohmiya A, Kishimoto S, Aida R, Yoshioka S, Sumitomo K (2006) Carotenoid cleavage dioxygenase (CmCCD4a) contributes to white color formation in chrysanthemum petals. Plant Physiol 142:1193–1201
Ortolani A (1999) Spots, stripes, tail tips and dark eyes: predicting the function of carnivore colour patterns using the comparative method. Biol J Linn Soc 67:433–476
Osorio D, Srinivasan MV (1991) Camouflage by edge enhancement in animal coloration patterns and its implications for visual mechanisms. Proc R Soc Lond B 244:81–85

Perevolotsky A, Seligman N (1998) Role of grazing in Mediterranean rangeland ecosystems. Inversion of a paradigm. BioScience 48:1007–1017
Pfister JA, Müller-Schwarze D, Balph DF (1990) Effects of predator fecal odors on feed selection by sheep and cattle. J Chem Ecol 16:573–583
Poethig RS (1987) Clonal analysis of cell lineage patterns in plant development. Am J Bot 74:581–594
Porter JK (1994) Chemical constituents of grass endophytes. In: Bacon CW, White JF (eds) Biotechnology of endophytic fungi of grasses. CRC, Boca Raton, FL, pp 103–123
Preisser EL (2009) The physiology of predator stress in free-ranging prey. J Anim Ecol 78:1103–1105
Prokopy RJ, Owens ED (1983) Visual detection of plants by herbivorous insects. Annu Rev Entomol 28:337–364
Rasmann S, Agrawal AA (2011) Latitudinal patterns in plant defense: evolution of cardenolides, their toxicity and induction following herbivory. Ecol Lett 14:476–483
Rausher MD (1978) Search image for leaf shape in a butterfly. Science 200:1071–1073
Reeves JL (2011) Vision should not be overlooked as an important sensory modality for finding host plants. Environ Entomol 40:855–863
Reznicek G, Jurenitsch J, Robien W, Kubelka W (1989) Saponins in *Cyclamen* species: configuration of cyclamiretin C and structure of isocyclamin. Phytochem 28:825–828
Ripple WJ, Beschta RL (2004) Wolves and the ecology of fear: can predation risk structure ecosystems? BioScience 54:755–766
Ronel M, Lev-Yadun S (2009) Spiny plants in the archaeological record of Israel. J Arid Environ 73:754–761
Ronel M, Lev-Yadun S (2012) The spiny, thorny and prickly plants in the flora of Israel. Bot J Linn Soc 168:344–352
Ronel M, Khateeb S, Lev-Yadun S (2009) Protective spiny modules in thistles of the Asteraceae in Israel. J Torrey Bot Soc 136:46–56
Ronel M, Ne'eman G, Lev-Yadun S (2010) Spiny east-Mediterranean plant species flower later and in a drier season than non-spiny species. Flora 205:276–281
Rosso D, Bode R, Li W, Krol M, Saccon D, Wang S, Schillaci LA, Rodermel SR, Maxwell DP, Hüner PA (2009) Photosynthetic redox imbalance governs leaf sectoring in the *Arabidopsis thaliana* variegation mutants *immutans*, *spotty*, *var1*, and *var2*. Plant Cell 21:3473–3492
Rothschild M (1986) The red smell of danger. New Sci 111:34–36
Rothschild M, Moore B (1987) Pyrazines as alerting signals in toxic plants and insects. In: Labeyrie V, Fabres G, Lachaise D (eds) Insects – plants. Dr W. Junk, Dordrecht, pp 97–101
Rowland HM, Mappes J, Ruxton GD, Speed MP (2010) Mimicry between unequally defended prey can be parasitic: evidence for quasi-Batesian mimicry. Ecol Lett 13:1494–1502
Rubino DL, McCarthy BC (2004) Presence of aposematic (warning) coloration in vascular plants of southeastern Ohio. J Torrey Bot Soc 131:252–256
Ruxton GD (2002) The possible fitness benefits of striped coat coloration for zebra. Mammal Rev 32:237–244
Ruxton GD, Sherratt TN, Speed MP (2004) Avoiding attack. The evolutionary ecology of crypsis, warning signals and mimicry. Oxford University Press, Oxford
Saikkonen K, Faeth SH, Helander M, Sullivan TJ (1998) Fungal endophytes: a continuum of interactions with host plants. Annu Rev Ecol Syst 29:319–343
Sakai WS, Hanson M, Jones RC (1972) Raphids with barbs and grooves in *Xanthosoma sagittifolium* (Araceae). Science 178:314–315
Salts D, Ward D (2000) Responding to a three-pronged attack: desert lilies subject to herbivory by dorcas gazelles. Plant Ecol 148:127–138
Scarchuk J, Lent JM (1965) The structure of mottled-leaf summer squash. J Hered 56:167–168
Schaefer HM, Ruxton GD (2009) Deception in plants: mimicry or perceptual exploitation? Trends Ecol Evol 24:676–685

Schaefer HM, Ruxton GD (2011) Plant-animal communication. Oxford University Press, New York, NY
Schaefer M, Stobbe N (2006) Disruptive coloration provides camouflage independent of background matching. Proc R Soc Lond B 273:2427–2432
Schemske DW, Mittelbach GG, Cornell HV, Sobel JM, Roy K (2009) Is there a latitudinal gradient in the importance of biotic interactions. Annu Rev Ecol Evol Syst 40:245–269
Sheriff MJ, Krebs CJ, Boonstra R (2009) The sensitive hare: sublethal effects of predator stress on reproduction in snowshoe hares. J Anim Ecol 78:1249–1258
Shifriss O (1981) Do *Cucurbita* plants with silvery leaves escape virus infection? Cucurbit Genetics Cooperative Report 4:42–43
Skelhorn J, Rowland HM, Ruxton GD (2010a) The evolution and ecology of masquerade. Biol J Linn Soc 99:1–8
Skelhorn J, Rowland HM, Speed MP, Ruxton GD (2010b) Masquerade: camouflage without crypsis. Science 327:51
Skelhorn J, Rowland HM, Speed MP, De Wert L, Quinn L, Delf J, Ruxton GD (2010c) Size-dependent misclassification of masqueradinf prey. Behav Ecol 21:1344–1348
Smith AP (1986) Ecology of leaf color polymorphism in a tropical forest species: habitat segregation and herbivory. Oecologia 69:283–287
Soltau U, Dötterl S, Liede-Schumann S (2009) Leaf variegation in *Caladium steudneriifolium* (Araceae): a case of mimicry? Evol Ecol 23:503–512
Speed MP, Ruxton GD (2005) Warning displays in spiny animals: one (more) evolutionary route to aposematism. Evolution 59:2499–2508
Stanley RM II (1998) To fool a glass eye. Camouflage versus photoreconnaissance in World War II. Smithsonian Institute Press, Washington, DC
Steiner MC (2005) The faunas of Hayonim Cave (Israel): a 200,000-year record of Paleolithic diet, demography and society. Peabody Museum of Archaeology and Ethnology, Cambridge
Stevens M, Cuthill IC, Windsor AMM, Walker HJ (2006) Disruptive contrast in animal camouflage. Proc R Soc B 273:2433–2438
Tchernov E (1979) Quaternary fauna. In: Horowitz A (ed) The quaternary of Israel. Academic, New York, NY, pp 257–290
Terlouw EMC, Boissy A, Blinet P (1998) Behavioural responses of cattle to the odours of blood and urine from conspecifics and to the odour of faeces from carnivores. Appl Anim Behav Sci 57:9–21
Tilney-Bassett RAE (1986) Plant chimeras. Edward Arnold, London
Tsukaya H, Okada H, Mohamed M (2004) A novel feature of structural variegation in leaves of the tropical plant *Schismatoglottis calyptrata*. J Plant Res 117:477–480
von Helversen B, Schooler LJ, Czienskowski U (2013) Are stripes beneficial? Dazzle camouflage influences perceived speed and hit rates. PLoS One 8:e61173
Waage JK (1981) How the zebra got its stripes – biting flies as selective agents in the evolution of zebra coloration. J Ent Soc S Afr 44:351–358
Wäli PP, Wäli PR, Saikkonen K, Tuomi J (2013) Is the pathogenic ergot fungus a conditional defensive mutualist for its host grass? PLoS One 8:e69249
Ward D, Spiegel M, Saltz D (1997) Gazelle herbivory and interpopulation differences in calcium oxalate content of leaves of a desert lily. J Chem Ecol 23:333–346
Wickler W (1968) Mimicry in plants and animals. Weidenfeld and Nicolson, London
Wiens D (1978) Mimicry in plants. Evol Biol 11:365–403
Wilkinson N (1969) A brush with life. Seeley Service & Co, London
Wilkinson D, Sherratt TN (2008) The art of concealment. Biologist 55:10–15
Williams D (2001) Naval camouflage 1914–1945. A complete visual reference. Naval Institute Press, Annapolis
Yamazaki K (2010) Leaf mines as visual defensive signals to herbivores. Oikos 119:796–801

Yu F, Fu A, Aluru M, Park S, Xu Y, Liu H, Liu X, Foudree A, Nambogga M, Rodermel S (2007) Variegation mutants and mechanisms of chloroplast biogenesis. Plant Cell Environ 30:350–365

Zheng S-J, Snoeren TAL, Hogewoning SW, van Loon JJA, Dicke M (2010) Disruption of plant carotenoid biosynthesis through virus-induced gene silencing affects oviposition behaviour of the butterfly *Pieris rapae*. New Phytol 186:733–745

Zohary M (1962) Plant life of Palestine. Israel and Jordan. Ronald Press, New York, NY

Zohary M (1973) Geobotanical foundations of the Middle East. Gustav Fischer, Stuttgart

Zohary M (1983) Man and vegetation in the Middle East. In: Holzner W, Werger MJA, Ikusima I (eds) Man's impact on vegetation. Dr W. Junk, The Hague, pp 287–295

Zohary D, Hopf M, Weiss E (2012) Domestication of plants in the Old World, 4th edn. Clarendon, Oxford

Part III
Ecology

Sunlight and Soil–Litter Mixing: Drivers of Litter Decomposition in Drylands

Paul W. Barnes, Heather L. Throop, Steven R. Archer, David D. Breshears, Rebecca L. McCulley, and Mark A. Tobler

Contents

P.W. Barnes (✉) • M.A. Tobler
Department of Biological Sciences, Loyola University New Orleans, New Orleans, LA 70118, USA
e-mail: pwbarnes@loyno.edu; tobler@loyno.edu

H.L. Throop
Department of Biology, New Mexico State University, Las Cruces, NM 88003, USA
e-mail: throop@nmsu.edu

S.R. Archer • D.D. Breshears
School of Natural Resources and the Environment, University of Arizona, Tucson, AZ 85721, USA
e-mail: sarcher@ag.arizona.edu; daveb@email.arizona.edu

R.L. McCulley
Department of Plant and Soil Science, University of Kentucky, Lexington, KY 40546, USA
e-mail: rebecca.mcculley@uky.edu

U. Lüttge, W. Beyschlag (eds.), *Progress in Botany*, Progress in Botany 76,
DOI 10.1007/978-3-319-08807-5_11

Abstract Decomposition of leaf litter is a key component of biogeochemical cycles but the mechanisms driving it in arid and semiarid ecosystems (drylands) remain unresolved. Here, we review recent findings that demonstrate dual roles of solar radiation (ultraviolet and photosynthetically active radiation) and soil–litter mixing as drivers of decomposition in drylands. We focus on the known and potential mechanisms by which these factors influence leaf litter decomposition, explore how the importance of these two drivers may shift over time, and propose possible avenues by which these factors may interact. Special attention is given to UV in sunlight, as this radiation is known to have multiple roles in influencing decomposition and has received considerable recent research attention. We also identify important uncertainties and challenges and offer a generalized conceptual model to guide future research aimed at enhancing our mechanistic understanding and quantitative modeling of the processes by which soil deposition and solar radiation together influence leaf litter decomposition rates in globally extensive dryland ecosystems.

1 Introduction

Decomposition of organic material strongly controls patterns of nutrient and carbon (C) retention and release in ecosystems. Although C and nutrients in the litter pool account for only a small portion of system-wide totals, the relatively rapid turnover of this pool makes leaf litter decomposition a key component of biogeochemical cycles (Aerts 1997; Berg and Laskowski 2005). Traditionally, the prevailing drivers of litter decomposition in terrestrial ecosystems have been viewed as a combination of abiotic (e.g., temperature, moisture) and biotic (e.g., litter quality) factors interacting to mediate decomposer community composition and metabolic activity, and considerable progress has been made in developing a mechanistic understanding of the controls over decomposition at local, regional, and global scales (Meentemeyer 1978; Couteaux et al. 1995; Aerts 1997; Hibbard et al. 2005; Cable et al. 2011). However, predicting decomposition dynamics in globally extensive arid and semi-arid systems (hereafter "drylands") has proven to be problematic, with models typically underestimating its rates (Whitford et al. 1981; Moorhead and Reynolds 1991; Kemp et al. 2003; Parton et al. 2007; Adair et al. 2008).

The disconnect between decomposition models and measurements suggests controls over decomposition in drylands differ fundamentally from those in wetter environments and that unique drivers may be operating in drylands (reviewed in Throop and Archer 2009; Austin 2011; King et al. 2012). Recently, several studies have shown that ultraviolet (UV; 280–400 nm) and photosynthetic active radiation (PAR; 400–700 nm) in ambient sunlight can accelerate litter mass loss in drylands via the process of photodegradation (Austin and Vivanco 2006; Brandt et al. 2007; Day et al. 2007; Brandt et al. 2010). Although the magnitudes and proposed mechanisms of photodegradation are variable and poorly understood (King et al. 2012; Song et al. 2013a), it may be an important, historically overlooked driver that could potentially explain, at least in part, why traditional models typically underpredict decomposition rates in drylands (Throop and Archer 2009; Austin 2011). Photodegradation results in the direct loss of a number of gases, including CO_2 (Brandt et al. 2009; Lee et al. 2012), CH_4 (McLeod et al. 2008; Bloom et al. 2010), CO (Schade and Crutzen 1999; Lee et al. 2012), and N_2O (Foereid et al. 2010), and recent analyses suggest that photodegradation of surface litter could have a measurable influence on landscape-level CO_2 flux rates, and ultimately C storage (Brandt et al. 2009; Rutledge et al. 2010).

While the climate and low and sparse vegetation cover of drylands create conditions of high solar radiation flux near ground level, these environments also favor considerable soil movement via wind and water transport (Breshears et al. 2003; Okin et al. 2009b), which can partially cover and eventually bury plant litter on the soil surface (Throop and Archer 2007). This combination of litter and the soil that covers it (the "soil–litter matrix") includes both loose soil mixed with litter and soil that, over time, adheres to leaf surfaces to form a complex aggregate of soil and microbial products (Throop and Archer 2009; Barnes et al. 2012; Hewins et al. 2013). Although the nature and development of this soil–litter matrix remains poorly understood, available evidence indicates that decomposition in drylands can be strongly correlated with levels of soil accumulation onto litter and hence the development of this soil–litter matrix (Throop and Archer 2007). The mechanisms underlying this response have yet to be fully elucidated; however, the mixing of soil and litter and the resultant soil–litter matrix appears to enhance microbial activity (Hewins et al. 2013) while simultaneously shielding litter from photodegradation (Barnes et al. 2012). Soil coverage may also shield sensitive microbes from potential detrimental effects of solar UV (Moody et al. 1999; Johnson 2003; Cockell et al. 2008). An increased understanding of the factors that influence soil deposition onto litter, the processes governing soil–litter matrix development, and the mechanisms by which solar radiation and soil–litter mixing interact to influence decomposition appears critical to understanding litter decomposition in drylands and resolving seemingly conflicting views on this process.

Here we examine the dual roles of sunlight and soil–litter mixing as drivers of litter decomposition in dryland ecosystems. Specifically, we focus on the known and potential mechanisms by which these factors influence leaf litter degradation, explore how the importance of these two drivers may shift over time, and propose possible avenues by which these factors may interact to influence decomposition. We give special attention to UV in sunlight, as this radiation is known to have multiple roles in influencing decomposition and has received considerable recent research attention. We also identify important uncertainties and challenges and offer a generalized conceptual model to guide future research aimed at enhancing our mechanistic understanding and quantitative modeling of the processes by which soil deposition and solar radiation together influence leaf litter decomposition rates in globally extensive dryland ecosystems.

2 Overview

Solar radiation (UV and PAR) and soil–litter mixing can influence decomposition of leaf litter in dryland ecosystems by multiple mechanisms, and interactions between these and other environmental factors can further modify these effects (Fig. 1; see Table 1 for definition of terms). The total radiation exposure of litter (Fig. 1B) will be influenced by a combination of climatic, landscape/vegetation, and species-specific factors (Fig. 1A) that determine the timing and duration of exposure. The effects of solar radiation (primarily UV) on biotic processes generally reduce rates of decomposition (−) though there is the potential for some beneficial effects (Fig. 1D). The effects of sunlight on abiotic processes (primarily photodegradation) tend to enhance (+) decomposition (Fig. 1E). Both of these processes exhibit distinct spectral sensitivities (i.e., action spectra) depending on the underlying chromophores and mechanisms involved (Fig. 1C). Solar radiation can also influence decomposition via its effects on leaf chemistry and structure

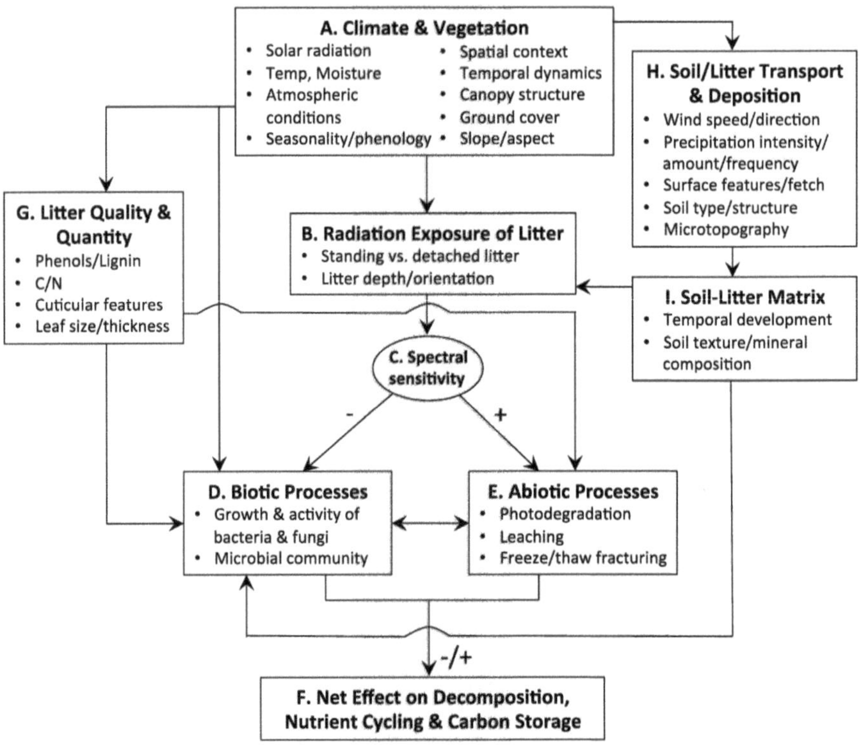

Fig. 1 Potential effects of solar radiation (UV and PAR) and soil–litter mixing on leaf litter decomposition in drylands, including interactions with other environmental factors. See Table 1 for definition of terms

Table 1 Terminology used to describe the various mechanisms by which solar radiation influences terrestrial litter decomposition. Although "photodegradation" is sometimes used broadly to include all radiation effects on mass loss, including microbial effects, we use it in a strict sense to refer only to abiotic effects on mass loss

Name	Mediation	Process	Material intercepting radiation	Outcome for plant litter decomposition	Examples
Primary photodegradation (Fig. 1E; aka photolysis; photooxidation[a])	Abiotic	Solar radiation breaks chemical bonds in litter via the direct absorption of radiation by photoreactive compounds	Litter	Efflux of gaseous compounds and mass loss	Austin and Vivanco (2006); Brandt et al. (2009); Lee et al. (2012)
Secondary photodegradation (Fig. 1E)	Abiotic	Breakdown of litter via reactive intermediates formed by primary photodegradation	Litter	Fragmentation and mass loss; increased potential for leaching, increased surface area for microbial attack	Austin and Vivanco (2006); Brandt et al. (2009); Lee et al. (2012)[b]
Thermal degradation	Abiotic	Low temperature (<100 °C) breakdown of chemical bonds in litter; may occur in absence of solar radiation; enhances photodegradation	Litter	Efflux of gaseous compounds and mass loss	Lee et al. (2012)
Photopriming (Fig. 1E→1D)	Abiotic/ biotic	Primary and/or secondary photodegradation that affects subsequent microbial decomposition	Litter	Often, but not always, enhanced microbial activity promoting mass loss	Foereid et al. (2010)
Microbially enhanced photodegradation (Fig. 1D→1E)	Biotic	Microbes modify litter which then influences photodegradation	Microbes	Increased mass loss	Ma et al. (2012)
Microbial photoinhibition (Fig. 1D)	Biotic	Solar radiation stress on microbial physiology/communities, affecting microbial decomposition	Microbes	Usually, but not always, decreased microbial activity and reduced mass loss	Duguay and Klironomos (2000); Verhoef et al. (2000); Newsham et al. (1997)

(continued)

Table 1 (continued)

Name	Mediation	Process	Material intercepting radiation	Outcome for plant litter decomposition	Examples
Plant photochemistry (Fig. 1G)	Abiotic/biotic	Solar radiation-induced changes in plant tissue structure and/or chemical composition that affect later microbial decomposition and/or photodegradation	Live plants including leaves and stems	Attenuated microbial activity leading to reduced mass loss; effects on photodegradation unknown at present	Gehrke et al. (1995); Rozema et al. (1997); Pancotto et al. (2003)

[a]Lee et al. (2012) found that O_2 was not required for this process; thus photo-oxidation is apparently one of several pathways for photodegradation of compounds in litter

[b]Field and laboratory methods to date do not differentiate between abiotic decomposition from primary or secondary photodegradation

(Fig. 1G), which subsequently influences both biotic and abiotic degradation processes when foliage dies (Fig. 1D, E).

Soil accumulation onto litter will be influenced by meteorological, vegetation, landscape, and edaphic factors (Fig. 1A) that influence the rate, magnitude, and direction of soil and litter transport and soil deposition (Fig. 1H). Over time, and depending on soil mineralogy and particle size composition, a complex mixture of soil, plant material, and microbial products can develop to form an adhesive soil–litter matrix (Fig. 1I) that can shield litter from solar radiation (Fig. 1B), reduce photodegradation (Fig. 1E), and enhance microbial processes (Fig. 1D).

Ultimately, the net effect of solar radiation and soil deposition on the rates of decomposition, nutrient cycling, and carbon storage (Fig. 1F) will depend on the weighted contribution of biotic and abiotic processes and may be positive, negative, or neutral depending on the relative strength of the individual effects. Subsequent sections review our current understanding of the factors itemized in Fig. 1, examine how interactions among them might play out under field conditions, and address some of the knowledge gaps and challenges associated with quantifying them.

3 Sunlight, UV Radiation and Decomposition

3.1 Brief History and Overview of Experimental Approaches

The solar spectrum at the Earth's surface consists of a mixture of UV, PAR, and near-infrared (IR) radiation, with the majority of the energy and photon flux coming from the latter two wavebands. Although the UV component of the spectrum comprises a small (<5 %) portion of the total surface solar irradiance, its influence on terrestrial plants and ecosystems can be significant (Day and Neale 2002; Ballaré et al. 2011; Paul et al. 2012; Wargent and Jordan 2013). Historically, research examining UV effects on decomposition was undertaken to evaluate potential ecological impacts of the changing solar ultraviolet-B (UV-B; 280–320 nm) regime associated with stratospheric ozone depletion, and field studies were therefore often conducted in high-latitude ecosystems where ozone loss was acute (Gehrke et al. 1995; Pancotto et al. 2003; Zaller et al. 2009). These studies typically employed plastic films to reduce ambient solar UV-B (i.e., UV-exclusion experiments) or filtered UV-emitting lamps to simulate elevated solar UV-B conditions (i.e., UV-B-enhancement experiments) associated with ozone depletion. More recently, efforts have shifted to explore the mechanisms and fundamental roles of UV-B, UV-A (320–400 nm), and PAR in influencing terrestrial decomposition and biogeochemistry using a combination of field radiation-attenuation experiments and controlled laboratory experiments with artificial light sources. Although the technical issues and uncertainties associated with the different experimental approaches to UV experiments are beyond the scope of this review (but see Caldwell et al. 1983a; Flint et al. 2003; Aphalo et al. 2013), it is worth noting that the detection of UV effects on decomposition appears to be influenced by the nature and type of experiments conducted (i.e., field UV-exclusion vs. lamp studies and field vs. laboratory studies, King et al. 2012; Song et al. 2013a). These findings suggest that experimental techniques used to manipulate UV exposure and the maintenance of proper spectral balances (i.e., UV-B:UV-A:PAR ratios) are important in interpreting both the quantitative and qualitative effects of UV radiation on decomposition (Fig. 1D, E), as has been well documented for UV studies on higher plants (e.g., Caldwell and Flint 1989; Flint et al. 2003; Krizek 2004).

3.2 Mechanisms for Solar Radiation Influence on Decomposition

Findings to date indicate that UV (and PAR), either at ambient or enhanced levels, can influence litter decomposition in terrestrial ecosystems via multiple mechanisms including effects on microbes (Fig. 1D) and abiotic photochemistry (Fig. 1E) as well as effects mediated through alterations in leaf chemistry (Fig. 1G and recent meta-analyses of King et al. 2012; Song et al. 2013a). These processes also interact with one another [e.g., abiotic processes such as photodegradation may enhance or retard biotic (microbial) process; Fig. 1E→D] and multiple pathways can occur within a given process (e.g., different pathways of photodegradation as described below). The terminology surrounding these processes and the mechanism underlying them is somewhat ambiguous in the literature, and interpretations are further complicated with respect to what constitutes "primary" vs. "secondary" and "direct" vs. "indirect" effects. Table 1 summarizes the definitions and interpretations used in this paper.

Photodegradation (Fig. 1E) is an abiotic process that occurs via photochemical mineralization of photo-reactive compounds (King et al. 2012), such as lignin (i.e., primary photodegradation; Table 1), and/or the transformation of compounds as a result of solar radiation-induced formation of reactive oxygen species and other intermediates (i.e., secondary photodegradation; Rozema et al. 1997; Anesio et al. 1999; Gallo et al. 2006; Day et al. 2007; Austin and Ballaré 2010; King et al. 2012). Photodegradation is enhanced in the presence of oxygen but also occurs under anoxic conditions, suggesting there are multiple chemical pathways involved (Lee et al. 2012). While photodegradation has long been viewed as an important mechanism influencing decomposition in aquatic ecosystems (e.g., Zepp et al. 1995), only recently has it been shown to be an important driver of decomposition in terrestrial ecosystems (Henry et al. 2008; Brandt et al. 2010; Song et al. 2012). In a semi-arid Patagonian steppe, Austin and Vivanco (2006) found that reducing solar radiation affected decomposition much more strongly than reducing microbial decomposition with a biocide treatment, and they attributed about 60 % of the observed litter mass loss to shortwave radiation. About half of this mass loss was due to UV-B. Similarly, 14–22 % of leaf mass loss was attributed to solar UV-B in a field litterbag experiment in the Sonoran Desert (Day et al. 2007). However, not all investigators have found significant photodegradation effects (e.g., Kirschbaum et al. 2011) and there is evidence that the degree of photodegradation may vary with litter chemical composition (Uselman et al. 2011; Lee et al. 2012) and moisture (Schade and Crutzen 1999; Andrady et al. 2003; Gallo et al. 2006). Photodegradation rates increase with increasing ambient temperature (Lee et al. 2012). Furthermore, thermal degradation, the thermal decay of litter compounds at relatively low temperatures (<100 °C; well below the ignition point), can account for a substantial component of measured trace gas fluxes in photodegradation experiments (Lee et al. 2012). Increases in temperature from solar radiation may therefore influence litter decay both through enhancing photodegradation and from thermal degradation alone.

The effects of UV radiation on bacteria and fungi (Fig. 1D) tend to be negative, with growth, survival, and the production and germination of spores generally inhibited, especially by UV-B (Table 1, Caldwell et al. 2007). These "microbial photoinhibition" effects of sunlight are generally thought to be manifestations of detrimental impacts on DNA and repair processes (Hughes et al. 2003; Johnson 2003; Jacobs et al. 2005; Gunasekera and Sundin 2006). However, species vary in their UV sensitivity (Moody et al. 1999; Braga et al. 2001; Ulevičius et al. 2004), resulting in shifts in microbial community composition when material is exposed to sunlight (Kadivar and Stapleton 2003; Rangel et al. 2004). The UV tolerance of microbes may be related to the solar UV environment of origin, with microbes from sites with low UV exposures being more sensitive to UV insult than those from sites experiencing higher UV fluxes (Gunasekera et al. 1997; Zucconi et al. 2002). Microbes isolated from deserts, where natural UV exposure is high (e.g., the Atacama, Gobi, and Negev Deserts), can be relatively tolerant to wide ranges of UV irradiation (including UV-C; <280 nm) (Paulino-Lima et al. 2013), especially when present as desiccated spores and associated with soil particles (Cockell et al. 2008; Osman et al. 2008).

Not all effects of UV on microbes are negative, however. UV (together with blue light) can stimulate spore production and hyphal development in some fungi (Gressel and Rau 1983; Nagahashi and Douds 2003) and benefit microbial growth. Also, the effects of UV on microbes will depend on prevailing environmental conditions (i.e., temperature, moisture, and substrate availability) that influence microbial activity (Rangel et al. 2004; Gunasekera and Paul 2007; Belnap et al. 2008). Consequently, the overall effect of solar UV (UV-B + UV-A) on the community composition and function of microbial decomposers may be complex (Denward et al. 2001; Johnson 2003; Kadivar and Stapleton 2003). In relatively wet ecosystems (e.g., forests, marshes, and bogs), solar UV-B has been shown to retard litter mass loss and microbial activity and change microbial community composition, but effects are often subtle and variable over time (Newsham et al. 1997; Moody et al. 2001; Pancotto et al. 2003; M. Tobler and P. Barnes, unpubl data). The effect of UV on microbial-driven decomposition is little understood in drylands. It is conceivable that UV effects would be less important and more temporally variable in drylands as compared to moist environments due to the more extreme temperatures and the sparse, intermittent nature of precipitation in drylands that govern microbial activity, and therefore potential sensitivity to UV (see for example Belnap et al. 2008). Alternatively, the often intense UV in drylands may be sufficient to exceed the UV tolerances of microbial decomposers, at least under certain conditions.

Solar radiation also influences decomposition by altering the chemistry and structure of live plant tissue ("plant photochemistry" effects; Table 1; Fig. 1G). Exposure to ambient or enhanced UV-B typically elevates levels of phenylpropanoid compounds (flavonoids and related phenolics) that serve as UV-absorbing compounds and free radical scavengers (Caldwell et al. 1983b; Day 1993; Searles et al. 2001; Agati and Tattini 2010). Changes in UV-B during plant growth has also been linked with changes in leaf C, N, P, K, and lignin concentrations (Song et al. 2013b). How these UV-induced changes in live leaf chemistry might influence subsequent litter photodegradation is unknown. It does appear that UV-absorbing compounds can persist for some time even in dried leaf tissue (Ryel et al. 2010) and this may protect inner mesophyll cells in litter from photodegradation effects. Nonetheless, the potential for these plant photochemistry effects on decomposition suggests that differences in decomposition rates may exist between sun- and shade leaves of the same plant as there can be significant variation in phenolics and other chemical constituents in leaves within plant canopies depending upon the light environment experienced during development (Barnes et al. 2013). Also, both UV and PAR can alter leaf structure (e.g., leaf size or area, thickness, and area/mass ratios (Fig. 1G) (Boardman 1977; Barnes et al. 2005), which may then influence subsequent photodegradation (e.g., Anesio et al. 1999).

Changes in leaf chemistry induced by UV exposure (Fig. 1G) can, in turn, influence decomposer organisms (Fig. 1D). For example, Gehrke et al. 1995) found significantly lower rates of microbial decomposition in *Vaccinium uliginosum* litter collected from plants growing in an arctic heathland exposed to enhanced UV-B and attributed the differences to increased polyphenol and reduced cellulose contents in the litter. Similar UV-induced increases in phenolics and changes in other chemical constituents have been linked to decreases in mass loss and/or microbial activity in decomposing leaves of *Calamagrostis epigeios* (Rozema et al. 1997), *Hordeum vulgare* (Pancotto et al. 2005), and *Alnus incana* (Kotilainen et al. 2009), but not all plant species exhibit these responses (Newsham et al. 2001; Kotilainen et al. 2009; Song et al. 2013b).

In addition to modifying microbial decomposition through changing live leaf chemistry, solar radiation may affect decomposition through photodegradation that then modifies subsequent microbial decomposition (Fig. 1E→D). The "photopriming" of litter (Table 1) may break down or partially degrade compounds, leaving litter with a higher proportion of constituents more susceptible to microbial decomposition. Thus, even when primary and secondary photodegradation may have little effect on litter mass loss, respiration rates in subsequent incubations with moisture and soil can be positively correlated with length of prior radiation exposure (Foereid et al. 2010). Photopriming may be of particular importance in the "conditioning" of standing litter prior to its detachment and incorporation into the soil (Fig. 1B). However, photopriming can also enhance C mineralization from surface soil organic matter (Mayer et al. 2012). Although some laboratory studies have not observed photopriming (Brandt et al. 2009; Kirschbaum et al. 2011), these may have been of insufficient time, radiation intensity, or incorrect wavelengths to produce measureable change. It is also likely that there will be considerable variation in plant species susceptibility to photopriming, with species most susceptible to mass loss through primary photodegradation also being the most affected by photopriming. Future photopriming experiments with multiple species in field situations are needed to assess whether this is a frequent or important facet of the photodegradation processes.

3.3 Evaluating the Role of Sunlight on Decomposition in Natural Settings

It is clear from field and laboratory studies to date that UV and PAR *can* play substantive roles in decomposition under experimental conditions, but it is likely that their effect will be attenuated by other factors under field conditions. The overall net effect of UV on litter decomposition under field conditions will reflect a balance between positive (e.g., photodegradation, photopriming, microbially enhanced photodegradation) and negative (e.g., microbial photoinhibition, plant photochemistry) effects (Table 1) such that decomposition may be increased, decreased, or unaffected by UV exposure depending on prevailing environmental conditions and litter chemistry (Fig. 1F; e.g., Rozema et al. 1997; Moody et al. 2001; Pancotto et al. 2005; Brandt et al. 2007; Smith et al. 2010; Uselman et al. 2011). This balance will also be influenced by the amount, wavebands, and timing of solar UV radiation received by litter (Fig. 1B, Song et al. 2013a). Understanding radiation loads that litter typically experiences under field conditions and assessing the impact of these exposures on decomposition is a crucial next step for advancing our understanding of the role of sunlight in influencing abiotic and biotic processes in natural systems.

Little is known of the precise nature of the dose–response relationships for the various mechanisms of UV-driven decomposition and whether there are differences in dose responses for abiotic and biotic mechanisms. Certainly, the UV exposure of standing and ground litter will vary over short (e.g., diurnal) and long (e.g., seasonal) time scales, and these patterns can be modified by cloud cover (Fig. 2a, b). Indeed, during the summer monsoon period in the North American Sonoran Desert (July–August), clouds can reduce daily UV-B levels by 50 % relative to seasonal maximum clear sky conditions (Fig. 2a). Due to the strong seasonality of solar UV, the timing of litter production is also important in influencing litter UV exposure, and differences in UV doses would be expected between dryland plant growth forms

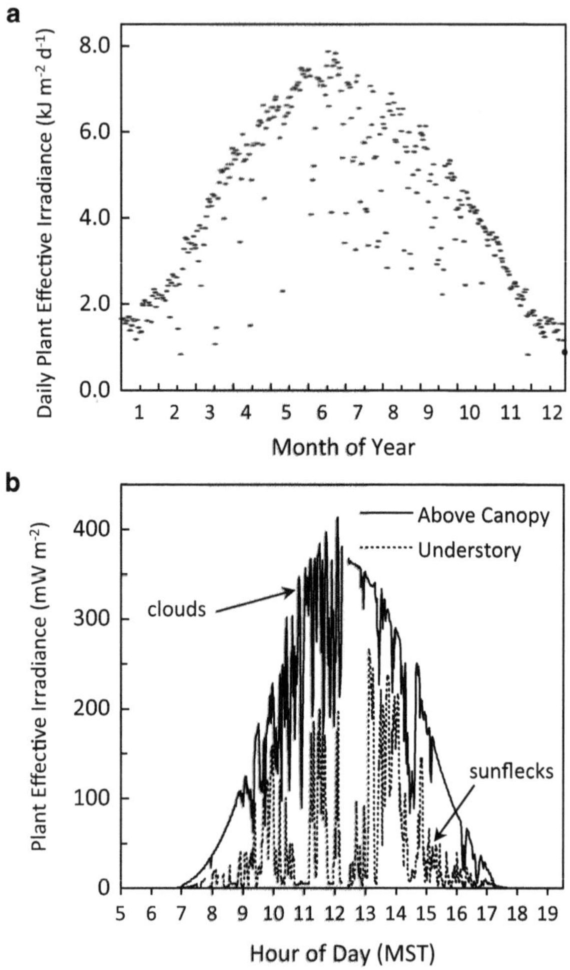

Fig. 2 The surface solar UV radiation environment in a semi-desert savanna [Sonoran Desert, Santa Rita Experimental Range (SRER), southeastern Arizona, USA; 31° 47′ 36″ N, 110° 53′ 4″ W; elevation ca. 1,000 m]. (**a**) Integrated daily UV-B irradiance [weighted according to the generalized plant action spectrum of Caldwell (1971) and normalized to unity at 300 nm] over 2012. UV measurements were made with a broadband UV sensor (UVB-1 pyranometer; Yankee Environmental Systems, Inc.; Turners Falls, MA, USA) calibrated against a double monochromator scanning UV spectroradiometer (OL 756; Gooch & Housego, Orlando, FL, USA). (**b**) Representative diurnal course of plant effective UV-B irradiance above and below the canopy of an isolated, mature *Prosopis velutina* shrub (Fig. 4) on 22 May 2013. Measurements were made with calibrated broadband UV-B sensors (SKU 430; Skye Instruments, Ltd., Powys, UK)

which differ in leaf chemistry (Sect. 3.2, Fig. 1G) and also in leaf phenology and leaf area duration (e.g. C_3 and C_4 grasses, grasses and shrubs, evergreen and deciduous shrubs; Figs. 1A and 4a). High photodegradation potential is likely in settings where the primary growing season occurs during wet, warm spring months, leaving large amounts of standing dead and surface litter exposed to solar radiation during dry, hot summer months when cloud cover is low. Photodegradation accounted for a substantial portion of the dry season ecosystem CO_2 flux in a California annual grassland (Rutledge et al. 2010)—ostensibly a consequence of the Mediterranean climate. Unfortunately, many of the field UV-exclusion decomposition studies conducted to date do not report UV or PAR irradiances, which makes it difficult to both interpret and compare results from studies conducted at different locations and times of year. At a minimum, total daily PAR and appropriate effective UV irradiances (UV-B and UV-A; see below) should be reported over the time period when decomposition data are collected.

The effectiveness of incident radiation in driving litter decomposition will be determined, in part, by the spectral sensitivity of the underlying decomposition processes (Fig. 1C). Action spectra represent the relative effectiveness of different wavelengths of radiation in causing biophysical responses and are typically developed under very controlled laboratory conditions (Holmes 1997). While few action spectra specific to decomposition have been developed, representative action spectra of related processes may yield insights. Potentially important action spectra for biotic and abiotic processes involved in litter decomposition are shown in Fig. 3, along with action spectra commonly used in UV photobiology studies. These indicate that UV effects on both biotic and abiotic processes are strongly wavelength-dependent, with shorter wavelengths showing greater quantum effectiveness than longer wavelengths (Fig. 3a). However, the slopes of these curves can vary considerably. For example, within the UV-B range (280–320 nm), the effectiveness of UV in damaging DNA can increase five orders of magnitude with decreasing wavelength. By comparison, UV-induced CO emission increases less than one order of magnitude over this same waveband. The relatively flat action spectrum for this aspect of photodegradation is consistent with experiments demonstrating that photodegradation can be caused by UV-A and visible (PAR) in addition to UV-B (Anesio et al. 1999; Austin and Ballaré 2010).

Action spectra are used to identify potential chromophores mediating photobiological responses and, in UV studies, as weighting functions to derive measures of biologically effective UV irradiance. In the case of weighting functions, measured raw spectral irradiances (Fig. 3b) are multiplied by relative effectiveness derived from the action spectrum, and then summed over the appropriate wavelength range, to give the *biologically effective radiation* (Fig. 3b, Caldwell and Flint 1997). Thus, the selection of the action spectrum can significantly influence the calculated biologically effective radiation. Steep action spectra (e.g., microbial DNA damage) amplify the importance of the shorter wavelengths (i.e., UV) to a greater degree than flatter action spectra (e.g., CO emissions; Fig. 3a). Because of this, differences in the spectral sensitivities of biotic and abiotic decomposition processes (Fig. 1C) would have important implications for experimental procedures and for interpreting the consequences of stratospheric ozone depletion and latitudinal UV gradients on decomposition. For example, using a relatively steep action spectrum, such as that

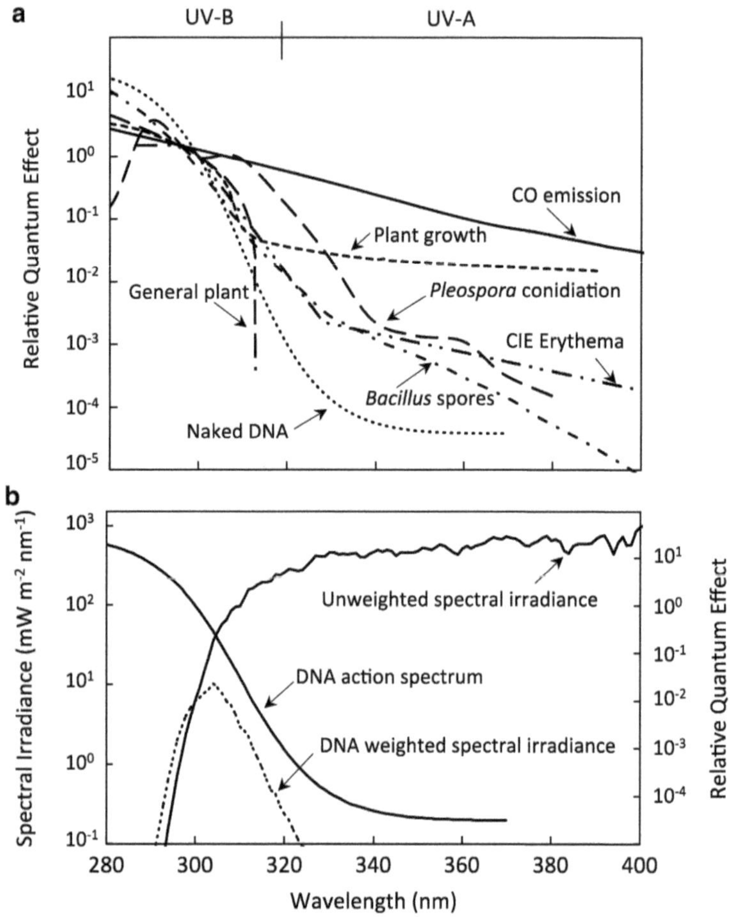

Fig. 3 Representative action spectra and biological weighting functions used in UV photobiology (note log scale on *y*-axes). (**a**) Relevant action spectra for biotic and abiotic processes associated with decomposition and common action spectra used as biological weighting functions in UV photobiology. All action spectra are normalized to unity at 300 nm and have been converted to quantum units if originally reported in energy units. The general plant action spectrum is from Caldwell (1971). The DNA action spectrum is for UV-induced damage to "naked" microbial DNA (Setlow 1974). The *Bacillus* action spectrum is based on the inactivation of spores as reported by Cockell et al. (2003). The *Pleospora* action spectrum is for UV-induced conidiation (asexual spore production) in the fungus *P. herbarum* originally described by Leach and Trione (1966) as reported by Ensminger (1993). CIE is the human erythemal action spectrum (McKinlay and Diffey 1987a, b), a widely used weighting function to report UV irradiances and the basis for the UV Index. The plant growth action spectrum describes the influence of UV on shoot elongation (Flint and Caldwell 2003). The CO emission action spectrum is for savanna grass (*Trachypogon* sp.) leaf litter (Schade et al. 1999). This action spectrum extends into the visible (>600 nm), but only the UV portion is shown here. (**b**) UV spectral irradiance at midday under clear skies on 7 June 2011 at the SRER as measured with a UV scanning spectroradiometer (see Fig. 1), the DNA damage action spectrum from panel **a**, and the calculated biologically effective UV irradiance weighted according to the DNA action spectrum. For this spectrum, the unweighted UV-B and UV-A irradiances are 2.5 and 43.1 W m^{-2}, respectively, and the DNA weighted UV irradiance is 0.10 W m^{-2}

for DNA, to report effective UV radiation may be inappropriate for photodegradation studies and could result in large errors if UV doses were applied using lamps that differ in spectral composition relative to natural sunlight (i.e., UV fluorescent bulbs, Flint and Caldwell 1996). Furthermore, the shift in the UV spectrum in favor of the shorter wavelengths as a result of ozone depletion becomes significant only if a relatively steep action spectrum exists (Caldwell et al. 1986). Finally, because of latitudinal variation in stratospheric ozone thickness and prevailing solar angles, there is a potential natural latitudinal gradient in ambient solar UV-B (Caldwell et al. 1980; Barnes et al. 1987). However, this latitudinal UV gradient would be trivial for responses exhibiting a relatively flat action spectrum, such as that for CO emissions. If other photodegradation processes exhibit a similarly flat action spectrum, this may explain, in part, why Brandt et al. (2010) found no strong differences in UV-driven photodegradation across a latitudinal gradient of grassland sites in North America. Because of the fundamental importance of action spectra in UV photobiology, additional studies are needed to develop and test appropriate action spectra/weighting functions for processes involved in litter decomposition.

The UV environment of litter in drylands also exhibits substantial spatial variability as a result of the pronounced discontinuous nature of vegetative cover in these ecosystems (i.e., herbaceous patches in a matrix of bare soil or tree/shrub patches in a matrix of bare soil and herbaceous plants, Noy-Meir 1979/80; West 1983; Evenari et al. 1985). In systems with discontinuous cover of woody plants, litter often accumulates in the understories of woody plants (Fig. 4b) and thus receives considerably less UV (and PAR) than that in the intercanopy zones (Fig. 2b). The UV exposure of litter will also depend on the depth of litter layer (Henry et al. 2008) and vertical position within the litter layer (Lin and King 2013) (Fig 1B). The angle at which litter is oriented would also have strong influences over its exposure to solar radiation, with vertically oriented standing dead (e.g., grasses; Fig. 4a) potentially intercepting less radiation than would detached litter resting horizontally on the soil surface. Orientation effects would, however, be less for UV than PAR because of the pronounced diffuse (isotropic) nature of solar UV radiation.

Once the litter falls to the soil surface, it may become covered with loose soil and tightly bound soil–litter films that can block solar radiation from hitting the litter and consequently negate photodegradation (Fig. 1I→B, Barnes et al. 2012). Because of these complexities, quantifying the actual UV exposure of litter in field environments is challenging. The use of inexpensive biological or synthetic UV dosimeters (e.g., Rahn and Lee 1998; Turner et al. 2009) deployed in a variety of habitats and conditions over varying time periods would aid in quantifying the patterns of UV exposure at spatial and temporal scales relevant to litter decomposition.

Fig. 4 Temporal and spatial aspects of litter production and distribution in a semi-desert savanna (Sonoran Desert, SRER; see Fig. 2 for location details). (**a**) End-of growing season standing litter of the C_4 grass *Heteropogon contortus* with the winter-deciduous shrub, *Prosopis velutina*, before leaf drop. Note the spatial heterogeneity in herbaceous cover. (**b**) Spatial variation in bare ground, surface litter accumulation, and light conditions under and near a *P. velutina* canopy after leaf drop and prior to the onset of the growing season (photos: S. Archer)

4 Soil–Litter Mixing and Decomposition

4.1 Soil Redistribution in Drylands

Dryland ecosystems are, by definition, water-limited, and this water limitation usually results in a mosaic of vegetation cover that is sparse and incomplete, with herbaceous patches in a matrix of bare soil or tree/shrub patches in a matrix of bare soil and herbaceous plants. As a consequence of sparse and patchy ground cover, soil erosion and associated processes of transport and deposition can be particularly pronounced (Fig. 1H, Kirkby 1980; Heathcote 1983; Fryrear 1985; Toy et al. 2002). Wind- and water-driven transport of soils is widely recognized as having a substantial influence on nutrient and vegetation distribution (e.g., Ludwig et al. 1997; Okin et al. 2006; Peters et al. 2006). Even so, the mechanisms by which plant community structure and ecosystem processes are influenced by wind and water transport of soils are poorly understood. Different physical forces promote movement of soil via wind and water, but these processes share three critical phases: detachment of soil particles from the soil surface, transport as overland flow or

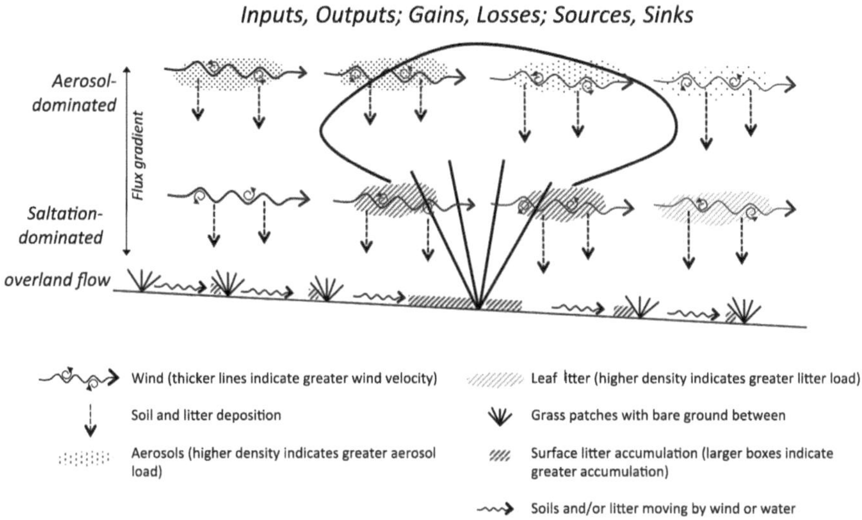

Fig. 5 Schematic representation of wind and water redistribution of soil and plant litter leading to the development of a soil–litter matrix (Fig. 1H, I) in drylands characterized by herbaceous patches in a matrix of bare soil or tree/shrub patches in a matrix of bare soil and herbaceous plants. In this figure a single shrub (large plant) is surrounded by grasses (small plants) and bare ground. Aerosols, saltating soil particles, and overland flow transfer soils and litter from areas of low vegetation cover to areas of higher vegetation cover. Net exchanges and source–sink relationships are mediated by the area, density, spacing, and stature of vegetated patches, the size and connectivity of bare gaps, topography, and disturbances such as fire and grazing

aerosols, or via saltation, and deposition at a location (Toy et al. 2002). These processes are interrelated and produce a net flux that can result in either an accumulation or erosional loss of soil at a given location (Fig. 5). Within sites, winderosion (net vertical dust flux) appears to be correlated with wind-driven transport (horizontal dust flux; Whicker et al. 2006). Evaluating the role of erosion on fine-scale processes such as decomposition requires coupled estimates of transport and deposition (Fig. 1H). Numerous studies have documented broad-scale or long-term manifestations of wind and water erosion, but few have focused on erosion and associated transport and especially redistribution at the finer spatial scales (cm–m) and the shorter time frames (weeks to a few years) relevant to litter decomposition (e.g., Whicker et al. 2002; Ludwig et al. 2005). Rates of wind erosion are poorly documented relative to those of water erosion, even though a recent evaluation that included major dryland ecosystem types (grassland, shrubland, woodland) found that annual rates of wind-driven soil transport could exceed those of water-driven transport by an order of magnitude or more (Breshears et al. 2003).

Soil transport by both wind and water is highly dependent on and sensitive to changes in woody plant cover (Fig. 1A→H, Bagnold 1941; Fryrear 1985; Reid et al. 1999; Wilcox et al. 2003; Warner 2004; Breshears et al. 2009). Notably, there has been a strong, directional increase in woody plant cover in drylands over the past century (Archer et al. 1995). This global-scale change has altered the quality and quantity of litter inputs (e.g. Hibbard et al. 2003) and the spatial and temporal patterns of erosion processes (Schlesinger et al. 1990; Okin et al. 2009a; Ravi et al. 2009a, 2010). These shifts in grass-woody plant ratios may potentially affect decomposition rates by mediating soil transport processes that determine rates of soil deposition into litter (Figs. 1H→I and 5). Recent research highlights two complementary aspects of horizontal dust flux: increased production with reduction in grass cover (Li et al. 2007) and capture by shrubs and grasses (Field et al. 2009). The later work highlights an important mechanism by which horizontal sediment flux and associated nutrients are likely to be deposited onto litter beneath plant canopies.

4.2 *Litter Redistribution in Drylands*

Wind and water promote the detachment and redistribution of plant litter, increasing spatial heterogeneity of litter and its nutrient constituents (Fig. 5). Although often observed and clearly evident (Fig. 6) the magnitude, patterns of litter redistribution, and the dynamics and ecological significance of litter mass and nutrient transfer have seldom been quantified in drylands. Surface water flows can redistribute detached litter and soil particles and concentrate them in ostensibly predictable locations related to microtopography and obstructions posed by rocks, animal disturbances, and other plants. In dryland plant communities with woody vegetation, coarse woody debris on the soil surface can trap and retain leaf and twig litter and soil. These accumulations presumably hasten the localized

Fig. 6 Localized accumulation of surface litter (*arrows*) in a semi-desert savanna (Sonoran Desert, SRER; see Fig. 2 for location details). (**a**) Litter accumulation in a bare patch as a result of microtopography. (**b**) Litter accumulation at the base of a small woody plant with coarse woody debris on soil surface. Note marking pen for scale (photos: S. Archer)

formation of soil–litter matrices. The self-facilitated burial of coarse woody debris would also accelerate its breakdown.

Redistribution of surface litter by overland flow of water is supplemented by wind-mediated transfers. Nutrient inputs can be substantially augmented by litter transferred from upwind to downwind communities (Shen et al. 2011). Redistribution of litter by wind from ridge tops to leeward locations in the Arctic can lead to increases in C and N inputs and subsequent increases in soil respiration in depositional locations (Fahnestock et al. 2000). As with water, litter transported by wind accumulates in predictable locations that likely vary depending on the size, shape, density, and mass of the litter and vegetation and landscape features that cause turbulence and alter wind speed and direction (Fig. 1H).

Assessing the biogeochemical consequences of litter redistribution and its subsequent decomposition at a given site requires quantifying inputs (gains) and outputs (losses). Most of the litter generated by a plant falls on the ground near the plant subsequent to detachment (input), but some litter is transported away from the plant by wind or overland flow (output) (Fig. 5). Litter deposited near a plant's canopy may be supplemented by litter transferred from other plant patches on the landscape (input). However, quantifying these litter inputs and outputs is challenging. Litter deposition has typically been quantified using litter traps, but the design of such traps is generally strongly biased toward the collection of gravity-deposited litter, and it is difficult to know what fraction of the litter in a trap, if any, is derived from external sources. Quantifying litter ground cover and its change through time offers alternative perspectives on the net outcomes of local litter gains and losses, but here too, it is difficult to know the amount of litter arriving from external sources. Furthermore, local surface litter cover reflects a hard-to-quantify combination of reductions (owing to burial by soil, comminution by arthropods and weathering) and increases (associated with inputs of new litter from local or external sources). Litter arriving from outside sources may also differ in quality relative to that of the locally produced litter and such differences may be pronounced in drylands consisting of heterogeneous patches of annuals, perennial grasses, and shrubs.

In locales where litter accumulates, the soil surfaces may be stabilized by formation of the soil–litter matrix, likely reducing ground surface temperatures and PAR and UV radiation levels. These changes could also promote the establishment of plants whose wind- and water-dispersed seeds would be likely to accumulate in the litter deposition zones. Locations where the soil–litter matrix forms via the processes outlined in Sect. 4.1 may therefore represent nutrient cycling "hot spots," a nascent phase in the formation of vegetated patches and a feedback mechanism reinforcing the persistence and expansion of vegetated patches. At the landscape scale, the dynamics of soil–litter patches will depend on the degree of bare gap connectivity (Okin et al. 2009b), and source–sink relationships governed by interactions among disturbance (e.g., grazing, fire), topography, and prevailing winds (e.g., Ravi et al. 2009b; Bestelmeyer et al. 2013). Multiple drivers interacting across scales probably combine with positive feedbacks to govern litter–soil distribution and redistribution (e.g., D'Odorico et al. 2012).

4.3 Soil–Litter Mixing and Decomposition

Once litter is on the ground it is inevitably covered with varying degrees of soil or other litter and, in some cases, fully buried (Fig. 6). Initially, litter is covered with loose soil that can be easily dislodged. Over time, soil films consisting of soil particles, microbes, and microbial exudates develop and adhere to the litter surface (Fig. 7; Barnes et al. 2012; Hewins et al. 2013). Unlike loose soil, these adhering soil films are more resistant to removal by rainfall and wind. At time scales of weeks to months, however, soil films are dynamic and may develop or degrade in response to temperature and moisture conditions (D. Hewins and H. Throop, unpublished). These soil films appear to be composed of inorganic and biological constituents with fungal hyphae and microbial exudates binding mineral particles to each other and to the leaf surface (Fig. 7). The specific nature of the abiotic and biotic components of these soil films and the degree and timing of soil film coverage will likely be influenced by site-specific edaphic and vegetation factors that influence local-scale differences in soil transport (Okin and Gillette 2001; Okin 2008).

While positive correlations have been found between rates of litter decomposition and the degree of soil–litter mixing (e.g., Throop and Archer 2007), the underlying mechanisms of this response have yet to be elucidated. Soil coverage of litter could potentially influence decomposition by several mechanisms, with the net effect ranging from positive to negative depending on conditions and the extent of coverage (Fig. 1I→B, I→D). Soil may serve as a vector for microbial colonization of litter. In a laboratory incubation study, soil–litter mixing led to differences in the quantity and composition of phospholipid fatty acids extracted from the soil–litter matrix following the first week of the incubation, suggesting that colonization

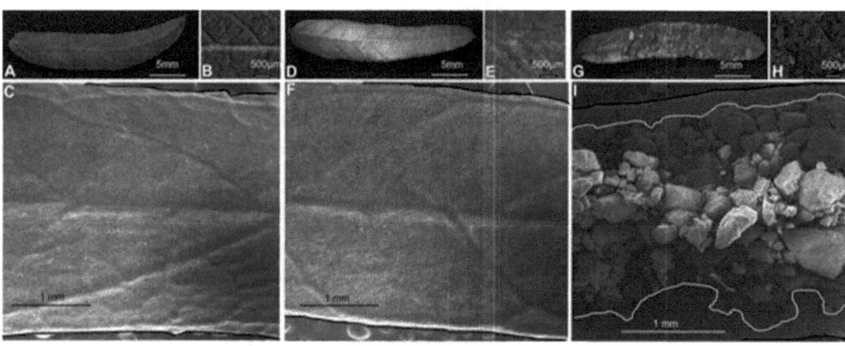

Fig. 7 Development of soil films on *Prosopis glandulosa* leaf litter over time (0, 30, and 180 days in **a–c**, **d–f**, and **g–i**, respectively) in a Chihuahuan Desert shrubland (Jornada Experimental Range, New Mexico, USA; 32°33′ N, 106°45′ W; elevation ca. 1,190 m) illustrated by low magnification (×1.6) stereo micrographs showing entire leaflets (**a**, **d**, **g**), high magnification (×3.2) stereo micrographs (**b**, **e**, **h**), and SEM micrographs (×25, ×31, and ×35 for **c**, **f**, and **i**, respectively). In the SEM micrographs, *black lines* denote leaflet margins and *white lines* denote the edge of the soil film (from Barnes et al. 2012)

may be influenced by soil–litter mixing (Lee et al. 2014). Soil–litter mixing may also buffer litter and resident microbes from the high temperatures and desiccation that commonly occur in drylands (Moorhead and Reynolds 1991). These effects could enhance decomposition by extending windows of opportunity for microbial activity following rainfall events (e.g., Cable et al. 2011). Indeed, soil–litter mixing strongly enhanced C mineralization in a laboratory experiment when the soil–litter matrix was subjected to wetting–drying cycles (Lee et al. 2014). The arrival of soil at the litter surface via saltating soil particles or the translocation of litter via overland flow may also promote surface abrasion and increase the surface area available to microbial colonization, leaching, or fragmentation (Throop and Archer 2009; Uselman et al. 2011). Enhanced microbial colonization of recently detached litter may be offset by the negative effects of solar UV on microbes (Sect. 3.2), but subsequent soil coverage, either as an adhering soil film or as loose soil, could partially and eventually fully shield litter from UV radiation and therefore ameliorate its adverse effects (Cockell et al. 2003; Barnes et al. 2012). Soil cover may therefore mediate photodegradation and other abiotic forces (Fig. 1D, E).

5 Integrated Conceptual Model of UV and Soil Mixing Effects on Dryland Decomposition

Based on findings from field and laboratory studies, we have proposed a generalized conceptual model for UV-soil mixing effects in dryland decomposition (Fig. 8, Barnes et al. 2012). Over a continuum of soil coverage of litter from none (e.g., standing dead) to partial (e.g., recently detached) to full burial, the mechanisms driving decomposition are predicted to shift from strongly abiotic (photodegradation of standing dead driven by UV together with PAR) to strongly biotic (microbial degradation of buried litter). Intermediate conditions consist of a combination of these processes whose influence varies depending on the extent of development of the soil–litter matrix, its biogeochemical constituency (e.g., litter quality, soil mineral composition, and organic matter content of soil [Fig. 1G, I]), the microbial community composition and activity (Fig. 1D), and the prevailing moisture/temperature conditions (Fig 1A). As the relative importance of photodegradation and microbial decomposition change through time, the overall rate of decomposition may approximate a unimodal curve that reflects the outcome of interactions between the speed of the concurrent drivers of decomposition and the recalcitrance of the chemical constituents present in the litter.

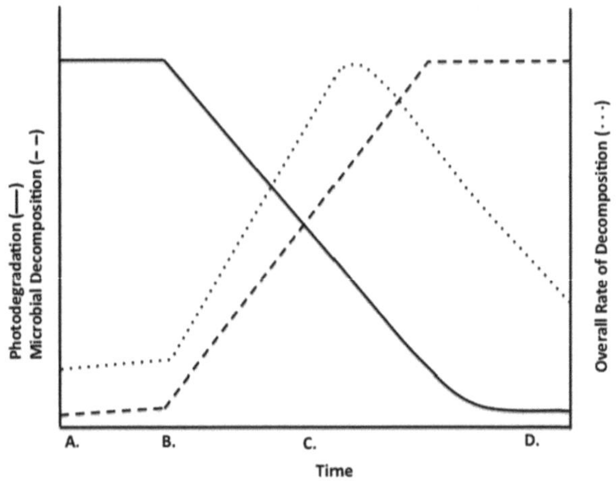

Fig. 8 Conceptual model of dryland decomposition following leaf senescence, illustrating the shifting relative importance of abiotic (photodegradation, Table 1) and biotic (microbial) processes through time and consequent changes in the overall rate of decomposition (Fig. 1F). Additional processes that may be important in decomposition, such as UV effects on microbes (Fig. 1D), leaching, fragmentation (Fig. 1E), or effects of UV on leaf chemistry/structure (Fig. 1G) are not illustrated. Recently senesced plant material is initially subject to high rates of photodegradation while it is standing dead (A). Limited microbial decomposition may occur on leaf surfaces at this time. While the majority of decomposition that occurs at this time is from photodegradation, the overall rates of decomposition remain low. When standing dead plant material falls to the soil surface (B), the soil–litter matrix develops (Figs. 1I and 6), gradually covering the litter (C). During this time the relative importance of photodegradation declines while microbial decomposition increases due to colonization opportunities, favorable microclimate, or abrasion afforded by the litter–soil matrix. Decomposition rates increase with microbial colonization, and overall rates of decomposition peak due to rapid losses of easily decomposable chemical constituents in the litter. Negative effects of UV on microbes are small and transient initially but increase over time in association with increased microbial biomass and activity until soil coverage negates these negative effects. Eventually nearly all the litter surface is cover by soil (D) and photodegradation accounts for a trivial portion of decomposition while microbial degradation prevails. The overall rate of decomposition is low as remaining litter is highly recalcitrant. From Barnes et al. (2012)

6 Summary and Conclusions

Over the past several decades, significant progress has been made in understanding the nature and importance of solar radiation in influencing litter decomposition in terrestrial ecosystems. Although a number of uncertainties remain, the information available indicates that solar UV (UV-B and UV-A) and PAR can have positive, negative, or minimal effects on decomposition depending on the balance of abiotic (photodegradation) and biotic (microbial) processes (Fig. 1D, E). In moisture-limited ecosystems (i.e., grasslands, savannas, and deserts), the net effects of sunlight/UV on decomposition are generally positive and photodegradation is now being considered as an important driver of decomposition that may account for the discrepancies between measurements and model predictions of decomposition rates. However, the majority of studies to date that have explored the effects of UV and PAR on decomposition in drylands have done so without explicitly considering soil–litter mixing. While such studies may reasonably ascertain decomposition of standing plant litter, their extrapolation to decomposition of detached plant litter on soil surfaces fails to take into account the formation of soil–litter complexes (Fig. 1I) that can strongly mediate or even negate these abiotic effects. Soil and litter movement and translocation are common in moisture-limited environments with low and patchy vegetation cover, and litter on the ground is frequently covered to varying degrees with soil and eventually buried. This mixing of soil and litter is associated with increased rates of decomposition. Although the mechanisms underlying these soil-mixing effects remain to be fully explored, it is likely that the formation of soil–litter–microbial complexes enhance microbial activity while simultaneously shielding litter from photodegradation. Thus, extrapolating the importance of photodegradation from measurements obtained in environments with either no soil or soil with restricted movement (e.g., litter boxes or glass jars) would overestimate the importance of photodegradation. Additional studies conducted under realistic field conditions are needed to fully explore how solar radiation and soil coverage interact through time to influence litter decomposition in dryland ecosystems characterized by soil movement and deposition. A greater understanding of the interactive effects of soil deposition and sunlight may aid, at least in part, in resolving the seemingly contradictory findings reported in photodegradation and soil deposition studies. Ongoing shifts in dryland life-form composition (e.g., from grass to shrub domination), driven by changes in land use and climate, will likely increase soil movement in these environments (Okin et al. 2009b). The role of soil deposition on litter decomposition in globally extensive dryland ecosystems may thus be magnified under future conditions.

Acknowledgments Financial support was provided by the US National Science Foundation (DEB 0815897 [Loyola University], DEB 0815808 [New Mexico State University], DEB 0816162 [University of Arizona], DEB 0814461 [University of Kentucky]), the Loyola University J.H. Mullahy Endowment for Environmental Biology, and the Jornada Basin LTER (DEB 0618210). Additional support was provided by the Arizona and Kentucky Agricultural Experiment Stations, and the University of Arizona – Universidad Nacional Autonoma de Mexico Drylands Research Consortium (#15), and US National Science Foundation (EAR 0724958 and 1331408 [University of Arizona]). We thank Jason P. Field, Daniel B. Hewins, Dean Keller, Darin J. Law, Hanna Lee, Eva Levi, Cianna Logie, Rachel Power, Katie Predick, Theresa Melhem, and Lee Chedister for related research and discussions.

References

Adair EC, Parton WJ, Del Grosso SJ, Silver WL, Harmon ME, Hall SA, Burke IC, Hart SC (2008) Simple three-pool model accurately describes patterns of long-term litter decomposition in diverse climates. Glob Change Biol 14:2636–2660

Aerts R (1997) Climate, leaf litter chemistry, and leaf litter decomposition in terrestrial ecosystems: a triangular relationship. Oikos 79:439–449

Agati G, Tattini M (2010) Multiple functional roles of flavonoids in photoprotection. New Phytol 186:786–793
Andrady AL, Hamid HS, Torikai A (2003) Effects of climate change and UV-B on materials. Photochem Photobiol Sci 2:68–72
Anesio AM, Tranvik LJ, Graneli W (1999) Production of inorganic carbon from aquatic macrophytes by solar radiation. Ecology 80:1852–1859
Aphalo PJ, Albert A, Björn LO, McLeod A, Robson TM, Rosenqvist E (eds) (2013) Beyond the visible: a handbook of best practice in plant UV photobiology. University of Helsinki, Department of Biosciences, Division of Plant Biology, Helsinki
Archer S, Schimel DS, Holland EA (1995) Mechanisms of shrubland expansion: land use, climate or CO_2? Clim Change 29:91–99
Austin AT (2011) Has water limited our imagination for aridland biogeochemistry? Trends Ecol Evol 26:229–235
Austin AT, Ballaré CL (2010) Dual role of lignin in plant litter decomposition in terrestrial ecosystems. Proc Natl Acad Sci U S A 107:4618–4622
Austin AT, Vivanco L (2006) Plant litter decomposition in a semi-arid ecosystem controlled by photodegradation. Nature 442:555–558
Bagnold RA (1941) The physics of blown sand and desert dunes. Chapman and Hall, London
Ballaré CL, Caldwell MM, Flint SD, Robinson SA, Bornman JF (2011) Effects of solar ultraviolet radiation on terrestrial ecosystems. Patterns, mechanisms, and interactions with climate change. Photochem Photobiol Sci 10:226–241
Barnes PW, Flint SD, Caldwell MM (1987) Photosynthesis damage and protective pigments in plants from a latitudinal arctic/alpine gradient exposed to supplemental UV-B radiation in the field. Arctic Alpine Res 19:21–27
Barnes PW, Shinkle JR, Flint SD, Ryel RJ (2005) UV-B radiation, photomorphogenesis and plant-plant interactions. Prog Bot 66:313–340
Barnes PW, Throop HL, Hewins DB, Abbene ML, Archer SR (2012) Soil coverage reduces photodegradation and promotes the development of soil-microbial films on dryland leaf litter. Ecosystems 15:311–321
Barnes PW, Kersting AR, Flint SD, Beyschlag W, Ryel RJ (2013) Adjustments in epidermal UV-transmittance of leaves in sun-shade transitions. Physiol Plant 149:200–213
Belnap J, Phillips SL, Flint SD, Money J, Caldwell MM (2008) Global change and biological soil crusts: effects of ultraviolet augmentation under altered precipitation regimes and nitrogen additions. Glob Change Biol 14:670–686
Berg B, Laskowski R (eds) (2005) Litter decomposition: a guide to carbon and nutrient turnover. Elsevier, London
Bestelmeyer BT, Duniway MC, James DK, Burkett LM, Havstad KM (2013) A test of critical thresholds and their indicators in a desertification-prone ecosystem: more resilience than we thought. Ecol Lett 16:339–345
Bloom AA, Lee-Taylor J, Madronich S, Messenger DJ, Palmer PI, Reay DS, McLeod AR (2010) Global methane emission estimates from ultraviolet irradiation of terrestrial plant foliage. New Phytol 187:417–425
Boardman NK (1977) Comparative photosynthesis of sun and shade plants. Annu Rev Plant Physiol Plant Mol Biol 28:355–377
Braga GUL, Flint SD, Miller CD, Anderson AJ, Roberts DW (2001) Variability in response to UV-B among species and strains of *Metarhizium* isolated from sites at latitudes from 61 degrees N to 54 degrees S. J Invertebr Pathol 78:98–108
Brandt LA, King JY, Milchunas DG (2007) Effects of ultraviolet radiation on litter decomposition depend on precipitation and litter chemistry in a shortgrass steppe ecosystem. Glob Change Biol 13:2193–2205
Brandt LA, Bohnet C, King JY (2009) Photochemically induced carbon dioxide production as a mechanism for carbon loss from plant litter in arid ecosystems. J Geophys Res Biogeosci 114: G02004. doi:10.1029/2008jg000772

Brandt LA, King JY, Hobbie SE, Milchunas DG, Sinsabaugh RL (2010) The role of photodegradation in surface litter decomposition across a grassland ecosystem precipitation gradient. Ecosystems 13:765–781

Breshears DD, Whicker JJ, Johansen MP, Pinder JE (2003) Wind and water erosion and transport in semi-arid shrubland, grassland and forest ecosystems: quantifying dominance of horizontal wind-driven transport. Earth Surf Process Landforms 28:1189–1209

Breshears DD, Whicker JJ, Zou CB, Field JP, Allen CD (2009) A conceptual framework for dryland aeolian sediment transport along the grassland–forest continuum: effects of woody plant canopy cover and disturbance. Geomorphology 105:28–38

Cable J, Ogle K, Lucas R, Huxman T, Loik M, Smith S, Tissue D, Ewers B, Pendall E, Welker J, Charlet T, Cleary M, Griffith A, Nowak R, Rogers M, Steltzer H, Sullivan P, Gestel N (2011) The temperature responses of soil respiration in deserts: a seven desert synthesis. Biogeochemistry 103:71–90

Caldwell MM (1971) Solar UV irradiation and the growth and development of higher plants. Photophysiology 6:131–177

Caldwell MM, Flint SD (1989) Plant response to UV-B radiation: comparing growth chamber and field environments. In: Payer HD, Pfirrmann T, Mathy P (eds) Environmental research with plants in closed chambers, Commission of the European Communities Report No. 26, Brussels, pp 264–270

Caldwell MM, Flint SD (1997) Uses of biological spectral weighting functions and the need of scaling for the ozone reduction problem. Plant Ecol 128:66–76

Caldwell MM, Robberecht R, Billings WD (1980) A steep latitudinal gradient of solar ultraviolet-B radiation in the arctic-alpine life zone. Ecology 61:600–611

Caldwell MM, Gold WG, Harris G, Ashurst CW (1983a) A modulated lamp system for solar UV-B (280-320 nm) supplementation studies in the field. Photochem Photobiol 37:479–485

Caldwell MM, Robberecht R, Flint SD (1983b) Internal filters: prospects for UV-acclimation in higher plants. Physiol Plant 58:445–450

Caldwell MM, Camp LB, Warner CW, Flint SD (1986) Action spectra and their key role in assessing biological consequences of solar UV-B radiation change. In: Worrest RC, Caldwell MM (eds) Stratospheric ozone reduction, solar ultraviolet radiation and plant life. Springer, Berlin, pp 87–111

Caldwell MM, Bornman JF, Ballaré CL, Flint SD, Kulandaivelu G (2007) Terrestrial ecosystems, increased solar ultraviolet radiation, and interactions with other climate change factors. Photochem Photobiol Sci 6:252–266

Cockell CS, Rettberg P, Horneck G, Scherer K, Stokes MD (2003) Measurements of microbial protection from ultraviolet radiation in polar terrestrial microhabitats. Polar Biol 26:62–69

Cockell CS, McKay CP, Warren-Rhodes K, Horneck G (2008) Ultraviolet radiation-induced limitation to epilithic microbial growth in and deserts – dosimetric experiments in the hyperarid core of the Atacama Desert. J Photochem Photobiol B Biol 90:79–87

Couteaux MM, Bottner P, Berg B (1995) Litter decomposition, climate and litter quality. Trends Ecol Evol 10:363–367

D'Odorico P, Okin GS, Bestelmeyer BT (2012) A synthetic review of feedbacks and drivers of shrub encroachment in arid grasslands. Ecohydrology 5:520–530

Day TA (1993) Relating UV-B radiation screening effectiveness of foliage to absorbing-compound concentration and anatomical characteristics in a diverse group of plants. Oecologia 95:542–550

Day TA, Neale PJ (2002) Effects of UV-B radiation on terrestrial and aquatic primary producers. Annu Rev Ecol Syst 33:371–396

Day TA, Zhang ET, Ruhland CT (2007) Exposure to solar UV-B radiation accelerates mass and lignin loss of *Larrea tridentata* litter in the Sonoran Desert. Plant Ecol 194:185–194

Denward CMT, Anesio AM, Granéli W, Tranvik LJ (2001) Solar radiation effects on decomposition of macrophyte litter in a lake littoral. Archiv fur Hydrobiologie 152:69–80

Duguay KJ, Klironomos JN (2000) Direct and indirect effects of enhanced UV-B radiation on the decomposing and competitive abilities of saprobic fungi. Appl Soil Ecol 14:157–164
Ensminger PA (1993) Control of development in plants and fungi by far-UV radiation. Physiol Plant 88:501–508
Evenari M, Noy-Meir I, Goodall DW (eds) (1985) Hot deserts and arid shrublands. Elsevier, Amsterdam
Fahnestock JT, Povirk KL, Welker JM (2000) Ecological significance of litter redistribution by wind and snow in arctic landscapes. Ecography 23:623–631
Field JP, Belnap J, Breshears DD, Neff JC, Okin GS, Whicker JJ, Painter TH, Ravi S, Reheis MC, Reynolds RL (2009) The ecology of dust. Front Ecol Environ 8:423–430
Flint SD, Caldwell MM (1996) Scaling plant ultraviolet spectral responses from laboratory action spectra to field spectral weighting factors. J Plant Physiol 148:107–114
Flint SD, Caldwell MM (2003) A biological spectral weighting function for ozone depletion research with higher plants. Physiol Plant 117:137–144
Flint SD, Ryel RJ, Caldwell MM (2003) Ecosystem UV-B experiments in terrestrial communities: a review of recent findings and methodologies. Agric Forest Meteorol 120:177–189
Foereid B, Bellarby J, Meier-Augenstein W, Kemp H (2010) Does light exposure make plant litter more degradable? Plant Soil 333:275–285
Fryrear DW (1985) Soil cover and wind erosion. Trans ASAE 28:781–784
Gallo ME, Sinsabaugh RL, Cabaniss SE (2006) The role of ultraviolet radiation in litter decomposition in arid ecosystems. Appl Soil Ecol 34:82–91
Gehrke C, Johanson U, Callaghan TV, Chadwick D, Robinson CH (1995) The impact of enhanced ultraviolet-B radiation on litter quality and decomposition processes in *Vaccinium* leaves from the subarctic. Oikos 72:213–222
Gressel J, Rau W (1983) Photocontrol of fungal development. In: Shropshire W Jr, Mohr H (eds) Photomorphogenesis, vol 16B, Encyclopedia of plant physiology. Springer, Heidelberg, pp 603–639
Gunasekera TS, Paul ND (2007) Ecological impact of solar ultraviolet-B (UV-B: 320-290 nm) radiation on *Corynebacterium aquaticum* and *Xanthomonas* sp. colonization on tea phyllosphere in relation to blister blight disease incidence in the field. Lett Appl Microbiol 44:513–519
Gunasekera TS, Sundin GW (2006) Role of nucleotide excision repair and photoreactivation in the solar UVB radiation survival of *Pseudomonas syringae* pv. syringae B728a. J Appl Microbiol 100:1073–1083
Gunasekera TS, Paul ND, Ayres PG (1997) Responses of phylloplane yeasts to UV-B (290–320 nm) radiation: interspecific differences in sensitivity. Mycol Res 101:779–785
Heathcote RL (1983) The arid lands: their use and abuse. Longman, New York
Henry HAL, Brizgys K, Field CB (2008) Litter decomposition in a California annual grassland: interactions between photodegradation and litter layer thickness. Ecosystems 11:545–554
Hewins DB, Archer SR, Okin GS, McCulley RL, Throop HL (2013) Soil–litter mixing accelerates decomposition in a Chihuahuan desert grassland. Ecosystems 16:183–195
Hibbard KA, Schimel D, Archer S, Ojima D, Parton W (2003) Grassland to woodland transitions: integrating changes in landscape structure and biogeochemistry. Ecol Appl 13:911–926
Hibbard KA, Law BE, Reichstein M, Sulzman J (2005) An analysis of soil respiration across northern hemisphere temperate ecosystems. Biogeochemistry 73:29–70
Holmes MG (1997) Action spectra for UV-B effects on plants: monochromatic and polychromatic approaches for analysing plant responses. In: Lumsden PJ (ed) Plants and UV-B: responses to environmental change. Cambridge University Press, Cambridge, pp 31–50
Hughes KA, Lawley B, Newsham KK (2003) Solar UV-B radiation inhibits the growth of Antarctic terrestrial fungi. Appl Environ Microbiol 69:1488–1491
Jacobs JL, Carroll TL, Sundin GW (2005) The role of pigmentation, ultraviolet radiation tolerance, and leaf colonization strategies in the epiphytic survival of phyllosphere bacteria. Microb Ecol 49:104–113

Johnson D (2003) Response of terrestrial microorganisms to ultraviolet-B radiation in ecosystems. Res Microbiol 154:315–320
Kadivar H, Stapleton AE (2003) Ultraviolet radiation alters maize phyllosphere bacterial diversity. Microb Ecol 45:353–361
Kemp PR, Reynolds JF, Virginia RA, Whitford WG (2003) Decomposition of leaf and root litter of Chihuahuan desert shrubs: effects of three years of summer drought. J Arid Environ 53:21–39
King JY, Brandt LA, Adair EC (2012) Shedding light on plant litter decomposition: advances, implications and new directions in understanding the role of photodegradation. Biogeochemistry 111:57–81
Kirkby MJ (1980) The problem. In: Kirkby MJ, Morgan RPC (eds) Soil erosion. Wiley, New York, pp 1–16
Kirschbaum MUF, Lambie SM, Zhou H (2011) No UV enhancement of litter decomposition observed on dry samples under controlled laboratory conditions. Soil Biol Biochem 43:1300–1307
Kotilainen T, Haimi J, Tegelberg R, Julkunen-Tiitto R, Vapaavuori E, Aphalo PJ (2009) Solar ultraviolet radiation alters alder and birch litter chemistry that in turn affects decomposers and soil respiration. Oecologia 161:719–728
Krizek DT (2004) Influence of PAR and UV-A in determining plant sensitivity and photomorphogenic responses to UV-B Radiation. Photochem Photobiol 79:307–315
Leach CM, Trione EJ (1966) An action spectrum for light-induced sporulation of *Pleospora herbarum* and *Alternaria dauci*. Photochem Photobiol 5:621–630
Lee H, Rahn T, Throop H (2012) An accounting of C-based trace gas release during abiotic plant litter degradation. Glob Change Biol 18:1185–1195
Lee H, Fitzgerald J, Hewins DB, McCulley RL, Archer SR, Rahn T, Throop HL (2014) Soil-litter mixing effects are limited to early stage decomposition and modified by soil moisture in a controlled environment study. Soil Biol Biochem 72:123–132
Li J, Okin GS, Alvarez L, Epstein H (2007) Quantitative effects of vegetation cover on wind erosion and soil nutrient loss in a desert grassland of southern New Mexico. Biogeochemistry 85:317–332
Lin Y, King JY (2013) Effects of UV exposure and litter position on decomposition in a California grassland. Ecosystems. doi:10.1007/s10021-013-9712-x
Ludwig JA, Tongway DJ, Freudenberger D, Noble J, Hodgkinson K (1997) Landscape ecology function and management: principles from Australia's rangelands. CSIRO, Collingwood, Australia
Ludwig JA, Wilcox BP, Breshears DD, Tongway DJ, Imeson AC (2005) Vegetation patches and runoff–erosion as interacting ecohydrological processes in semiarid landscapes. Ecology 86:288–297
Ma S, Baldocchi D, Hatala J, Detto JM, Yuste J (2012) Are rain-induced ecosystem respiration pulses enhanced by legacies of antecedent photodegradation in semi-arid environments? Agric Forest Meteorol 154–155:203–213
Mayer LM, Thornton KR, Schick LL, Jastrow JD, Harden JW (2012) Photodissolution of soil organic matter. Geoderma 170:314–321
McKinlay AF, Diffey BL (1987a) A reference action spectrum for ultra-violet induced erythema in human skin. In: Passchier WF, Bosnjakovic BF (eds) Human exposure to ultraviolet radiation: risks and regulations. Elsevier, Amsterdam, pp 83–87
McKinlay AF, Diffey BL (1987b) A reference action spectrum for ultraviolet induced erythemal in human skin. CIE J 6:17–22
McLeod AR, Fry SC, Loake GJ, Messenger DJ, Reay DS, Smith KA, Yun BW (2008) Ultraviolet radiation drives methane emissions from terrestrial plant pectins. New Phytol 180:124–132
Meentemeyer V (1978) Macroclimate and lignin control of litter decomposition rates. Ecology 59:465–472
Moody SA, Newsham KK, Ayres PG, Paul ND (1999) Variation in the responses of litter and phylloplane fungi to UV-B radiation (290-315 nm). Mycol Res 103:1469–1477

Moody SA, Paul ND, Björn LO, Callaghan TV, Lee JA, Manetas Y, Rozema J, GwynnJones D, Johanson U, Kyparissis A, Oudejans AMC (2001) The direct effects of UV-B radiation on Betula pubescens litter decomposing at four European field sites. Plant Ecol 154:27–36

Moorhead DL, Reynolds JF (1991) A general model of litter decomposition in the northern Chihuahuan Desert. Ecol Model 56:197–219

Nagahashi G, Douds DD (2003) Action spectrum for the induction of hyphal branches of an arbuscular mycorrhizal fungus: exposure sites versus branching sites. Mycol Res 107:1075–1082

Newsham KK, McLeod AR, Roberts JD, Greenslade PD, Emmett BA (1997) Direct effects of elevated UV-B radiation on the decomposition of *Quercus robur* leaf litter. Oikos 79:592–602

Newsham KK, Anderson JM, Sparks TH, Splatt P, Woods C, McLeod AR (2001) UV-B effect on *Quercus robur* leaf litter decomposition persists over four years. Glob Change Biol 7:479–483

Noy-Meir I (1979/80) Structure and function of desert ecosystems. Isr J Bot 28:1–19

Okin GS (2008) A new model of wind erosion in the presence of vegetation. J Geophys Res 113: F02S10. doi:10.1029/2007JF000758

Okin GS, Gillette DA (2001) Distribution of vegetation in wind-dominated landscapes: implications for wind erosion modeling and landscape processes. J Geophys Res 106:9673–9684

Okin GS, Herrick JE, Gillette DA (2006) Multi-scale controls on and consequences of aeolian processes in landscape change in arid and semiarid environments. J Arid Environ 65:253–275

Okin GS, D'Odorico P, Archer SR (2009a) Impact of feedbacks on Chihuahuan desert grasslands: transience and metastability. J Geophys Res Biogeosci 114:G01004. doi:10.1029/2008JG000833

Okin GS, Parsons AJ, Wainwright J, Herrick JE, Bestelmeyer BT, Peters DC, Fredrickson EL (2009b) Do changes in connectivity explain desertification? Bioscience 59:237–244

Osman S, Peeters Z, LaDuc MT, Mancinelli R, Ehrenfreund P, Venkateswaran K (2008) Effect of shadowing on survival of bacteria under conditions simulating the Martian atmosphere and UV radiation. Appl Environ Microbiol 74:959–970

Pancotto VA, Sala OE, Cabello M, Lopez NI, Robson TM, Ballaré CL, Caldwell MM, Scopel AL (2003) Solar UV-B decreases decomposition in herbaceous plant litter in Tierra del Fuego, Argentina: potential role of an altered decomposer community. Glob Change Biol 9:1465–1474

Pancotto VA, Sala OE, Robson TM, Caldwell MM, Scopel AL (2005) Direct and indirect effects of solar ultraviolet-B radiation on long-term decomposition. Glob Change Biol 11:1982–1989

Parton W, Silver WL, Burke IC, Grassens L, Harmon ME, Currie WS, King JY, Adair EC, Brandt LA, Hart SC, Fasth B (2007) Global-scale similarities in nitrogen release patterns during long-term decomposition. Science 315:361–364

Paul ND, Moore JP, McPherson M, Lambourne C, Croft P, Heaton JC, Wargent JJ (2012) Ecological responses to UV radiation: interactions between the biological effects of UV on plants and on associated organisms. Physiol Plant 145:565–581

Paulino-Lima IG, Azua-Bustos A, Vicuña R, González-Silva C, Salas L, Teixeira L, Rosado A, da Costa Leitao AA, Lage C (2013) Isolation of UVC-tolerant bacteria from the Hyperarid Atacama Desert, Chile. Microb Ecol 65:325–335

Peters DPC, Bestelmeyer BT, Herrick JE, Fredrickson EL, Monger HC, Havstad KM (2006) Disentangling complex landscapes: new insights into arid and semiarid system dynamics. Bioscience 56:491–501

Rahn R, Lee MA (1998) Iodouracil as a personal dosimeter for solar UVB. Photochem Photobiol 68:173–178

Rangel DEN, Braga GUL, Flint SD, Anderson AJ, Roberts DW (2004) Variations in UV-B tolerance and germination speed of *Metarhizium anisopliae* conidia produced on insects and artificial substrates. J Invertebr Pathol 87:77–83

Ravi S, D'Odorico P, Collins SL, Huxman TE (2009a) Can biological invasions induce desertification? New Phytol 181:512–515

Ravi S, D'Odorico P, Wang L, White C, Okin G, Macko S, Collins S (2009b) Post-fire resource redistribution in desert grasslands: a possible negative feedback on land degradation. Ecosystems 12:434–444
Ravi S, Breshears DD, Huxman TE, D'Odorico P (2010) Land degradation in drylands: interactions among hydrologic, aeolian erosion and vegetation dynamics. Geomorphology 116:236–245
Reid KD, Wilcox BP, Breshears DD, MacDonald L (1999) Runoff and erosion in a piñon-juniper woodland: influence of vegetation patches. Soil Sci Soc Am J 63:1869–1879
Rozema J, Tosserams M, Nelissen HJM, Vanheerwaarden L, Broekman RA, Flierman N (1997) Stratospheric ozone reduction and ecosystem processes: enhanced UV-B radiation affects chemical quality and decomposition of leaves of the dune grassland species *Calamagrostis epigeios*. Plant Ecol 128:284–294
Rutledge S, Campbell DI, Baldocchi D, Schipper LA (2010) Photodegradation leads to increased carbon dioxide losses from terrestrial organic matter. Glob Change Biol 16:3065–3074
Ryel RJ, Flint SD, Barnes PW (2010) Solar UV-B radiation and global dimming: effects on plant growth and UV-shielding. In: Gao W, Schmoldt DL, Slusser J (eds) UV radiation in global climate change. Measurements, modeling and effects on ecosystems. Springer, New York, pp 370–394
Schade GW, Crutzen PJ (1999) CO emissions from degrading plant matter (II). Estimate of a global source strength. Tellus Ser B Chem Phys Meteorol 51:909–918
Schade GW, Hofmann RM, Crutzen PJ (1999) CO emissions from degrading plant matter (I). Measurements. Tellus Ser B Chem Phys Meteorol 51:889–908
Schlesinger WH, Reynolds JF, Cunningham GL, Huenneke LF, Jarrell WM, Virginia RA, Whitford WG (1990) Biological feedbacks in global desertification. Science 24:1043–1048
Searles PS, Flint SD, Caldwell MM (2001) A meta-analysis of plant field studies simulating stratospheric ozone depletion. Oecologia 127:1–10
Setlow RB (1974) The wavelengths in sunlight effective in producing skin cancer: a theoretical analysis. Proc Natl Acad Sci U S A 71:3363–3366
Shen W, Lin Y, Jenerette G, Wu J (2011) Blowing litter across a landscape: effects on ecosystem nutrient flux and implications for landscape management. Landsc Ecol 26:629–644
Smith WK, Gao W, Steltzer H, Wallenstein MD, Tree R (2010) Moisture availability influences the effect of ultraviolet-B radiation on leaf litter decomposition. Glob Change Biol 16:484–495
Song X, Jiang H, Zhang HL, Peng CH, Yu SQ (2011) Elevated UV-B radiation did not affect decomposition rates of needles of two coniferous species in subtropical China. Eur J Soil Biol 47:343–348
Song XZZ, Zhang HLL, Chang SX, Jiang H, Peng CHH, Yu SQQ (2012) Elevated UV-B radiation increased the decomposition of Cinnamomum camphora and Cyclobalanopsis glauca leaf litter in subtropical China. J Soils Sediments 12:307–311
Song X, Peng C, Jiang H, Zhu Q, Wang W (2013a) Direct and indirect effects of UV-B exposure on litter decomposition: a meta-analysis. PLoS One 8(6):e68858. doi:10.1371/journal.pone.0068858
Song XZ, Zhang HL, Jiang H, Donaldson LA, Wang HL (2013b) Influence of elevated UV-B radiation on leaf litter chemistry and subsequent decomposition in humid subtropical China. J Soils Sediments 13:846–853
Throop HL, Archer SR (2007) Interrelationships among shrub encroachment, land management, and litter decomposition in a semidesert grassland. Ecol Appl 17:1809–1823
Throop HL, Archer SR (2009) Resolving the dryland decomposition conundrum: some new perspectives on potential drivers. Prog Bot 70:171–194
Toy TJ, Foster GR, Reynard KG (2002) Soil erosion: processes, prediction, measurement and control. Wiley, New York
Turner J, Parisi AV, Turnbull DJ (2009) Dosimeter for the measurement of plant damaging solar UV exposures. Agric Forest Meteorol 149:1301–1306
Ulevičius V, Pečiulyte D, Lugauskas A, Andriejauskiene J (2004) Field study on changes in viability of airborne fungal propagules exposed to UV radiation. Environ Toxicol 19:437–441

Uselman SM, Snyder KA, Blank RR, Jones TJ (2011) UVB exposure does not accelerate rates of litter decomposition in a semi-arid riparian ecosystem. Soil Biol Biochem 43:1254–1265
Verhoef HA, Verspagen JMH, Zoomer HR (2000) Direct and indirect effects of ultraviolet-B radiation on soil biota, decomposition and nutrient fluxes in dune grassland soil systems. Biol Fertil Soils 31:366–371
Wargent JJ, Jordan BR (2013) From ozone depletion to agriculture: understanding the role of UV radiation in sustainable crop production. New Phytol 197:1058–1076
Warner TT (2004) Desert meteorology. Cambridge University Press, New York
West N (ed) (1983) Temperate deserts and semi-deserts, vol 5, Ecosystems of the world. Elsevier, Amsterdam
Whicker JJ, Breshears DD, Wasiolek PT, Kirchner TB, Tavani RA, Schoep DA, Rodgers JC (2002) Temporal and spatial variation of episodic wind erosion in unburned and burned semiarid shrubland. J Environ Qual 31:599–612
Whicker JJ, Pinder JE, Breshears DD (2006) Increased wind erosion from forest wildfire: implications for contaminant-related risks. J Environ Qual 35:468–478
Whitford WG, Meentemeyer V, Seastedt TR, Cromack K, Crossley DA, Santos P, Todd RL, Waide JB (1981) Exceptions to the AET model – deserts and clear-cut forest. Ecology 62:275–277
Wilcox BP, Breshears DD, Allen CD (2003) Ecohydrology of a resource-conserving semiarid woodland: effects of scale and disturbance. Ecol Monogr 73:223–239
Zaller JG, Caldwell MM, Flint SD, Ballare CL, Scopel AL, Sala OE (2009) Solar UVB and warming affect decomposition and earthworms in a fen ecosystem in Tierra del Fuego, Argentina. Glob Change Biol 15:2493–2502
Zepp RG, Callaghan T, Erickson D (1995) Effects of increased solar ultraviolet radiation on biogeochemical cycles. Ambio 24:181–187
Zucconi L, Ripa C, Selbmann L, Onofri S (2002) Effects of UV on the spores of the fungal species *Arthrobotrys oligospora* and *A. ferox*. Polar Biol 25:500–505

Interspecific Competition in *Arabidopsis thaliana*: A Knowledge Gap Is Starting to Close

Maik Bartelheimer, Christoph Schmid, Joana Storf, Katharina Hell, and Sibylle Bauer

Contents

Abstract The model species *Arabidopsis thaliana* offers an interesting ecological background as a non-mycorrhizal annual species and it can be analysed by outstanding molecular tools. However, its interspecific interactions are scarcely analysed, especially its competitive effect, which is found to be strong despite the species' small size. *A. thaliana*'s competitive response has received more attention during the last few years. Mechanisms were found to be multi-faceted and to involve resource competition for shifting limiting resources, impacts of environmental factors including environmental stress, perception of neighbours as well as responses to allelopathy and neighbour-associated mycorrhiza. Most mechanisms underlying *A. thaliana* interspecific interactions still require clarification and offer research perspectives both for plant molecular biology and plant ecology.

M. Bartelheimer (✉) • C. Schmid • J. Storf • K. Hell • S. Bauer
Institute of Botany, Faculty of Biology and Preclinical Medicine, University of Regensburg, 93040 Regensburg, Germany
e-mail: maik.bartelheimer@biologie.uni-regensburg.de; christoph.schmid@biologie.uni-regensburg.de; joana.storf@stud.uni-regensburg.de; katharina.hell@stud.uni-regensburg.de; sibylle.bauer@biologie.uni-regensburg.de

U. Lüttge, W. Beyschlag (eds.), *Progress in Botany*, Progress in Botany 76,
DOI 10.1007/978-3-319-08807-5_12

1 Why Study Interspecific Interactions in *Arabidopsis thaliana*

Plant–plant interactions can either occur between individuals of the same species (intraspecific) or between individuals of different species (interspecific). In natural ecosystems up to 89 species can co-occur on a single square metre (Wilson et al. 2012), meaning that interspecific encounters are very common in vascular plants. Different views exist on the relative intensity of intraspecific competition and interspecific competition. One common notion is that competition between individuals of the same species should be most intense, because they have exactly the same resource requirements and their niche overlap should therefore be maximal (e.g. MacArthur and Levins 1967). However, practically all plants have a common requirement for a basic set of resources, whether they are conspecifics or not, since they all need light, water, and mineral nutrients. Hence, the intensity of resource competition is not so much ruled by the degree of niche overlap, but by the question which resource is limiting plant growth at a site and which species is most efficient in its uptake (Tilman 1982). Whether a species can occur at a site is therefore strongly co-determined by its adaptations and its interspecific interac tions. Ecological research has intensely investigated interspecific interactions and some species are especially favoured. For instance, a keyword based search by the ISI Web of Knowledge search engine (accessed December 2013) produces 62 hits for the term combination 'Lolium perenne' AND 'interspecific competition' and it produced 56 hits, when the species name is replaced by 'Trifolium repens'. However, the else most commonly used model species *A. thaliana* is almost neglected in the topic of interspecific interaction research (23 hits, of which many only used '*Arabidopsis thaliana*' as a 'KeyWord Plus'). This finding contrasts to the obvious advantages the species has to offer considering its interesting ecology as a non-mycorrhizal, (winter-)annual species and the molecular toolbox that has been developed to facilitate its use in research.

1.1 Ecological Characteristics of Arabidopsis thaliana

Arabidopsis thaliana (L.) Heynh. (mouse-ear cress) is a small flowering plant in the mustard family (Brassicaceae) widely used as a model organism in plant molecular genetics (see below). It is a rosette annual (rarely biennial) native to Europe, North Africa and North America (Mitchell-Olds 2001), with a maximum height of 40 (-70) cm (Tutin et al. 1993). *A. thaliana* mainly occurs in open or disturbed rocky, sandy or loamy habitats, for example riversides, roadsides or rocky hills (Al-Shehbaz and O'Kane 2002). In Europe it belongs to the plant association of Sedo-Scleranthetea with persistent, succulent or summer annual species (*Sempervivum arachnoideum*, *S. soboliferum* and *S. tectorum*, *Sedum acre* and *S. annuum*, *Potentilla neumanniana*, *Myosotis ramosissima*, *Cerastium glutinosum*). It also occurs in the association of Aperion spicae-venti (*Apera spica-venti*, *Centaurea cyanus*, *Vicia hirsuta* and *V. angustifolia*) on nutrient poor fields and Digitario-Setarion (*Erodium cicutarium*, *Galinsoga parviflora* and *G. ciliata*, *Setaria viridis*, *S. glauca*, *S. digitaria* and *S. ischaemum*, *Echinochloa crus-galli*, *Stachys arvensis*, *Anchusa arvensis*, *Chrysanthemum segetum*) on dry sandy and loamy soils (Oberdorfer 2001; Pott 1995). *A. thaliana* is classified as a ruderal/stress-tolerant species (Grime et al. 2007) and exhibits rapid growth and is able to complete its life cycle in less than 6 weeks with prolific seed production and high rates of germination. Considering that competition is often assumed to be of lesser relevance for short-lived and prolific seed producing species occurring in disturbed habitats (Grime et al. 2007), the ecology of *A. thaliana* and most of its plant properties would not suggest a species of strong competitive ability, nor that competition is of major importance during its life cycle. Yet, *A. thaliana* usually co-occurs with 15–23 species per m^2 (Grime et al. 2007), making interspecific encounters likely. Furthermore, a number of studies find strong competitive ability in the species (e.g. Bauer 2010; Müller and Bartelheimer 2013), describing its extraordinarily fine roots as a potential reason for this. *A. thaliana*'s high reproductive efficiency could give a hint to its high competitive ability, because reproductive efficiency has been proposed as a hypothesis for explaining evolutionary success of small species (Aarssen et al. 2006). By producing a higher number of offspring, genetic variants are generated faster and could allow for better adaptation to changing environments with the opportunity for emergence of new ecotypes. Also, Bonser and Ladd (2011) point out that competition has been important in the evolution of strategies in many *A. thaliana* genotypes. Over 750 natural accession of *A. thaliana* are available from seed stock centres like ABRC and NASC. Beyond morphology they vary for example in flowering time, seed dormancy, resistance to pathogens, enzyme activity, DNA methylation and gene expression level (Koornneef et al. 2004). The species is non-mycorrhizal, but it does have symbiotic endophytic fungi like *Piriformospora indica* (Peškan-Berghöfer et al. 2004; Sherameti et al. 2008). The bacterial flora hosted by *Arabidopsis* roots is in part dependent on soil type, but notably it is in part also typical of the plant species (Bulgarelli et al. 2012).

1.2 Advantages Offered by the Arabidopsis *Molecular Toolbox*

The possibilities offered by the use of the model species *A. thaliana* for answering ecological questions are highly diverse and valuable. One advantage is its small genome of 125 Mbp consisting of five chromosomes, which was fully sequenced in the year 2000 (Arabidopsis Genome Initiative 2000). This makes the species most suitable for combining morphological studies with molecular analyses. The knowledge of the plant's genome, proteome and metabolome greatly facilitates working procedures, starting from simple in silico work, through availability of suitable hard- and software (e.g. for micro-array analyses), to data interpretation based on current information on gene function. For most genes (77 %) such functional annotations have already been assigned (Lamesch et al. 2012). These can be used for the identification of regulatory mechanisms reacting to the applied manipulation. Not only do so-called "omics studies" provide the opportunity to gain insights into the physiological background of any given experimental manipulation. The existence and availability of a wide range of mutant lines (cf. TAIR; Lamesch et al. 2012) and well-established laboratory techniques, such as quantitative PCR techniques, RNAi, GC/MS analysis or immunolabelling, allow for studying such mechanisms in any detail both in vitro and in vivo. Furthermore, many characterised natural *Arabidopsis* ecotypes can be purchased, offering the opportunity to study their differential reactions to the intended manipulation. Additionally, data can be transferred to species related to *A. thaliana* quite easily. For many ecological topics these possibilities have already been adopted with good success. Proteome studies already have provided valuable information on several topics in plant ecology, e.g. in plant–microbe interactions (Kav et al. 2007; Cheng et al. 2010). In pollination biology the combined informations gained from metabolomics, proteomics and genomics led to a quite detailed understanding of the underlying mechanisms for both plants and insects (Heil 2011). As a very versatile set of tools without the need of sequence information, metabolomic studies have a broad range of applications (Macel et al. 2010). Successful applications include the chemical ecology of plant–herbivore interactions (e.g. Ramadan et al. 2011). A variety of transcriptome studies on the reactions to abiotic stressors has already been conducted. One example here is the identification of a conserved set of genes in iron deficiency responses of several *A. thaliana* ecotypes (e.g. Stein and Waters 2012). Concerning interactions with other organisms, much progress has been made, e.g. in plant–pathogen interactions (e.g. Attard et al. 2010) or in using *Medicago truncatula* as a model species in mutualistic communities (e.g. Küster et al. 2007). Even for plant–plant interactions some exciting research including transcriptome analysis has been conducted. However, the latter is true only for intraspecific competition (Biedrzycki et al. 2011; Masclaux et al. 2012). So far research on interspecific competition has scarcely made use of the advantages brought about by the model species in experimental research. What is more, there are only a very limited number of studies on *A. thaliana* that deal with its characteristics in interspecific competition. To our knowledge there are currently only 13 publications that are in one way or another related to the topic in question (see below). Some of these involve the use of different ecotypes in order to investigate their differences in competitive effect and response, and few make attempts to profit from the model species as a molecular toolbox.

For these reasons, we believe that interspecific interactions of *A. thaliana* will be interesting for at least two large groups of researchers with (1) interaction ecologists wishing to make use of the molecular possibilities offered by a model plant's molecular toolbox and (2) plant scientists working with *A. thaliana* and wishing to elaborate on the (interaction-)ecological relevance of their findings.

2 Literature Survey

Existing literature was screened using the Web of Knowledge v.5.12 search engine (assessed in December 2013) with the following keyword combinations: 'Arabidopsis thaliana AND interspecific AND competition', 'Arabidopsis thaliana AND interspecific AND interaction', 'Arabidopsis thaliana AND allelopathy'. There are but 13 studies examining interspecific interactions of *A. thaliana*, 10 of which were published only during the last 5 years (Table 1).

We structure these studies following one of the most important distinctions in plant–plant interaction ecology, competitive effect and competitive response (Goldberg and Fleetwood 1987). The competitive effect is one component of competitive ability describing how strongly a plant or species of interest affects its neighbour's performance. The competitive response is the second component of competitive ability describing how strongly the plant or species of interest is affected by its neighbours, i.e. its tolerance to competition.

2.1 *Studies Examining Competitive Effects of* A. thaliana

In a greenhouse pot experiment, Bauer (2010) used *A. thaliana* wild type from field collections to compare its competitive effect on a phytometer, *Hieracium pilosella*. The latter is known to be a sensitive indicator of competitive pressure (Weigelt et al. 2007). *A. thaliana* was found to inflict stronger competitive effects than any other of the tested sandy grassland species including two forbs (*Cerastium semidecandrum* and *H. pilosella*) and two grasses (*Deschampsia flexuosa* and *Koeleria glauca*). The astonishing finding in this was that *A. thaliana* had the lowest biomass of all species (together with *D. flexuosa*), meaning that the smallest species had the largest competitive effect. Plant size (or biomass as its surrogate) is often seen as a prime plant property lending competitive ability, since large plants are especially capable of extensive resource uptake. Here, *A. thaliana* represents an example of a strong effect competitor of small plant size, posing the question by which mechanism this competitive pressure is inflicted. The following synopsis is intended to give an overview of current knowledge on this topic (also see Table 1).

Apparently, these mechanisms do not involve negative allelopathic effects. To the contrary, in a study by Tomita-Yokotani et al. (2003) germination and growth of seeds of *Celosia cristata* exposed to germinating seeds of *A. thaliana* wild type on agar petri dishes were promoted as compared to the control (i.e. in absence of *A. thaliana*). In a second part experiment, this promotive effect could be tracked to seedling exudates. While this one single study on *A. thaliana* allelopathic effects is insufficient for a sound statement, negative allelopathic interactions are an unlikely explanation for *A. thaliana* competitive effects. In fact, two manipulative studies by Bossdorf et al. (2009) and Müller and Bartelheimer (2013) suggest that the background to *A. thaliana* competitive ability is a multi-faceted matter. Bossdorf

Table 1 Overview of studies examining *A. thaliana* interactions with other species

Ecotype of *A. thaliana*	Competing species	Examined feature	Outcome of interaction	
Wild type; field collection (Bayreuth, Germany)	*H. pilosella*	Competitive **effect**	Competitive effect of *A. thaliana* was stronger than that of four other species, despite *A. thaliana* had the lowest biomass	Bauer (2010)
Shokei (wild collection at Shokei University, Japan)	*Celosia cristata*	Allelopathic **effect** of *A. thaliana* seeds	Growth of *Celosia cristata* seedlings promoted when *A. thaliana* seeds were present in the petri dish	Tomita-Yokotani et al. (2003)
Various wild-type genotypes collected from across Europe	*Senecio vulgaris*/*Anagallis arvensis* (densities 0 or 1 per pot)	Competitive **effect**/competitive **response**	Clear ranking of competitive ability (*Senecio* > *Arabidopsis* > *Anagallis*). Competitive effect (especially on *Anagallis*) depended on genotype Competitive response ability was highly context-dependent (diverse responses of genotypes to the different neighbours)	Bossdorf et al. (2009)
Col-0	*H. pilosella* for Competitive effect. Species mixture of *Aira caryophyllea*, *Conyza canadensis*, *Deschampsia flexuosa*, *Hieracium pilosella*, *Rumex acetosella* for competitive response	Competitive **effect**/competitive **response**	Aberrant root phenotypes with differential impact on competitive ability in competitive effect/competitive response	Müller and Bartelheimer (2013)
Col-0	*Brachypodium distachyon* (densities 0 or 1 per pot)	**Response** of *A. thaliana* in biomass and metabolic fingerprinting. **Effect** of *A. thaliana* on its neighbour's biomass and metabolic fingerprint	Biomass in *A. thaliana* sig. reduced by neighbour. Neighbour biomass unaffected by *A. thaliana*, but allocation measures responded Metabolic fingerprint: no changes in *A. thaliana*, significant changes in *B. distachyon*	Gidman et al. (2003)
n.a.	*Trifolium repens*	**Effect** and **response** concerning metabolites	Interaction increases 34 metabolites in *A. thaliana* and decreases another 54; conversely it increases 20 metabolites in *T. repens* and decreases another 33	Pedersen et al. (2013)

Inbred lines of Col-0 and Ri-0	*Poa annua* (density 0, 4, or 8 per pot)	Competitive **response**	Competition delays anthesis, reduces flowering time and the duration of the period of reproduction. It also reduces seed output, though this was not statistically significant	Brachi et al. (2012)
Mix of various genotypes	Artificial communities dominated by forbs, grasses or legumes	Competitive **response**	Seedling emergence reduced in forb-dominated as compared to grass-dominated communities	Hovick et al. (2012)
Various ecotypes from Europe and North America	*Bromus inermis*/*Andropogon gerardii* (densities 0 or 2 per pot)	Competitive **response**	Competitive pressure from both grasses reduced in elevated CO_2	Lau et al. (2010)
Col-0	*H. pilosella* (densities 0 or 1 per pot)	Competitive **response** and transcriptomic reactions	Strong transcriptomic reactions induced by neighbour. Similarity between neighbour perception and pathogen response	Schmid et al. (2013)
Col-0	*Capsella rubella* (densities 0 or 1 per magenta box)	Competitive **response** in root secreted proteins	Levels of root excreted defense-related proteins significantly increased with *C. rubella* At the same time, the number of individual proteins identified was reduced	Badri et al. (2012)
Col-0	*Phragmites australis*, genotypes BB and P38 Exp. 1 densities 0 or 1 per pot. Exp. 2 densities 0 or 1 per pot; with/without activated charcoal	Competitive **response**	Survival drastically reduced with *P. australis*. Charcoal strongly increased seed survival to match survival under control conditions	Rudrappa et al. (2007)
Exp. 1: Col-0; Exp. 2: Col-0; myb72-1; jin1-2	Exp. 1. Arbuscular mycorrhiza (AM) supported by *Trifolium pratense* (densities 0 or 3 per pot) Exp. 2. AM supported by *Lolium multiflorum* (densities 0 or 3 per pot)	Competitive **response**	Presence of a neighbour, albeit divided from *A. thaliana* by nylon mesh, significantly reduced *A. thaliana* growth. With AM inoculums, this effect was strongly reinforced	Veiga et al. (2013)

et al. (2009) found a clear competition hierarchy in three forbs when examining *Senecio vulgaris*, *Anagallis arvensis* and *A. thaliana* in a pot experiment with a presence/absence approach. Interestingly, competitive effect of *A. thaliana* varied between its genotypes. In addition, the competitive effect of *A. thaliana* on the inferior *A. arvensis* also varied with the nutrient or clipping treatment the *Arabidopsis* maternal generation had previously received (maternal effects). A part experiment with activated carbon showed that allelopathy was not involved in *A. thaliana*'s competitive effect. In a recent study, Müller and Bartelheimer (2013) focused on a single plant property (root hairs) and its role for competitive ability. Their pot experiment with phosphorus-deficient soil examined competitive effects of wild type or of aberrant root phenotypes either in intraspecific competition or on the above-mentioned phytometer *H. pilosella*. Differences between root phenotypes were strongest in intraspecific competition, but they also showed in interspecific interaction, where wild type was stronger than the aberrant phenotypes. Following a metabolomic approach, Gidman et al. (2003) used an additive design in a greenhouse experiment and cultivated *Brachypodium distachyon* and *A. thaliana* either as solitary plants or in 1:1 mixtures. While *Arabidopsis*' competitive effect on *B. distachyon* was negligible in terms of total biomass, it did impact on *B. distachyon*'s allocation and metabolism. Root–shoot allocation was significantly shifted towards higher root biomass, possibly indicating the importance of below ground processes in this experiment. Likewise, the general metabolomic fingerprint was significantly altered, though specific patterns were indistinct. Clearly, to date only incomplete knowledge exists on what determines the often considerable competitive effect of *A. thaliana* on other species. While it appears that plant size and allelopathy can be excluded, an array of other factors concerning resource competition remains to be explored. The above-mentioned notion that *A. thaliana*'s competitive effect does not depend on allelopathy by far does not imply that chemical compounds are not important in its interactions with other species. To the contrary it is clear that beyond resource competition plant–plant communication involves root exudates. This has been demonstrated repeatedly for intraspecific interactions of *A. thaliana* (e.g. Biedrzycki et al. 2010; Caffaro et al. 2011, 2013). However, merely one study has, to date, investigated *Arabidopsis*' interaction effects on another species, when resource competition itself is prevented. Pedersen et al. (2013) co-cultivated *A. thaliana* and *Trifolium repens* in nutrient medium for 2 weeks and found that *A. thaliana* impacted on *T. repens* metabolic profile by significantly affecting 53 of its metabolites, of which 20 were increased. This might indicate the strong potential of *A. thaliana*'s non-toxic exudates to influence other species as potential interaction partners.

2.2 *Studies Examining Competitive Response of* A. thaliana

Four of the studies mentioned above (Bossdorf et al. 2009; Müller and Bartelheimer 2013; Pedersen et al. 2013; Gidman et al. 2003) not only investigated competitive effects but also competitive responses. A further seven studies (Table 1) plus two very recent studies outlined below investigated competitive response exclusively. First we will summarize the findings on competitive response from the four studies already mentioned in Sect. 2.1 (see above) and then continue with the remaining ones.

Similar to its ranking in competitive effect, *A. thaliana* takes an intermediate position in competitive response in the experiments by Bossdorf et al. (2009). It was shown that to some extent, *A. thaliana* responded negatively to allelopathy (manipulated by activated carbon) from both *S. vulgaris* as superior competitor and *A. arvensis* as inferior competitor. On a general basis, *A. thaliana* suffered a biomass loss of 57 % when competing with *S. vulgaris* and 18 % when competing with *A. arvensis*. Different *A. thaliana* genotypes responded differently to competition, i.e. both their reactions to the two competing species and reactions to allelopathy depended on genotypes. The authors point out that such variation in competitive ability between genotypes is known for other species as well and that this might reflect the potential of evolution of competitive ability on short temporal scales (Bossdorf et al. 2009). It is tempting to speculate that this potential might be especially high in a species with a short life cycle and tremendous potential seed output like *A. thaliana*. Interestingly, maternal effects, which had been found to be important for competitive effect, were unimportant for competitive response.

In contrast to the findings by Bossdorf et al. (2009), Müller and Bartelheimer (2013) found competitive response to not depend on geno- nor phenotype. As mentioned in Sect. 2.1 they compared different root phenotypes in a pot experiment with phosphorus deficiency. All phenotypes were shown to be unaffected by a community of six sandy grassland species, though among these six species competition was intense. This example might well represent a case where growing conditions resemble a natural situation favourable to *A. thaliana*: a combination of relatively resource-poor soil, slow-growing neighbours and a low degree of canopy closure. It is conceivable that the species' competitive response would be stronger, and likely with more pronounced differences between phenotypes, when conditions match less well with the species realized habitat niche (see introduction for a brief description). In fact, our gap of knowledge is starting to close as far as *Arabidopsis*' different morphological, phenological and physiological responses to competition are concerned. Current knowledge can be compiled as follows.

In the study by Gidman et al. (2003) growth of *A. thaliana* is significantly reduced, though impacts on its metabolomic fingerprint were not detected. Pedersen et al. (2013) found that different metabolites were altered by the presence of *T. repens*. Brachi et al. (2012) used *A. thaliana* plants in standard culture soil in the presence of 0, 4, or 8 plants of naturally co-occurring *Poa annua* to test for different phenological traits. They observed that seed production (represented by cumulative silique length) was reduced by competition. Likewise, anthesis was delayed and so was the duration of the reproductive period. In an experiment on *A. thaliana* colonization success in four types of constructed plant communities, *A. thaliana* seedling emergence was reduced in forb-dominated communities relative to grass-dominated ones (Hovick et al. 2012). The authors concluded that forbs, being most productive in that experiment, brought about higher plant cover, to which *A. thaliana* responded negatively. This conclusion is interesting, because it suggests that competition for light is of high relevance for *A. thaliana* competitive response. While for competitive effect it is somewhat obvious that this rosette species can hardly inflict severe shading on its neighbours, preconditions are different for competitive response. Once *Arabidopsis* switches to reproductive state, it does produce stems with considerable amounts of green tissue above the leaf rosette that might cover the plant's energy demand. However, the study by Hovick et al. (2012) suggests that this species has a weak competitive response ability when competition takes place aboveground. There are two other studies that investigated how *A. thaliana*'s ability to respond to competition changes with altered environmental conditions. One is a study by Lau et al. (2010) that varies presence/absence of intraspecific competition or competition from grasses (the C3-grass *Bromus inermis* and the C4-grass *Andropogon gerardii*) in pots subjected to ambient or elevated CO_2. Elevated CO_2 led to an increase of intraspecific competition intensity, but to a decrease of interspecific competition intensity. The authors discuss that with elevated CO_2 a shift might have occurred concerning the resource most strongly limiting growth of *A. thaliana* (e.g. *A. thaliana* being limited by water under ambient CO_2 but limited by nitrogen under elevated CO_2). This might have led to the named shifts in competitive rankings. While competitive rankings shifting with environmental factors is by no means an extraordinary finding in itself (Goldberg 1996), this illustrates how strongly the competitive response of *A. thaliana* depends on various factors and that it is not much of a species' inherent property in itself.

Additionally, a study by Hell (2014) finds the intensity of *A. thaliana* competitive response to vary with a manipulated environmental factor, in this case water table depth (Fig. 1). The applied additive design compared solitary *A. thaliana* plants to those growing together with plants from six other Brassiceaen species (ranging from indicators of drought to indicators of waterlogging). Water table depth was varied in three factor levels (Fig. 1a). Combined effects of water table depth and competition are evident from Fig. 1b. On the one hand, competition does severely affect *A. thaliana* and reduces its growth tremendously, especially under dry to intermediate water conditions. On the other hand, waterlogging has strongly negative effects by itself, reducing growth to a fraction of what is found for solitary plants under dry conditions. The combined effects of waterlogging and competition did result in the lowest total dry weight, but this was not significantly different from most other treatments. Consequently, once *Arabidopsis*' growth was strongly impeded by waterlogging, competition did not have as much extra effect as under drier conditions. It appears that the observed differences in competition intensity were not primarily due to shifts in the limiting resource (as in Lau et al. 2010), but to environmental stress (in this case anoxia) resulting in overall reduced biomass. Besides *Arabidopsis*' general preference for drier conditions, this study illustrates how severely competition can affect *Arabidopsis*' growth (up to 97 % reduction under dry conditions) (Fig. 1b).

A study by Storf (2014) examined *A. thaliana* vertical root distribution in response to competition from *Vulpia bromoides* and varied water availability (well-watered vs. drought treatments, Fig. 2). Drought resulted in pronounced growth reductions both in *A. thaliana* and *V. bromoides*. *A. thaliana* competitive response (in terms of growth reduction relative to the control) was very severe and similar between water treatments. The calculated Relative Neighbour Effect (RNE, an indicator of competition intensity with max. values of 1.0, see Markham and Chanway 1996) was $RNE_{well\text{-}watered} = 0.89 \pm 0.03$ and $RNE_{drought} = 0.87 \pm 0.05$, respectively (not shown). However, root distribution patterns responded differentially (Fig. 2b, c). In the well-watered treatment (Fig. 2b), *A. thaliana* formed the great majority of its roots in the uppermost soil layer, with near to zero roots in the bottom soil layer. In the presence of competition root amounts were drastically reduced, but the deepest soil layer did contain some deep reaching roots. In the drought treatment (Fig. 2c) control plants distributed their roots quite evenly along the soil layers, while competition resulted in such small root systems that no roots could even reach the deepest soil layer. Clearly, strong biomass effects on root distribution were evident, here. Additionally, at least in the well-watered treatment (Fig. 2b), it appears that the *A. thaliana* root system is capable of plastically avoiding competition by favouring deeper soil layers. Similar rooting strategies have been observed before (Tosti and Thorup-Kristensen 2010), where an alternative interpretation was suggested and the occupation of certain soil layers during early stages of competition led to dominance in these layers later on.

A different approach to test for *A. thaliana* root reactions was taken by Schmid et al. (2013). An additive design with presence/absence of *H. pilosella* was used to investigate *A. thaliana* horizontal root distribution and for transcriptomic response to the neighbour. Similar to the above study by Storf (2014) it was found that the *A. thaliana* root system is capable of plastic reactions, avoiding the neighbour by preferentially placing roots in the adverse direction (root segregation). The examined transcriptomic response made use of the advantages of *A. thaliana* as a model plant of molecular biology. It was found that even though *A. thaliana* biomass was not significantly reduced by the neighbour, there was a far reaching transcriptomic reaction. Analyses concerning the identity of responding transcripts revealed induction especially of those transcripts usually responding to pathogen attacks. A possible explanation brought up by the authors was that neighbour roots, their exudates, or their associated microbiological flora are perceived by *A. thaliana* with mechanisms else active during pathogen attacks.

A recent study by Badri et al. (2012) also investigates neighbour recognition. The *A. thaliana* ecotype Col-0 was cultured on MS medium as a solitary plant or with either different *A. thaliana* ecotypes or with *Capsella rubella*. Short term reactions in root exudates were examined. The observed patterns of excreted proteins were complex, but there was a distinct increase in the amount of excreted defence proteins in the plant combination Col-0 vs. *C. rubella* as compared to Col-0 as solitary plant. At the same time the secretion of stress-related proteins decreased in the named comparison. The authors conclude that some proteins are specifically

Fig. 1 Method and result illustrations of a competition experiment (Hell 2014) with *A. thaliana* (Col-0) and six other species from the family of Brassicaceae with (**a**) picture of an experimental unit containing a mixture of competing *A. thaliana*, *Arabis hirsuta*, *Arabis nemorensis*, *Barbarea vulgaris*, *Berteroa incana*, *Cochlearia pyrenaica*, *Erysimum odoratum*. Cache pots were used to set water tables. (**b**) Total dry weight (means ± SE) in *A. thaliana* in reaction to water table depth and interspecific interaction. Different *minor letters* indicate statistically significant differences in Welch-ANOVA with post-hoc Tamhane ($n = 6–7$)

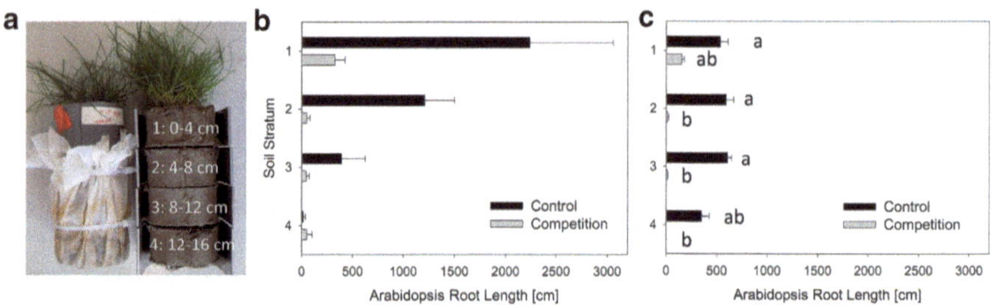

Fig. 2 Method and result illustrations of a competition and root distribution experiment (Storf 2014) with *A. thaliana* (Col-0) as focal species and *Vulpia bromoides* as competitor. (**a**) Picture of an experimental unit (*left*) consisting of a PVC tube filled with sandy loam and planted with *A. thaliana* (Col-0) as focal plant (not visible) in the centre of six plants of *V. bromoides*. Upon harvest, the soil was stratified in 4 cm strata (*right*) and roots were washed out and sorted by species. Control treatments consisted of one solitary plant of *A. thaliana* without *V. bromoides*. (**b**) *A. thaliana* root distribution (means ± SE; $n = 3–5$) in the *well-watered* treatment with and without competition, respectively. (**c**) *A. thaliana* root distribution (means ± SE; $n = 7–8$) in the *dry* treatment with and without competition, respectively. Different *minor letters* indicate statistically significant differences in Kruskal–Wallis Test with corrected *p*-values

secreted depending on the identity of the neighbour and that some level of neighbour recognition is possible even before resource competition is taking place.

Allelopathy also appears to be a potentially important factor for *A. thaliana* competitive response (also see above for Bossdorf et al. 2009). Due to its sensitivity to allelochemicals *A. thaliana* is progressively gaining importance for bioassays, gradually replacing *Lepidium sativum* as prime test species (e.g. Qin et al. 2007;

Abhilasha et al. 2008). However, such bioassays are usually carried out by challenging *A. thaliana* with plant extracts, meaning that actual interspecific interaction is not examined in these studies. As an exception Rudrappa et al. (2007) tested *A. thaliana* response to *Phragmites australis* and its allelopathy. Pots were seeded with *A. thaliana* and presence/absence of 40-day-old *P. australis* was varied. The application of activated charcoal was used to absorb allelochemicals. *A. thaliana* survival and growth was largely reduced by *P. australis*, but restored almost to control condition when activated charcoal was applied. Together with the findings by Bossdorf et al. (2009) (see above), it is clear that allelopathy can play an important role in *A. thaliana*'s competitive response.

A newly emerging research area is how *A. thaliana* competitive response is influenced by mycorrhiza. *A. thaliana* is known to be non-mycorrhizal, but it can still interact with mycorrhizal plants. An exciting finding in that respect was presented by Veiga et al. (2013). They grew *A. thaliana* in dual compartment systems as solitary plants or with either *Trifolium pratense* or *Lolium perenne*. Root systems of neighbour plants were separated by mesh, which can be penetrated by hyphae but not by roots. It was found that arbuscular mycorrhiza (AM) inoculum caused 50 % growth reduction in *A. thaliana*, but only when *P. pratense* or *L. perenne* were present to support the AM network. In the latter case, roots of *A. thaliana* were found to be colonized by AM. The authors discussed that *A. thaliana* likely suffered from nutrient removal by AM hyphae that were further allocated to the neighbouring plant.

3 Conclusions

The term 'underused' is seldom associated with *A. thaliana*, but it is appropriate in connection with this species when it comes to interspecific competition. Especially our current knowledge on competitive effect is scarce. It would be rewarding to further investigate this in *A. thaliana*, as this is a rare example of a species that has strong competitive effects despite being comparatively small. Its competitive ability appears to be brought about by (so far unknown) morphological or physiological plant properties beyond plant size. For the sake of a clearer picture, investigations on *A. thaliana* effects on more than just a handful of species will be necessary.

As far as the competitive response is concerned, the above synopsis shows that especially over the last few years *A. thaliana* is becoming more recognized as a valuable species for studying interspecific interaction. The matter is complex and involves resource competition for shifting limiting resources, impacts of environmental factors including environmental stress, perception of neighbours as well as responses to allelopathy and neighbour-associated mycorrhiza. Consequently, it appears important to continue to make full use of the possibilities offered by this model plant.

Generally speaking there are almost no field experiments on interspecific interactions of *A. thaliana* (but see Lau et al. 2010), and, to our knowledge, also just one field experiment on intraspecific interactions (Fitter et al. 2002). Considering the high level of complexity of biotic interactions in the field, the use of a model plant to identify mechanisms suggests itself. Moreover, it can be taken for granted that plant interactions in the field are strongly influenced, if not mediated, by microbiota. *A. thaliana*, being at the forefront of research on plant-associated microorganisms suggests itself even in this area. For field experiments the use of natural genotypes is one possibility, while the use of genetically modified knock-down or knock-out mutants is usually prohibited. However, there is still a very good possibility available using chemically mutagenized material screened by 'targeting induced local lesions in genomes' (TILLING, McCullum et al. 2000). The whole screening process can be completed within 4 months (Bush and Krysan 2010) and is therefore well within a reasonable time frame for many projects.

Also concerning the near-to-natural situation, it would be worthwhile to have more studies of *A. thaliana* interacting with species it naturally co-occurs with. One way to prove the ecological relevance of many results from molecular plant biology could be to grow the respective knock-down or overexpression mutants together with species from their natural habitat and to show thereby the impacts on competitive ability.

Plant ecologists, especially those interested in ecophysiological mechanisms, might find powerful approaches in the *Arabidopsis* molecular toolbox. Access to methods and know-how are nowadays often facilitated by service units, companies and collaboration partners. Molecular methods continue to become more powerful and more and more available also for non-model species.

References

Aarssen LW, Schamp B, Pither J (2006) Why are there so many small plants? Implications for species coexistence. J Ecol 94:569–580

Abhilasha D, Quintana N, Vivanco JM, Joshi J (2008) Do allelopathic compounds in invasive *Solidago canadensis* s.l. restrain the native European flora? J Ecol 96:933–1001

Al-Shehbaz IA, O'Kane SL Jr (2002) Taxonomy and phylogeny of *Arabidopsis* (Brassicaceae). Arabidopsis Book 1:e0001, American Society of Plant Biologists

Arabidopsis Genome Initiative (2000) Analysis of the genome sequence of the flowering plant *Arabidopsis thaliana*. Nature 408:796

Attard A, Gourgues M, Callemeyn-Torre N, Keller H (2010) The immediate activation of defense responses in *Arabidopsis* roots is not sufficient to prevent *Phytophthora parasitica* infection. New Phytol 187:449–460

Badri DV, De-la-Peña C, Lei Z, Manter DK, Chaparro JM, Guimarães RL, Sumner LW, Vivanco JM (2012) Root secreted metabolites and proteins are involved in the early events of plant-plant recognition prior to competition. PLoS One 7:e46640

Bauer S (2010) Experimentell-ökologische Untersuchungen zur Bedeutung der Aluminiumtoxizität für die Konkurrenzkraft von Pflanzenarten aus Sandökosystemen. Bachelor thesis, University of Regensburg

Biedrzycki ML, Jilany TA, Dudley SA, Bais HP (2010) Root exudates mediate kin recognition in plants. Commun Integr Biol 3:28–35

Biedrzycki ML, Venkatachalam L, Bais HP (2011) Transcriptome analysis of *Arabidopsis thaliana* plants in response to kin and stranger recognition. Plant Signal Behav 6:1515–1524

Bonser SP, Ladd B (2011) The evolution of competitive strategies in annual plants. Plant Ecol 212:1441–1449

Bossdorf O, Shuja Z, Banta JA (2009) Genotype and maternal environment affect belowground interactions between *Arabidopsis thaliana* and its competitors. OIKOS 118:1541–1551

Brachi B, Aimé C, Glorieux C, Roux F (2012) Adaptive value of phenological traits in stressful environments: predictions based on seed production and laboratory natural selection. PLoS One 7:e32069

Bulgarelli D, Rott M, Schlaeppi K, van Theemat EVL, Ahmadinejad N, Assenza F, Rauf P, Huettel B, Reinhardt R, Schmelzer E, Peplies J, Gloeckner FO, Amann R, Eickhorst T, Schulze-Lefert P (2012) Revealing structure and assembly cues for *Arabidopsis* root-inhabiting microbiota. Nature 488:91–95

Bush SM, Krysan PJ (2010) iTILLING: a personalized approach to the identification of induced mutations in *Arabidopsis*. Plant Physiol 154:25–35

Caffaro MM, Vivanco JM, Gutierrez-Boem FH, Rubio G (2011) The effect of root exudates on root architecture in Arabidopsis thaliana. Plant Growth Regul 64:241–249

Caffaro MM, Vivanco JM, Botto J, Rubio G (2013) Root architecture of *Arabidopsis* is affected by competition with neighbouring plants. Plant Growth Regul 70:141–147

Cheng Z, Woody ZO, Glick RB, McConkey JB (2010) Characterization of plant-bacterial interactions using proteomic approaches. Curr Proteomics 7:244–257

Fitter AH, Williamson L, Linkohr B, Leyser O (2002) Root system architecture determines fitness in an Arabidopsis mutant in competition for immobile phosphate ions but not for nitrate ions. Proc R Soc Lond 269:2017–2022

Gidman E, Goodacre R, Emmett B, Smith AR, Gwynn-Jones D (2003) Investigating plant-plant interference by metabolic fingerprinting. Phytochemistry 63:705–710

Goldberg DE (1996) Competitive ability: definitions, contingency and correlated traits. Philos Trans R Soc Lond B 351:1377–1385

Goldberg DE, Fleetwood L (1987) Competitive effect and response in four annual plants. J Ecol 75:1131–1143

Grime JP, Hodgson JG, Hunt R (2007) Comparative plant biology—a functional approach to common British species. Castlepoint, Colvend

Heil M (2011) Nectar: generation, regulation and ecological functions. Trends Plant Sci 16:191–200

Hell C (2014) Differenzierung hydrologischer Nischen bei Vertretern der Familie der Brassicaceen. Exam thesis, University of Regensburg

Hovick SM, Gümüşer ED, Whitney KD (2012) Community dominance patterns, not colonizer genetic diversity, drive colonization success in a test using grassland species. Plant Ecol 213:1365–1380

Kav NN, Srivastava S, Yajima W, Sharma N (2007) Application of proteomics to investigate plant-microbe interactions. Curr Proteomics 4:28–43

Koornneef M, Alonso-Blanco C, Vreugdenhil D (2004) Naturally occurring genetic variation in *Arabidopsis thaliana*. Annu Rev Plant Biol 55:141–172

Küster H, Vieweg MF, Manthey K, Baier MC, Hohnjec N, Perlick AM (2007) Identification and expression regulation of symbiotically activated legume genes. Phytochemistry 68:8–18

Lamesch P, Berardini TZ, Li D, Swarbreck D, Wilks C, Sasidharan R, Muller R, Dreher K, Alexander DL, Garcia-Hernandez M, Karthikeyan AS, Lee CH, Nelson WD, Ploetz L, Singh S, Wensel A, Huala E (2012) The *Arabidopsis* Information Resource (TAIR): improved gene annotation and new tools. Nucleic Acids Res 40:D1202–D1210

Lau JA, Shaw RG, Reich PB, Tiffin P (2010) Species interactions in a changing environment: elevated CO2 alters the ecological and potential evolutionary consequences of competition. Evol Ecol Res 12:435–455

MacArthur R, Levins R (1967) The limiting similarity, convergence, and divergence of coexisting species. Am Nat 101:377–385

Macel M, Van Dam NM, Keurentjes JJB (2010) Metabolomics: the chemistry between ecology and genetics. Mol Ecol Resour 10:583–593

Markham JH, Chanway CP (1996) Measuring plant neighbour effects. Funct Ecol 10:548–549

Masclaux F, Bruessow F, Schweizer F, Gouhier-Darimont C, Keller L, Reymond P (2012) Transcriptome analysis of intraspecific competition in *Arabidopsis thaliana* reveals organ-specific signatures related to nutrient acquisition and general stress response pathways. BMC Plant Biol 12:227

McCullum CM, Comai L, Greene EA, Henikoff S (2000) Targeted screening for induced mutations. Nat Biotechnol 18:455–457

Mitchell-Olds T (2001) *Arabidopsis thaliana* and its wild relatives: a model system for ecology and evolution. Trends Ecol Evol 16:693–700

Müller B, Bartelheimer M (2013) Interspecific competition in *Arabidopsis thaliana*: root hairs are important for competitive effect, but not for competitive response. Plant Soil 371:167–177

Oberdorfer E (2001) Pflanzensoziologische Exkursionsflora für Deutschland und angrenzende Gebiete. Ulmer, Stuttgart

Pedersen HA, Kudsk P, Fiehn O, Fomsgaard IS (2013) The response of *Arabidopsis* to co-cultivation with clover investigating plant-plant interactions with metabolomics. In: Beck JJ, Coats RJ, Duke SO, Koivunen ME (eds) Pest management with natural products. American Chemical Society, Washington, DC, pp 189–201

Peškan-Berghöfer T, Shahollari B, Giong PH, Hehl S, Markert C, Blanke V, Kost G, Varma A, Oelmüller R (2004) Association of *Piriformospora indica* with *Arabidopsis thaliana* roots represents a novel system to study beneficial plant–microbe interactions and involves early plant protein modifications in the endoplasmic reticulum and at the plasma membrane. Physiol Plant 122:465–477

Pott R (1995) Die Pflanzengesellschaften Deutschlands. Ulmer, Stuttgart

Qin B, Lau JA, Kopshever J, Callaway RM, McGray H, Perry LG, Weir TL, Paschke MW, Hierro JL, Yoder J, Vivanco JM, Strauss S (2007) No evidence for root-mediated allelopathy *in Centaurea solstitialis*, a species in a commonly allelopathic genus. Biol Invasions 9:897–907

Ramadan A, Muroi A, Arimura G-i (2011) Herbivore-induced maize volatiles serve as priming cues for resistance against post-attack by the specialist armyworm *Mythimna separata*. J Plant Interact 6:155–158

Rudrappa T, Bonsall J, Gallagher JL, Seliskar DM, Bais HP (2007) Root-secreted allelochemicals in the noxious weed *Phragmites australis* deploys a reactive oxygen species response and microtubule assembly disruption to execute rhizotoxicity. J Chem Ecol 33:1898–1918

Schmid C, Bauer S, Müller B, Bartelheimer M (2013) Belowground neighbor perception in *Arabidopsis thaliana* studied by transcriptome analysis: roots of *Hieracium pilosella* cause biotic stress. Front Plant Sci 4:296

Sherameti I, Tripathi S, Varma A, Oelmüller R (2008) The root-colonizing endophyte *Piriformospora indica* confers drought tolerance in *Arabidopsis* by stimulating the expression of drought stress-related genes in leaves. Mol Plant Microbe Interact 21:799–807

Stein RJ, Waters BM (2012) Use of natural variation reveals core genes in the transcriptome of iron-deficient *Arabidopsis thaliana* roots. J Exp Bot 63:1039–1055

Storf J (2014) Die Rolle des hydrotropischen Wurzelwachstums für die Konkurrenzfähigkeit von *Arabidopsis thaliana*. Exam thesis, University of Regensburg

Tilman D (1982) Resource competition and community structure. Princeton University Press, Princeton, NJ

Tomita-Yokotani K, Kato T, Masud Parvez M, Mori Y, Goto N, Hasegawa K (2003) Approach of allelopathy study with *Arabidopsis thaliana* (L.) Hevnh. and Neurospora crassa. Weed Biol Manage 3:93–97

Tosti G, Thorup-Kristensen K (2010) Using coloured roots to study root interaction and competition in intercropped legumes and non-legumes. J Plant Ecol 3:191–199

Tutin G, Heywood VH, Burges NA, Valentine DH, Moore DM (1993) Flora Europaea. Cambridge University Press, Cambridge, UK

Veiga RSL, Faccio A, Genre A, Pieterse CMJ, Bonfante P, van der Heijden MGA (2013) Arbuscular mycorrhizal fungi reduce growth and infect roots of the non-host plant *Arabidopsis thaliana*. Plant Cell Environ 36:1926–1937

Weigelt A, Schumacher J, Walther T, Bartelheimer M, Steinlein T, Beyschlag W (2007) Identifying mechanisms of competition in multi-species communities. J Ecol 95:53–64

Wilson JB, Peet RK, Dengler J, Pärtel M (2012) Plant species richness: the world records. J Veg Sci 23:796–802

Carbon Reserves as Indicators for Carbon Limitation in Trees

Günter Hoch

Contents

Abstract In view of the current increase of atmospheric CO_2 concentrations, the question if carbon is a limiting resource for tree growth or not gained large attention over the last decades. This review summarizes how tissue concentrations of nonstructural carbon (C) reserves compounds can be used to assess the C-supply status of trees. Studies that investigated the tissue concentrations of C-reserves and their seasonal variations in trees growing under natural conditions suggested that tree growth and reproduction are currently not limited by photosynthesis under benign or non-stressful climatic conditions. The comparative analysis of C-reserves in trees exposed to environmental stresses like cold temperatures and drought revealed that against previous assumption, the stress-induced decline of growth is also not caused by insufficient C-assimilation. However, recent studies on the C-relation in dying trees exposed to sustained drought indicated organ-specific

G. Hoch (✉)
Institute of Botany, University of Basel, Schönbeinstrasse 6, 4056 Basel, Switzerland
e-mail: guenter.hoch@unibas.ch

U. Lüttge, W. Beyschlag (eds.), *Progress in Botany*, Progress in Botany 76,
DOI 10.1007/978-3-319-08807-5_13

different reactions of tissue C-reserve concentrations, probably as a result of impaired C-transport and reserve re-mobilization under drought stress.

1 Introduction

Like all C-autotrophic life forms, trees depend on the continuous supply of C that is bound from atmospheric CO_2. On average, about half of all dry biomass consists of C, rendering C the quantitatively most important element in plants. Historically, the photosynthetic fixation of inorganic C has thus been received as the physiological key process, which ultimately drives all other plant functions, especially growth. As it has been previously put forward by Körner (2006), the perception of C as the limiting resource for plant growth gained further acceptance by the development of mobile and easy to use CO_2 gas exchange analyzing systems, and the resulting finding that the net-photosynthesis of most terrestrial plants is yet not saturated at ambient atmospheric CO_2 concentrations (close to 400 ppm at the time of writing of this article). Mature trees in particular were supposed to be limited in terms of C-supply, since the majority of tissues are heterotrophic and dependent on the import of assimilated C from green leaves that make up only a small fraction of the entire biomass in this plant growth form (Yoda et al. 1965). However, over the last decades there has been increasing observational and experimental evidence that growth and reproduction of trees living under natural (un-fertilized) conditions are most likely not directly controlled by photosynthetic C-uptake, but by other factors like the availability of soil nutrients, water, or space (Bader et al. 2013; Körner 2003; Norby et al. 2010). Despite the likelihood of sufficient C-supply for growth under benign conditions at current atmospheric CO_2 concentrations, trees might still face C-limitation under certain environmental or biological stresses, like cold temperatures, drought, or leaf-loss by herbivores. To assess the C-balance of large trees is a methodological challenge, and apart from expensive large-scale CO_2-enrichment studies, the prediction of the absence or presence of C-limitation has to rely on indirect clues and models. Here, I will discuss the usefulness of mobile C-reserve compound concentrations as a proxy for a tree C-source-sink-balance, and summarize our current knowledge about mobile C-reserves in trees under potential C-limiting situations.

2 How to Assess the Carbon Supply Status of Tall Trees?

Unlike for single leaves, small plants but also entire ecosystems, the C-exchange cannot be directly measured as a whole for individual tall trees. Hence, predictions for the net C-balance of trees have to (1) either rely on models that integrate selective measurements of photosynthesis and respiration with climatic records (temperature and light) as well as estimates for the proportional biomass of different organs (e.g., leaves, sapwood, fine roots, etc.) or (2) use the comparative analyses of the tissue concentrations of mobile C-reserves. Gas-exchange measurements can be achieved precisely at an areal base for single tissue types (e.g., leaves, bark) and short periods (i.e., seconds to minutes). However, the uncertainty for the modeled net exchange of an entire tree increases markedly with spatial and temporal upscaling (e.g., from cuvette area to the whole organ, from single measurements to whole season integrals). Even more important, a model for the C-supply status of a plant has to account also for all C-sink activities apart from respiration, like growth, reproduction, exudation, etc., which again can at best be approximated from temporally and spatially punctual measurements in tall trees. Obviously, precise determinations of all C-source and -sink activities over longer periods are laborious and time-consuming, thus limiting the number of trees that can be assessed in a single study. Hence, except for very obvious situations like deep shade, direct measurements of C-source and sinks will generally result in more or less rough estimates about whether or not the availability of C is driving specific plant processes, like growth.

An alternative approach is the comparative analysis of the concentrations of carbon-reserve compounds in tree tissues. As photosynthesis and C-sink processes (and thus the ratio between C-source and -sink activities) are not constant over time, but vary diurnally, seasonally, or ontogenetically, plants are experiencing periods where C-assimilation exceeds C-usage alternating with periods where the demand for C is higher than the photosynthetic supply. Mobile C-reserves (e.g., starch or low molecular weight sugars) are metabolites that are produced during periods when C is assimilated faster than used for the diverse C-sinks within a plant, while they are dissipated and serve as a C-source whenever the demand for C exceeds the rate of C-assimilation (Chapin et al. 1990; Dickson 1991; Sala et al. 2011). A simple "bucket-model" for C-reserves (Fig. 1a) thus assumes expanding reserve

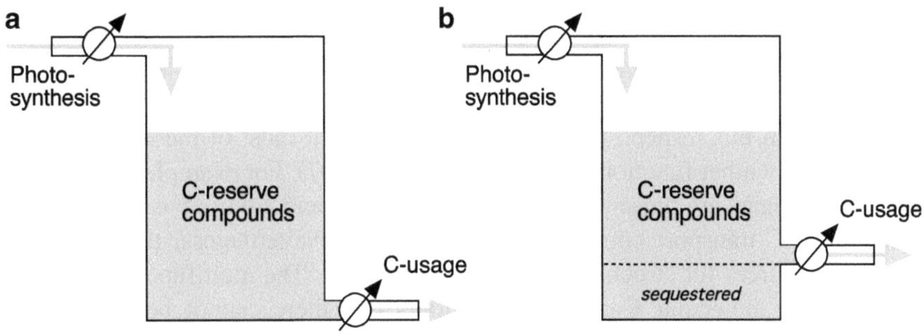

Fig. 1 Simplified "bucket-model" for the size of a plant's C-reserve store in dependency of its C-source-sink-relations. C-sources (photosynthesis) and C-usage (all C-sinks) can be actively controlled. (**a**) If all C-reserve compounds can be used as C-sources, the C-reserve pool would be completely depleted at C-starvation. (**b**) Not all compounds that can serve as C-reserves can be completely depleted at C-limitation, forming a de facto sequestered fraction within the C-reserve pool

concentrations whenever the C-influx from photosynthesis outbalance the net usage of C, and shrinking C-reserves if the demand for C-assimilates is higher than the current photosynthetic activity. In this respect it is important to note that in view of the indicative nature of C-reserves for the C-supply status of a plant, they are the actual tissue concentrations rather than the total reserve pool per plant, which are the decisive factor. A large C-limited tree will show lower tissue concentrations of C-reserves compared to a small well C-supplied sapling, but it will have the larger total C-reserve pool. Based on these assumptions, the comparison of the sizes of C-reserve concentrations among species, dates, or treatments has been used as an integrative proxy for a plant's C-supply status. The sensitivity of C-reserve concentrations to changed C-supplies could be demonstrated in an experiment that exposed plants from different functional groups (including trees) to artificially very low or high CO_2 concentrations (Schädel et al. 2010). A recent study by Hartmann et al. (2013a) showed an almost complete depletion of starch and very low concentrations of low molecular weight sugars in different tissues of *Picea abies* seedlings treated with extremely low atmospheric CO_2 concentrations. Also, trees exposed to deep shade exhibited very strong declines of their C-reserve concentrations (e.g., Piper et al. 2009; Sevanto et al. 2014). On the other hand, C-reserve pools might be formed against a prevailing C-sink demand during certain phenological or developmental stages, especially for C storage before dormant periods in seasonal climates, and probably also under environmental stress (Wiley and Helliker 2012; but see also Palacio et al. 2014). Nevertheless, trees exposed to C-starvation should use up these reserves before death. Otherwise, these C-compounds should be considered as "sequestered" (in terms of not retrievable) rather than "stored" (Millard et al. 2007).

3 Which Carbon Reserves?

In principal, all organic compounds within a plant that can be reintroduced into primary metabolism (recycled) can serve as C-reserves for C-sink activities if the requirement for C exceeds the C-supply by current photosynthesis (Chapin et al. 1990). In this respect, it is important to note that most of these compounds serve primarily other functions than storage (Hoch 2007). For example, low molecular weight sugars are present in all living plant cells, and are used as intermediate metabolites, C-transport compounds, or osmolytes. Nevertheless, they can also serve as C-sources for structural growth or respiration. The multifunctional nature of many low molecular C-compounds entails that there cannot be a complete depletion of all potential C-reserves in living plant tissue (Hoch 2007), comparable with the concentration of blood sugar in mammals, which cannot decrease to zero even if individuals should starve to death. This is especially important if C-reserve concentrations are used as proxies for the C-supply status of plants, since there are C-reserves present even in severely C-limited plants, detectible as a lower minimum concentration of C-reserves (e.g., Piper et al. 2009; Schädel et al. 2010;

Veneklaas and den Ouden 2005; Fig. 1b). The identification of this lower C-reserve threshold in different species and tissues certainly will be an important future research task, in order to improve the diagnostic potential of C-reserve concentrations for a plant's C-supply status.

Besides the large number of C-compounds that can principally be used as reserves but primarily serve other functions, only two C-compound classes are synthesized exclusively as storage compounds, namely, nonstructural polysaccharides (starch and fructans) and neutral lipids (triacylglycerols). These "strict" C-reserves are also expected to react strongest to C-source-sink imbalances. On one hand, there is principally no lower concentration limit for starch or storage lipids that is determined by immediate cell-physiological needs like, for example, for free sugars (see above). On the other hand, polysaccharides as well as neutral lipids are mostly osmotically inactive, which means that they can accumulate to high concentrations without interfering with other cell-biological processes. Indeed, it could be shown that changes in the nonstructural carbohydrate concentrations of plants in response to different atmospheric CO_2 supplies are largely due to changes in starch concentrations while tissue concentrations of low molecular weight sugars stayed relatively constant (Schädel et al. 2010). Interestingly, to my knowledge, there are no reports of a complete depletion of starch reserves, even under sever C-limitation. This might indicate the presence of a lower minimum concentration for starch, which is likely not related to direct cell-physiological needs like for low molecular sugars, but due to steric constraints for the complete degradation of starch granules (e.g., Srichuwong and Jane 2007). Nevertheless, compared to low molecular weight sugars, starch shows generally much stronger variability towards C-source sink imbalances. With respect to the usage of C-reserve concentrations as a proxy for the C-supply status of plants, our main concern might thus be with nonstructural polysaccharides and storage lipids, with the latter being of quantitative importance in only a limited number of species (Hoch et al. 2003). However, the majority of studies dealing with C-reserves in plants reported the sum of low molecular weight sugars plus starch, which is generally referred to as either as "nonstructural carbohydrates" (NSC) or "total nonstructural carbohydrates" (TNC).

4 How Likely Is Carbon Limitation in Trees?

Photosynthesis is the quantitatively most significant assimilation process in plants. However, the central role of C-assimilation for plant growth does not automatically mean that productivity is determined by the photosynthetic activity of plants. The differentiation between C-assimilation and C-usage, and the fact that the former does not necessarily drive the latter has been extensively reviewed previously (e.g., Körner 2003, 2006; Körner 2013; Millard et al. 2007). The question, if and under which conditions plant growth is limited by the supply of photoassimilates, is of special interest in view of the future increase of atmospheric CO_2 concentrations.

Because globally about 90 % of the C fixed in biomass is sequestered in forests (Körner 2003), the dependency of tree growth on C-assimilation is of prime importance. In the following, I will provide an overview of the likelihood of C-limitation in trees as it can be derived from comparative analyses of tissue C-reserve concentrations. First, I will focus on trees growing under more or less benign conditions, and thereafter, I will address two situations of climatic stresses, which potentially can lead to C-starvation, namely cold temperatures and drought. Many of the reported data are results of my own studies during the last decade, which will be compared to findings on C-reserves in trees from other research groups. In addition, these results will be put in a broader plant-physiological context, to better assess the existence or absence of C-limitation independent from the comparative analyses of C-reserves.

4.1 Carbon Relations Under Benign Climatic Conditions

Research over the last two decades has largely indicated that tree growth is driven by other factors than C-supply under non-stressful conditions. Traditionally it has been assumed that trees from seasonal climates (especially deciduous species) rely strongly on stored C-reserves for spring bud break and the consecutive early season growth, implying that the amount of C-reserves that can be stored throughout one season is significantly affecting the growth potential in the following season. In contrast, more recent studies that investigated the C-relations in temperate trees could not confirm the assumed strong dependency of deciduous trees on stored C-reserves.

4.1.1 Independency of Growth from Stored Carbon Reserves

A study at the Swiss Canopy Crane facility investigated the seasonal variations of NSC stores in seven deciduous and three evergreen temperate forest tree species (Hoch et al. 2003). This survey revealed only moderate changes of NSC from before bud break to the end of the growing season, indicating that early season growth does not strain the stored C-reserves in these trees, except for reserves in very young branches during spring bud break (Schädel et al. 2009). On average, deciduous trees showed a seasonal variation of NSC of about 30 % in branch wood and only 10 % in stem sapwood. Surprisingly, the magnitude of the seasonal variations of the trees' C-reserve pools were similar between deciduous and evergreen species (but the seasonal patterns are in opposite directions, see Hoch et al. 2003), clearly contradicting the old paradigm that deciduous trees rely to a larger extend on stored C than evergreen species. In addition, the seasonal NSC variations occurred on top of large total C-reserve pools. Averaged across all broad-leaved deciduous species, the estimated nonstructural carbohydrate reserves stored only in the aboveground wood of these 100-year old trees accounted for over four times the entire C incorporated in all leaves. Additionally, some of the species investigated in Hoch et al. (2003), e.g., *Tilia platyphyllos* and *Pinus sylvestris*, had high sapwood concentrations of storage lipids that did not change significantly across the entire growing season. Similar to this study in temperate trees, year-round relatively high levels of C-reserve concentrations had been documented for trees at other temperate sites (e.g., Barbaroux and Breda 2002; Fischer and Höll 1992; Landhäusser and Lieffers 2003) and also in tropical forest trees (Würth et al. 2005). In view of the overall large C-reserve pools of trees and the fact that, unlike other nutrients like nitrogen, nonstructural C-compounds are often not recovered from senescent tissues, Millard et al. (2007) proposed that trees are sufficiently supplied with C up to a point where considerable portions of a trees' C-reserves are never remobilized and de facto sequestered.

The often weak seasonal fluctuations of nonstructural C-reserves in mature trees led to the conclusion that the productivity of forests is likely not limited by photosynthesis under current atmospheric CO_2 concentrations (Körner 2003). This notion was supported by CO_2 enrichment experiments that were applied in natural or near-natural forests. An 8-year, 550 ppm CO_2, Free Air CO_2 Experiment (FACE) with mature deciduous trees at the Swiss Canopy Crane facility revealed no sustained increase in tree biomass production (Bader et al. 2013; Körner et al. 2005), despite higher leaf-level net-photosynthesis, which was on average 40 % higher in trees exposed to high atmospheric CO_2 compared to ambient CO_2 control trees (Bader et al. 2010; Zotz et al. 2005). Hence, the extra C assimilated at high CO_2 conditions likely increased the C-flux through these trees, without enhancing growth (Körner et al. 2005). Within a North-American long-term FACE experiment (550 ppm CO_2), trees of a sweetgum (*Liquidambar styraciflua*) forest reacted to increased CO_2 supply with a significant +24 % enhancement of NPP during the initial 6 years of the experiment, but their productivity declined to the level of ambient CO_2 control trees during the following 5 years, most likely due to declining N availability under elevated CO_2 at this site (Norby et al. 2010). A third long-term FACE experiment with larger trees investigated *Pinus taeda* trees in an afforestation 11-year-old at the beginning of the experiment. Although this study on young trees found a sustained enhancement of tree productivity at elevated CO_2 over the course of 10 consecutive years of CO_2 enrichment, it could be shown that the effect of elevated CO_2 was primarily driven by the availability of soil nitrogen and water (McCarthy et al. 2010). Hence, in the long term, soil nutrients and water rather than C are the limiting resource for tree growth in mature forests under current atmospheric CO_2 concentrations.

There is additional evidence for abundant C-supply for trees under non-stressed growth conditions from yet another field of research: the analyses of radiocarbon (^{14}C) bomb spike signals as tracers for the age of C used for growth or metabolism in trees. By cross-dating ^{14}C-signals in CO_2 from stem respiration, in tree ring cellulose, as well as in stem wood NSC reserves in maple trees (*Acer rubrum*), Carbone et al. (2013) demonstrated that metabolism and growth largely rely on C that is younger than 1 year, whereas the average age of stored NSC was found to be 10 years, and the old C-reserves are readily available and not sequestered if needed

(Richardson et al. 2013). As a consequence, C-reserve pools are likely to increase with tree age under benign conditions (Carbone et al. 2013), again signifying the abundant C-supply of trees under current atmospheric CO_2 concentrations.

4.1.2 Independency of Mast Fruiting from Stored Carbon Reserves

While the above-mentioned studies delivered clear evidence that vegetative tree growth is currently not under C-limitation in unfertilized sites, the phenomenon of mast seeding (i.e., the production of high fruit loads in some years intermitted by no to very low fruit production in all other years; Silvertown 1980) might be related to a restricted availability of C-reserves in trees. It has been hypothesized that the formation of high fruit loads in one season might deplete tree-internal pools of a resource, and it therefore takes one or more seasons to refill this pool before the next masting event can take place (Isagi et al. 1997; Kelly 1994). Traditionally, C-reserves are thought to be likely candidates for this limiting resource in plants with masting reproduction behavior (e.g., Crone et al. 2009; Kozlowski et al. 1991). However, the analysis of NSC reserves in stem wood of *Fagus sylvatica* and *Quercus petraea* did not show significant concentration differences between masting and non-masting years (Körner 2003). Obeso (1998) and Hoch (2005) tested the dependency of fruit production on stored C-reserves by applying ring-girdling (i.e., the removal of a stripe of phloem without interrupting the xylem sap flow) treatments on fruiting shoots of different broad-leaved tree species in order to cut off the terminal, fruit-bearing part of the shoot from C-reserves stored in other parts of the tree. These experiments showed uniformly that gridling alone had no effect on the final fruit-number and -biomass, suggesting that fruit production can be supplied exclusively from current photosynthesis by leaves and photosynthetically active fruit tissue on the girdled branches. Even a 50 % defoliation on girdled branches led to no or a relatively small reduction in final fruit biomass, while girdling and complete defoliation massively reduced fruit number and biomass. Accordingly, NSC reserve concentrations in the xylem of girdled branches were also found to be unaffected relative to controls in undefoliated and 50 %-defoliated branches, but were significantly reduced in girdled and completely defoliated branches (Hoch 2005), reflecting the branches' negative C-balance. The exclusive usage of current photoassimilates for fruiting were recently confirmed in a study that estimated the age of C used for fruit production in ten temperate deciduous tree species by radiocarbon (^{14}C) analyses (Ichie et al. 2013). Finally, an unambiguous indication for the independency of fruiting from stored C-reserves in masting tree species was revealed within a study at the Swiss Canopy Carne site. In this study, adult trees that had been labeled with a low $\delta^{13}C$ signal for eight consecutive seasons, produced fruits in the first season after the labeling that carried exclusively the ambient $\delta^{13}C$ of control trees but no traces of the $\delta^{13}C$-label of old C-reserves (Hoch et al. 2013).

Overall, studies performed over the last two decades clearly contradicted the classical view that mast fruiting depletes C-reserves in trees. However, this does not inevitably suggest that there is no constraint on the C-supply for growth for the entire tree. Indeed, several studies indicated that high fruit loads constrain the vegetative growth of either the fruit-bearing shoot (Han et al. 2011) or the entire tree (Mund et al. 2010), although others did not find a pronounced effect of mast seeding on tree stem increments (Sala et al. 2012a; Yasumura et al. 2006). Probably, lower vegetative growth in masting seasons might indicate that, although fruit growth is not directly constraining stored reserves, the increased C-sink strength from fruits for currently assimilated C might compete with allocation of C to growth of stems or roots. This would also explain observations of lower C-reserve concentrations in wood of fruiting branches (Miyazaki et al. 2002). In contrast, the reduced vegetative growth in masting trees could be unrelated to C, but driven by a limitation of other nutrients, since high fruit production has also been described to decrease pools of nitrogen and/or phosphorus on the fruiting shoots as well as on the whole tree level (Han et al. 2014; Ichie and Nakagawa 2013; McDowell et al. 2000; Miyazaki et al. 2002; Sala et al. 2012a).

4.2 Carbon Relations Under Environmental Stress

Perhaps, the reason for the abundant C-supply of trees today is due to the fact that most of the currently existing tree species evolved in atmospheres where CO_2 concentrations were below 300 ppm CO_2 (compared to currently about 400 ppm), with all physiological processes optimized to operate under much lower atmospheric CO_2 than today (Körner 2006). While this finding has far-reaching consequences for all models on tree growth under future climates, there might be specific situations, where abiotic or biotic stress results in C-limitation for trees, not ameliorated by today's increased atmospheric CO_2 concentrations.

Environmental stress can theoretically have different effects on a plant's C-balance (Fig. 2). First, the stressor can affect primarily photosynthetis, thus leading to C-limitation for all C-sinks (direct C-limitation; arrow "a" in Fig. 2). Second, the stressful (limiting) environmental factor can affect C-sinks, like growth, before or stronger than photosynthesis (direct C-sink limitation; arrow "b" in Fig. 2). Third, plants might also react with the active accumulation of C-reserves as an immediate stress response (Wiley and Helliker 2012). This accumulation of reserves could be in competition with other C-sink activities of the plant (indirect C-limitation via enhanced C allocation to storage; arrow "c" in Fig. 2). With respect to the usage of C-reserve concentrations as proxies for a plant's C-supply status, it can be assumed that C-reserve concentrations are declining relative to unstressed plants, in the case of a direct C-limitation. At direct sink limitation, but also if C-reserves are actively formed against a prevailing C-sink strength, C-reserve concentrations are assumed to increase in stressed trees. In this case, the comparative analyses of C-reserve concentrations alone cannot predict the absence or presence of C-limitation, but further physiological information is needed to decide on the dominant biological mechanism behind the limitation of growth

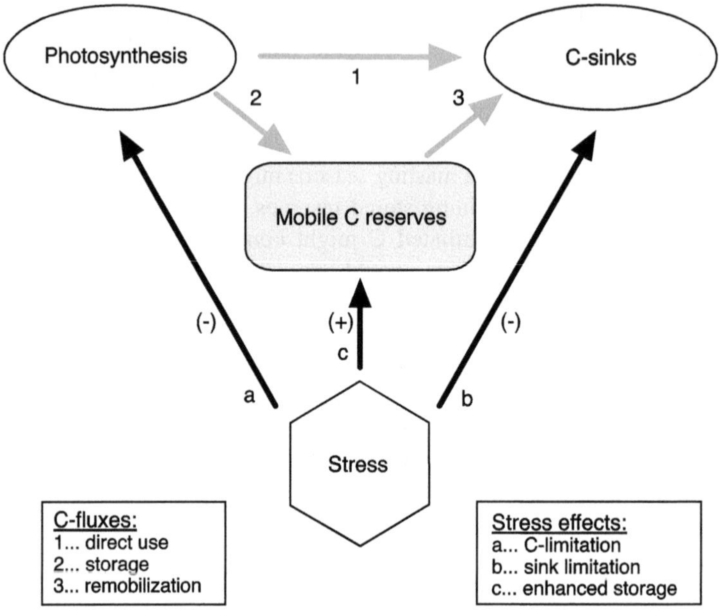

Fig. 2 Schematic presentation of plant internal C-fluxes (1, 2, 3) and the potential effects of environmental stressors on these fluxes (a, b, c). The direction of the stress effect on C-source activity, C-sink-activity, or C-reserve pools are indicated by (−) decrease or (+) increase. See text for further explanations

(Palacio et al. 2014). In the following, I will give an overview of our current state of knowledge about the C-supply status of trees exposed to two environmental stressors that potentially can induce reduced productivity and tree mortality via C-limitation, namely, cold temperature and hydraulic constraints.

4.2.1 Cold Limits of Tree Growth

Along elevational or latitudinal gradients from warmer to colder sites, different tree species are reaching their species-specific cold limit at very different temperature isoclines. However, if only alpine treelines (i.e., the cold limits of the life form tree) are considered, their elevational situations exhibit a remarkable climatic similarity, despite the fact that treelines are a world-wide phenomenon occurring at elevations between close to sea level to almost 5,000 m a.s.l. Independent of the length of the growing season, the average temperature across the local growing season at alpine treelines was found to fall within a narrow temperature window of around 6.5 °C (Körner and Paulsen 2004). Considering that the actual tree species which form these different treelines are from phylogenetically very different branches (including conifers and angiosperms), and that besides the mean growing season temperatures, all environmental factors (including soil properties, precipitation, and partial pressure) can be very different among treelines; this uniform isocline suggests one common process behind the formation of treelines worldwide. As discussed in detail in Körner (1998), there are basically two mechanisms that might be responsible for the decline of tree growth at the treeline: (1) C-limitation via insufficient photosynthesis or (2) a direct restriction of meristematic growth at cold temperatures. Within a global series of studies along gradients from closed montane forests to the uppermost occurrences of trees, the question, if C-limitation can be the cause for the decline of tree growth at alpine treelines, was addressed by investigating tissue concentrations of NSC in 14 different tree species at 13 locations between 68 °N and 46 °S (Hoch and Körner 2012). These surveys showed that NSC did not decline with elevation, but revealed a clear increase towards treeline, indicating that a direct C-limitation is unlikely the physiological cause for treeline formation. Across all species and sites, the elevational increase of NSC was mainly driven by higher concentrations of starch (+40 % at treeline), while there was only an insignificant increase in the concentrations of low molecular weight sugars. Hence, the observed increase of NSC is not an immediate physiological adjustment to cold temperatures (e.g., by increasing cytoplasmatic concentrations of osmotically active sugars), but an accumulation of osmotically inactive C-reserves. In addition to starch, storage lipids in sapwood were also found to increase significantly in three different pine species along treeline ecotones, and lipids concentrations of pines at alpine treelines were found to be about twice that of pines from lowland sites (Hoch and Körner 2003). The elevational patterns found in the global survey by Hoch and Körner (2012) were also described in several other studies (e.g., Piper et al. 2006; Shi et al. 2006), and the increase of NSC could be shown to be of a similar magnitude between evergreen and deciduous species (Fajardo et al. 2013). To my knowledge, there is so far only one single study that investigated the C-balance along elevational gradients in trees that do not reach the alpine treeline. Within a recent survey, NSC concentrations in stem wood of eight European broad-leaved tree species were investigated along elevational gradients from the species' upper distribution limit (which differed by almost 1,000 elevational meter among species) downslope to about 1,200 m below the species' upper limit (A. Lenz, *unpublished data*). Across all investigated species, this survey revealed no or increasing trends of NSC concentrations with elevation, indicating that also for tree species that do not reach the alpine treeline, C-limitation is unlikely the growth limiting mechanism at their respective cold limits.

Investigation of trees along different elevations have to deal with a shift in phenology along these gradients (e.g., an earlier start of the growing season at lower elevations, Fig. 3). Thus, comparisons of NSC between sites at different elevations should be either conducted with trees at the same phenological stage (i.e., at different dates for different elevations) or at a period during the year at which trees are at the same phenological stage across the whole elevational gradient (e.g., at the end of the season or during the dormant period). Consequently, studies that investigated NSC trends in evergreen trees across the alpine treeline ecotone at different dates during the growing season (Li et al. 2008; Sveinbjornsson et al. 2010) reported also a decline of NSC from lower to higher sites during the

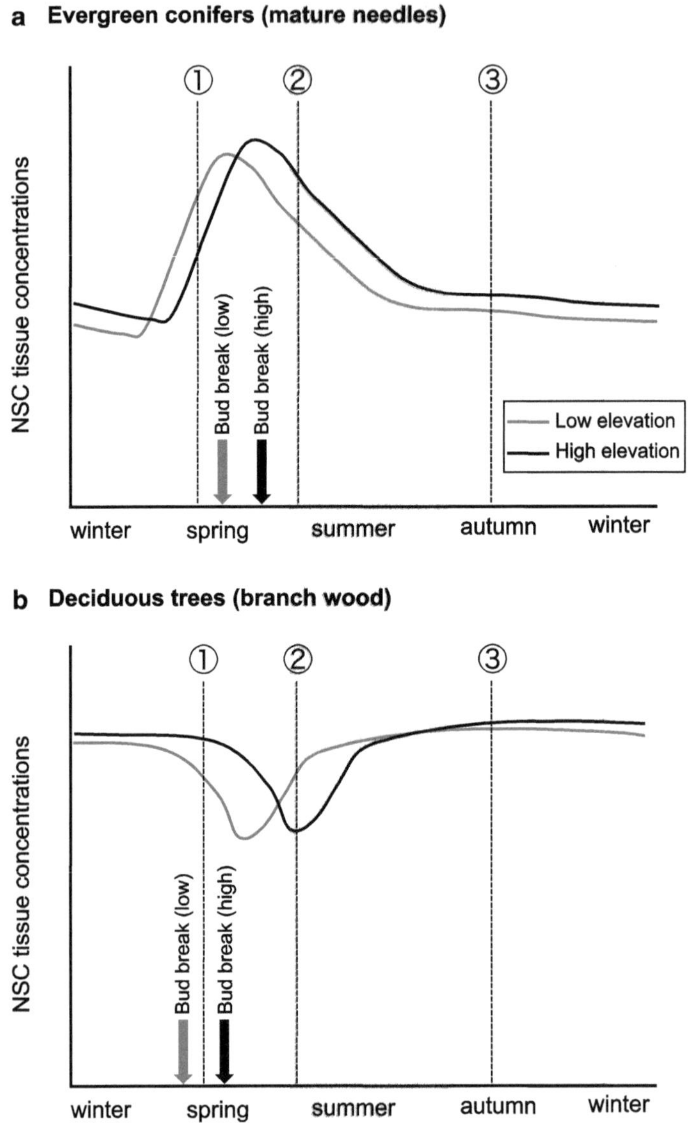

Fig. 3 The phenological shift in bud break dates between low and high elevation tree stands leads to an elevational shift in the seasonal NSC patterns in evergreen conifer needles (**a**) and deciduous branch wood (**b**). 1, 2, and 3 show three different sampling dates for NSC analyses along the growing season, indicating the stronger phenological bias on comparative NSC analyses along elevational gradients at sampling dates 1 and 2 compared to date 3 at the end of the growing season

first half of the growing season. Most likely, this is not an indication of reduced C-supply at higher elevations, but a bias from sampling trees at different phenophases around bud break, a period where temporal changes in NSC concentrations are most pronounced and fast (Fig. 3; Fischer and Höll 1991; Hoch et al. 2002, 2003). For example, a sampling campaign on date 1 in Fig. 3 would reveal higher reserve concentrations at lower elevations in evergreen species and lower concentrations at low elevation species of deciduous species, while the situation would be reversed at sampling date 2 (Fig. 3).

The observed increase of tissues C-reserve concentrations towards treeline has been interpreted as an indication of an oversupply of C resulting from a disproportional decrease of growth relative to photosynthesis (Hoch and Körner 2012; Körner 1998). In contrast, it has been also assumed more recently that this elevational pattern might be due to an internal (in the sense of "intentional") upregulation of C-stores in trees facing cold stress (Wiley and Helliker 2012), in which case the increased investment into C-storage might be even in competition with growth. However, there is unequivocal physiological evidence that cell growth in cold adapted trees ceases at considerable warmer temperature than net-photosynthesis (e.g., Grace et al. 2002; Palacio et al. 2014 and references therein). Tree growth becomes negligible below ca. 5 °C (Alvarez-Uria and Körner 2007; Rossi et al. 2007) and comes to a complete stop between 2 and 3 °C depending on species (Halter et al. 1997; Solfjeld and Johnsen 2006; Schenker et al. 2014), while in cold adapted trees considerable positive net-photosynthesis has been reported near 0 °C and below (Körner 2012; Wieser and Tausz 2007). Hence, there is substantial evidence apart from the analyses of C-reserve concentrations that points at a direct limitation of growth processes at cold treeline temperatures, rendering the hypothesis of a cold-induced reserve formation on the expense of growth rather unlikely.

It has been also suggested that the higher NSC concentrations found in trees at alpine treelines are ecotypical (genetic) adaptations to the cold and often harsher environment (Smith et al. 2003; Sveinbjörnsson 2000), resulting in tree populations at treeline that differ from trees at lower elevations by intrinsically higher C-reserve stores (Monson et al. 2006). However, recent investigations could largely rebut this hypothesis. A study that investigated dwarfed *Picea abies* trees growing on montane permafrost patches in the Swiss Jura mountains showed significantly enhanced NSC and lipid concentrations in different tissues of these trees compared to tall growing trees of the same age in the immediate neighborhood (Hoch 2008). The fact that the trees on these permafrost sites are most likely recruited from the surrounding tall forests, suggests that the NSC increase on permafrost cannot be a genetically fixed trait. An even stronger evidence for the absence of ecotypic adaptations with respect to higher NSC concentrations at treeline was delivered recently by Fajardo et al. (2012), who found consistently increasing NSC concentrations in wood and leaves in afforested trees (*Larix decidua*, *Pinus cembra*, *Pinus sylvestris*) from single seed sources that were planted along elevational gradients from lower montane sites up to the alpine treeline. These studies together with ex situ experiments (see below) that also used trees from single provenances, clearly suggest that the observed increase of NSC at cold, growth limiting temperatures is an immediate physiological effect and not an evolved ecotypic adaptation.

The increasing C-reserve concentrations in tree tissues towards alpine treelines and the presence of direct growth limitation at cold growing season temperatures typically found at these sites were largely confirmed in experimental studies. Within a phyotoron experiment, seedlings of *Pinus mugo* and *Larix decidua* were treated with 20 weeks growing seasons with either mean temperatures of 6 °C (corresponding to the mean season temperature at natural treelines) or of 12 °C. Seedlings treated with low temperatures showed significantly enhanced NSC concentrations while their growth was severely reduced compared to warm treated controls (Hoch and Körner 2009). The same study tested also the effect of seasonally variable temperatures versus completely constant temperatures on growth and C-relations. Interestingly, this comparison revealed very similar reaction of whole-season growth and NSC concentrations, independent of if temperatures varied diurnally and seasonally or not (Hoch and Körner 2009). Higher tissue concentrations of C-reserves were also found in seedlings of the broad-leaved species *Betula pendula* treated with growth limiting cold root zone temperatures (Solfjeld and Johnsen 2006) and in first year seedlings of *Abies lasiocarpa* and *Pseudotsuga menziesii* transplanted to treeline compared to seedlings grown 550 elevational meters below (Bansal and Germino 2010).

Up to date, there has been only one single study that tested the growth-limitation hypothesis at treeline by directly applying elevated CO_2 concentrations (Dawes et al. 2011, 2013; Handa et al. 2005). In this study that investigated afforested *Larix decidua* and *Pinus uncinata* which were exposed to elevated CO_2 by a FACE system for 9 consecutive years at the alpine treeline ecotone, both species reacted to high CO_2 with a significant increase of their NSC tissue concentrations. *P. uncinata* showed no stimulation of growth under elevated CO_2 throughout the experiment, suggesting a strong direct limitation of growth in this species. In contrast, *L. decidua* treated with high CO_2 exhibited a significant increase in aboveground wood production (average of 33 % larger tree rings over all 9 years, Dawes et al. 2013). However, while the CO_2 stimulation of growth in *L. decidua* was high during the first 7 years of the experiment, it seemed to have leveled off towards the end of the 9-year study period (Dawes et al. 2011). Hence, the initially very positive CO_2 effect on growth might have been transient (similar to the findings for temperate forests by Norby et al. (2010) discussed above) and was probably also influenced by a series of unusually warm growing seasons at the beginning of the experiment, including the extreme European heat-wave summer of 2003 (Jolly et al. 2005).

4.2.2 Hydraulic Limits of Tree Growth

Hydraulic constraints for water transport in trees induces a reduction of stomatal conductance, which in turn might lead to C-limitation for growth or even C-starvation as the cause for tree die-off. Hydraulic limitation on tree growth and drought-induced tree mortality have gained growing attention over the last few years, mainly because large-scale tree mortality that is very likely related to water shortage has increased globally during the last two decades (Allen et al. 2010). Global climate models predict significant reductions of precipitation within the next century (IPCC 2013) in many regions worldwide, including areas that are already currently facing water shortage, like the south-west of the USA or south-western Europe. In addition the future increase of temperatures will enhance the atmospheric evaporative demands, leading to an increasing vapor pressure deficit (VPD), thus potentially inducing water limitation also in regions that will not experience a direct decrease of precipitation in the years to come (Breshears et al. 2013; Will et al. 2013; Williams et al. 2013). Finally, warmer temperatures will probably further burden the C-balance of trees by an increase of respiration (Adams et al. 2009; Zhao et al. 2013). Drought-induced tree mortality on larger areas have dramatic effects for ecological and biogeochemical processes, including the ecosystem water balance, soil erosion, and ecosystem C-sequestration, but it might also have drastic feedback effects on the global climate (Bonan 2008). In order to better understand and assess these scenarios, precise models that predict the development of forests facing drought are needed. A thorough understanding of the underlying physiological processes responsible for the observed effects of drought is necessary to improve the current, largely correlative vegetation models (Powell et al. 2013). However, there are still large uncertainties about the actual biological mechanisms that lead to reduced growth and mortality in drought-stressed trees. Here I will summarize, how the analyses of C-reserve concentrations in tree tissues are used to investigate these mechanisms. Whenever hydraulic constraints in trees are discussed, it is important to consider the initial nonlethal drought stress that limits growth separately form severe and lethal drought stress. Likely, the physiological mechanisms behind growth limitation are different from the processes involved in tree mortality under drought. Of course, growth limitation and tree mortality are occurring not independently with the first occurring before the latter on a continuous gradient of drought stress severity (Fig. 4). In the following, I will first consider the effects and physiological mechanisms of hydraulic constraints on growth and then discuss the current state of knowledge about the mechanisms behind drought-induced tree mortality in context of a tree's C-relation.

Hydraulic Constraints on Tree Growth

It is well established in forestry that tree productivity decreases with decreasing site water availability. Ryan and Yoder (1997) hypothesized that also the decline of productivity in trees reaching their maximum height is due to increasing hydraulic constraints, since growing trees have to move water against gravity to even higher positions. Like in plants that are facing soil water shortage, the decreasing water potential in tree crowns that approach their maximum height enforce earlier stomatal closure (Woodruff et al. 2004), which in turn results in declining photosynthetic rates at the leaf level (Koch et al. 2004). Consequently, it has been proposed that the decline of productivity in trees on dry sites as well as in old trees is caused by C-limitation, resulting from the disproportional decrease of net-photosynthesis over C-sink demands ("hydraulic limitation hypothesis," Ryan and Yoder 1997).

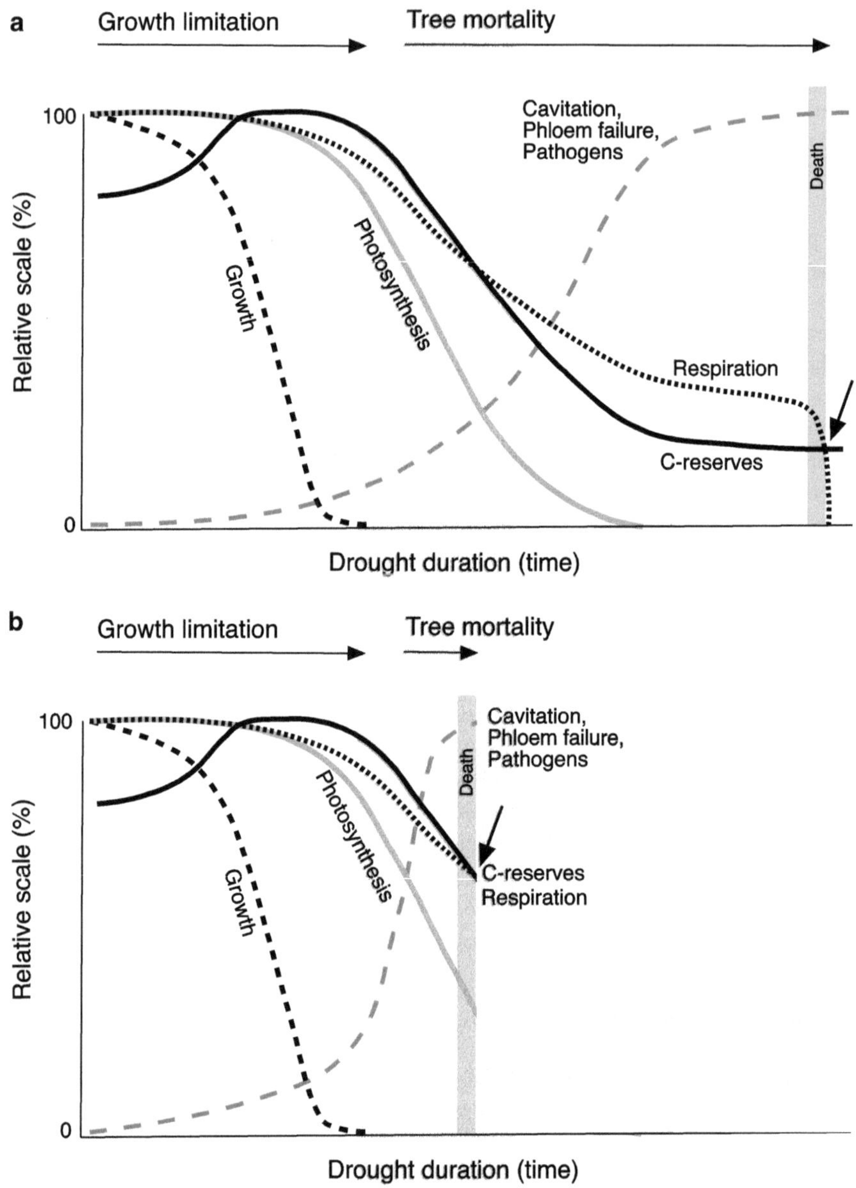

Fig. 4 Schematic concept for the interdependency of photosynthesis, respiration, and growth on a tree's C-reserve pool at sustained drought stress, and the interrelated effect of xylem cavitation, phloem failure, and pathogen infestation on the size of the C-reserve pool at the time of tree death. In terms of C-relations, growth limitation needs to be considered separately from tree mortality under drought, indicated by the *two arrows* at the top of the figures. (**a**) Situation in trees with sustained hydraulic conductivity and pathogen resistance at drought. (**b**) Situation in trees with fast hydraulic conductivity loss and high vulnerability against pathogens at drought. The point of tree death is indicated by the *vertical grey bar*; the *arrow* indicates the level of the C-reserve pool at tree die-off. This scheme is a simplified synthesis of a previous concept presented in McDowell et al. (2011)

To investigate this hypothesis, the analysis of C-reserve concentrations in tree tissue is a promising approach.

The hydraulic limitation hypothesis was first evaluated by comparative NSC analyses in a study that investigated *Pinus ponderosa* trees growing on two sites differing in soil moisture (Sala and Hoch 2009). Against the initial hypothesis by Ryan and Yoder (1997), this study revealed significantly increasing concentrations of NSC in sapwood and needles, and of storage lipids in branch wood as the trees approach their site-specific maximum heights. Importantly, the average concentrations of C-reserves, as well as their relative increase with tree height, were found to be higher in trees at a dry stand with smaller trees compared to a moister site with larger trees (Sala and Hoch 2009). This study suggested clearly that the hydraulic constraints in tall growing trees do not exert negative C-balances, and C-limitation can thus unlikely be the responsible physiological mechanism behind declining growth. The results described in Sala and Hoch (2009) were confirmed by Woodruff and Meinzer (2011), who analyzed the seasonal course of NSC in wood and needles of *Pseudotsuga menziesii* trees of different heights between 2 and 57 m. Cross-seasonal average NSC concentrations were higher and seasonal variations of NSC were smaller the higher the trees. The study by Woodruff and Meinzer (2011) could further show a very strong negative correlation between NSC concentrations in branch wood and shoot elongation, since smaller trees had stronger shoot growth and lower NSC concentrations. The same study found also a negative correlation between NSC concentration and the branch water potential (i.e., higher NSC concentrations at lower water potentials). In conclusion, these results point to a growth limiting process, which, in contrast to the original hydraulic limitation hypothesis, leads to a relative oversupply with photo-assimilates, despite the observed reduced stomatal conductance. Thus, under hydraulic constraints, meristematic growth (cell division and expansion) declines significantly earlier than photosynthesis (Fig. 4), and in analogy to the processes at the cold growth limit of trees, this direct limitation of growth processes is indicated by higher NSC tissue concentrations in taller trees and drier sites (see also the initial increase in NSC in Fig. 4). In fact, the higher sensitivity of growth compared to photosynthesis to decreasing plant water potentials has been documented already more than 40 years ago in different plants (e.g., Boyer 1970; Hsiao and Acevedo 1974) and has been recently reviewed by Muller et al. (2011). Hence, NSC analyses are in line with predictions from independent physiological measurements of growth rates and photosynthesis, and the observed growth declines are not caused by C-limitation but by a direct growth limitation, likely induced by increasing difficulties to establish the necessary turgor for cell division and expansion within growing meristems (Woodruff et al. 2004).

Drought-Induced Tree Mortality

Although C-limitation is most likely not decisive for growth reduction under hydraulic constraints, it might be a significant process for tree mortality under severe drought. Hypothesized first in the seminal review by McDowell et al. (2008), and exhaustively discussed thereafter in several papers (e.g., McDowell et al. 2011; Sala et al. 2010; Sevanto et al. 2014), drought-induced tree mortality might occur by three processes that likely are not acting separately, but in combination (1) hydraulic failure (i.e., a lethal loss of conductivity in the xylem), (2) C-starvation (i.e., the depletion of C-reserves to a point where they cannot supply basic C-needs), and (3) biotic attack (i.e., the infestation of pathogenic insects or fungal infection). While biotic attacks might often follow the weakening of trees by either (1) or (2) or both, the debate if and under which conditions, either hydraulic failure or C-starvation is the initial cause for tree mortality is ongoing. In addition, hydraulic limitations of xylem water transport are probably also impairing the phloem movement of photoassimilates as well as the remobilization of stored C, thereby linking hydraulic constraints and C-relations beyond the effect of drought on stomatal conductance (e.g., McDowell et al. 2011; Sala et al. 2010; Sevanto 2014). Finally, trees depend on the presence of a certain amount of carbohydrates for the osmoregulation of C-transport and the maintenance of the hydraulic conductivity in the sapwood, which likely limits the degree to which living trees can degrade C-reserve stores under drought (Sala et al. 2012b).

The comparative analysis of C-reserve concentrations in tree tissues has been used in a number of recent studies to test the absence or presence of C-starvation in drought-stressed trees. So far, these tests revealed very different results ranging from a significant increase of NSC concentrations, over no changes to an almost complete depletion of C-reserves under drought. For example, Anderegg (2012) reported increasing NSC concentrations in drought-stressed *Populus tremuloides*, with the strongest increase occurring in roots. Similarly, a very strong increase of starch concentrations in roots was found in the same species treated with experimental drought under green house conditions (Galvez et al. 2011). Anderegg and Anderegg (2013) found no significant changes in seedlings of two conifers (*Juniperus osteosperma* and *Pinus edulis*) at drought-induced mortality, but severe losses of hydraulic conductivity. A study on two *Nothofagus* species revealed opposite NSC reactions in response to experimentally applied drought (Piper 2011). While NSC concentrations increased in the more drought-tolerant *N. dombeyi*, it decreased in the more drought-sensitive *N. nitida*. Other experimental drought studies reported changes in C-reserve concentrations that differed among organs. A recent study on *Populus balsamifera* and *Populus tremuloides* seedlings showed decreased C-reserve concentrations especially in roots, but to a lesser extend in leaves and stems in drought-stressed seedlings under outdoor conditions (Galvez et al. 2013). The lowered NSC concentrations in roots in the study by Galvez et al. (2013) were probably responsible for higher rates of winter mortality following the dry season. A drought experiment by Hartmann et al. (2013b) in *Picea abies* trees showed a very strong decline of NSC (especially of starch) reserves in roots in drought treated trees at the time of death, while starch and sugar concentrations in aboveground organs were similar to watered controls. The same pattern was reported in Hartmann et al. (2013a), where it was additionally shown that *P. abies* decreased C-reserves much faster under CO_2-depletion than under drought, indicating that hydraulic failure rather than C-starvation was the cause for drought-induced mortality in trees at ambient CO_2 in that study. Finally, significantly lower starch concentrations in stem collar sapwood of *Quercus robur* in autumn after the extreme European heat- and drought-summer of 2003 were reported for trees that exhibited high branch die-back in the following spring (Bréda et al. 2006), and a survey of heavily drought-stressed individuals of mature *Pinus sylvestris* trees showed almost a complete depletion of NSC in their stem sapwood the season before they died (Galiano et al. 2011).

The contrasting results for tissue NSC concentrations in response to lethal drought among the above-mentioned studies, deliver important information about the effect of drought on the tree C-balance. There are three potential explanations that are likely causes for these varying results: (1) species-specific differences (i.e., different tree species show different sensitivities to cavitation and C-starvation), (2) differences in the experimentally applied drought (i.e., severity and/or duration of drought among studies), and (3) differences among tree organs (i.e., C-starvation does not occur simultaneously throughout the whole tree, but is restricted to one organ). Certainly, the species-specific vulnerability to conductivity loss is an important factor determining, if a tree can survive long enough under drought to deplete its C-reserves. Cavitation is supposed to occur earlier in so-called anisohydric species (Fig. 4b) than in isohydric species, which avoid critical negative tree water potentials by earlier stomata closure at drought (Fig. 4a; McDowell et al. 2008). In addition, the speed with which cavitation occurs depends on the severity of the applied drought. A slow decline of soil water potential will more likely lead to C-starvation in trees than an abrupt change in soil water availability, as could be experimentally shown recently in *Pinus edulis* trees (Sevanto et al. 2014), where fast dying trees had higher NSC branch reserve concentrations than trees that survived longer under drought. Several studies that investigated above- and belowground NSC concentrations found no decline, or even an increase of NSC in leaves and branch wood, but sometimes significant declines of NSC in stem and especially in roots. Such a local decline of C-reserves in tissue furthest away from leaves might be indicative of an insufficient downward transport of C-assimilates in the phloem as a consequence of hydraulic constraints to the upward water movement in the xylem (Sala et al. 2010; Sevanto 2014) and reduced translocation of recent carbon form leaves in drought-stressed trees (Ruehr et al. 2009). Hence, while the drought-induced reduction of productivity is unlikely driven by C-limitation, C-starvation might occur locally (i.e., restricted to specific organs) in dying trees experiencing sustained and mild drought. However, the majority of the existing studies on drought-induced tree mortality suggest that in most species and under most conditions, hydraulic conductivity loss by cavitation generally precedes carbon starvation (e.g., Hartmann et al. 2013a; Sevanto et al. 2014).

5 Concluding Remarks

Over the last decades, the comparative analysis of C-reserve concentrations in plant tissue has been established as a valuable tool to assess the C-supply status of trees. The principal responsiveness of the C-reserves to changes of a plant's C-source-sink-balance could be demonstrated unequivocally in previous experiments. Most importantly, environmental stressors that affect exclusively photosynthesis, like deep shade and low CO_2, lead to a significant depletion of C-reserves in plant tissue. Hence, the increasing tissue concentrations of NSC and storage lipids found in trees that experience growth restriction at cold temperatures and under hydraulic constraints are very likely indicative for a direct limitation of growth processes and the absence of C-limitation under these adverse conditions (Palacio et al. 2014). However, to improve the usability of quantitative C-reserve analyses for predictions of a plant's C-supply status, further research in the line of Schädel et al. (2009) and Schädel et al. (2010) will be needed to clarify which cell compounds beside NSC and lipids can serve as C-sources. In addition, because C-limitation will never lead to a complete depletion of all C-reserves in living cells, experimental studies and meta-analyses over existing studies should identify the minimum and maximum concentrations of C-reserve compounds in different plant groups and tissues. Finally, we are still lacking a conclusive picture about how and to which extent intrinsic reserve formation, like C storage before the dormant season, does interfere with the simple "bucket-model" of C-reserve dynamics in trees by allocating photoassimilates to storage against other C-sink demands. Such information is needed to unequivocally predict C-starvation in plants by quantitative C-reserve analyses.

The temporal dynamics of C-reserves within entire trees will be only understood once the long-distance C-transport via the phloem can be measured quantitatively and continuously. Up to date the phloem C-flux has been mainly inferred from pulse labeling experiments with either ^{13}C or ^{14}C labeled photoassimilates (e.g., Dannoura et al. 2011; Hansen and Grauslun 1973; Streit et al. 2013). However, in order to accurately interpret the local depletion of C-reserves found in some of the experimental studies that investigated drought-induced tree mortality, a continuous and nondestructive monitoring of phloem transport would be needed. In this respect two new techniques appear promising: the concurrent and continuous measurement of bark and xylem diameter changes (e.g., Mencuccini et al. 2013; Sevanto et al. 2003) and the usage of nuclear magnetic resonance imaging (e.g., Windt et al. 2006). Combining quantitative analyses of C-reserve concentrations with gas-exchange and new techniques of sap-flux measurements in field studies and ex situ experiments will certainly help to further improve our understanding of the C-relations of trees under environmental stress.

Acknowledgments I would like to thank all colleagues, students, and field helpers who contributed to my research over the last years. Especially I thank Christian Körner, Anna Sala, Frida Piper, Alex Fajardo, and Qingmin Han for the fruitful collaborations in experiments and field surveys about the C-supply status of trees. During the time of writing this manuscript, I received funding from the European Research Council (ERC) grant No. 233399 (project 'TREELIM' to Christian Körner).

References

Adams HD, Guardiola-Claramonte M, Barron-Gafford GA, Villegas JC, Breshears DD, Zou CB, Troch PA, Huxman TE (2009) Temperature sensitivity of drought-induced tree mortality portends increased regional die-off under global-change-type drought. Proc Natl Acad Sci U S A 106(17):7063–7066. doi:10.1073/pnas.0901438106

Allen CD, Macalady AK, Chenchouni H, Bachelet D, McDowell N, Vennetier M, Kitzberger T, Rigling A, Breshears DD, Hogg EH, Gonzalez P, Fensham R, Zhang Z, Castro J, Demidova N, Lim JH, Allard G, Running SW, Semerci A, Cobb N (2010) A global overview of drought and heat-induced tree mortality reveals emerging climate change risks for forests. Forest Ecol Manage 259(4):660–684. doi:10.1016/j.foreco.2009.09.001

Alvarez-Uria P, Körner C (2007) Low temperature limits of root growth in deciduous and evergreen temperate tree species. Funct Ecol 21(2):211–218

Anderegg WRL (2012) Complex aspen forest carbon and root dynamics during drought A letter. Clim Change 111(3–4):983–991. doi:10.1007/s10584-012-0421-9

Anderegg WRL, Anderegg LDL (2013) Hydraulic and carbohydrate changes in experimental drought-induced mortality of saplings in two conifer species. Tree Physiol 33(3):252–260. doi:10.1093/treephys/tpt016

Bader MKF, Siegwolf R, Körner C (2010) Sustained enhancement of photosynthesis in mature deciduous forest trees after 8 years of free air CO_2 enrichment. Planta 232(5):1115–1125. doi:10.1007/s00425-010-1240-8

Bader MKF, Leuzinger S, Keel SG, Siegwolf RTW, Hagedorn F, Schleppi P, Körner C (2013) Central European hardwood trees in a high-CO_2 future: synthesis of an 8-year forest canopy CO_2 enrichment project. J Ecol. doi:10.1111/1365-2745.12149

Bansal S, Germino MJ (2010) Unique responses of respiration, growth, and non-structural carbohydrate storage in sink tissue of conifer seedlings to an elevation gradient at timberline. Environ Exp Bot 69:313–319

Barbaroux C, Breda N (2002) Contrasting distribution and seasonal dynamics of carbohydrate reserves in stem wood of adult ring-porous sessile oak and diffuse-porous beech trees. Tree Physiol 22(17):1201–1210

Bonan GB (2008) Forests and climate change: forcings, feedbacks, and the climate benefits of forests. Science 320(5882):1444–1449. doi:10.1126/science.1155121

Boyer JS (1970) Leaf enlargement and metabolic rates in corn, soybean, and sunflower at various leaf water potentials. Plant Physiol 46(2):233–235. doi:10.1104/pp. 46.2.233

Bréda N, Huc R, Granier A, Dreyer E (2006) Temperate forest trees and stands under severe drought: a review of ecophysiological responses, adaptation processes and long-term consequences. Ann Forest Sci 63:625–644

Breshears DD, Adams HD, Eamus D, McDowell N, Law DJ, Will RE, Williams AP, Zou CB (2013) The critical amplifying role of increasing atmospheric moisture demand on tree mortality and associated regional die-off. Front Plant Sci 4:266

Carbone MS, Czimczik CI, Keenan TF, Murakami PF, Pederson N, Schaberg PG, Xu X, Richardson AD (2013) Age, allocation and availability of nonstructural carbon in mature red maple trees. New Phytol 200:1145–1155. doi:10.1111/nph.12448

Chapin FS, Schulze ED, Mooney HA (1990) The ecology and economics of storage in plants. Annu Rev Ecol Syst 21:423–447

Crone EE, Miller E, Sala A (2009) How do plants know when other plants are flowering? Resource depletion, pollen limitation and mast-seeding in a perennial wildflower. Ecol Lett 12 (11):1119–1126. doi:10.1111/j.1461-0248.2009.01365.x

Dannoura M, Maillard P, Fresneau C, Plain C, Berveiller D, Gerant D, Chipeaux C, Bosc A, Ngao J, Damesin C, Loustau D, Epron D (2011) In situ assessment of the velocity of carbon transfer by tracing ^{13}C in trunk CO_2 efflux after pulse labelling: variations among tree species and seasons. New Phytol 190(1):181–192. doi:10.1111/j.1469-8137.2010.03599.x

Dawes MA, Hättenschwiler S, Bebi P, Hagedorn F, Handa IT, Körner C, Rixen C (2011) Species-specific tree growth responses to 9 years of CO_2 enrichment at the alpine treeline. J Ecol 99 (2):383–394. doi:10.1111/j.1365-2745.2010.01764.x

Dawes MA, Hagedorn F, Handa IT, Streit K, Ekblad A, Rixen C, Körner C, Hättenschwiler S (2013) An alpine treeline in a carbon dioxid-rich world: synthesis of a nine-year free-air carbon dioxide enrichment study. Oecologia 171:623–637. doi:10.1007/s00442-012-2576-5

Dickson RE (1991) Assimilate distribution and storage. In: Raghavendra AS (ed) Physiology of trees. Wiley, New York, pp 51–85

Fajardo A, Piper FI, Pfund L, Körner C, Hoch G (2012) Variation of mobile carbon reserves in trees at the alpine treeline ecotone is under environmental control. New Phytol 195:794–802

Fajardo A, Piper FI, Hoch G (2013) Similar variation in carbon storage between deciduous and evergreen treeline species across elevational gradients. Ann Bot 112:623–631

Fischer C, Höll W (1991) Food reserves of scots pine (*Pinus sylvestris* L.) I. Seasonal changes in the carbohydrate and fat reserves of pine needles. Trees 5:187–195

Fischer C, Höll W (1992) Food reserves of scots pine (*Pinus sylvestris* L.) II. Seasonal changes and radial distribution of carbohydrate and fat reserves in pine wood. Trees 6:147–155

Galiano L, Martínez-Vilalta J, Lloret F (2011) Carbon reserves and canopy defoliation determine the recovery of Scots pine 4 yr after a drought episode. New Phytol 190(3):750–759. doi:10.1111/j.1469-8137.2010.03628.x

Galvez DA, Landhäusser SM, Tyree MT (2011) Root carbon reserve dynamics in aspen seedlings: does simulated drought induce reserve limitation? Tree Physiol 31(3):250–257. doi:10.1093/treephys/tpr012

Galvez DA, Landhäusser SM, Tyree MT (2013) Low root reserve accumulation during drought may lead to winter mortality in poplar seedlings. New Phytol 198(1):139–148. doi:10.1111/nph.12129

Grace J, Berninger F, Nagy L (2002) Impacts of climate change on the tree line. Ann Bot 90:537–544

Halter R, Sands R, Ashton DH, Nambiar EKS (1997) Root growth of subalpine and montane Eucalyptus seedlings at low soil temperatures. Trees 12(1):35–41

Han Q, Kabeya D, Hoch G (2011) Leaf traits, shoot growth and seed production in mature Fagus sylvatica trees after 8 years of CO_2 enrichment. Ann Bot 107(8):1405–1411. doi:10.1093/aob/mcr082

Han Q, Kabeya D, Iio A, Inagaki Y, Kakubari Y (2014) Nitrogen storage dynamics are affected by masting events in *Fagus crenata*. Oecologia 174:679–687. doi:10.1007/s00442-013-2824-3

Handa IT, Körner C, Hattenschwiler S (2005) A test of the tree-line carbon limitation hypothesis by in situ CO_2 enrichment and defoliation. Ecology 86(5):1288–1300

Hansen P, Grauslun J (1973) ^{14}C-studies on apple-trees. 8. Seasonal-variation and nature of reserves. Physiol Plant 28(1):24–32

Hartmann H, Ziegler W, Kolle O, Trumbore S (2013a) Thirst beats hunger – declining hydration during drought prevents carbon starvation in Norway spruce saplings. New Phytol 200(2):340–349. doi:10.1111/nph.12331

Hartmann H, Ziegler W, Trumbore S (2013b) Lethal drought leads to reduction in nonstructural carbohydrates in Norway spruce tree roots but not in the canopy. Funct Ecol 27(2):413–427. doi:10.1111/1365-2435.12046

Hoch G (2005) Fruit-bearing branchlets are carbon autonomous in mature broad-leaved temperate forest trees. Plant Cell Environ 28(5):651–659

Hoch G (2007) Cell wall hemicelluloses as mobile carbon stores in non-reproductive plant tissues. Funct Ecol 21(5):823–834

Hoch G (2008) The carbon supply of *Picea abies* trees at a Swiss montane permafrost site. Plant Ecol Divers 1(1):13–20

Hoch G, Körner C (2003) The carbon charging of pines at the climatic treeline: a global comparison. Oecologia 135(1):10–21

Hoch G, Körner C (2009) Growth and carbon relations of tree line forming conifers at constant vs. variable low temperatures. J Ecol 97(1):57–66. doi:10.1111/j.1365-2745.2008.01447.x
Hoch G, Körner C (2012) Global patterns of mobile carbon stores in trees at the high-elevation tree line. Glob Ecol Biogeogr 21(8):861–871. doi:10.1111/j.1466-8238.2011.00731.x
Hoch G, Popp M, Körner C (2002) Altitudinal increase of mobile carbon pools in *Pinus cembra* suggests sink limitation of growth at the Swiss treeline. Oikos 98(3):361–374
Hoch G, Richter A, Körner C (2003) Non-structural carbon compounds in temperate forest trees. Plant Cell Environ 26:1067–1081
Hoch G, Siegwolf RTW, Keel SG, Körner C, Han Q (2013) Fruit production in three masting tree species does not rely on stored carbon reserves. Oecologia 171:653–662
Hsiao TC, Acevedo E (1974) Plant responses to water deficits, water-use efficiency, and drought resistance. Agric Meteorol 14:59–84
Ichie T, Nakagawa M (2013) Dynamics of mineral nutrient storage for mast reproduction in the tropical emergent tree Dryobalanops aromatica. Ecol Res 28:151–158. doi:10.1007/s11284-011-0836-1
Ichie T, Igarashi S, Yoshida S, Kenzo T, Masaki T, Tayasu I (2013) Are stored carbohydrates necessary for seed production in temperate deciduous trees? J Ecol 101(2):525–531. doi:10.1111/1365-2745.12038
IPCC (2013) Climate change 2013: The Physical Science Basis. Working Group I Contribution to the 5th Assessment Report of the Intergovernmental Panel on Climate Change. Cambridge University Press, New York
Isagi Y, Sugimura K, Sumida A, Ito H (1997) How does masting happen and synchronize? J Theor Biol 187(2):231–239. doi:10.1006/jtbi.1997.0442
Jolly WM, Dobbertin M, Zimmermann NE, Reichstein M (2005) Divergent vegetation growth responses to the 2003 heat wave in the Swiss Alps. Geophys Res Lett 32(18). doi:10.1029/2005GL023252
Kelly D (1994) The evolutionary ecology of mast seeding. Trends Ecol Evol 9(12):465–470
Koch GW, Sillett SC, Jennings GM, Davis SD (2004) The limits to tree height. Nature 428(6985):851–854
Körner C (1998) A re-assessment of high elevation treeline positions and their explanation. Oecologia 115(4):445–459
Körner C (2003) Carbon limitation in trees. J Ecol 91:4–17
Körner C (2006) Plant CO_2 responses: an issue of definition, time and resource supply. New Phytol 172(3):393–411
Körner C (2012) Alpine treelines: functional ecology of the global high elevation tree limits. Springer, Berlin
Körner C (2013) Growth controls photosynthesis – mostly. Nova Acta Leopoldina 31(5):713–732
Körner C, Paulsen J (2004) A world-wide study of high altitude treeline temperatures. J Biogeogr 31(5):713–732
Körner C, Asshoff R, Bignucolo O, Hättenschwiler S, Keel SG, Pelaez-Riedl S, Pepin S, Siegwolf RTW, Zotz G (2005) Carbon flux and growth in mature deciduous forest trees exposed to elevated CO_2. Science 309(5739):1360–1362
Kozlowski TT, Davis WJ, Pallardy SG (1991) The physiological ecology of woody plants. Academic, San Diego
Landhäusser SM, Lieffers VJ (2003) Seasonal changes in carbohydrate reserves in mature northern *Populus tremuloides* clones. Trees 17:471–476
Li MH, Xiao WF, Wang SG, Cheng GW, Cherubini P, Cai XH, Liu XL, Wang XD, Zhu WZ (2008) Mobile carbohydrates in Himalayan treeline trees I. Evidence for carbon gain limitation but not for growth limitation. Tree Physiol 28(8):1287–1296
McCarthy HR, Oren R, Johnsen KH, Gallet-Budynek A, Pritchard SG, Cook CW, LaDeau SL, Jackson RB, Finzi AC (2010) Re-assessment of plant carbon dynamics at the Duke free-air CO_2 enrichment site: interactions of atmospheric [CO_2] with nitrogen and water availability over stand development. New Phytol 185(2):514–528. doi:10.1111/j.1469-8137.2009.03078.x

McDowell SCL, McDowell NG, Marshall JD, Hultine K (2000) Carbon and nitrogen allocation to male and female reproduction in Rocky Mountain Douglas-fir (*Pseudotsuga menziesii* var. *glauca*, Pinaceae). Am J Bot 87(4):539–546. doi:10.2307/2656598

McDowell N, Pockman WT, Allen CD, Breshears DD, Cobb N, Kolb T, Plaut J, Sperry J, West A, Williams DG, Yepez EA (2008) Mechanisms of plant survival and mortality during drought: why do some plants survive while others succumb to drought? New Phytol 178(4):719–739. doi:10.1111/j.1469-8137.2008.02436.x

McDowell NG, Beerling DJ, Breshears DD, Fisher RA, Raffa KF, Stitt M (2011) The interdependence of mechanisms underlying climate-driven vegetation mortality. Trends Ecol Evol 26(10):523–532. doi:10.1016/j.tree.2011.06.003

Mencuccini M, Holtta T, Sevanto S, Nikinmaa E (2013) Concurrent measurements of change in the bark and xylem diameters of trees reveal a phloem-generated turgor signal. New Phytol 198 (4):1143–1154. doi:10.1111/nph.12224

Millard P, Sommerkorn M, Grelet GA (2007) Environmental change and carbon limitation in trees: a biochemical, ecophysiological and ecosystem appraisal. New Phytol 175(1):11–28

Miyazaki Y, Hiura T, Kato E, Funada R (2002) Allocation of resources to reproduction in *Styrax obassia* in a masting year. Ann Bot 89(6):767–772

Monson RK, Rosenstiel TN, Forbis TA, Lipson DA, Jaeger CH (2006) Nitrogen and carbon storage in alpine plants. Integr Comp Biol 46(1):35–48

Muller B, Pantin F, Genard M, Turc O, Freixes S, Piques M, Gibon Y (2011) Water deficits uncouple growth from photosynthesis, increase C content, and modify the relationships between C and growth in sink organs. J Exp Bot 62(6):1715–1729. doi:10.1093/jxb/erq438

Mund M, Kutsch WL, Wirth C, Kahl T, Knohl A, Skomarkova MV, Schulze ED (2010) The influence of climate and fructification on the inter-annual variability of stem growth and net primary productivity in an old-growth, mixed beech forest. Tree Physiol 30(6):689–704. doi:10.1093/treephys/tpq027

Norby RJ, Warren JM, Iversen CM, Medlyn BE, McMurtrie RE (2010) CO_2 enhancement of forest productivity constrained by limited nitrogen availability. Proc Natl Acad Sci U S A 107 (45):19368–19373. doi:10.1073/pnas.1006463107

Obeso JR (1998) Effects of defoliation and girdling on fruit production in *Ilex aquifolium*. Funct Ecol 12(3):486–491

Palacio S, Hoch G, Sala A, Körner C, Millard P (2014) Does carbon storage limit tree growth? New Phytol 201:1096–1100

Piper FI (2011) Drought induces opposite changes in the concentration of non-structural carbohydrates of two evergreen *Nothofagus* species of differential drought resistance. Ann Forest Sci 68(2):415–424

Piper FI, Cavieres LA, Reyes-Diaz M, Corcuera LJ (2006) Carbon sink limitation and frost tolerance control performance of the tree *Kageneckia angustifolia* D. Don (Rosaceae) at the treeline in central Chile. Plant Ecol 185(1):29–39

Piper FI, Reyes-Diaz M, Corucera LJ, Lusk CH (2009) Carbohydrate storage, survival, and growth of two evergreen *Nothofagus* species in two contrasting light environments. Ecol Res 24:1233–1241

Powell TL, Galbraith DR, Christoffersen BO, Harper A, Imbuzeiro HMA, Rowland L, Almeida S, Brando PM, da Costa ACL, Costa MH, Levine NM, Malhi Y, Saleska SR, Sotta E, Williams M, Meir P, Moorcroft PR (2013) Confronting model predictions of carbon fluxes with measurements of Amazon forests subjected to experimental drought. New Phytol 200(2):350–365. doi:10.1111/nph.12390

Richardson AD, Carbone MS, Keenan TF, Czimczik CI, Hollinger DY, Murakami P, Schaberg PG, Xu XM (2013) Seasonal dynamics and age of stemwood nonstructural carbohydrates in temperate forest trees. New Phytol 197(3):850–861. doi:10.1111/nph.12042

Rossi S, Deslauriers A, Anfodillo T, Carraro V (2007) Evidence of threshold temperatures for xylogenesis in conifers at high altitudes. Oecologia 152(1):1–12

Ruehr NK, Offermann CA, Gessler A, Winkler JB, Ferrio JP, Buchmann N, Barnard R (2009) Drought effects on allocation of recent carbon: from beech leaves to soil CO_2 efflux. New Phytol 184:950–961. doi:10.1111/j.1469-8137.2009.03044.x

Ryan MG, Yoder BJ (1997) Hydraulic limits to tree height and tree growth. Bioscience 47(4):235–242
Sala A, Hoch G (2009) Height-related growth declines in ponderosa pine are not due to carbon limitation. Plant Cell Environ 32(1):22–30. doi:10.1111/j.1365-3040.2008.01896.x
Sala A, Piper FI, Hoch G (2010) Physiological mechanisms of drought-induced tree mortality are far from being resolved. New Phytol 186:274–281
Sala A, Fouts W, Hoch G (2011) Carbon storage in trees: does relative carbon supply decrease with tree size? In: Meinzer FC, Dawson TE, Ladenbruch B (eds) Size- and age-related changes in tree structure and function. Springer, Heidelberg, pp 287–306
Sala A, Hopping K, McIntire EJB, Delzon S, Crone EE (2012a) Masting in whitebark pine (*Pinus albicaulis*) depletes stored nutrients. New Phytol 196(1):189–199. doi:10.1111/j.1469-8137.2012.04257.x
Sala A, Woodruff DR, Meinzer FC (2012b) Carbon dynamics in trees: feast or famine? Tree Physiol 32:764–775
Schädel C, Blöchl A, Richter A, Hoch G (2009) Short-term dynamics of nonstructural carbohydrates and hemicelluloses in young branches of temperate forest trees during bud break. Tree Physiol 29(7):901–911. doi:10.1093/treephys/tpp034
Schädel C, Richter A, Blöchl A, Hoch G (2010) Hemicellulose concentration and composition in plant cell walls under extreme carbon source-sink imbalances. Physiol Plant 139(3):241–255. doi:10.1111/j.1399-3054.2010.01360.x
Schenker G, Lenz A, Körner C, Hoch G (2014) Physiological minimum temperatures for root growth in seven common European broad-leaved tree species. Tree Physiol 34(3):302–313. doi:10.1093/treephys/tpu003
Sevanto S (2014) Phloem transport and drought. J Exp Bot 65:1751–1759. doi:10.1093/jxb/ert467
Sevanto S, Vesala T, Peramaki M, Nikinmaa E (2003) Sugar transport together with environmental conditions controls time lags between xylem and stem diameter changes. Plant Cell Environ 26(8):1257–1265. doi:10.1046/j.1365-3040.2003.01049.x
Sevanto S, McDowell NG, Dickman LT, Pangle R, Pockman WT (2014) How do trees die? A test of the hydraulic failure and carbon starvation hypotheses. Plant Cell Environ 37:153–161. doi:10.1111/pce.12141
Shi P, Körner C, Hoch G (2006) End of season carbon supply status of woody species near the treeline in western China. Basic Appl Ecol 7(4):370–377
Silvertown JW (1980) The evolutionary ecology of mast seeding in trees. Biol J Linean Soc 14:235–250
Smith WK, Germino MJ, Hancock TE, Johnson DM (2003) Another perspective on altitudinal limits of alpine timberlines. Tree Physiol 23(16):1101–1112
Solfjeld I, Johnsen O (2006) The influence of root-zone temperature on growth of *Betula pendula* Roth. Trees 20(3):320–328
Srichuwong S, Jane JL (2007) Physicochemical properties of starch affected by molecular composition and structures: a review. Food Sci Biotechnol 16(5):663–674
Streit K, Rinne KT, Hagedorn F, Dawes MA, Saurer M, Hoch G, Werner RA, Buchmann N, Siegwolf RTW (2013) Tracing fresh assimilates through *Larix decidua* exposed to elevated CO_2 and soil warming at the alpine treeline using compound-specific stable isotope analysis. New Phytol 197(3):838–849. doi:10.1111/nph.12074
Sveinbjörnsson B (2000) North American and European treelines: external forces and internal processes controlling position. Ambio 29(7):388–395
Sveinbjornsson B, Smith M, Traustason T, Ruess RW, Sullivan PF (2010) Variation in carbohydrate source-sink relations of forest and treeline white spruce in southern, interior and northern Alaska. Oecologia 163(4):833–843. doi:10.1007/s00442-010-1597-1
Veneklaas EJ, den Ouden F (2005) Dynamics of non-structural carbohydrates in two *Ficus* species after transfer to deep shade. Environ Exp Bot 54:148–154. doi:10.1016/j.envexpbot.2004.06.010

Wieser G, Tausz M (2007) Current concepts for treelife limitation at the upper timberline. In: Wieser G, Tausz M (eds) Trees at their upper limit: treelife limitation at the alpine timberline. Springer, Berlin, pp 1–10
Wiley E, Helliker B (2012) A re-evaluation of carbon storage in trees lends greater support for carbon limitation to growth. New Phytol 195:285–289
Will RE, Wilson SM, Zou CB, Hennessey TC (2013) Increased vapor pressure deficit due to higher temperature leads to greater transpiration and faster mortality during drought for tree seedlings common to the forest–grassland ecotone. New Phytol 200(2):366–374. doi:10.1111/nph.12321
Williams AP, Allen CD, Macalady AK, Griffin D, Woodhouse CA, Meko DM, Swetnam TW, Rauscher SA, Seager R, Grissino-Mayer HD, Dean JS, Cook ER, Gangodagamage C, Cai M, McDowell NG (2013) Temperature as a potent driver of regional forest drought stress and tree mortality. Nat Clim Chang 3(3):292–297. doi:10.1038/nclimate1693
Windt CW, Vergeldt FJ, De Jager PA, Van As H (2006) MRI of long-distance water transport: a comparison of the phloem and xylem flow characteristics and dynamics in poplar, castor bean, tomato and tobacco. Plant Cell Environ 29(9):1715–1729. doi:10.1111/j.1365-3040.2006.01544.x
Woodruff DR, Meinzer FC (2011) Water stress, shoot growth and storage of non-structural carbohydrates along a tree height gradient in a tall conifer. Plant Cell Environ 34(11):1920–1930. doi:10.1111/j.1365-3040.2011.02388.x
Woodruff DR, Bond BJ, Meinzer FC (2004) Does turgor limit growth in tall trees? Plant Cell Environ 27(2):229–236
Würth MKR, Pelaez-Riedl S, Wright SJ, Körner C (2005) Non-structural carbohydrate pools in a tropical forest. Oecologia 143(1):11–24
Yasumura Y, Hikosaka K, Hirose T (2006) Resource allocation to vegetative and reproductive growth in relation to mast seeding in *Fagus crenata*. Forest Ecol Manage 229(1–3):228–233
Yoda K, Shinozaki K, Ogawa H, Hozumi K, Kira T (1965) Estimation of the total amount of respiration in woody organs of trees and forest communities. J Biol Osaka City Univ 16:15–26
Zhao J, Hartmann H, Trumbore S, Ziegler W, Zhang Y (2013) High temperature causes negative whole-plant carbon balance under mild drought. New Phytol 200(2):330–339. doi:10.1111/nph.12400
Zotz G, Pepin S, Körner C (2005) No down-regulation of leaf photosynthesis in mature forest trees after three years of exposure to elevated CO_2. Plant Biol 7(4):369–374

Consequences of Changing Precipitation Patterns for Ecosystem Functioning in Grasslands: A Review

Stephan Unger and Marjan Jongen

Contents

Abstract Grassland ecosystems worldwide provide agricultural goods and important ecosystem services. Productivity and other ecosystem processes in grasslands are, in most cases, strongly linked to the ecosystems' water status, a factor that is predicted to experience major alterations with global climate change. Future predictions include changes in the amount, distribution, frequency, and intensity of

S. Unger (✉)
Department for Experimental and Systems Ecology, University of Bielefeld, Universitätsstr. 25, 33615 Bielefeld, Germany
e-mail: stephan.unger@uni-bielefeld.de

M. Jongen
Instituto Superior de Agronomia, Universidade de Lisboa, Tapada da Ajuda, 1349-017 Lisboa, Portugal

U. Lüttge, W. Beyschlag (eds.), *Progress in Botany*, Progress in Botany 76,
DOI 10.1007/978-3-319-08807-5_14

precipitation, which, particularly in grasslands, may have important consequences for ecosystem state and functioning. This review analyses the effects of experimental precipitation manipulation on plant productivity, species diversity, soil/ecosystem respiration, and soil nitrogen in grassland-type ecosystems over a wide range of climate types, synthesising the results from 72 studies.

We found that sensitivity of ecosystem processes to changes in precipitation amounts increased with aridity. In addition, ecosystem processes were more responsive to precipitation addition than to precipitation reduction. However, we did observe high resilience of grassland ecosystems to both changing precipitation amounts and variability, which may be explained by the fact that the applied manipulation scenarios often lie within the range of the natural inter-annual precipitation variability experienced by ecosystems, and by evolutionary adaptation of grassland ecosystems to these natural inter-annual differences. Long-term effects of altered precipitation regimes on ecosystem processes, i.e. by changes in species composition and soil properties, are rarely covered within the time frame of most studies and thus cannot be ruled out as a possible consequence of a gradually changing climate.

Increasing the comparability between individual precipitation manipulation studies is needed to facilitate the evaluation of ecosystem responses to altered precipitation regimes. We recommend future precipitation manipulation studies to aim at capturing possible long-term effects with comparable designs and standardised data compilation.

1 Introduction

1.1 Climate Change Influence on Precipitation Regimes

Anthropogenic fossil fuel emissions continue to impinge on global climate change, resulting in ecosystems worldwide being subjected to altered temperature and precipitation regimes. Basic theory and empirical evidence suggest that state, functioning and service provision of ecosystems worldwide are increasingly influenced by this development. A major goal in current research is to achieve a comprehensive understanding of the possible consequences of climatic changes for ecosystem processes and subsequently develop adequate mitigation strategies. However, in view of the complex tangle of co-dependent ecosystem processes (e.g. productivity, biodiversity, mineralisation, soil carbon and nutrient cycling and storage), all potentially exhibiting differential responses to environmental change, the scientific community is still far from achieving this aim.

One of the major factors affected by climate change is the water status of terrestrial ecosystems all over the globe. With the ongoing temperature increase, and a 7 % higher water holding capacity of the atmosphere per 1 °C warming (Wentz et al. 2007; Trenberth 2011), overall evaporation and atmospheric water vapour concentrations will increase. The latter will promote cloud formation in humid regions (Meehl et al. 2007; Trenberth 2011), leading to increases in

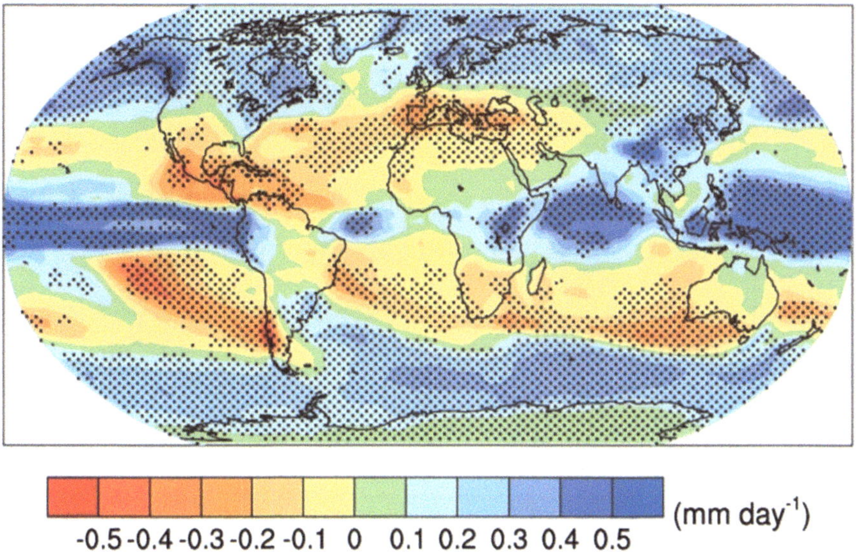

Fig. 1 Multi-model mean changes in precipitation (mm day^{-1}). Regions are stippled where at least 80 % of the climate models agree on the sign of the mean change. Changes are annual means for the *SRES A1B* scenario for the period 2088–2099 relative to 1988–1999. Taken from Meehl et al. (2007, their Fig. 10.12)

precipitation (Fig. 1). In contrast, arid regions are expected to experience a decrease in precipitation (Fig. 1), as here adequate surface moisture, a determinant of increased evaporation, is lacking (Meehl et al. 2007; Trenberth 2011). Indeed, long-term observations for the period between 1900 and 2005 have already demonstrated significant increases in precipitation in eastern North and South America, northern Europe, and northern and central Asia, and significant decreases in precipitation in the Sahel, southern Africa, the Mediterranean, and southern Asia (Trenberth et al. 2007). In addition to the expected changes in precipitation amount, the type of precipitation might change in temperate regions, with much of the precipitation usually falling as snow increasingly falling as winter rain, thus reducing water storage in snow packs, and concomitantly reducing water availability in summer (Trenberth et al. 2007).

While alterations in the amount of precipitation may have large implications for ecosystem functioning in the affected regions, precipitation distribution, frequency and intensity are regarded as nonetheless important (Easterling et al. 2000). Over the twentieth century, estimates suggest that atmospheric water vapour concentrations have increased by 5 %, which has generally increased the intensity of precipitation events (Trenberth et al. 2007). Scenarios predict an almost universal increase in precipitation intensity (Fig. 2a), although particularly at middle and high latitude regions where mean precipitation (Fig. 1) is also projected to increase (Meehl et al. 2007). However, both increases and decreases in consecutive dry days between precipitation events can be found (Fig. 2b), with regions located in the

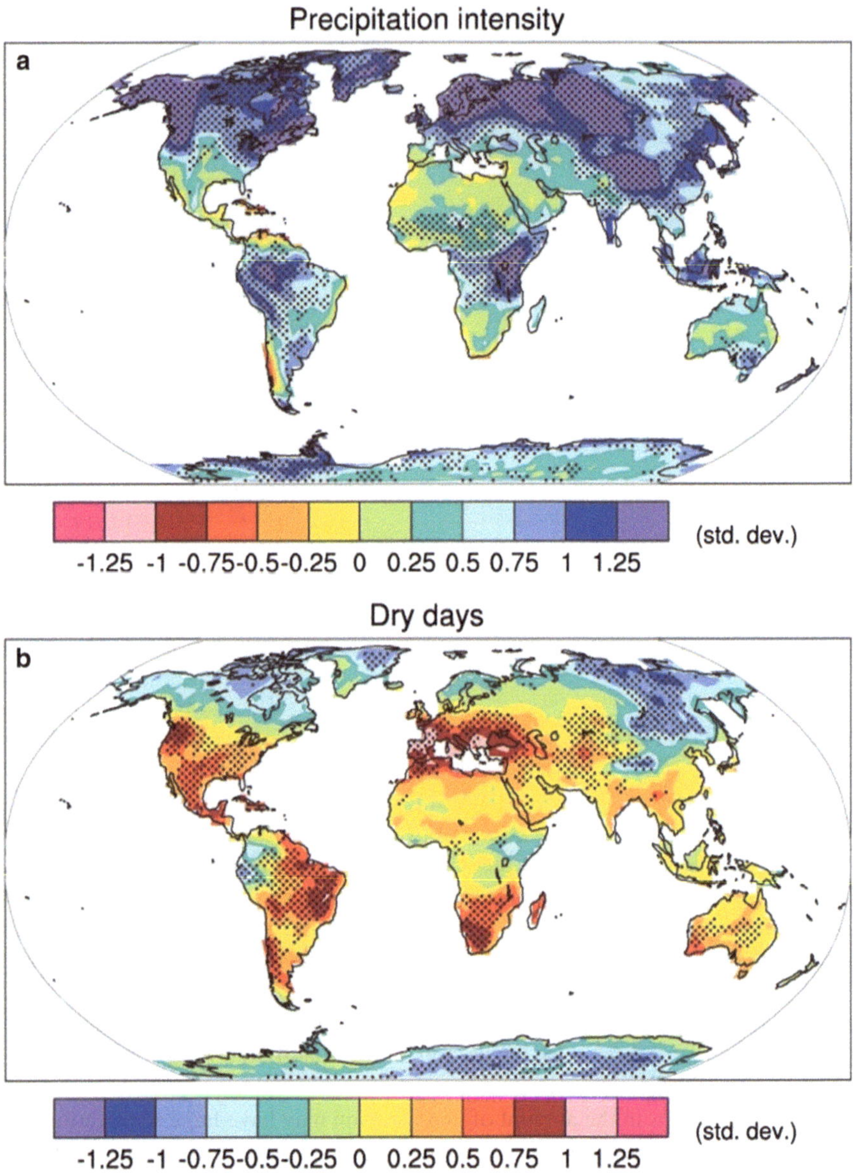

Fig. 2 Changes in spatial patterns of simulated precipitation intensity and dry days between two 20-year means (2080–2099 minus 1980–1999) for the *A1B* scenario. *Stippling* denotes areas where at least five of the nine models concur in determining that the change is statistically significant. Changes are given in units of standard deviation, following Frich et al. (2002). Taken from Meehl et al. (2007, their Fig. 10.18)

subtropics and lower mid-latitudes exhibiting an increased run of dry days, thus having a concomitant greater risk of drought (Meehl et al. 2007).

In this review we will concentrate on understanding the effects of altered precipitation regimes on one of the most important terrestrial ecosystem types worldwide: grasslands. With precipitation being one of the most important determinants of the majority of ecosystem processes, the observed and predicted changes in amount, intensity and frequency of precipitation are particularly threatening for future stability and functioning of these ecosystems.

1.2 Grassland Ecosystems

Grasslands are disturbance-dependent terrestrial ecosystems dominated by graminoid and herbaceous vegetation in climates with a distinct seasonality of productivity (Smith 1973), which are maintained by fire, grazing, drought, and/or freezing temperatures (Anderson 1982). These factors provide selective pressure for a short ruderal life cycle involving early reproduction with a high number of seeds, a high turnover of aboveground biomass, high belowground carbon investment and the location of perennating organs near the soil surface (Sala et al. 1996), which promotes the dominance of graminoids and forbs. However, grasslands encompass not only non-woody systems but also savannas, woodlands, shrublands, and tundra (White et al. 2000). Estimates of the extent of the earth's land area in grasslands (excluding Greenland and Antarctica), depending on land cover characterisation, range from ~42 to 56 million km^2, or ~31 to 43 % (Whittaker and Likens 1975; Atjay et al. 1979; Olson et al. 1983). Figure 3 gives the global distribution of grasslands, with the largest areas found in central and southern Asia, southern South America, Africa and central North America (Sala et al. 1996).

Grassland ecosystems are of high economic importance for provisioning of agricultural goods as, together with the livestock they sustain, they constitute one

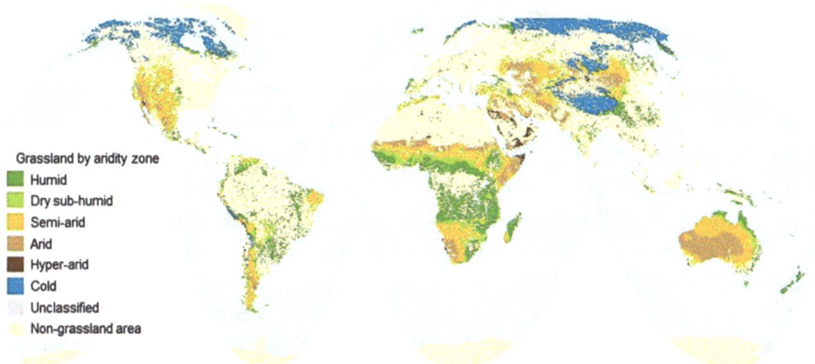

Fig. 3 Grasslands and climate zones. From White et al. (2000) and UNEP (1992)

of the earth's major food resources (Singh et al. 1983). Apart from their vital role in food production, grasslands goods and services include wildlife and biodiversity conservation, resource storage, prevention of soil degradation as well as supporting tourism and recreational activities, and offering aesthetic and spiritual gratification (White et al. 2000). The provision of these ecosystem services depends on the maintenance of grassland ecosystem state (Miller et al. 2011). As the extent of drought, fire and grazing determines the state transition of grasslands into deserts or shrublands and forest ecosystems (Sala et al. 1996), state shifts may easily occur with climate and land use change, afforestation, eutrophication, or the invasion of neophytes (Faber-Langendoen and Josse 2010), with restoration of previous conditions being difficult, costly, or effectively unfeasible (Miller et al. 2011). In relation to changing temperature and precipitation patterns, this problem has increased the interest of grassland ecosystem research on biotic and abiotic attributes conferring ecosystem resilience with changing environmental conditions and thus subsequently reducing system susceptibility to state shifts.

1.3 Drivers of Ecosystem Processes in Grasslands

There are a range of factors controlling ecosystem processes in grassland ecosystems, of which precipitation and temperature are thought to exhibit the strongest ties to grassland functioning (Sala et al. 1996). Accordingly, a wide range of studies describe the relationship of these factors with important ecosystem traits such as productivity, biodiversity, soil and ecosystem carbon cycling, and soil nutrient dynamics. Precipitation and consequently soil moisture have been shown to modulate the carbon cycle of grassland ecosystems, with arid systems generally exhibiting significantly lower plant productivity (e.g. Huxman et al. 2004a), biodiversity (Sala et al. 1996), soil and ecosystem carbon fluxes (e.g. Merbold et al. 2009), and soil nutrient cycling (Aranibar et al. 2004), as compared to semi-arid or mesic systems. However, negative effects of precipitation on the carbon and nitrogen cycles in grasslands have been observed, particularly with large precipitation pulses interrupting long dry periods (e.g. Kim et al. 2012; Unger et al. 2012). These precipitation pulses can lead to large carbon and nitrogen losses through high soil respiration rates (Kim et al. 2012) and leaching of nitrogen, particularly nitrate, below the rooting zone (Austin et al. 2004). In contrast to arid and semi-arid grasslands, ecosystem processes in mesic grasslands, where moisture is not limiting, are controlled by temperature, with increasing temperatures generally enhancing productivity (e.g. Flanagan and Adkinson 2011) and soil microbial activity (e.g. Davidson et al. 2006), thus resulting in higher carbon and nitrogen cycling (Aranibar et al. 2004).

Apart from the climatic controls on ecosystem functioning in grasslands, natural disturbance factors in the form of fire and grazing or land use factors like fertilisation, mowing, and tillage are known to affect ecosystem functioning. While disturbance is necessary to maintain ecosystem state in grasslands, particularly in the more humid regions, where natural succession favours tree growth (Anderson 1982), an increase in disturbance with land use intensification or even transformation of grasslands into croplands has a negative effect on ecosystem functioning and stability leading to diversity loss, increasing ecosystem vulnerability to climate change or irreversible ecosystem collapse (MacDougall et al. 2013).

1.4 Experimental Designs to Study Precipitation Impacts at the Ecosystem Scale

Knowledge of the impacts of alterations in precipitation on ecosystem processes can be gained from multi-year observational records. For example, over a 20-year period (1986–2005), inter-annual variation in aboveground productivity in a Mediterranean grassland was found to correlate with annual precipitation (Vázquez-de-Aldana et al. 2008). However, a 22-year measurements series (1982–2003) from an Inner Mongolia grassland revealed a correlation of aboveground productivity with previous-year precipitation (Ma et al. 2010).

More recently, several studies have interpreted multi-year eddy flux data to explain productivity–precipitation relationships. For example, on an annual basis, ecosystem CO_2 exchange correlated with annual precipitation in a Mediterranean grassland in Portugal (Jongen et al. 2011), whereas the inter-annual variation in ecosystem CO_2 exchange in a Mediterranean grassland in California depended primarily on the timing of precipitation, rather than total annual precipitation (Ma et al. 2007).

However, analysis of impacts of precipitation on ecosystem state using both of the above-mentioned approaches, observational records or eddy flux measurements, may be confounded by other co-varying factors. Precipitation manipulation experiments enable replication, control for confounding factors, and allow for multiple scenarios to be studied simultaneously. They can therefore contribute to our understanding of impacts of precipitation on ecosystem state. Water manipulation experiments at the pot and mesocosm scale are widely conducted (e.g. Li et al. 2011; Nagy et al. 2013; O'Brien et al. 2013). However, they do not necessarily reflect natural conditions, often are of short duration, and are limited in the number of parameters studied (i.e. through destructive sampling). Extending the outcome of these studies to the field or ecosystem scale or incorporating their results into modelling approaches is therefore in most cases not appropriate. To achieve a thorough picture of potential effects of changes in precipitation on ecosystem functioning, field studies with experimentally manipulated precipitation in natural ecosystems with global representation are needed.

Although not always a requirement (e.g. in those studies with only simulation of addition scenarios), precipitation manipulation experiments often involve the use of

Fig. 4 Experimental designs to study the effects of precipitation manipulation: (**a**) closed shelters after Jongen et al. (2013b); (**b**) throughfall shelters, after Yahdjian and Sala (2002), picture from K. Tielbörger; and (**c**) movable shelters, after Báez et al. (2013), picture from S. Collins

rainout shelters, which enables control of the amount and/or timing of precipitation received by the vegetation. To date three different designs have been employed:

1. 'Closed shelters' (Fig. 4a), covered by a complete roof, which excludes almost all precipitation from the experimental plots. Consequently, the researcher has full control over the precipitation regime, generally through fixed sprinklers. Although shelter sides and ends remain open to maximise air movement and minimise temperature and relative humidity artefacts, closed shelters have the disadvantage of altering the microclimate, in particular solar radiation (Fay et al. 2000). Comparison to unsheltered control plots to evaluate the impacts of the shelter is thus desirable (Owens 2003). However, closed shelters have the advantage of being easily constructed and dismantled and comparatively inexpensive, allowing for a necessary number of replications. There are a range of shelter designs with different sizes, supports, and roofing materials, the choice of which depends largely on site-specific demands and research objectives (e.g. short-term versus long-term investigations). Nevertheless, shelters should be sufficiently large to allow for the exclusion of edge effects. The RaMPS (Rainfall Manipulation Plot Study) installed at the Konza Prairie Biological station in northeast Kansas, USA, is an example of this experimental set-up, with timing and quantity of precipitation being experimentally manipulated since 1997.
2. 'Throughfall shelters' (Fig. 4b), originally based on the design of Yahdjian and Sala (2002), have angled roofs composed of bands of transparent acrylic, blocking a certain amount of precipitation. This design gives the possibility to collect the intercepted rain and subsequently use it to irrigate other experimental plots (e.g. Holub et al. 2013), making throughfall shelters useful for manipulation experiments where a combination of addition/reduction scenarios is investigated. However, the throughfall shelter design does not allow for a full control of the amount and timing of precipitation received by the experimental plots, with vegetation being subjected to intra- and inter-annual variations in the amount of precipitation. In comparison to closed shelters, alterations in microclimate are reduced, with wind convection, temperature, and solar radiation less affected (Gherardi and Sala 2013).

3. 'Movable shelters' (Fig. 4c) have a design with roofs or curtains that slide diagonally along rollers to cover the experimental plots, activated by a rain sensor (e.g. Báez et al. 2013). These shelters have the advantage that the effect of the shelter on the microclimate is small. Nevertheless, care must be taken to ensure that the parked shelter does not create shade on a part of the experimental plot (Owens 2003). The disadvantages of movable shelters are higher costs for acquisition and maintenance.

1.5 Problems Associated with Precipitation Manipulation Experiments

In order to adapt current models on future ecosystem behaviour with climate change, manipulation experiments should cover the global range of ecosystems under representative current and future climate scenarios (Beier et al. 2012). In a comprehensive review, Beier et al. (2012) describe that, whereas precipitation manipulation experiments in grasslands and forest ecosystems are widely conducted, other important ecosystem types, such as arable lands and tundra, are underrepresented. In addition, experiments representing the Southern hemisphere, in particular Africa, and experiments conducted in high rainfall zones (>1.500 mm) were scarce. Additionally, the chosen scenarios are often relatively conservative and related to historical or current conditions, while in the future many ecosystems are likely to be exposed to climates exceeding historical and current climatic variations (Beier et al. 2012; IPCC 2012). In many studies, the applied precipitation regimes lie within the range of natural year-to-year climatic variability, explaining the often-found resilience of many ecosystem processes. Therefore, a major issue with precipitation manipulation experiments is to choose the right climate change scenarios for the ecosystems under study. Further issues identified by Beier et al. (2012) are the (1) relatively short duration of most studies, which therefore do not reflect long-term integrated effects on ecosystem performance that might be expected from a future climate scenario, (2) the lack of appropriate reference conditions (i.e. controls being subjected to large inter-annual variation), (3) the unequal distribution of precipitation, caused by some irrigation set-ups or the presence of slopes and soil type gradients, (4) the lack of grazing in fenced off areas altering plant communities and productivity measures, and (5) the often limited plot size constraining sampling strategy and exacerbating problems associated with edge effects and disturbance caused during sampling and measurements. However, many of these issues are hard to avoid and have to be accounted for when incorporating data into modelling approaches.

2 Methods

2.1 Data Compilation

In this review on the effects of changing precipitation patterns on ecosystem processes, data collection was restricted to studies in which precipitation was experimentally manipulated in the field, as they represent the best way to explore cause–effect relationships between water availability and ecosystem functioning. We incorporated studies conducted in grassland ecosystems in the broad sense, i.e. grasslands, savannas, woodlands, shrublands, and tundras (White et al. 2000). In addition, we only incorporated those studies with a precipitation manipulation for a period of at least two consecutive months. For each selected study, we collected information on ecosystem type, latitude, longitude, mean annual temperature (MAT), mean annual precipitation (MAP), data on experimental duration and set-up, and magnitude of manipulation. In the case of multifactor experiments (e.g. precipitation in combination with warming, nutrients, or CO_2), we only used data of the precipitation treatments, in relation to the controls, with the other factors kept at ambient levels. In relation to ecosystem type we used the following criteria: ecosystems with MAP < 300 mm were denoted as arid, MAP between 300 and 600 mm as semi-arid, and MAP > 600 mm as mesic. Studies in which the amount of precipitation was manipulated (addition or reduction) are assembled in Table 1, while studies aiming at altering the variability of precipitation, without changing the absolute amount of rainfall received by the vegetation, are assembled in Table 2.

2.2 Data Analysis

Collected data on the magnitude of change with precipitation manipulation was grouped into three categories of response variables: (1) productivity, including aboveground net primary productivity (ANPP), belowground net primary productivity (BNPP), absolute growth rate (AGR), gross ecosystem photosynthesis (GEP), vegetation cover, and additional growth parameters (e.g. shoot length, leaf length, number of leaves), (2) diversity, including data on species richness (i.e. number of species), species composition, and diversity indices such as the Shannon-Wiener and Simpson's indices, and (3) soil respiration (R_S), ecosystem respiration (R_{ECO}) and responses of soil nitrogen (N-availability, N-mineralisation or concentration of nitrate, ammonium, or total N). For multi-year studies, we calculated the average response for inclusion in the analysis. The aridity index (AI) was calculated according to Köppen (1923), with AI = MAP/(MAT + 33). To normalise the responses to the magnitude of the treatment imposed, we calculated the sensitivity index as the % change in response variable divided by % change in precipitation following the manipulation. The sensitivity index yields positive values if the response is unidirectional with the treatment imposed (i.e. productivity increase

Table 1 Summary of precipitation manipulation studies with rainfall addition or reduction—effects on productivity, biodiversity, respiration, and soil nitrogen

Location	Ecosystem	MAT (°C)	MAP (mm)	Experimental design	Study period	Precipitation manipulation	Productivity	Diversity	Respiration, soil nitrogen	References
Minqin County, China N38°34′, E102°58′	Arid: desert shrubland, dominated by *Nitraria tangutorum*	7.8	115	No shelter	2009	May–Sept.: (1) ↑25 % (2) ↑50 % (3) ↑75 % (4) ↑100 %	ANPP: (1) ↑ 7 % (2) ↑ 44 % (3) ↑ 83 % (4) ↑ 266 %	ND	R_S: ↑ 31–59 %	Song et al. (2012)
DREX, Bayan Unjuul, Mongolia, China N47°02′, E105°57′	Arid: Steppe	0.1	163	Closed shelter	2005	May–August: ↓100 %	ANPP: ↓~75 % BNPP: 0	ND	ND	Shinoda et al. (2010)
Cabo de Gata natural park, Spain N36°49′, W02°15′	Arid: coastal sand dune	19	200	Closed shelter	Oct. 2005–June 2006	(1) ↓25 % (2) ↓50 %	ANPP: (1) and (2): 0	Diversity: (1) and (2): 0	ND	Miranda et al. (2009)
Kytalyk, NE-Siberia, Russia N70°49′, E147°28′	Arid: Arctic shrub tundra, dominated by *Betula nana*	−13.9	205	No shelter	2007–2009	Summer: ↑~100 %	ANPP: 0 Length increment *B. nana*: ↑~30 %	ND	ND	Keuper et al. (2012)
El Cautivo, Tabernas basin, Spain N37°00′, W02°26′	Arid: Grassland	18	230	Closed shelter	Oct. 2005–June 2006	(1) ↓25 % (2) ↓50 %	ANPP: (1) 0 (2) ↓60 %	Diversity: (1) 0 (2) ↓~60 %	ND	Miranda et al. (2009)
PSA (site, Tabernas basin, Spain N37°05′, W02°21′	Arid: grasslands and shrublands	18	242	Closed shelter	2006–2009	↓30 %	ANPP: 2006 + 2007: 0 2009: ↓~50 %	ND	R_S: 2007: 0 2009: ↓~65 % Soil N '07: 0	Miranda et al. (2011)
PSA site, Tabernas basin, Spain N37°05′, W02°21′	Arid: Grassland	17	250	Closed shelter	Oct. 2005–June 2006	(1) ↓25 % (2) ↓50 %	ANPP: (1) ↓35 % (ns) (2) ↓60 % (ns)	Diversity: (1) 0 (2) ↓25–30 %	ND	Miranda et al. (2009)
LTER Sevilleta, New Mexico, USA N34°20′, W106°43′	Arid: native grassland, shrub-land, grass-shrub ecotone	13.2	250	Movable shelter	2002–2008	April–Nov.: (1) ↓~50 % May–Oct.: (2) ↑42 %	Vegetation cover: (1) *B. eriopoda* (grass) ↓ (2) *B. eriopoda* (grass) ↑ Shrub cover: no consistent effect	Species composition: shifts in (1) and (2)	ND	Báez et al. (2013)[a]
LTER Sevilleta, New Mexico, USA N34°20′, W106°43′	Arid: grassland	13.2	250	No shelter	2007–2008	July–Sept.: ↑55 %	ANPP: ↑86 %	ND	R_S: ↑32 %	Thomey et al. (2011)
Eagle Summit, Alaska, USA N65°26′, W145°30′	Arid: montane tundra, dominated by *Dryas octopetala*	−2.8	268	No shelter	1979–1980	Summer: ↑~100 %	Leaf number per shoot: 0 Shoot growth rate: 0	ND	ND	McGraw (1985)

(continued)

Table 1 (continued)

Location	Ecosystem	MAT (°C)	MAP (mm)	Experimental design	Study period	Precipitation manipulation	Productivity	Diversity	Respiration, soil nitrogen	References
Ny Ålesund, Svalbard, Norway N78°56′, E11°50′	Arid: subarctic tundra dominated by *Dryas octopetala*	−4	271	No shelter	1991–1995	June–August: ↑50 %	Cover vegetation groups and several species: 0	ND	ND	Robinson et al. (1998)
Isdammen, Svalbard, Norway N78°122′, E15°44′	Arid: Arctic tundra, dominated by *Cassiope tetragona*	−4	271	No shelter	2005–2008	Mid-June to late August: ↑100 %	Growth parameters (shoot and leaf length, number of leaves): 0	ND	ND	Weijers et al. (2013)
Abisko Scientific Research Station, Sweden N68°21′, E18°49′	Semi-arid: subarctic heath ecosystem	−0.8	~300	No shelter	1991–1994	Summer: ↑50 %	ANPP: 0	ND	ND	E.g. Press et al. (1998)
Eshkolot, Israel N31°23′, E34°54′	Semi-arid: shrubland	18.4	300	Throughfall shelter	2003–2010 R_S: 2006–2007	(1) ↓30 % (2) ↑30 %	ANPP: (1) 0 (2) ↑~80 %	Richness and diversity: (1) and (2): 0	R_S: (1) ↓10 % (2) ↑31 %	Sternberg (2011), Talmon et al. (2011)
Stordalen, Sweden N68°21′, E19°03′	Semi-arid: *Sphagnum fuscum*-dominated bog	−0.5	300	No shelter	2007–2009	Summer: ↑~100 %	ANPP: 0	ND	ND	Keuper et al. (2012)
Irvine Range Land Reserve, Orange County, California, USA N33°62′, W117°78′	Semi-arid: Mediterranean grassland	17	325	No shelters	2006	March–May: ↑~85 %	ANPP: 0 BNPP: 0	Species richness and abundance of forbs: 0	ND	Harpole et al. (2007)
Irvine Range Conservancy, Irvine, California, USA N33°44′, W117°42′	Semi-arid: Mediterranean grassland	17	325	Closed shelters in reduction plots	2006–2007 growing season Note: severe drought with ambient of 79.4 mm	Nov.–June: (1) ↓23 % (2) ↑52 %	ANPP: (1) ↓~70 % (2) ↑~275 %	ND	R_{ECO}: (1) ↓~50 % (2) ↑~30 % N avail: (1) ↓~65 % (2) ↑~60 %	Potts et al. (2012)
CPER Nunn, Colorado, USA N40°49′, W104°46′	Semi-arid: shortgrass steppe	8.2	341	Movable shelter	1999–2009 (11-year drought experiment)	April–Oct.: (1) ↓50 % (2) ↓75 %	Vegetation cover: (1) '03: ↓~45 % '09: ↓~55 % Other years: 0 (2) '99–'01: 0 '02→: ↓35–80 %	Species composition: 2005–2009: *B. gracilis*: ↓~12–38 % ruderals: ↑~1–20 %	ND	Evans et al. (2011)
CPER Nunn, Colorado, USA N40°49′, W104°46′	Semi-arid: shortgrass steppe	8.2	341	Movable shelter	2008–2009 (end of 11-year drought experiment)	April–Oct.: (1) ↓50 % (2) ↓75 %	(1) ANPP: ↓~26 % BNPP: 0 (2) ANPP: ↓~42 % BNPP: ↓~16 %	ND	(1) R_S: ↓25 % (2) R_S: ↓33 % In (1) and (2) N avail.: 0 2009: NO_3^- and NH_4^+ ↑	Evans and Burke (2013)

CPER LTER Nunn, Colorado, USA N40°49′, W104°46′	Semi-arid: shortgrass steppe	8.2	341	Throughfall shelters	2008–2010	Growing season: (1) ↓60 % (2) ↑~70 %	(1) ANPP: 2008 + 2009: 0 2010: ↑~27 % BNPP: 0 (2) ANPP: 0 BNPP: 2010: 0 2009: ↓~59 %	ND	ND	Byrne et al. (2013)
CPER Nunn, Colorado, USA N40°49′, W104°46′	Semi-arid: shortgrass steppe	7.2	342	Throughfall shelters	2007–2008	(1) ↓50 % (2) ↓80 %	ANPP: (1) ↓~17–20 % (2) ↓~25–42 %	ND	ND	Cherwin and Knapp (2012)
Duolun County, Inner Mongolia, China N42°27′, E116°41′	Semi-arid: temperate steppe	3.3	382	No shelter	2006–2007	July–August: ↑~30 %	ANPP:↑24–63 % BNPP:↑33–42 % GEP: ↑11–90 %	ND	R_S: ↑19–61 % R_{ECO}: ↑11–71 %	Liu et al. (2009), Yan et al. (2010, 2011a)
Duolun County, Inner Mongolia, China N42°02′, E116°16'	Semi-arid: temperate steppe	3.3	383	No shelter	2008–2009	June–August: ↑~30 %	ANPP:↑40–60 % BNPP: 0	ND	R_S: ↑15–43 %	Yan et al. (2011b)
HPGRS, Cheyenne, Wyoming, USA N41°20′, W104°89′	Semi-arid: mixed-grass prairie	8	384	No shelter	2006	May–Aug.: ↑115 %	Vegetation cover: ↑38 % GEP: ↑121 %	ND	R_{ECO}: ↑90 %	Bachman et al. (2010)
Sand Creek NHS, Eads, Colorado, USA N38°32′, W102°31'	Semi-arid: shortgrass steppe	8.3	385	Throughfall shelters	2007–2008	(1) ↓50 % (2) ↓80 %	ANPP: (1) ↓~0–35 % (2) ↓~0–51 %	ND	ND	Cherwin and Knapp (2012)
Lethbridge, Alberta, Canada N49°47′, W112°94′	Semi-arid: short/mixed grassland	~5.8	386	Throughfall shelter	2011	May–Oct.: (1) ↓50 % (2) ↑50 %	(1) and (2): ANPP: 0 BNPP: 0	ND	(1) and (2): R_S: 0	Flanagan et al. (2013)
Duolun County, Inner Mongolia, China N42°02′, E116°17′	Semi-arid: temperate steppe	2.1	386	No shelter	2005–2006	July–August: ↑~30 %	GEP: ↑7–34 %	ND	R_{ECO}: ↑11–26 %	Niu et al. (2008)
Kevo Subarctic Research Station, Lapland, Finland N69°45′, E27°01′	Semi-arid: subarctic forest-tundra with *Empetrum nigrum* and *Vaccinium vitis-idaea*	−2	400	No shelter	1992–1994	June–August: ↑13–69 %	Growth parameters *E. nigrum* ↓~25 % *V. vitis-idaea* ↑~45 %	ND	ND	Shevtsova et al. (1997)[a]
Fort Union NM, New Mexico, USA N35°91′, W105°01′	Semi-arid: shortgrass steppe	9.4	425	Throughfall shelters	2007–2008	(1) ↓50 % (2) ↓80 %	(1) and (2): ANPP: 0	ND	ND	Cherwin and Knapp (2012)
CLIMAITE, Garraf, Spain N41°18′, E01°49′	Semi-arid: Mediterranean shrubland	15.1	455	Movable shelters	1999–2005	Two-month period in growing season: ↓65–90 %	ANPP: 0	Species richness: ↓	R_S: 0 N-mineralisation: 0	Emmett et al. (2004), Peñuelas et al. (2007)
CLIMAITE Kiskunság, Hungary N46°88′, E19°38′	Semi-arid: Mediterranean heathland	10.4	504	Movable shelters	1999–2005	Two-month period in growing season: ↓65–90 %	ANPP: 0	Species richness: 0	ND	Peñuelas et al. (2007)

(continued)

Table 1 (continued)

Location	Ecosystem	MAT (°C)	MAP (mm)	Experimental design	Study period	Precipitation manipulation	Productivity	Diversity	Respiration, soil nitrogen	References
Podyjí National Park, Znojmo, Czech Republic N48°49′, E16°00′	Semi-arid: lowland *Festuca* grassland	9.6	529	Throughfall shelters	2006–2010	(1) ↓50 % (2) ↑50 %	(1) ANPP:↓21 % BNPP: 0 root incr.:↓ 21 % (2) ANPP:↑36 % BNPP: 0 root incr.: ↑38 %	Species composition: shifts in both (1) and (2), grasses ↓ and forbs ↑	Soil properties (a.o. N) in (1) and (2): 0	Fiala et al. (2009, 2012), Tůma et al. (2009), Holub et al. (2013)
Mata, Israel N31°42′, E35°03′	Semi-arid: Mediterranean shrubland	17.7	540	Throughfall shelter	2003–2010 R_S: 2006–2007	(1) ↓30 % (2) ↑30 %	ANPP: (1) and (2): 0	Richness and diversity: (1) and (2): 0	R_S: (1) ↓5 % (2) ↑12 %	Sternberg (2011), Talmon et al. (2011)
Satara, Kruger National Park, South Africa S24°24′, E31°44′	Semi-arid: understorey herbaceous vegetation	22.9	547	Throughfall shelter	2004–2008	(1) ↓50 % (2) ↑50 %	ANPP (DPM measurements): (1) ↓25–30 % (2) ↑20–25 %	ND	ND	February et al. (2013)
Saline experimental range, Hays, Kansas, USA N38°52′, W99°23′	Semi-arid: mixed grass prairie	12.1	583	Throughfall shelters	2008–2010	Growing season: (1) ↓50 % (2) ↑~60 %	(1) ANPP: 0 BNPP:0–↓~35 % (2) ANPP: ↑31–55 % BNPP:0–↓~40 %	ND	ND	Byrne et al. (2013)
McLaughlin Reserve, Lower Lake, California, USA N38°51, W123°30′	Mesic: serpentine grassland	~16.5	620	No shelters	2010	Spring: ↑50 %	ANPP: 0	ND	ND	Fernandez-Going and Harrison (2013)
McLaughlin Reserve, Lower Lake, California, USA N38°51′, W123°30′	Mesic: non-serpentine grassland	~16.5	620	No shelters	2010	Spring: ↑50 %	ANPP: ↑~50 %	ND	ND	Fernandez-Going and Harrison (2013)
Los Yébenes, Toledo, Spain N39°25′, W04°04′	Mesic: Meditteranean shrubland dominated by *Cistus ladanifer*	14.9	622	Movable shelter	2009	March–Sept.: (1) ↓25 % (2) ↓45 %	AGR *C. ladanifer*: (1) 0 (2) ↓ 40 %	ND	ND	Parra et al. (2012)
Porto Conte Capo Caccia, Sardinia, Italy N40°37′, E08°10′	Mesic: shrubland	16.8	640	Movable shelter	2002–2004	April–May and Oct.–Nov.: ↓100 %	ND	ND	R_S: 0	de Dato et al. (2010)
JRBP, Woodside, California, USA N37°24′, W122°14′	Mesic: Mediterranean grassland dominated by annual species	14.6	655	No shelters	1999–2003	Nov.–June: ↑50 %	ANPP: 0 BNPP: 0	ND	ND	Dukes et al. (2005)
Hohenheim Climate Change experiment, Heidfeldhof, Germany N48°42′, E09°11′	Mesic: agricultural crop 2009: *T. aestivum* 2010: *H. vulgare*	8.7	679	Closed shelter	2009–2010	June–August: ↓25 %	ANPP: 0	ND	R_S: 0	Poll et al. (2013)

Wytham, Oxford, UK N51°46′, W01°30′	Mesic: calcareous grassland	10	680	Movable shelter in (1)	1994–1998	(1) July–August:- ↓100 % (2) June–Sept.: ↑20 %	ANPP: (1) ↓~7 % (2) ↑~38 %	ND	ND	Grime et al. (2000)
Herdade da Giblaceira, Montemor-O-Novo, Portugal N38°36′, W08°10′	Mesic: woodland understorey vegetation	15.5	700	Closed shelter	2008	Growing season: ↓32 %	ANPP: ↓44 % BNPP: 0	Diversity: ↓17 %	R_S: ↓12 % Soil N: 0	Jongen et al. (2009)
La Copita Research Area, Central, Rio Grande Plains, Texas, USA N27°40′, E98°12′	Mesic: grass- to woodland	22.4	716	No shelters	1996–1997	↑~300 %	ND	ND	R_S: ↑48–131 %	McCulley et al. (2007)
Pretoriuskop, Kruger National Park, South Africa S25°07′, E31°13′	Mesic: understorey herbaceous vegetation	21.9	737	Throughfall shelter	2004–2008	(1) ↓50 % (2) ↑50 %	ANPP (DPM measurements): (1) ↓25–30 % (2) ↑20–25 %	ND	ND	February et al. (2013)
SFREC, Browns Valley, California, USA N39°15′, W121°17′	Mesic: annual grassland	16.0	750	No shelters	2003–2007	↑50 %	ANPP: ↑~35 % BNPP: 0	Species composition: 0	R_S: ↑~10 %	Chou et al. (2008)
CLIMAITE, Mols, Denmark N56°23′, E10°57′	Mesic: heath land	9.4	758	Movable shelters	1999–2005	Two-month period in growing season: ↓65–90 %	ANPP: 0	Species richness: 0	R_S: 0 N-mineralisation.: 0	Emmett et al. (2004), Peñuelas et al. (2007)
Kameničky, Hlinsko, Czech Republic N49°45′, E15°54′	Mesic: highland *Cirsium* grassland	7	762	Throughfall shelters	2006–2010	(1) ↓50 % (2) ↑50 %	(1) BNPP:↓18 % root incr.: 0 (2) BNPP:↑17 % root incr.:↑15 %	ND	Soil properties (a.o. N) in (1) and (2): 0	Fiala et al. (2009, 2012), Tůma et al. (2009)
Clermont Ferrard, France N45°43′, E03°01′	Mesic: grassland	12.4	780	Closed shelters	2005–2007	June–Aug.: ↓20 %	ANPP: 0	Diversity: 0 Species comp.: 0	ND	Bloor et al. (2010)
La Cortijuela, Sierra Nevada NP, Granada, Spain N37°05′, W03°28′	Mesic: herbaceous vegetation, shrubland and forest	14.3	816	Throughfall shelters in (1)	2007	April–Sept.: (1) ↓35 % (2) ↑50 %	ND	ND	R_S: (1) ↓19.2 % (2) ↑14.6 %	Matías et al. (2012)
Konza Prairie, Kansas, USA N39°05′, W96°35′	Mesic: upland tallgrass prairie	13	835	No shelters	1991–2009	↑31 %	Cover: ↑~22 %	Richness and diversity: 0	ND	Collins et al. (2012)
Konza Prairie, Kansas, USA N39°05′, W96°35′	Mesic: lowland tallgrass prairie	13	835	No shelters	1991–2009	↑31 %	ANPP: ↑51 % Cover: ↑~9 % (ns)	Richness and diversity: 0	ND	Collins et al. (2012), Knapp et al. (2012)

(continued)

Table 1 (continued)

Location	Ecosystem	MAT (°C)	MAP (mm)	Experimental design	Study period	Precipitation manipulation	Productivity	Diversity	Respiration, soil nitrogen	References
RaMPs facility, Konza Prairie, Kansas, USA N39°05′, W96°35′	Mesic: tallgrass prairie	13	835	Closed shelters	1998–2000	↓30 %	ANPP: ↓10–15 %	ND	R_S: ↓7–8 %	Fay et al. (2003, 2011), Harper et al. (2005)
CLIMAITE, Porto Conte Capo Caccia, Italy N40°62′, E08°17′	Mesic: Mediterranean: forest-steppe	6.8	840	Movable shelters	1999–2005	Two-month period in growing season: ↓65–90 %	ANPP: 0	Species richness: 0	ND	Peñuelas et al. (2007)
KAEFS, McClain County, Oklahoma, USA N34°59′, W97°31′	Mesic: mixed-grass prairie	16.3	914	Throughfall shelters	2010–2011	(1) ↓50 % (2) ↑100 %	(1) and (2): ANPP: 0 BNPP: 0	ND	ND	Xu et al. (2013)
Alp Weissenstein, Switzerland N46°34′, E09°47′	Mesic: Alpine grassland	2.3	918	Closed shelters	2006–2007	Two to three-month exclusion ↓100 %	ANPP:↓65–69 % BNPP: 0	ND	ND	Gilgen and Buchmann (2009)
KFFL, McClain County, Oklahoma, USA N34°59′, W97°31′	Mesic: tallgrass prairie	16.3	967	No shelters	2003	↑100 %	ANPP: ↑~80 % BNPP: 0	ND	R_S: ↑ 9 % N availability: 0	Zhou et al. (2006, 2012), Sherry et al. (2008)
Yellowstone National Park, Wyoming, USA N44°50′, W110°33′	Mesic: grasslands (5 sites)	~1	site-dependent	No shelters	2005	April–Sept.: ↑~50 %	GEP: 0	ND	R_S: 0 R_{ECO}: 0	Risch and Frank (2007)
CLIMAITE Oldebroek, Netherlands N52°24′, E05°55′	Mesic: heathland/grassland	10.1	1042	Movable shelters	1999–2005	Two-month period in growing season: ↓65–90 %	ANPP: 0	Species richness: 0	R_S: 0 N-mineralisation: ↓	Emmett et al. (2004), Peñuelas et al. (2007)
BACE, Waltham, Massachusetts, USA N42°23′, W71°13′	Mesic: old-field grassland	10.3	1063	Throughfall shelters	2009–2010	(1) ↓50 % (2) ↑50 %	(1) and (2) ANPP: 0 ENPP: 0	Diversity: 0	R_H: (1) ↓21 % (2) 0	Hoeppner and Dukes (2012), Suseela et al. (2012)
Chamau, Switzerland N47°12′, E08°24′	Mesic: planted meadow	9.8	1179	Closed shelters	2005–2007	Two- to three-month in spring/summer ↓100 %	ANPP: 0 BNPP: 0	ND	ND	Gilgen and Buchmann (2009)

BCCIL, Buxton, UK N53°20′, W02°00′	Mesic: calcareous grassland	8	1300	Movable shelter in (1)	1994–1998	(1) July–August:- ↓100 % (2) June–Sept.: ↑20 %	ANPP: (1) ↓~5 % (2) ↑~20 %	ND	ND	Grime et al. (2000)
BCCIL, Buxton, UK N53°20′, W02°00′	Mesic: calcareous grassland	8	1300	Movable shelter in (1)	2004 (ANPP) 2006 (species richness) after >11 year manipulation	(1) July–August:- ↓100 % (2) June–Sept.: ↑20 %	ANPP: (1) ↓~35 % (2) 0	Species richness: (1) ↓~25 % (2) 0	ND	Grime et al. (2008)
Monte Rondinaio, northern Apennines, Italy N44°08′, E10°35′	Mesic: dwarf- shrub heathland	2	1300	No shelter	1999–2003	June–Sept.: ↑~35 %	ANPP: 0	Species richness: 0	Soil N: 0	Brancaleoni et al. (2007)
Bîly Kříž, Moravian-Silesian Beskydy mountain, Czech Republic N49°30′, E18°32′	Mesic: mountain *Nardus* grassland	6.8	1312	Throughfall shelters	2006–2010	(1) ↓50 % (2) ↑50 %	(1) ANPP: 0 BNPP: ↓11 % root incr.: 0 (2) ANPP: 0 BNPP: 0 root incr.:↑54 %	Species richness and composition in (1) and (2): 0	Soil properties (a.o. N) in (1) and (2): 0	Fiala et al. (2009, 2010, 2012), Tůma et al. (2009), Holub et al. (2013)
Früebüel, Switzerland N47°06′, E08°31′	Mesic: managed pasture	7.7	1632	Closed shelters	2005–2007	Two- to three-month in spring/ summer ↓100 %	ANPP: 0 BNPP: 0	ND	ND	Gilgen and Buchmann (2009)
CLIMAITE Clocaenog, UK N53°03′, W03°28′	Mesic: shrubland	8.2	1741	Movable shelters	1999–2005	Two-month period in growing season: ↓65–90 %	ANPP: 0	Species richness: 0	R_S: 0 N-mineralisation: 0	Emmett et al. (2004), Peñuelas et al. (2007)

ANPP aboveground net primary productivity, *BNPP* belowground net primary productivity, *AGR* absolute growth rate, *GEP* gross ecosystem photosynthesis, R_S soil respiration, R_H heterotrophic soil respiration, R_{ECO} ecosystem respiration, *MAT* mean annual temperature, *MAP* mean annual precipitation, ND not determined. Effect is indicated as positive (↑), negative (↓), or no effect (0)

[a]Studies excluded from analysis, as no numerical results given, or results depended on species

Table 2 Summary of precipitation manipulation studies with changes in the distribution of precipitation

Location	Ecosystem	MAT (°C)	MAP (mm)	Experimental design	Study period	Precipitation manipulation	Productivity	Diversity	Respiration, soil nitrogen	References
Río Mayo, Chubut, Argentina S45°41′, W70°16′	Arid: Patagonian steppe	8.4	168	No shelters	?	Precipitation addition with altered distribution: Oct., Dec., and Jan.: 1 × 15 mm or 3 × 5 mm	ND	ND	N-mineralisation ↓30 %, NO_3^- leaching ↑~15 % with large events	Yahdjian and Sala (2010)
Cabo de Gata natural park, Spain N36°49′, W02°15′	Arid: coastal sand dune	19	200	Closed shelters	Oct. 2005–June 2006	Altered distribution (dry period of 1, 2, or 4 weeks), total: 224 mm year^{-1}	ANPP: 0	Diversity: 0	ND	Miranda et al. (2009)
El Cautivo, Tabernas basin, Spain N37°0′, W02°26′	Arid: grassland	18	230	Closed shelters	Oct. 2005–June 2006	Altered distribution (dry period of 1, 2, or 4 weeks), total: 224 mm year^{-1}	ANPP: ↑~50 % with biweekly watering	Diversity: 0	ND	Miranda et al. (2009)
PSA site, Tabernas basin, Spain N37°5′, W02°21′	Arid: grassland	17	250	Closed shelters	Oct. 2005–June 2006	Altered distribution (dry period of 1, 2, or 4 weeks), total: 224 mm y^{-1}	ANPP: 0	Diversity: 0	ND	Miranda et al. (2009)
PSA site, Tabernas basin, Spain N37°5′, W02°21′	Arid: grasslands and shrublands	18	242	Closed shelters	2006–2009	Seasonal distribution changed (autumn and spring −15 %, winter +30 %)	ANPP: 0	ND	R_S: 0 soil N: 0	Miranda et al. (2011)
LTER Sevilleta, New Mexico, USA N34°20′, W106°43′	Arid: grassland	13.2	~250	No shelter	July–Sept. 2007 and 2008, during monsoon season	Precipitation addition with altered distribution: weekly (12 × 5 mm) or monthly (3 × 20 mm)	ANPP:↑28 % (ns) ANPP *Bouteloua eriopoda*: ↑54 % with large infrequent rainfall events	ND	R_S: ↑15–30 % with large infrequent rainfall events	Thomey et al. (2011), Vargas et al. (2012)
Northern Great Basin Experimental Range, Oregon, USA N43°29′, W119°34′	Semi-arid: sagebrush steppe	7.6	300	Closed shelters	1994–2000	Altered distribution (1) 80 % of precipitation between Oct. and March (2) 80 % of precipitation between April and July	Density and cover of *A. tridentata*: 0 ANPP herba-ceous vegetation: in (1) ↓~50 %	ND	ND	Bates et al. (2006)
CPER LTER Nunn, Colorado, USA N40°49′, W104°46′	Semi-arid: shortgrass steppe	8.6	321	Closed shelters	2006	Altered distribution during May–Sept. (dry period of 10, 20, or 30 days), total: 191 mm	ANPP: ↑30 % with large infrequent rainfall events	Species richness and composition: 0	Soil N: ↑~60 % with large infrequent rainfall events	Heisler-White et al. (2009)
CPER, LTER Nunn, Colorado, USA N40°49′, W104°46′	Semi-arid: shortgrass steppe	8.6	321	Closed shelters	2005	Altered distribution during May–Sept. (dry period of 10, 20, or 30 days), total: 190 mm	ANPP: ↑70–75 % with larger infrequent rainfall events	ND	ND	Heisler-White et al. (2008)

Saline experimental range, Hays, Kansas, USA N38°53', W99°23′	Semi-arid: mixed grass prairie	11.9	576	Closed shelters	2006	Altered distribution during May-Sept. (dry period of 10, 20, or 30 days), total: 340 mm	ANPP: ↑70 % with larger infrequent rainfall events	Species richness and composition: 0	Soil N: 0	Heisler-White et al. (2009)
LTER montado, Coruche, Portugal Nd°8′, W8°20′	Mesic: woodland understorey herbaceous vegetation	15.9	680	Closed shelters	Oct. 2010–June 2012	Altered distribution, with dry period of 1 and 3 weeks (Nov. 2010–June 2011) and 3 and 6 weeks (Nov. 2011–June 2012), total: 580 mm	ANPP: 0 BNPP: 0	Species composition: 0	R_S: 0 soil N: 0	Jongen et al. (2013a, b, c)
Hohenheim Climate Change experiment, Heidfeldhof, Germany N48°42′, E09°11′	Mesic: agricultural crop (2009: *T. aestivum*, 2010: *H. vulgare*)	8.7	679	Closed shelters	2009–2010	Altered distribution during June–August with 50 % increase in dry period	ANPP: 0	ND	R_S: 0	Poll et al. (2013)
Konza Prairie, Kansas, USA N39°05′, W96°35′	Mesic: tallgrass prairie	13	835	Closed shelters	2006	Altered distribution during the period of May–Sept. (dry period of 10, 20, or 30 days), total: 450 mm	ANPP: ↓18 % with large infrequent rainfall events	Species richness and composition: 0	Soil N: ↑~100 % with large infrequent rainfall events	Heisler-White et al. (2009)
RaMPs facility, Konza Prairie, Kansas, USA N39°05′, W96°35′	Mesic: tallgrass prairie	13	835	Closed shelters	1998–2000	Altered distribution with 50 % increase in dry period	ANPP: ↓10–15 %	Diversity: ↑~15 %	R_S: ↓13–16 %	Knapp et al. (2002), Fay et al. (2003, 2011), Harper et al. (2005)
RaMPs facility, Konza Prairie, Kansas, USA N39°05′, W96°35′	Mesic: tallgrass prairie	13	835	Closed shelters	2003–2007	Altered distribution with 50 % increase in dry period	ANPP: 0	ND	R_S: 0	Fay et al. (2011)
RaMPs facility, Konza Prairie, Kansas, USA N39°05′, W96°35′	Mesic: tallgrass prairie	13	835	Closed shelters	2007–2011	Altered distribution with 50 % increase in dry period	ANPP: 0	Diversity: 0	ND	Koerner et al. (2014)
Echo Bay, Ontario, Canada N46°25′, W8402′	Mesic: grassland	4.4	906	Closed shelters	1998	Altered distribution during the period of June–Sept. (dry period of 2, 4, 8, 14, or 28 days), total: 252 mm	ANPP: ↓31 % (ns) with large infrequent rainfall events	ND	R_S: ↓44 % with large infrequent rainfall events	Laporte et al. (2002)

ANPP aboveground net primary productivity, *BNPP* belowground net primary productivity, R_S soil respiration, *MAT* mean annual temperature, *MAP* mean annual precipitation, *ND* not determined. Effect is indicated as positive (↑), negative (↓), or no effect (0) with water addition, or productivity decrease with water reduction). Finally, for the Partial Least Squares Regression (PLSR) analysis we differentiated the data according to ecosystem (arid, semi-arid, and mesic), with an additional climate classification: cold (MAT < 5 °C), temperate (5 °C < MAT < 15 °C), and warm (MAT > 15 °C), thus giving a set of nine different biomes. PLSR (Wold et al. 1983, 2001) was applied to model percentage change in aboveground productivity (response variable, Y data) using MAT, MAP and percent manipulation as explaining variables (X data). By visually inspecting the scores and loading plots, the main factors determining the percent change in productivity can be assessed depending on ecosystem types, and relationships between potential explaining variables can be evaluated. X and Y data were mean centred and weighted by 1/(standard deviation). NIPALS algorithm was used. PLSR calculations were performed using the software package The Unscrambler X 10.3 (CAMO Software AS, Oslo, Norway).

3 The Impact of Changes in the Amount of Precipitation

3.1 *Precipitation Addition*

A total of 61 publications were included in the analysis for effects of changing precipitation quantity on ecosystem processes, 44 of which describe results of rainfall addition experiments, with a total number of 34 ecosystems (7 arid, 13 semi-arid, and 14 mesic systems) and 45 addition scenarios studied (Table 1). Addition experiments usually do not require shelters, and are often conducted in combination with precipitation exclusion, using the intercepted rainfall of a throughfall manipulation experiment (e.g. Talmon et al. 2011; Hoeppner and Dukes 2012; Byrne et al. 2013; Flanagan et al. 2013). The applied addition scenarios increased precipitation between 20 and 115 %, with the exception of McCulley et al. (2007), adding 300 % of rainfall. Some of these addition studies were conducted in ecosystems where climate change scenarios predict an increase in precipitation, such as in northern Europe (Press et al. 1998; Robinson et al. 1998; Keuper et al. 2012; Weijers et al. 2013), central Asia (Niu et al. 2008; Liu et al. 2009; Yan et al. 2010, 2011a, 2011b), or cold tundra regions in northern Russia (Keuper et al. 2012) and North America (McGraw 1985). However, several other addition studies, conducted in warm and temperate arid and semi-arid systems (e.g. Harpole et al. 2007; Evans et al. 2011; Talmon et al. 2011; Potts et al. 2012; Báez et al. 2013; Xu et al. 2013), investigated the effect of an increase in precipitation, although this is in contrast to future climate change scenarios, with these studies generally done by mere opportunity (e.g. using intercepted precipitation from throughfall shelters) or aiming at studying the effects of inter-annual variation in precipitation. The main parameters studied in precipitation addition experiments in the field were aboveground productivity (AP), totaling 40 observations, and soil/

ecosystem respiration ($R_{S/ECO}$), with 16 observations; less attention was given to effects on biodiversity (9 observations) and soil nitrogen properties (6 observations).

Figure 5 shows the variation of percentage response in the above-mentioned parameters to precipitation addition scenarios. None of the addition studies reported a negative response to any of the studied parameters. Both AP and $R_{S/ECO}$ showed the largest response to precipitation addition in arid regions, with average increases of 54 % and 46 %, respectively. Parameter increases in semi-arid (40 % for AP, and 31 % for $R_{S/ECO}$) and mesic regions (20 % for AP and 21 % for $R_{S/ECO}$) were lower. A 0-response was shown for diversity (Fig. 5d) in semi-arid and mesic ecosystems, a result which has to be interpreted carefully due to the low observation count. Similarly, 5 out of 6 studies reporting on soil N properties in semi-arid and mesic ecosystems gave a 0-response (Fig. 5h). In arid ecosystems, none of the studies reported on diversity and soil N properties (Fig. 5d, h). The highest percentage responses to precipitation addition were found in an arid shrubland in central China, with maximum increases of 266 % in AP and 59 % in R_S (Song et al. 2012), and a semi-arid Mediterranean grassland in California, with maximum increases of 275 %, 30 %, and 60 % in AP, R_{ECO}, and soil N, respectively (Potts et al. 2012).

Relating the percentage of manipulation to the percentage of productivity response (Fig. 6) or respiration response (Fig. 7) for all addition studies revealed a higher responsiveness of arid and semi-arid ecosystems to precipitation addition, as compared to mesic ecosystems. A 50 % increase in the amount of precipitation would result, on average, in a productivity increase of 37 %, 27 %, and 23 % in arid, semi-arid, and mesic ecosystems, respectively (Fig. 6). In addition, respiration would increase by 29 %, 35 %, and 13 % in arid, semi-arid, and mesic ecosystems, respectively, with a 50 % water addition (Fig. 7). However, the respiration results have to be regarded with care, due to the low number of replicate studies. Further, site-specific differences in other well-known drivers of soil respiration, such as soil nutrient and carbon availability, can be expected to affect these results.

Considering the strong moisture relationship with plant productivity generally found in terrestrial ecosystems (e.g. Nippert et al. 2006), it is remarkable that many of the applied addition scenarios did not result in significant increases in ecosystem processes. In total, 18 out of the 40 studied scenarios did not find increases in productivity responses, with approximately equal representation of arid and semi-arid ecosystems (44 % and 40 %, respectively), while 8 out of the 16 addition scenarios in mesic systems did not result in a productivity response. Notably, the four studies finding no response in arid systems were all performed in cold Tundra climates (McGraw 1985; Robinson et al. 1998; Keuper et al. 2012; Weijers et al. 2013). Respiration was generally found to be more responsive to water addition than productivity, with only 3 out of 16 studied scenarios reporting no effects.

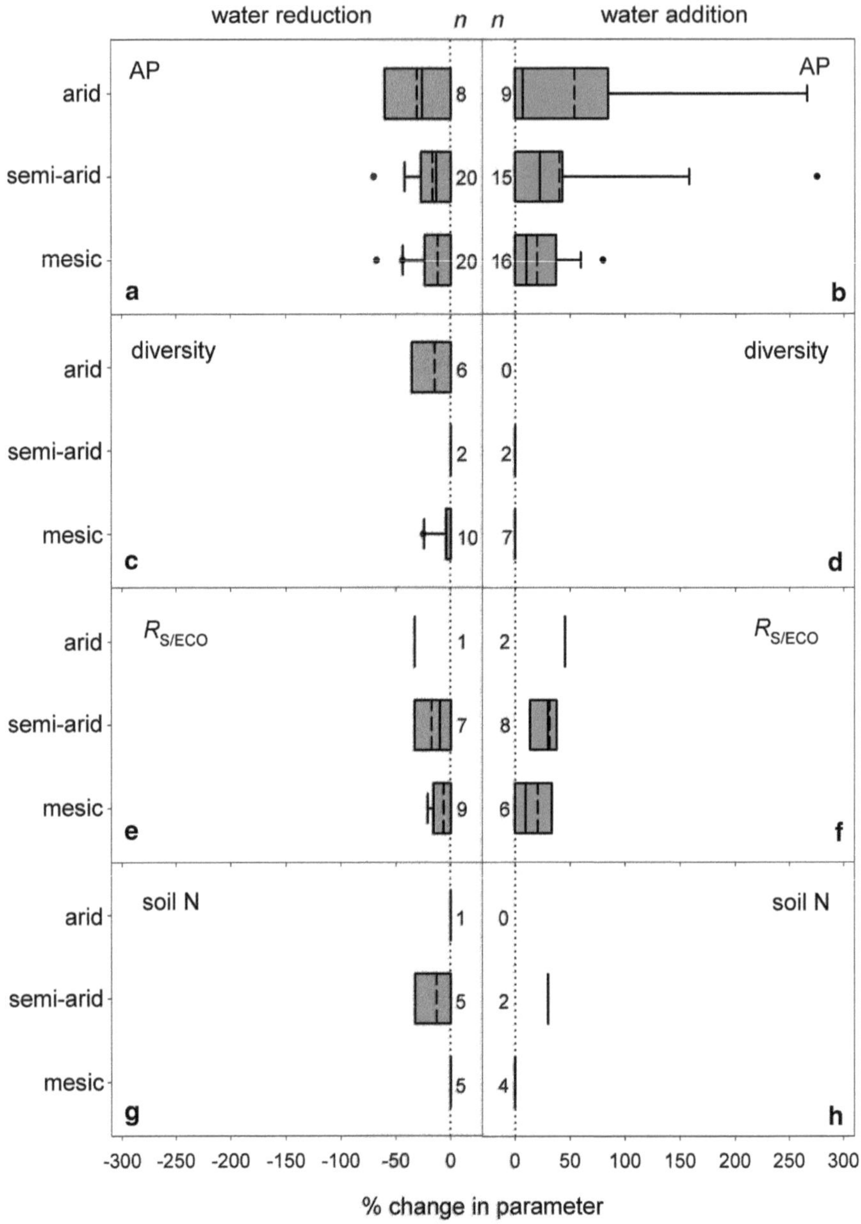

Fig. 5 Boxplot of variation in the range of percentage change in (**a**, **b**) aboveground productivity (AP), (**c**, **d**) diversity, (**e**, **f**) soil/ecosystem respiration ($R_{S/ECO}$), and (**g**, **h**) soil N properties observed in arid, semi-arid, and mesic ecosystems (see Table 1) in response to precipitation addition (**a**, **c**, **e**, **g**) or reduction (**b**, **d**, **f**, **h**). Number (*n*) refers to sample size. Boxplots visualise the first, second, and third quartile, and the mean values (*dotted lines*), with whiskers indicating the maximum and minimum. Outliers are shown as *dots*

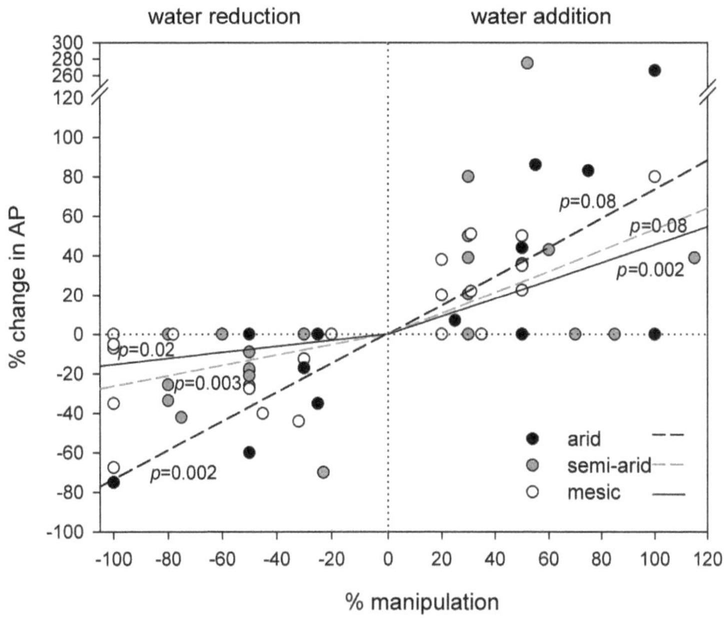

Fig. 6 Correlation between the percentage manipulation versus the percentage change in aboveground productivity (AP), for arid, semi-arid, and mesic ecosystems

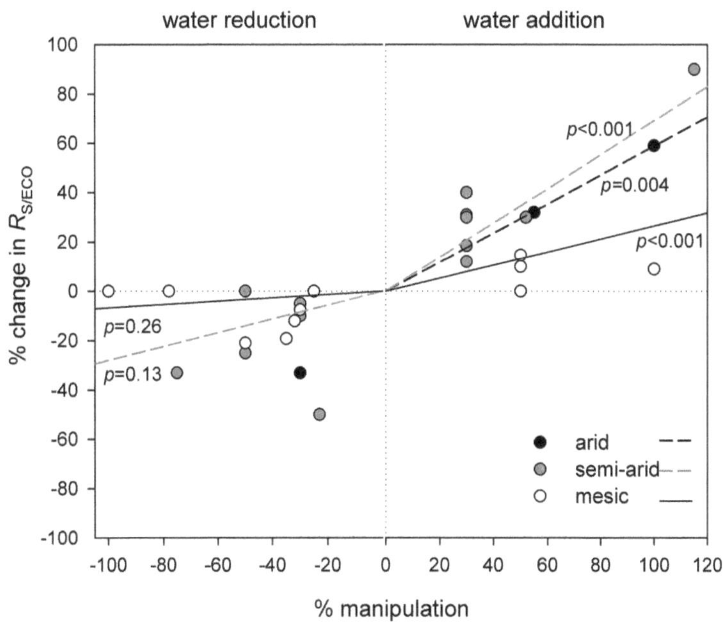

Fig. 7 Correlation between the percentage manipulation versus the percentage change in soil/ecosystem respiration ($R_{S/ECO}$), for arid, semi-arid, and mesic ecosystems

3.2 Precipitation Reduction

Of the 61 publications that were included in the analysis for effects of changing precipitation quantity on ecosystem processes, 41 described results of rainfall reduction experiments (Table 1). In total, these studies covered 38 ecosystem types (5 arid, 12 semi-arid, and 21 mesic systems) with 52 reduction scenarios applied. Experiments were performed using any type of the shelter designs described in Sect. 1.4, with the applied scenarios reducing precipitation between 20 and 100 % of natural precipitation during the experimental period. Most studies were conducted in ecosystems with future climate change prognoses indicating decreasing amounts of precipitation (e.g. Miranda et al. 2009, 2011; Evans et al. 2011; Talmon et al. 2011; Cherwin and Knapp 2012; Potts et al. 2012; Báez et al. 2013; Byrne et al. 2013). However, some studies, especially those conducted in northern European mesic ecosystems, chose to study precipitation reduction without a clear consensus of climate model predictions (e.g. CLIMAITE project studies of Emmett et al. 2004; Peñuelas et al. 2007).

AP, totaling 48 observations, was the most studied parameter, followed by biodiversity (18 observations), $R_{S/ECO}$ (17 observations), and soil N properties (11 observations). None of the featured reduction studies reported a positive response of either parameter with a decrease of precipitation (Fig. 5). Both biodiversity and soil N were found to be resilient to water reduction (Fig. 5c, g), with the exception of biodiversity in two arid ecosystems, exhibiting a negative response of 60 % and 27 % (Fig. 5c; Miranda et al. 2009), and a semi-arid grassland giving a 65 % decrease in soil N availability (Potts et al. 2012).

AP and $R_{S/ECO}$ showed the highest responsiveness to precipitation reduction in arid ecosystems, with average decreases of 31 and 33 %, respectively. Parameter decreases in semi-arid (17 % for AP, 18 % for $R_{S/ECO}$, respectively) and mesic ecosystems (12 % for AP, 7 % for $R_{S/ECO}$, respectively) were considerably lower (Fig. 5). Most responsive to precipitation reduction was an arid steppe in Mongolia, with a decrease in AP of ~75 % (Shinoda et al. 2010), and a semi-arid warm grassland in California (Potts et al. 2012), where AP decreased by 70 % in response to a mere 23 % water reduction. Surprisingly large reductions in AP were reported for an Alpine grassland (mesic ecosystem), where a 2–3 month exclusion of precipitation resulted in a decrease in AP of ~67 % (Gilgen and Buchmann 2009). In addition, in an arid warm grassland ecosystem in Spain (Miranda et al. 2009), exposed to a 50 % precipitation reduction, AP was reduced by 60 %.

Relating the percentage of manipulation to the percentage of productivity response (Fig. 6) or respiration response (Fig. 7) for all reduction studies revealed a higher responsiveness of arid ecosystems to precipitation reduction, as compared to semi-arid and mesic ecosystems. On average, a 50 % precipitation reduction would lead to productivity decreases of 37 %, 13 %, and 8 % in arid, semi-arid, and mesic grassland ecosystems, respectively (Fig. 6). However, the magnitude of responses of $R_{S/ECO}$ to precipitation reduction was substantially lower, with average decreases of 14 % and 4 % in semi-arid and mesic ecosystems, respectively, with a

50 % precipitation reduction (Fig. 7). The response of arid systems could not be estimated due to the lack of observations.

In total, 24 out of the 48 studied precipitation reduction scenarios found no significant treatment responses in AP (38, 45, and 69 % of the studies in arid, semi-arid, and mesic systems, respectively). With regard to $R_{S/ECO}$, 7 out of the 17 studied scenarios reported the absence of a significant treatment effect, with this non-responsiveness particularly pronounced in mesic ecosystems (5 out of 9 studies).

3.3 Sensitivity to Changes in the Amount of Precipitation

Our findings (Figs. 5, 6, and 7) indicate a hierarchy in the responsiveness of ecosystems to water manipulation, with largest responses to changing precipitation amounts in arid ecosystems, immediately followed by semi-arid ecosystems, while most mesic ecosystems were resilient to both water addition and reduction. In addition, in all three ecosystem types, responsiveness to water addition was higher as compared to water reduction.

The magnitude of the changes in a response parameter likely depends on the magnitude of the exposed addition/reduction scenario. For example, Song et al. (2012), being the only study including more than one addition scenario, reported a progressively increasing productivity response (from 7 to 266 %), with increasing precipitation addition (from 25 to 100 %). Similarly, several studies, including two reduction scenarios (e.g. Evans et al. 2011; Cherwin and Knapp 2012; Evans and Burke 2013), reported higher parameter reductions with increasing precipitation reduction. Thus, the higher responsiveness of arid ecosystems to precipitation may be due to a coincidentally higher amount of manipulation in these ecosystems. To estimate the possible impact of this effect, we calculated a sensitivity index, by weighing the ecosystem parameter response to the relative magnitude of exposed manipulation. Subsequently, we evaluated the effect of aridity index (after Köppen 1923), which accounts for both temperature and precipitation, by combining MAP and MAT, on the sensitivity index. Note that by definition a high aridity index means a humid climate while a low aridity index means an arid climate.

Figure 8 demonstrates this cause–effect relationship, with sensitivity of ecosystem parameters to changes in the amount of precipitation strongly decreasing with the aridity index. The average sensitivity index to water addition was 0.70, 0.77, and 0.18, while reduction scenarios gave values of 0.44, 0.40, and 0.14 for arid, semi-arid, and mesic ecosystems, respectively.

Deviating from the sensitivity–aridity relationship are the results reported in Potts et al. (2012). In this study, a semi-arid Mediterranean-type grassland, subjected to relatively small manipulations of the precipitation amount, gave exceptionally high sensitivity indices in both addition scenarios (5.3, 0.6, and 1.1 for AP, R_{ECO}, and soil N, respectively) and reduction scenarios (3.0, 2.2, and 2.8 for

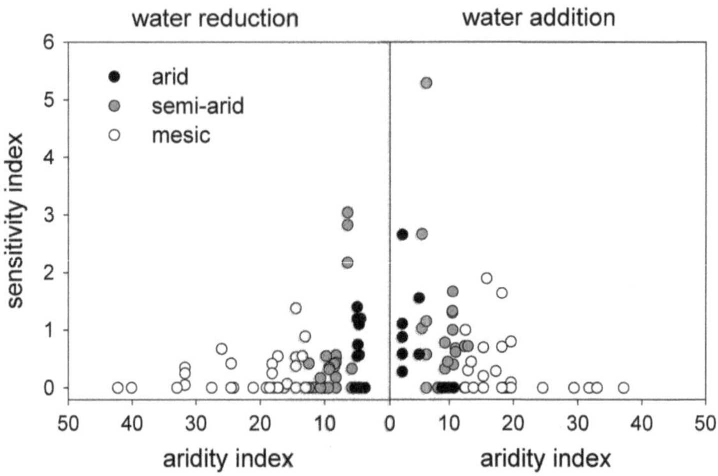

Fig. 8 Sensitivity index of arid, semi-arid, and mesic ecosystems with precipitation reduction or addition versus Köppen aridity index

AP, R_{ECO}, and soil N, respectively). However, the study was performed during a year characterised by natural drought, with severe consequences for the ambient treatment, which highlights the importance of considering climate conditions in the controls with precipitation manipulation studies. Omitting this study, the sensitivity indices for semi-arid ecosystems are smaller, i.e. 0.57 with water addition and 0.18 with reduction. In short, arid ecosystems showed the highest sensitivity to water manipulation, followed by semi-arid and mesic ecosystems. Thus, the observed hierarchy of ecosystem sensitivity to changes in the amount of precipitation with aridity is not an artefact of either the magnitude of manipulation or MAT.

Overall, ecosystem processes were more sensitive to water addition than to water reduction, with sensitivity indices of 0.45 and 0.28, respectively (Fig. 8). This effect was particularly pronounced in arid (0.70 with addition, 0.44 with reduction) and semi-arid regions (0.77 versus 0.40, respectively), while mesic ecosystems did not differ much in their responsiveness to either water addition or reduction (0.18 with addition, 0.14 with reduction). However, the trend towards higher sensitivity with aridity was maintained (Fig. 8).

4 The Impact of Changes in Precipitation Variability

In many ecosystems, a consequence of future climate change will be increasing intra-annual precipitation variability, with heavier rainfall events but longer intervening dry periods (see Sect. 1.1; Fig. 2). In recent years, several precipitation manipulation experiments have been conducted to study these scenarios. While some studies (e.g. Fay et al. 2003; Miranda et al. 2009) include a scenario of decreased precipitation amount (see Sect. 3.2), others try to single out the sole effect of altered precipitation distribution, frequency, and intensity on ecosystem performance (Knapp et al. 2002; Laporte et al. 2002; Heisler-White et al. 2008, 2009; Jongen et al. 2013a, b, c; Koerner et al. 2014). The experimental design usually involves closed shelters, with the collection and subsequent redistribution of natural precipitation according to the implemented scenario.

To date, the effects of precipitation variability, without altering total annual precipitation inputs, are still rarely studied, with only 18 publications included in the analysis. These studies were performed in 12 different ecosystem types (5 arid, 3 semi-arid, and 4 mesic systems) and 28 variability scenarios were studied (Table 2). The applied variability scenarios differed widely, with 0.5- to 14-fold increases in the both the extent of the normal dry period and the applied precipitation intensity (on average ~2-fold). Two studies given in Table 2 (i.e. Bates et al. 2006; Miranda et al. 2011) followed a different approach by altering the seasonal distribution of precipitation between the winter and the spring/summer periods. Main records made are on AP (27 observations), whereas observations on biodiversity (16), R_S (11), and soil N properties (10) are reported upon with less frequency. Figure 9a shows the variation of responses in AP to all applied precipitation variability scenarios in arid, semi-arid, and mesic systems. Observations in arid ecosystems consistently reported that AP was non-responsive to changes in the distribution pattern of precipitation (Miranda et al. 2009; Thomey et al. 2011; Vargas et al. 2012). However, Thomey et al. (2011) did report significant productivity increases in the dominant grass species with large infrequent rainfall events. In marked contrast, Heisler-White et al. (2008; 2009) with 6 observations in two different semi-arid ecosystems found a strong positive effect of larger infrequent precipitation pulses, with an increase in AP of on average 59 %, whereas negative effects on AP (−10 to −31 %) have been reported for mesic systems (Knapp et al. 2002; Laporte et al. 2002; Fay et al. 2003; Harper et al. 2005; Heisler-White et al. 2009). However, 9 out of 12 observations in mesic ecosystems report no significant response of AP with precipitation variability (Laporte et al. 2002; Fay et al. 2011; Jongen et al. 2013b, c; Poll et al. 2013; Koerner et al. 2014).

In a conceptual model ('bucket model') Knapp et al. (2008) predicted the responses of terrestrial ecosystems to more extreme intra-annual precipitation patterns. The model is based on the expected change in the amplification of the fluctuations of soil water content with respect to certain stress thresholds (e.g. field capacity and permanent wilting point) with increasing precipitation intensities and longer intervening dry periods. According to the 'bucket model', in mesic ecosystems, the altered fluctuations will increase the duration of soil moisture exceeding the stress thresholds, leading to more frequent and higher water stress with increasing precipitation variability, thus resulting in future negative responses in AP. However, arid ecosystems will experience the opposite effect, with a decrease in seasonal water stress with increasing precipitation variability, as the amplification of soil water dynamics would result in deeper soil water infiltration, thereby permitting soil moisture to be maintained above drought stress thresholds for longer periods (Knapp et al. 2008). The results, with a 0- or an on-average positive

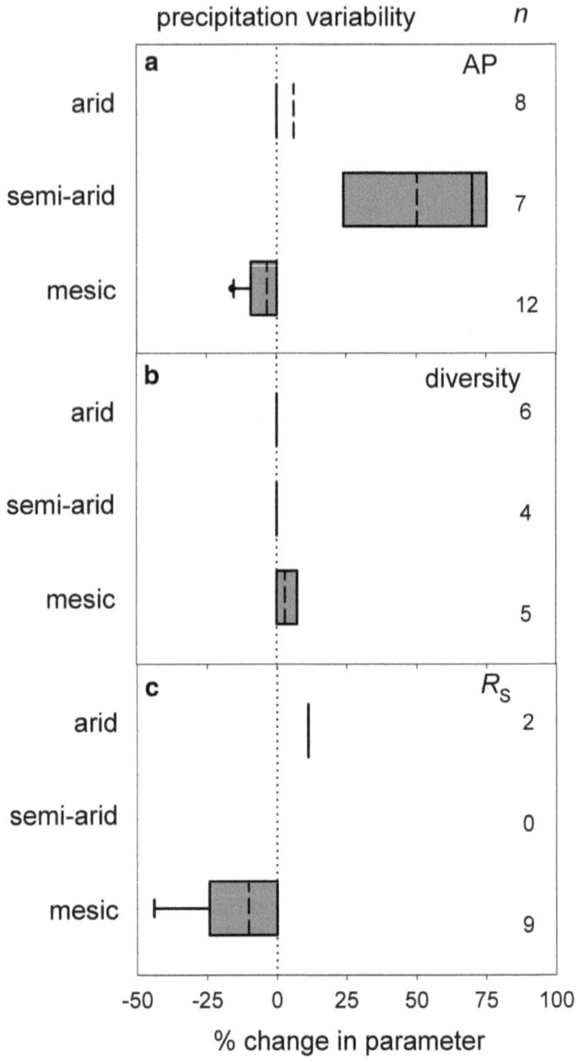

Fig. 9 Boxplot of variation in the range of percentage change, with mean values separately indicated (*dotted lines*) in measures of (**a**) productivity (AP), (**b**) diversity, and (**c**) soil respiration (R_S) observed in arid, semi-arid, and mesic ecosystems (see Table 2) in response to precipitation variability. Number (*n*) refers to sample size. Boxplots visualise the first, second, and third quartile, and the mean values (*dotted lines*), with whiskers indicating the maximum

response of AP in arid and semi-arid ecosystems, respectively, and an on-average negative response in mesic systems, support the predictions made in the 'bucket model'. However, the majority of the observations (19 out of 27) report no response of AP to altered precipitation variability, similar to the studies on the impacts of precipitation addition or reduction on AP (see Sects. 3.1 and 3.2).

In contrast to AP responses, the changes in biodiversity parameters with increased precipitation variability were very small (Fig. 9b). Arid and semi-arid ecosystems proved to be non-responsive, while mesic ecosystems exhibited a slight increase in biodiversity with decreasing precipitation frequency (Fig. 9b). This increase, however, was due to a single observation (Knapp et al. 2002), with none of the other studies in mesic ecosystems reporting significant biodiversity responses.

Changes in soil respiration were found to reflect the responses of AP to increased precipitation variability, with an 11 % increase (only 2 observations) and a 10 % decrease in arid and mesic ecosystems, respectively (Fig. 9c). Unfortunately, none of the studies in semi-arid ecosystems report results on R_S. Regarding soil nitrogen, general conclusions are difficult due to the small number of studies reporting on these parameters (10 observations). Nevertheless, soil nitrogen responses to decreasing precipitation frequency were extremely variable, with 7 observations finding no significant changes, while three observations report strong responses of −30 %, 60 %, and 100 % in arid, semi-arid, and mesic ecosystems, respectively (Heisler-White et al. 2009; Yahdjian and Sala 2010). As increasing AP and R_S in the arid/semi-arid regions should promote soil nitrogen mineralisation and turnover, the 0- and negative responses of soil N are unexpected. However, as not only precipitation frequency but also pulse size was altered, the increased precipitation intensity could enhance nitrogen losses from soils, which might counteract the expected positive effects of increased AP and R_S (see Sect. 4). Further, high temperatures in warm arid and semi-arid ecosystems can be expected to facilitate NH_3 volatilisation from soils (e.g. Fan et al. 2011).

What remains is the question why the majority of the observations (45 out of 63) report a non-responsiveness of ecosystem processes with increasing precipitation variability. Of major importance is the question whether an increase in precipitation variability actually causes an increase in variability of soil water content. As the fluctuations in soil moisture can easily be expressed as the coefficient of variation of daily average soil water content (CV_{SWC}) (e.g. Fay et al. 2011; Jongen et al. 2013b, c), we suggest the inclusion of this parameter in future studies, as non-responsiveness can be explained by the lack of a significant change in CV_{SWC}. In addition, the lack of responsiveness to increased precipitation variability can be explained by the resilience of grassland vegetation to short-term decreases in soil moisture during the growing period (Miranda et al. 2011; Jongen et al. 2013c). Indeed, plants have the ability to cope with irregularities of precipitation patterns through a high degree of phenotypic plasticity (Jump and Peñuelas 2005) and the possibility to employ strategies that improve water uptake and reduce water consumption (Moreno et al. 2008). Also, manipulation of precipitation variability during sensitive periods such as germination, seedling establishment, or the peak growing period can cause strong limitations on AP and dependent parameters that would not occur when studying less sensitive times, such as the end of the growing season. Finally, the length of the period that ecosystems experienced an increase in precipitation variability could be another reason for the non-responsiveness observed in many cases (Fay et al. 2011; Beier et al. 2012; Poll et al. 2013). Long-term effects of precipitation manipulation, including the loss of resilience, the possibility for adaptation, or a steady change in soil properties caused by the manipulation, might result in different ecosystem responses than those found in short-term studies (Beier et al. 2012). For example, the studies conducted in Konza Prairie reported no effects on AP and R_S in the long term (Fay et al. 2011; Koerner et al. 2014), while negative effects were found in the first 2 years of manipulation (Knapp et al. 2002; Fay et al. 2003; Harper et al. 2005).

5 The Impact of Changes in Precipitation Intensity

With future climate scenarios predicting a change in precipitation distribution (Sect. 1.1), the effects of extended dry periods between precipitation events (Fig. 2b) on ecosystem processes will be accompanied by effects of increased precipitation intensity (Fig. 2a). Studies addressing this increase in precipitation variability (Sect. 4, Table 2) often show a lack of responsiveness of grassland ecosystems. However, the presented analysis only included studies with a minimum duration of 2 months of precipitation manipulation, with parameter responses taken for each year or growing season. Thus, the effects of individual precipitation pulses on parameter responses on the short term, and ultimately on ecosystem functioning in the long term, should be considered.

For arid and semi-arid ecosystems, it has been hypothesised that discrete precipitation pulses stimulate brief but important episodes of biological activity (Huxman et al. 2004b), with pulse size and frequency differentially affecting above- and belowground biota. With the response thresholds of plants and microbes being determined by the ability of the organisms to utilise precipitation events of different infiltration depth, there is a hierarchical view of precipitation pulse patterns and their effect on ecosystem processes (Huxman et al. 2004b; Schwinning and Sala 2004). This suggests that small rainfall events will only stimulate soil microbes in the uppermost soil layer, subsequently increasing microbial respiration, while the stimulation of carbon assimilation in higher plants requires larger rainfall events, with concomitant increase in infiltration depth. Indeed, most studies show a positive effect of increased precipitation intensity on short-term plant performance (e.g. photosynthetic carbon uptake, phenology, reproductive traits) after a precipitation pulse, which, however, is seldom reflected in productivity increases (e.g. Patrick et al. 2007; Chen et al. 2009; Angert et al. 2010).

In relation to aboveground biological activity, several studies have shown that the magnitude and duration of the photosynthetic response were related to the size of the precipitation event (Chen et al. 2009; Jongen et al. 2013d) and that photosynthetic assimilation responds with a delay to precipitation pulses (Ogle and Reynolds 2004; Potts et al. 2006b). In addition, the plants' response differs across species and functional types, and depends on phenology, morphology, and physiological status (Huxman et al. 2004b; Ogle and Reynolds 2004; Potts et al. 2006a; Patrick et al. 2007; Angert et al. 2010).

In relation to belowground activity, a rapid increase in soil CO_2 efflux following precipitation pulses has been observed in various ecosystems (Mariko et al. 2007; Chen et al. 2009; Unger et al. 2010; Fan et al. 2012; Jongen et al. 2013a). This phenomenon of increased carbon and nitrogen losses after rewetting of dry soils, commonly termed 'Birch effect', has become an important subject in ecological studies. Although a complete understanding of the processes underlying the Birch effect has not yet been achieved (Borken and Matzner 2009), it is commonly accepted to be a direct response of the soil microbial and fungal community to changing moisture conditions (Borken and Matzner 2009; Inglima et al. 2009; Unger et al. 2010, 2012; Kim et al. 2012). Many studies agree on the theory of a positive effect of rewetting on soil microbial performance through a stimulation of microbial growth and matter transformation (e.g. Austin et al. 2004). However, several studies favour the hypothesis of a negative effect of precipitation pulses, with the large carbon and nitrogen losses being triggered by a hypo-osmotic stress response of soil microbes and fungi (Fierer and Schimel 2003; Xu and Baldocchi 2004; Jarvis et al. 2007; Unger et al. 2010, 2012). The majority of studies show that length and severity of the dry period prior to rewetting play a key role in the soil microbial response to such sudden changes in soil water status (Kieft et al. 1987; Fierer and Schimel 2003; Xu and Baldocchi 2004; Cable et al. 2008; Unger et al. 2010), with large and transient carbon pulses corresponding to high microbial stress (Unger et al. 2010; Meisner et al. 2013). Furthermore, wetting rate and infiltration to a soil seem to affect the microbial response to changing soil water potentials (Unger et al. 2012). Thus, the magnitude of change in soil water content related to a precipitation pulse is probably a key factor determining pulse effects on microbial activity and matter turnover.

Additionally, precipitation pulses may cause higher losses of nutrients, in particular the easily soluble components such as nitrate, due to higher water infiltration and leaching below the rooting zone (Yahdjian and Sala 2010; Jongen et al. 2013a).

As both increased Birch effects and leaching are to be expected with an increase of precipitation intensities, overall negative effects on soil microbial activity, nutrient availability, and hence productivity are a likely consequence. This effect might be exacerbated by larger precipitation pulses causing temporarily anoxic soil environments, impacting negatively on root and microbial performance (water logging, e.g. Jackson and Colmer 2005). However, the potential effects of increased precipitation intensity are masked by the concomitant effects of increasingly pulsed water availability and the longer intervening dry periods and therefore difficult to disentangle. Further, the impact of effects of increased leaching and soil nitrogen and carbon losses on ecosystem functioning is likely to be significant in the long term, and thus seldom detected in short-term observations.

For example, Patrick et al. (2007) found decreased soil respiration rates with concomitantly increased leaf level photosynthesis with a large precipitation pulse in summer during a 7-day observation period, resulting in an increase in carbon sequestration. However, the reduced soil respiration rates are likely to indicate substrate limitation or stress for soil microbes after the pulse, which could, in the long term, hamper the observed increase in carbon sequestration.

Although grassland ecosystems often show a lack of responsiveness to an increase in precipitation variability (Sect. 4, Table 2), this finding was particularly noticeable in arid and mesic ecosystems. In semi-arid ecosystems, overall positive effects of increased precipitation variability on productivity have been reported (Fig. 9), a finding that does not support the hypothesis of increased precipitation intensity causing negative effects due to the increased leaching and Birch effects in these ecosystems. Probably, the duration of most studies was not long enough to account for these long-term effects. In addition, it can be argued that the positive effect of longer periods with soil moisture being above stress thresholds, as hypothesised in the 'bucket model' for water-limited ecosystems (Knapp et al. 2008), counteracts negative effects of increased precipitation intensity on ecosystem functioning. However, Jongen et al. (2013c), increasing precipitation variability within the same experimental site for two subsequent years, reported lower soil nitrogen and productivity in the second year of manipulation, which could be related to negative effects of increased leaching and Birch pulses in the first year. Nevertheless, the results highlight a need for more long-term observation studies, ideally across several years, to assess the impact of precipitation intensity changes on ecosystem functioning.

6 Effects of Changing Precipitation Patterns on Synchronicity of the Matter Cycles

Due to the co-dependence of plant productivity and soil microbial mineralisation, there is a close linkage of ecosystem cycling of carbon and mineral nutrients. On the one hand, the temporal availability of carbon provided by plants, mainly through litter fall, root turnover, and exudation of organic substances to the soil, is a prerequisite for soil microbial growth (e.g. Kuzyakov and Gavrichkova 2010), while on the other hand, soil organic matter turnover and the mineralisation of nutrients, particularly of nitrogen, by soil microbes are vital for plant performance (e.g. Vitousek and Howarth 1991). Temporal synchronicity between both supply and demand of carbon and mineral nutrients is therefore crucial for ecosystem functioning (Augustine and McNaughton 2004).

The lack of soil moisture limitation in mesic ecosystems generally assures a tight coupling between microbial nutrient supply and plant nutrient demand (e.g. Vitousek et al. 1998; McCulley et al. 2009; Bobbink et al. 2010), with synchronicity between microbial and plant processes being mediated by intra-annual temperature variation, thereby minimising the loss of available nitrogen through leaching and gaseous emissions. However, in water-limited arid and semi-arid ecosystems, several studies observed an asynchronicity between nitrogen supply by microbes and nitrogen demand by plants (Jackson et al. 1988; Augustine and McNaughton 2004), due to the large fluctuations in soil moisture, with differences in the hierarchy, intricately linked to differences in thresholds, of the responses of plants and microbes to changes in soil water (Schwinning and Sala 2004; Collins et al. 2008). Such decoupling between peaks of mineral nutrient supply and plant growth can lead to substantial losses of mineral nutrients from the system, and result in a shift from a closed internal nitrogen cycle to an 'open' cycle, with the excess nitrogen being leached and/or emitted from the ecosystem (de Schrijver et al. 2008). For example, Yahdjian et al. (2006) suggested that net

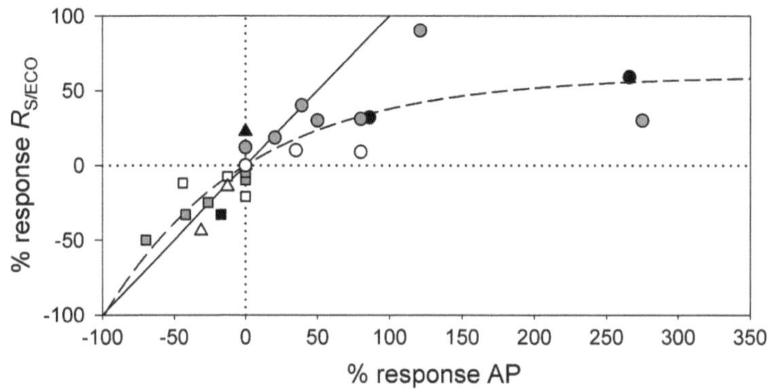

Fig. 10 Percentage change in AP versus percentage change of $R_{S/ECO}$ for precipitation addition (*circles*), reduction (*squares*), and variability (*triangles*) studies in arid (*black*), semi-arid (*grey*), and mesic (*white*) ecosystems. The *dashed line* indicates an exponential function ($y = 59.7 (1 - e^{(-0.01x)})$, $r^2 = 0.73$, $p < 0.0001$), whereas the 1:1 relationship is indicated by a *solid line*

mineralisation of nitrogen during long dry periods is less affected than plant and microbial nitrogen immobilisation, resulting in an accumulation of nitrate in the soil that is subsequently lost by leaching during first precipitation events, thereby leading to nitrogen limitation and decreased productivity.

With altered precipitation amount, frequency, and intensity, the response thresholds of plants and soil microbes might experience a larger deviation, amplifying the asynchronicity in nitrogen processes, and resulting in extended periods of decoupling between nitrogen and carbon cycles throughout the year.

Figure 10 shows the observed percentage responses of $R_{S/ECO}$ versus AP in arid, semi-arid, and mesic ecosystems, considering all reduction, addition, and variability scenarios. Only those studies with simultaneous measurements of $R_{S/ECO}$ and AP responses were included. Regardless of the ecosystem type, negative responses in AP, as found in precipitation reduction scenarios, were accompanied by negative responses of the same magnitude in $R_{S/ECO}$, indicating that a decrease in precipitation does not result in asynchronicity of plant and microbe performances (Fig. 10). This finding is unexpected, as moisture limitation has previously been found to enhance the possibility of asynchronicity (e.g. Evans and Burke 2013). However, although Evans and Burke (2013) found significant increases in soil inorganic nitrogen pools with a simulated long-term drought in a semi-arid grassland in California, the decoupling of AP and soil respiration due to different drought sensitivities of ecosystem processes was small (−42 % and −33 %, respectively).

In contrast to the observed synchronicity between AP and $R_{S/ECO}$ in studies with precipitation reduction, the different ecosystem compartments (AP and $R_{S/ECO}$) tended to show larger residuals from the 1:1 line with precipitation addition. In those precipitation addition studies with positive responses in AP and $R_{S/ECO}$, the AP response was often much stronger than the $R_{S/ECO}$ response. This effect became more pronounced with increasingly positive responses in AP and was observed in 2 arid (Thomey et al. 2011; Song et al. 2012), 2 semi-arid (Sternberg 2011; Talmon et al. 2011; Potts et al. 2012), and 1 mesic ecosystem (Zhou et al. 2006, 2012; Sherry et al. 2008). Thus, if precipitation manipulation increased plant growth, this was not always reflected in the same order of magnitude in soil or ecosystem respiration, which is indicative of a decoupling between carbon production and nitrogen mineralisation, potentially leading to a lagged nutrient deficiency of plants. As this effect was most pronounced in arid and semi-arid ecosystems, where asynchronicity of the matter cycles is reflected in nitrogen accumulation during longer dry periods, a higher response of plant productivity as compared to soil microbial activity might be expected, as additional water enables plants to better assimilate the readily available nitrogen pool. While this explains how, in the short term, precipitation addition accelerates plant growth in soils without nutrient limitation, the lagged effect of smaller nutrient supply by microbes with larger nutrient fixation by plants is not considered. However, the studies showing the most extreme deviations from the 1:1 line in Fig. 10 can be regarded as exceptions, with Potts et al. (2012) comparing precipitation addition to a control subjected to severe natural drought, and Song et al. (2012) studying an extreme addition scenario in the most arid ecosystem reported upon. Under these extreme conditions, a higher response of AP as compared to the response of $R_{S/ECO}$ with precipitation addition might be expected, as (1) nutrient accumulation during drought periods is expected to be higher, and (2) high soil water potential changes might cause greater stress to soil microbes (Sect. 5) than to plants.

Reports on both AP and $R_{S/ECO}$ in studies manipulating precipitation variability, without altering total precipitation inputs (Table 2), did not find pronounced differences between soil and plant treatment responses (Fig. 10). Dijkstra et al. (2012) showed that nitrogen release by soil microbes was enhanced as compared to plant nitrogen uptake with large and infrequent precipitation pulses, indicating that changes in precipitation event sizes could exacerbate losses of nitrogen in a semi-arid system. However, such short-term effects were not reproduced in any of the longer-term manipulation studies, most of them finding a near 1:1 response in AP and $R_{S/ECO}$ (Fig. 10).

Thus, in general, we found that in most ecosystems reported upon, both AP and $R_{S/ECO}$ did respond synchronously to precipitation manipulation scenarios. This is supported by the general lack of significant treatment effects on soil nitrogen availability or mineralisation with precipitation manipulation (Tables 1 and 2), with the exception of only four studies (Heisler White et al. 2009; Yahdjian and Sala 2010; Potts et al. 2012; Evans and Burke 2013). Therefore, increased asynchronicity of the matter cycles will not likely be a threat with changing precipitation patterns as predicted with future climate scenarios.

7 Synthesis

Experimentally manipulating precipitation patterns is an indispensable tool to describe and model future climate change impacts on ecosystem processes (Reichstein et al. 2013; Reyer et al. 2013; Vicca et al. 2013). During the last two decades, a growing body of studies has emerged reporting on manipulative experiments in a variety of ecosystems, with some of these studies including the effects of changing precipitation patterns. In this review, the findings of these individual studies were synthesised in order to make inferences from the combined results and to identify and refine strategies for future research. An earlier meta-analysis of precipitation manipulation studies, carried out by Wu et al. (2011), synthesised 39 studies, conducted in 34 ecosystems with different vegetation types, their analysis focusing on plant growth parameters and ecosystem carbon balance. They found that supplemental precipitation stimulated plant productivity and ecosystem carbon fluxes, whereas reduced precipitation suppressed these parameters (Wu et al. 2011). Our review analyses the effects of in situ precipitation manipulation on plant productivity, species diversity, soil/ecosystem respiration, and soil nitrogen in grassland type ecosystems over a wide range of climate types (MAT range: −13.9 °C to 22.9 °C; MAP range: 115–1,741 mm), synthesising the results from 72 studies. Our analysis showed a hierarchy in the responsiveness of grassland ecosystems to changing precipitation quantity, with largest responses found in arid ecosystems, immediately followed by semi-arid ecosystems, while the majority of mesic ecosystems did not respond to either water addition or reduction. Furthermore, our analysis indicated that, independent of ecosystem type, ecosystem processes were more responsive to water addition than to water reduction, which agrees with Wu et al (2011), reporting higher sensitivity to increased precipitation than to decreased precipitation. In a review on precipitation reduction experiments, Vicca et al. (2013) reported no evidence of differential effects of experimental drought over sites with different MAP, which is in contrast to our results for grassland type ecosystems, where responsiveness to manipulation was higher in arid regions than in mesic regions. This, however, is plausible, as productivity losses through water reduction in shallow-rooted grassland ecosystems are potentially stronger in arid regions, which might not be the case with deeper-rooted forest ecosystems (Reichstein et al. 2013). In agreement, Knapp and Smith (2001) concluded that systems dominated by herbaceous vegetation may be more responsive to future precipitation regimes, as their productivity shows generally larger inter-annual variability than forests. Hsu et al. (2012) reported higher sensitivity of aboveground productivity to changes in mean annual precipitation in semi-arid ecosystems as compared to arid ecosystems, concluding that at the driest sites, sensitivities may be lower due to low relative growth rates, density limitations, and high evaporation rates. This is in contrast to our results, with grassland ecosystems showing higher sensitivity with increasing aridity. However, the conclusion of Hsu et al. (2012) is based on two semi-arid sites with exceptionally high sensitivity.

Our analysis indicated that a large part of the studies reported a resilience of grassland ecosystems to changes in precipitation patterns. However, Wu et al. (2011) do not mention resilience in their findings. The resilience to changing precipitation amounts was most evident in mesic ecosystems, followed by semi-arid and arid ecosystems, for both addition and reduction scenarios. Furthermore, resilience was more evident for diversity, soil nitrogen, and productivity as compared to respiration processes. This resilience might be because the applied manipulation scenarios often lie within the range of the natural inter-annual precipitation variability experienced by ecosystems, and a long-term evolutionary adaptation of ecosystem components to these natural differences (Sardans and Peñuelas 2013).

To test the observed hierarchy in responsiveness of ecosystems towards changing precipitation amounts with increasing aridity, we performed a Partial Least Squares Regression analysis (PLSR) of the productivity responses with changing precipitation quantity. The x-score plot of the factors MAP, MAT, and manipulation amount revealed a separation between grassland responses towards manipulation in different biomes (Fig. 11). PLSR showed that x-scores at factor 1 (amount of manipulation) explained 29 % of the y-variance (response variable), while x-scores at factor 2 (MAP and MAT) explained only 2 % of the y-variance. Thus, although high loadings of MAT and MAP on factor 2 are observed, the variance in AP response is mainly explained by the amount of manipulation or other unknown factors, which is indicated in similar directions of the manipulation and response vectors. MAP and MAT influences on ecosystem response variance cannot be

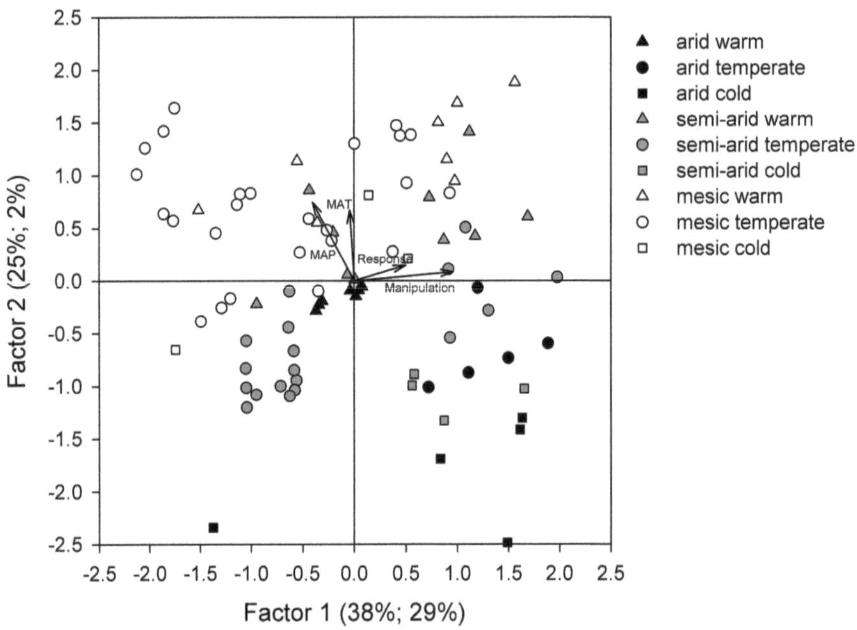

Fig. 11 Scores plot of the first two factors of a Partial Least Squares Regression (PLSR) analysis to model percentage change in aboveground productivity in grassland ecosystems (response variable, y-data) using MAT, MAP, and precipitation addition and reduction scenarios (percent manipulation) as explaining variables (x-data). Samples originating from nine different biomes are sorted by MAP (arid, semi-arid, mesic) and MAT (cold, temperate, warm). *Arrows* indicate mean x-and y-loadings of the predictor variables and the response. Explained variance in x and y, respectively, is given in *brackets* in the axes labels. $n = 87$

separated as they showed similar loading vectors; however, neither parameter contributed significantly to ecosystem responses towards manipulation.

The poor fit of the PLSR model, with only 38 % and 25 % of x-variance being explained by factors 1 and 2, respectively, can be an indication for data distribution being prejudiced by the resilience of many systems (0-responses). Thus, resilience, as indeed observed in most studies, overall seemed to have more impact than the hierarchy of different climate types in those studies that did observe ecosystem responses with precipitation manipulation. However, it has to be considered that the PLSR model is not only driven by MAT, MAP, and percentage manipulation, but also influenced by site-specific factors (e.g. soil properties, species composition, herbivory, management, climatic and site history) and means of experimental conduct (e.g. length and timing of experimental manipulation, seasonality, inter-annual variation in precipitation, shelter effects), which hampers generalisation and direct comparability of results in between sites.

In relation to changes in precipitation variability, resilience was prominent in arid and mesic grassland ecosystems, while semi-arid ecosystems showed an increase in productivity with increasing variability, although this result is based on a single study by Heisler-White et al. (2009). The hierarchy with a 0- or an on-average positive response of AP in arid and semi-arid ecosystems, respectively, and an on-average negative response in mesic systems in studies reporting effects towards precipitation variability experiments (Fig. 9), supports the predictions made in the 'bucket model' by Knapp et al. (2008). However, it must be considered that this conclusion is only based on a small amount of observations and, similar to the findings for changing precipitation amounts, is overshadowed by the resilience found in most studies. In addition, the hypothesised negative effects of increased leaching and gaseous carbon and nitrogen losses with higher precipitation pulse intensity in arid and semi-arid ecosystems could not be confirmed. This could be due to two reasons: (1) either the precipitation variability studies conducted are not long enough to account for negative long-term effects of soil nitrogen and carbon depletion or (2) the positive effects of soil moisture being above stress thresholds for longer periods counteract negative effects of increased precipitation intensity on ecosystem functioning.

For those studies that reported upon both AP and $R_{S/ECO}$, synchronous responses of both parameters towards precipitation manipulation scenarios were found, with the exception of some addition studies, with higher responses of productivity as compared to $R_{S/ECO}$. Therefore, increased asynchronicity of the matter cycles will not likely be a threat with the changing precipitation patterns that are predicted with future climate scenarios.

In a recent publication, Vicca et al. (2012) highlighted the necessity of a common metric to increase comparability of precipitation manipulation experiments in different ecosystems, this metric combining an index of both stress duration and stress intensity, thereby reflecting the actual treatments as experienced by plants. Indeed, it has to be considered that altering precipitation amounts might not always result in an equivalent change in available soil water for plants and microorganisms, as processes that distribute precipitation in the soil are complex, with interception, infiltration, run-off, seepage below the rooting zone, soil evaporation, plant water use, and hydraulic redistribution strongly differing between soil and ecosystem types (Loik et al. 2004) and differing between season and quantity of applied precipitation (Parton et al. 2012). Thus, these factors, by determining the available water for plants, might play a crucial role in determining whether ecosystems display either resilience or responsiveness to changing precipitation patterns. In addition, Sala et al. (2012) found that the responsiveness of ecosystem processes to changing water availability is influenced by a lagged effect of the previous year's situation, pointing to the need to incorporate information about climate legacies when interpreting results from precipitation manipulation experiments. Furthermore, most of the studies included in our review were conducted over short time periods. Shifts in species composition that can occur in the long term are thus rarely observed in precipitation manipulation studies (Weltzin et al. 2003). Considering climate change as a directional process, ecosystem responses will depend on both the magnitude of change and the time frame being considered (Sala et al. 2012).

The problem when comparing the results of a large number of individual studies to address large-scale ecological questions has recently been raised by Fraser et al. (2013), recommending the use of coordinated distributed experiments (CDEs), a possible solution to increase the level of in-between study comparability. CDEs provide a collaborative, coordinated, and hypothesis-driven approach for standardised experimental conduct and data analysis on an international level, thus controlling for site and study effects on both spatial and temporal scales, allowing to address important large-scale ecological issues, that would otherwise be difficult to resolve (Fraser et al. 2013).

We strongly support both the necessity to introduce a common metric to improve inter-study comparability, as proposed by Vicca et al. (2012), and the use of ecological CDEs, proposed by Fraser et al. (2013). In this review, we tried to improve inter-study comparability by relating the relative amount of ecosystem response with the relative amount of manipulation (sensitivity index, Sect. 2.2) and attempting to assess whether differences in ecosystem responses were driven by MAT and MAP. In addition, particularly for those manipulation experiments investigating the effects of precipitation variability, we suggest the use of the coefficient of variation of daily average soil water content (CV_{SWC}) to be included in future studies.

Although this review of precipitation manipulation experiments finds a general resilience of grassland ecosystems towards a range of manipulation scenarios, the question how future precipitation changes will affect ecosystem processes in global grasslands is far from answered. Thus, coordinated precipitation manipulation experiments with long-term field observations and increased comparability are desirable to capture and compare possible long-term effects (e.g. through changes in species composition and soil properties) on ecosystem state and functioning.

Acknowledgements The authors wish to thank Christine Hellmann for help and advice with the PLRS, and Peter Fay and Scott Collins for helpful suggestions on the manuscript. The authors gratefully acknowledge the financial support by FCT (Fundação para a Ciência e Tecnologia), through a postdoctoral fellowship to Marjan Jongen (SFRH/BPD/79662/2011).

References

Anderson RC (1982) An evolutionary model summarizing the roles of fire, climate and grazing animals in the origin and maintenance of grasslands: an end paper. In: Estes J, Tyrl R, Brunken J (eds) Grasses and grasslands: systematics and ecology. University of Oklahoma Press, Norman, OK, pp 297–308

Angert AL, Horst JL, Huxman TE, Venable DL (2010) Phenotypic plasticity and precipitation response in Sonoran Desert winter annuals. Am J Bot 97:405–411

Aranibar JN, Otter L, Macko SA, Feral CJW, Epstein HE, Dowty PR, Eckardt F, Shugart HH, Swap RJ (2004) Nitrogen cycling in the soil-plant system along a precipitation gradient in the Kalahari sands. Glob Change Biol 10:359–373

Atjay GL, Ketner P, Duvigneaud P (1979) Terrestrial primary production and phytomass. In: Bolin B, Degens ET, Kempe S, Ketner P (eds) The global carbon cycle. Wiley, Chichester, pp 129–181

Augustine DJ, McNaughton SJ (2004) Temporal asynchrony in soil nutrient dynamics and plant production in a semiarid ecosystem. Ecosystems 7:829–840

Austin AT, Yahdjian L, Stark JM, Belnap J, Porporato A, Norton U, Ravetta DA, Schaeffer SM (2004) Water pulses and biogeochemical cycles in arid and semiarid ecosystems. Oecologia 141:221–235

Bachman S, Heisler-White JL, Pendall E, Williams DG, Morgan JA, Newcomb J (2010) Elevated carbon dioxide alters impacts of precipitation pulses on ecosystem photosynthesis and respiration in a semi-arid grassland. Oecologia 162:791–802

Báez S, Collins SL, Pockman WT, Johnson JE, Small EE (2013) Effects of experimental rainfall manipulations on Chihuahuan Desert grassland and shrubland plant communities. Oecologia 172:1117–1127

Bates JD, Svejcar T, Miller RF, Angell RA (2006) The effects of precipitation timing on sagebrush steppe vegetation. J Arid Environ 64:670–697

Beier C, Beierkuhnlein C, Wohlgemuth T, Peñuelas J, Emmett B, Körner C, de Boeck H, Christensen JH, Leuzinger S, Janssens IA, Hansen K (2012) Precipitation manipulation experiments—challenges and recommendations for the future. Ecol Lett 15:899–911

Bloor JMG, Pichon P, Falcimagne R, Leadley P, Soussana J-F (2010) Effects of warming, summer drought, and CO_2 enrichment on aboveground biomass production, flowering phenology, and community structure in an upland grassland ecosystem. Ecosystems 13:888–900

Bobbink R, Hicks K, Galloway J, Spranger T, Alkemade R, Ashmore M, Bustamante M, Cinderby S, Davidson E, Dentener F, Emmett B, Erisman J-W, Fenn M, Gilliam F, Nordin A, Pardo L, de Vries W (2010) Global assessment of nitrogen deposition effects on terrestrial plant diversity: a synthesis. Ecol Appl 20:30–59

Borken W, Matzner E (2009) Reappraisal of drying and wetting effects on C and N mineralization and fluxes in soils. Glob Change Biol 15:808–824

Brancaleoni L, Gualmini M, Tomaselli M, Gerdol R (2007) Responses of subalpine dwarf-shrub heath to irrigation and fertilization. J Veg Sci 18:337–344

Byrne KM, Lauenroth WK, Adler PB (2013) Contrasting effects of precipitation manipulations on production in two sites within the Central Grassland Region, USA. Ecosystems 16:1039–1051

Cable JM, Ogle K, Williams DG, Weltzin JF, Huxman TE (2008) Soil texture drives responses of soil respiration to precipitation pulses in the Sonoran Desert: implications for climate change. Ecosystems 11:961–979

Chen S, Lin G, Huang J, Jenerette GD (2009) Dependence of carbon sequestration on the differential responses of ecosystem photosynthesis and respiration to rain pulses in a semiarid steppe. Glob Change Biol 15:2450–2461

Cherwin K, Knapp AK (2012) Unexpected patterns of sensitivity to drought in three semi-arid grasslands. Oecologia 169:845–852

Chou WW, Silver WL, Jackson RD, Thompson AW, Allen-Diaz B (2008) The sensitivity of annual grassland carbon cycling to the quantity and timing of rainfall. Glob Change Biol 14:1382–1394
Collins SL, Sinsabaugh RL, Crenshaw C, Green L, Porras-Alfaro A, Stursova M, Zeglin LH (2008) Pulse dynamics and microbial processes in aridland ecosystems. J Ecol 96:413–420
Collins SL, Koerner SE, Plaut JA, Okie JG, Brese D, Calabrese LB, Carvajal A, Evansen RJ, Nonaka E (2012) Stability of tallgrass prairie during a 19-year increase in growing season precipitation. Funct Ecol 26:1450–1459
Davidson EA, Janssens IA, Luo YQ (2006) On the variability of respiration in terrestrial ecosystems: moving beyond Q_{10}. Glob Change Biol 12:154–164
de Dato GD, De Angelis P, Sirca C, Beier C (2010) Impact of drought and increasing temperatures on soil CO_2 emissions in a Mediterranean shrubland (gariga). Plant Soil 327:153–166
de Schrijver A, Verheyen K, Mertens J, Staelens J, Wuyts K, Muys B (2008) Nitrogen saturation and net ecosystem production. Nature 451:E1. doi:10.1038/nature06578
Dijkstra FA, Augustine DJ, Brewer P, von Fischer JC (2012) Nitrogen cycling and water pulses in semiarid grasslands: are microbial and plant processes temporally asynchronus? Oecologia 170:799–808
Dukes JS, Chiariello NR, Cleland EE, Moore LA, Shaw MR, Thayer S, Tobeck T, Mooney HA, Field CB (2005) Responses of grassland production to single and multiple global environmental changes. PLoS Biol 3:1829–1837
Easterling DR, Evans JL, Groisman PY, Karl TR, Kunkel KE, Ambenje P (2000) Observed variability and trends in extreme climate events: a brief review. Bull Am Meteorol Soc 81:417–442
Emmett BA, Beier C, Estiarte M, Tietema A, Kristensen HL, Williams D, Peñuelas J, Schmidt I, Sowerby A (2004) The response of soil processes to climate change: results from manipulation studies of shrublands across an environmental gradient. Ecosystems 7:625–637
Evans SE, Burke IC (2013) Carbon and nitrogen decoupling under an 11-year drought in the shortgrass steppe. Ecosystems 16:20–33
Evans SE, Byrne KM, Lauenroth WK, Burke IC (2011) Defining the limit to resistance in a drought-tolerant grassland: long-term severe drought significantly reduces the dominant species and increases ruderals. J Ecol 99:1500–1507
Faber-Langendoen D, Josse C (2010) World grasslands and biodiversity patterns. NatureServe, Arlington, VA
Fan XH, Li YC, Alva AK (2011) Effects of temperature and soil type on ammonia volatilization from slow-release nitrogen fertilizers. Commun Soil Sci Plant Anal 42:1111–1122
Fan J, Jones SB, Qi LB, Wang QJ, Huang MB (2012) Effects of precipitation pulses on water and carbon dioxide fluxes in two semiarid ecosystems: measurement and modeling. Environ Earth Sci 67:2315–2324
Fay PA, Carlisle JD, Knapp AK, Blair JM, Collins SL (2000) Altering rainfall timing and quantity in a mesic grassland ecosystem: design and performance of rainfall manipulation shelters. Ecosystems 3:308–319
Fay PA, Carlisle JD, Knapp AK, Blair JM, Collins SL (2003) Productivity responses to altered rainfall patterns in a C4-dominated grassland. Oecologia 137:245–251
Fay PA, Blair JM, Smith MD, Nippert JB, Carlisle JD, Knapp AK (2011) Relative effects of precipitation variability and warming on tallgrass prairie ecosystem function. Biogeosciences 8:3052–3068
February EC, Higgins SI, Bond WJ, Swemmer L (2013) Influence of competition and rainfall manipulation on the growth responses of savanna trees and grasses. Ecology 94:1155–1164
Fernandez-Going BM, Harrison S (2013) Effects of experimental water addition depend on grassland community characteristics. Plant Ecol 214:777–786
Fiala K, Tůma I, Holub P (2009) Effect of manipulated rainfall on root production and plant belowground dry mass of different grassland ecosystems. Ecosystems 12:906–914

Fiala K, Tůma I, Holub P (2010) Interannual variability in plant below-ground biomass in a mountain grassland affected by manipulated rainfall. Beskydy 3:133–138
Fiala K, Tůma I, Holub P (2012) Interannual variation in root production in grasslands affected by artificially modified amount of rainfall. ScientificWorldJournal 2012:805298
Fierer N, Schimel JP (2003) A proposed mechanism for the pulse in carbon dioxide production commonly observed following the rapid rewetting of a dry soil. Soil Sci Soc Am J 67:798–805
Flanagan LB, Adkinson AC (2011) Interacting controls on productivity in a northern Great Plains grassland and implications for response to ENSO events. Glob Change Biol 17:3293–3311
Flanagan LB, Sharpa EJ, Letts MG (2013) Response of plant biomass and soil respiration to experimental warming and precipitation manipulation in a Northern Great Plains grassland. Agric For Meteorol 173:40–52
Fraser LH, Henry HAL, Carlyle CN, White SR, Beierkuhnlein C, Cahill JF, Casper BB, Cleland E, Collins SL, Dukes JS, Knapp AK, Lind E, Long R, Luo Y, Reich PB, Smith MD, Sternberg M, Turkington R (2013) Coordinated distributed experiments: an emerging tool for testing global hypothesis in ecology and environmental science. Front Ecol Environ 11:147–155
Frich P, Alexander LV, Della-Marta P, Gleason B, Haylock M, Klein Tank AMG, Peterson T (2002) Observed coherent changes in climatic extremes during the second half of the twentieth century. Clim Res 19:193–212
Gherardi LA, Sala OE (2013) Automated rainfall manipulation system: a reliable and inexpensive tool for ecologists. Ecosphere 4:article 18
Gilgen AK, Buchmann N (2009) Response of temperate grasslands at different altitudes to simulated summer drought differed but scaled with annual precipitation. Biogeosciences 6:2525–2539
Grime JP, Brown VK, Thompson K, Masters GJ, Hillier SH, Clarke IP, Askew AP, Corker D, Kielty JP (2000) The response of two contrasting limestone grasslands to simulated climate change. Science 289:762–765
Grime JP, Fridley JD, Askew AP, Thompson K, Hodgson JG, Bennett CR (2008) Long-term resistance to simulated climate change in an infertile grassland. PNAS 105:10028–10032
Harper CW, Blair JM, Fay PA, Knapp AK, Carlisle JD (2005) Increased rainfall variability and reduced rainfall amount decreases soil CO_2 flux in a grassland ecosystem. Glob Change Biol 11:322–334
Harpole WS, Potts DL, Suding KN (2007) Ecosystem responses to water and nitrogen amendment in a California grassland. Glob Change Biol 13:2341–2348
Heisler-White JL, Knapp AK, Kelly EF (2008) Increasing precipitation event size increases aboveground net primary productivity in a semi-arid grassland. Oecologia 158:129–140
Heisler-White JL, Blair JM, Kelly EF, Harmoney K, Knapp AK (2009) Contingent productivity responses to more extreme rainfall regimes across a grassland biome. Glob Change Biol 15:2894–2904
Hoeppner SS, Dukes JS (2012) Interactive responses of old-field plant growth and composition to warming and precipitation. Glob Change Biol 18:1754–1768
Holub P, Fabšičová M, Tůma I, Záhora J, Fiala K (2013) Effects of artificially varying amounts of rainfall on two semi-natural grassland types. J Veg Sci 24:518–529
Hsu JS, Powell J, Adler PB (2012) Sensitivity of mean annual primary production to precipitation. Glob Change Biol 18:2246–2255
Huxman TE, Smith MD, Fay PA, Knapp AK, Shaw MR, Loik ME, Smith SD, Tissue DT, Zak JC, Weltzin JF, Pockman WT, Sala OE, Haddad BM, Harte J, Koch GW, Schwinning S, Small EE, Williams DG (2004a) Convergence across biomes to a common rain-use efficiency. Nature 429:651–654
Huxman TE, Snyder KA, Tissue D, Leffler AJ, Pockman W, Ogle K, Sandquist D, Potts DL, Schwinning S (2004b) Precipitation pulses and carbon balance in semi-arid and arid ecosystems. Oecologia 141:254–268
Inglima I, Alberti G, Bertoloni T, Vaccari FP, Giolo B, Miglietta F, Cortufo MF, Peressortti A (2009) Precipitation pulse enhance respiration of Mediterranean ecosystems: the balance between organic and inorganic components of increased soil CO_2 efflux. Glob Change Biol 15:1289–1301

IPCC (2012) Summary for policymakers. In: Field CB, Barros V, Stocker TF et al (eds) Managing the risks of extreme events and disasters to advance climate change adaptation: a special report of working groups I and II of the intergovernmental panel on climate change. Cambridge University Press, Cambridge, pp 1–19

Jackson MB, Colmer TD (2005) Response and adaptation by plants to flooding stress. Ann Bot 96:501–505

Jackson LE, Strauss RB, Firestone MK, Bartolome JW (1988) Plant and soil nitrogen dynamics in California annual grassland. Plant Soil 110:9–17

Jarvis PG, Rey A, Petsikos C, Wingate L, Rayment M, Pereira JS, Banza J, David JS, Miglietta F, Borgetti M, Manca G, Valentini R (2007) Drying and wetting of Mediterranean soils stimulates decomposition and carbon dioxide emission: the "Birch effect". Tree Physiol 27:929–940

Jongen M, Caldeira MC, Pereira JS (2009) The effect of drought and subsequent precipitation pulse on productivity, species composition, and carbon fluxes of the herbaceous understorey in a cork oak woodland. Nat Prec. doi:10.1038/npre.2009.3706.1

Jongen M, Pereira JS, Aires LM, Pio CA (2011) The effects of drought and timing of precipitation on the inter-annual variation in ecosystem-atmosphere exchange in a Mediterranean grassland. Agric For Meteorol 151:595–606

Jongen M, Lecomte X, Unger S, Fangueiro D, Pereira JS (2013a) Precipitation variability does not affect soil respiration and nitrogen dynamics in the understorey of a Mediterranean oak woodland. Plant Soil 372:235–251

Jongen M, Lecomte X, Unger S, Pintó-Marijuan M, Pereira JS (2013b) The impact of changes in the timing of precipitation on the herbaceous understorey of Mediterranean oak woodlands. Agric For Meteorol 171–172:163–173

Jongen M, Unger S, Fangueiro D, Cerasoli S, Silva JMN, Pereira JS (2013c) Resilience of montado understorey to experimental precipitation variability fails under severe natural drought. Agric Ecosyst Environ 178:18–30

Jongen M, Unger S, Pereira JS (2013d) Effects of precipitation variability on carbon exchange in the understorey of a nitrogen-limited montado ecosystem. Oecologia, submitted

Jump AS, Peñuelas J (2005) Running to stand still: adaptation and the response of plants to rapid climate change. Ecol Lett 8:1010–1020

Keuper F, Parmentier F-JW, Blok D, van Bodegom PM, Dorrepaal E, van Hal J, van Logtestijn RSP, Aerts R (2012) Tundra in the rain: differential vegetation responses to three years of experimentally doubled summer-precipitation in Siberian shrub and Swedish bog tundra. Ambio 41:269–280

Kieft TL, Soroker E, Firestone MK (1987) Microbial biomass response to a rapid increase in water potential when dry soil is wetted. Soil Biol Biochem 19:119–126

Kim D-G, Vargas R, Bond-Lamberty B, Turetsky MR (2012) Effects of soil rewetting and thawing on soil gas fluxes: a review of current literature and suggestions for future research. Biogeosciences 9:2459–2483

Knapp AK, Smith MD (2001) Variations among biomes in temporal dynamics of aboveground primary productivity. Science 291:481–484

Knapp AK, Fay PA, Blair JM, Collins SL, Smith MD, Carlisle JD, Harper CW, Danner BT, Lett MS, McCarron JK (2002) Rainfall variability, carbon cycling, and plant species diversity in a mesic grassland. Science 298:2202–2205

Knapp AK, Beier C, Briske DD, Classen AT, Luo Y, Reichstein M, Smith MD, Smith SD, Bell JE, Fay PA, Heisler JL, Leavitt SW, Sherry RA, Smith B, Weng E (2008) Consequences of more extreme precipitation regimes for terrestrial ecosystems. Bioscience 58:811–821

Knapp AK, Briggs JM, Smith MD (2012) Community stability does not preclude ecosystem sensitivity to chronic resource alteration. Funct Ecol 26:1231–1233

Koerner SE, Collins SL, Blair JM, Knapp AK, Smith MD (2014) Rainfall variability has minimal effects on grassland recovery from repeated grazing. J Veg Sci 25:36–44

Köppen W (1923) Die Klimate der Erde. Walter de Gruyter, Berlin

Kuzyakov Y, Gavrichkova O (2010) Time lag between photosynthesis and carbon dioxide efflux from soil: a review of mechanisms and controls. Glob Change Biol 16:3386–3406
Laporte MF, Duchesne LC, Wetzel S (2002) Effect of rainfall patterns on soil surface CO_2 efflux, soil moisture, soil temperature and plant growth in a grassland ecosystem of northern Ontario, Canada: implications for climate change. BMC Ecol 2:article 10
Li Y, Yang H, Xia J, Zhang W, Wan S, Li L (2011) Effects of increased nitrogen deposition and precipitation on seed and seedling production of *Potentilla tanacetifolia* in a temperate steppe ecosystem. PLoS One 6:e28601
Liu W, Zhang Z, Wan S (2009) Predominant role of water in regulating soil and microbial respiration and their responses to climate change in a semiarid grassland. Glob Change Biol 15:184–195
Loik ME, Breshears DD, Lauenroth WK, Belnap J (2004) A multi-scale perspective of water pulses in dryland ecosystems: climatology and ecohydrology of the western USA. Oecologia 141:269–281
Ma S, Baldocchi DD, Xu L, Hehn T (2007) Inter-annual variability in carbon dioxide exchange of an oak/grass savanna and open grassland in California. Agric For Meteorol 147:157–171
Ma W, Liu Z, Wang Z, Wang W, Liang C, Tang Y, He J-S, Fang J (2010) Climate change alters interannual variation of grassland aboveground productivity: evidence from a 22-year measurement series in the Inner Mongolian grassland. J Plant Res 123:509–517
MacDougall AS, McCann KS, Gellner G, Turkington R (2013) Diversity loss with persistent human disturbance increases vulnerability to ecosystem collapse. Nature 494:86–89
Mariko S, Urano T, Asanuma J (2007) Effects of irrigation on CO_2 and CH_4 fluxes from Mongolian steppe soil. J Hydrol 333:118–123
Matías L, Castro J, Zamora R (2012) Effect of simulated climate-change on soil respiration in a Mediterranean-type ecosystem: rainfall and habitat-type are more important than temperature or the soil carbon pool. Ecosystems 15:299–310
McCulley RL, Boutton TW, Archer SR (2007) Soil respiration in a subtropical savanna parkland: response to water additions. Soil Sci Soc Am J 71:820–828
McCulley RL, Burke IC, Lauenroth WK (2009) Conservation of nitrogen increases with precipitation across a major grassland gradient in the Central Great Plains of North America. Oecologia 159:571–581
McGraw J (1985) Experimental ecology of *Dryas octopetala* ecotypes, III. Environmental factors and plant growth. Arct Alp Res 17:229–239
Meehl GA, Stocker TF, Collins WD, Friedlingstein P, Gaye AT, Gregory JM, Kitoh A, Knutti R, Murphy JM, Noda A, Raper SCB, Watterson IG, Weaver AJ, Zhao Z-C (2007) Global climate projections. In: Solomon S, Qin D, Manning M et al (eds) Climate change 2007: the physical science basis. Contribution of working group I to the fourth assessment report of the intergovernmental panel on climate change. Cambridge University Press, Cambridge, pp 747–846
Meisner A, Bååth E, Rousk J (2013) Microbial growth responses upon rewetting soil dried for four days or one year. Soil Biol Biochem 66:188–192
Merbold L, Ardo J, Arneth A, Scholes RJ, Nouvellon Y, de Grandcourt A, Archibald S, Bonnefond JM, Boulain N, Brueggemann N, Bruemmer C, Cappelaere B, Ceschia E, El-Khidir HAM, El-Tahir BA, Falk U, Lloyd J, Kergoat L, Le Dantec V, Mougin E, Muchinda M, Mukelabai MM, Ramier D, Roupsard O, Timouk F, Veenendahl EM, Kutsch WL (2009) Precipitation as driver of carbon fluxes in 11 African ecosystems. Biogeosciences 6:1027–1041
Miller ME, Belote RT, Bowker MA, Garman S (2011) Alternative states of a semiarid grassland ecosystem: implications for ecosystem services. Ecosphere 2:article 55
Miranda JD, Padilla FM, Lázaro R, Pugnaire FI (2009) Do changes in rainfall patterns affect semiarid annual plant communities? J Veg Sci 20:269–276
Miranda JD, Armas C, Padilla FM, Pugnaire FI (2011) Climatic change and rainfall patterns: effects on semi-arid plant communities of the Iberian Southeast. J Arid Environ 75:1302–1309
Moreno M, Gulías J, Lazaridou M, Medrano H, Cifre J (2008) Ecophysiological strategies to overcome water deficit in herbaceous species under Mediterranean conditions. Opt Médit 79:247–256

Nagy L, Kreyling J, Gellesch E, Beierkuhnlein C, Jentsch A (2013) Recurring weather extremes alter the flowering phenology of two common temperate shrubs. Int J Biometeorol 57:579–588

Nippert JB, Knapp AK, Briggs JM (2006) Intra-annual rainfall variability and grassland productivity: can the past predict the future? Plant Ecol 184:65–74

Niu S, Wu M, Han Y, Xia J, Li L, Wan S (2008) Water-mediated responses of ecosystem carbon fluxes to climatic change in a temperate steppe. New Phytol 177:209–219

O'Brien MJ, Philipson CD, Tay J, Hector A (2013) The influence of variable rainfall frequency on germination and early growth of shade-tolerant Dipterocarp seedlings in Borneo. PLoS One 8: e70287

Ogle K, Reynolds FJ (2004) Plant responses to precipitation in desert ecosystems: integrating functional types, pulses, thresholds, and delays. Oecologia 141:282–294

Olson JS, Watts JA, Allison LJ (1983) Carbon in live vegetation of major world ecosystems. Report ORNL-5862. Environmental Sciences Division, Oak Ridge National Laboratory, Oak Ridge, TN

Owens MK (2003) Approaches and techniques of rainfall manipulation. In: Weltzin JF, McPherson GR (eds) Changing precipitation regimes and terrestrial ecosystems: a North American perspective. University of Arizona Press, Tucson, AZ, pp 72–89

Parra A, Ramírez DA, Resco V, Velasco Á, Moreno JM (2012) Modifying rainfall patterns in a Mediterranean shrubland: system design, plant responses, and experimental burning. Int J Biometeorol 56:1033–1043

Parton W, Morgan J, Smith D, Del Grosso S, Prihodko L, Lecain D, Kelly R, Lutz S (2012) Impact of precipitation dynamics on net ecosystem productivity. Glob Change Biol 18:915–927

Patrick L, Cable J, Potts D, Ignace D, Barron-Gafford G, Griffith A, Alpert H, van Gestel N, Robertson T, Huxman TE, Zak J, Loik ME, Tissue D (2007) Effects of an increase in summer precipitation on leaf, soil, and ecosystem fluxes of CO_2 and H_2O in a sotol grassland in Big Bend National Park, Texas. Oecologia 151:704–718

Peñuelas J, Prieto P, Beier C, Cesaraccio C, de Angelis P, de Dato G, Emmett BA, Estiarte M, Garadnai J, Gorissen A, Kovács Láng E, Kröel-Dulay G, Llorens L, Pellizzaro G, Riis-Nielsen T, Schmidt IK, Sirca C, Sowerby A, Spano D, Tietema A (2007) Response of plant species richness and primary productivity in shrublands along a north-south gradient in Europe to seven years of experimental warming and drought. Reductions in primary productivity in the heat and drought year of 2003. Glob Change Biol 13:2563–2581

Poll C, Marhan S, Back F, Niklaus PA, Kandeler E (2013) Field-scale manipulation of soil temperature and precipitation change soil CO_2 flux in a temperate agricultural ecosystem. Agric Ecosyst Environ 165:88–97

Potts DL, Huxman TE, Cable JM, English NB, Ignace DD, Eilts JA, Mason MJ, Weltzin JF, Williams DG (2006a) Antecedent moisture and seasonal precipitation influence the response of canopy-scale carbon and water exchange to rainfall pulses in a semi-arid grassland. New Phytol 170:849–860

Potts DL, Huxman TE, Enquist BJ, Weltzin JF, Williams DG (2006b) Resilience and resistande of ecosystem functional response to a precipitation pulse in a semi-arid grassland. J Ecol 94:23–30

Potts DL, Suding KN, Winston GC, Rocha AV, Goulden ML (2012) Ecological effects of experimental drought and prescribed fire in a southern California coastal grassland. J Arid Environ 81:59–66

Press MC, Potter JA, Burke MJW, Callaghan TV, Lee JA (1998) Responses of a subarctic dwarf shrub heath community to simulated environmental change. J Ecol 86:315–327

Reichstein M, Bahn M, Ciais P, Frank D, Mahecha MD, Seneviratne SI, Zscheischler J, Beer C, Buchmann N, Frank DC, Papale D, Rammig A, Smith P, Thonicke K, van der Velde M, Vicca S, Walz A, Wattenbach M (2013) Climate extremes and the carbon cycle. Nature 500:287–295

Reyer CPO, Leuzinger S, Rammig A, Wolf A, Bartholomeus RP, Bonfante A, de Lorenzi F, Dury M, Gloning P, Abou Jaoudé R, Klein T, Kuster TM, Martins M, Niedrist G, Riccardi M, Wohlfahrt G, de Angelis P, de Dato G, François L, Menzel A, Pereira M (2013) A plant's perspective of extremes: terrestrial plant responses to changing climatic variability. Glob Change Biol 19:75–89

Risch AC, Frank DA (2007) Effects of increased soil water availability on grassland ecosystem carbon dioxide fluxes. Biogeochemistry 86:91–103

Robinson CH, Wookey PA, Lee JA, Callaghan TV, Press MC (1998) Plant community responses to simulated environmental change at a high arctic polar semi-desert. Ecology 79:856–866

Sala OE, Lauenroth WK, McNaughton SJ, Rusch G, Zhang X (1996) Biodiversity and ecosystem functioning in grasslands. In: Mooney HA, Cushman JH, Medina E, Sala OE, Schulze E-D (eds) Functional roles of biodiversity: a global perspective. Wiley, Chichester, pp 129–149

Sala OE, Gherardi LA, Reichmann L, Jobbágy E, Peters D (2012) Legacies of precipitation fluctuations on primary productivity: theory and data synthesis. Philos Trans R Soc B 367:3135–3144

Sardans J, Peñuelas J (2013) Plant-soil interactions in Mediterranean forest and shrublands: impacts of climatic change. Plant Soil 365:1–33

Schwinning S, Sala OE (2004) Hierarchy of responses to resource pulses in arid and semi-arid ecosystems. Oecologia 141:211–220

Sherry RA, Weng E, Arnone JA, Johnson DW, Verburg PS, Wallace LL, Luo Y (2008) Lagged effects of experimental warming and doubled precipitation on annual and seasonal aboveground biomass production in a tallgrass prairie. Glob Change Biol 14:2923–2936

Shevtsova A, Haukioja E, Ojala A (1997) Growth response of subarctic dwarf shrubs, *Empetrum nigrum* and *Vaccinium vitis-idaea*, to manipulated environmental conditions and species removal. Oikos 78:440–458

Shinoda M, Nachinshonhor GU, Nemoto M (2010) Impact of drought on vegetation dynamics of the Mongolian steppe: a field experiment. J Arid Environ 74:63–69

Singh JS, Lauenroth WK, Milchunas DG (1983) Geography of grassland ecosystems. Progr Phys Geogr 7:46–80

Smith FM (1973) Abiotic functions in grassland ecosystems. In: van Dyne GM (ed) Analysis of structure, function, and utilization of grassland ecosystems, vol 2, A progress report. Colorado State University, Fort Collins, CO, pp 29–96

Song W, Chen S, Wu B, Zhu Y, Zhou Y, Li Y, Cao Y, Lu Q, Lin G (2012) Vegetation cover and rain timing co-regulate the responses of soil CO_2 efflux to rain increase in an arid desert ecosystem. Soil Biol Biochem 49:114–123

Sternberg M (2011) Climate change in unpredictable terrestrial ecosystems: an integrative approach along an aridity gradient in Israel. Paper presented at the 1st Interface workshop: how do we improve earth system models: integrating earth system models, ecosystem models, experiments and long-term data, February 28–March 3, Captiva Island, FL

Suseela VA, Conant RT, Wallenstein MD, Dukes JS (2012) Effects of soil moisture on the temperature sensitivity of heterotrophic respiration vary seasonally in an old-field climate change experiment. Glob Change Biol 18:336–348

Talmon Y, Sternberg M, Grünzweig JM (2011) Impact of rainfall manipulations and biotic controls on soil respiration in Mediterranean and desert ecosystems along an aridity gradient. Glob Change Biol 17:1108–1118

Thomey ML, Collins SL, Vargas R, Johnson JE, Brown RF, Natvig DO, Friggens MT (2011) Effect of precipitation variability on net primary production and soil respiration in a Chihuahuan Desert grassland. Glob Change Biol 17:1505–1515

Trenberth KE (2011) Changes in precipitation with climate change. Clim Res 47:123–138

Trenberth KE, Jones PD, Ambenje P, Bojariu R, Easterling D, Klein Tank A, Parker D, Rahimzadeh F, Renwick JA, Rusticucci M, Soden B, Zhai P (2007) Observations: surface and atmospheric climate change. In: Solomon S, Qin D, Manning M et al (eds) Climate change 2007: the physical science basis. Contribution of working group I to the fourth assessment report of the intergovernmental panel on climate change. Cambridge University Press, Cambridge, pp 235–336

Tůma I, Fiala K, Holub P, Záhora J (2009) Can soil properties in different grasslands be altered after three years of experimentally manipulated rain? Beskydy 2:71–76

UNEP United Nations Environment Programme (1992) World Atlas of Desertification. Arnold A (ed). UNEP, London, pp 69

Unger S, Máguas C, Pereira JS, David TS, Werner C (2010) The influence of precipitation pulses on soil respiration—Assessing the "Birch effect" by stable carbon isotopes. Soil Biol Biochem 42:1800–1810

Unger S, Máguas C, Pereira JS, David TS, Werner C (2012) Interpreting postdrought rewetting effects on soil and ecosystem carbon dynamics in a Mediterranean oak savannah. Agric For Meteorol 154–155:9–18

Vargas R, Collins SL, Thomey ML, Johnson JE, Brown RF, Natvig DO, Friggens MT (2012) Precipitation variability and fire influence the temporal dynamics of soil CO_2 efflux in an arid grassland. Glob Change Biol 18:1401–1411

Vázquez-de-Aldana BR, García-Ciudad A, García-Criado B (2008) Interannual variations of above-ground biomass and nutritional quality of Mediterranean grasslands in Western Spain over a 20-year period. Aust J Agric Res 59:767–779

Vicca S, Gilgen AK, Camino Serrano M, Dreesen FE, Dukes JS, Estiarte M, Gray SB, Guidolotti G, Hoeppner SS, Leakey ADB, Ogaya R, Ort DR, Ostrogovic MZ, Rambal S, Sardans J, Schmitt M, Siebers M, van der Linden L, van Straaten O, Granier A (2012) Urgent need for a common metric to make precipitation manipulation experiments comparable. New Phytol 195:518–522

Vicca S, Estiarte M, Bahn M, Peñuelas J, Janssens I (2013) Synthesizing effects of precipitation manipulation on plant production and soil respiration—results and challenges. Geophys Res Abstr 15:EGU2013-2363-2011

Vitousek PM, Howarth RW (1991) Nitrogen limitation on land and in the see: how can it occur? Biogeochemistry 13:87–115

Vitousek PM, Hedin LO, Matson PA, Fownes JH, Neff J (1998) Within-system element cycles, input-output budgets, and nutrient limitation. In: Pace ML, Groffmann PM (eds) Successes, limitations, and frontiers in ecosystem science. Springer, New York, pp 432–451

Weijers S, Auliaherliaty L, van Logtestijn R, Rozema J (2013) Effects of manipulated precipitation and shading on *Cassiope tetragona* growth and carbon isotope discrimination: a high arctic field study. Arct Antarc Alp Res 45:1–11

Weltzin JF, Loik ME, Schwinning S, Williams DG, Fay PA, Haddad BM, Harte J, Huxman TE, Knapp AK, Lin G, Pockman WT, Shaw MR, Small EE, Smith MD, Smith SD, Tissue DT, Zak JC (2003) Assessing the response of terrestrial ecosystems to potential changes in precipitation. Bioscience 53:941–952

Wentz FJ, Ricciardulli L, Hilburn K, Mears C (2007) How much more rain will global warming bring? Science 317:233–235

White RP, Murray S, Rohweder M (2000) Pilot analysis of global ecosystems: grassland ecosystems. World Resources Institute, Washington, DC

Whittaker RH, Likens E (1975) The biosphere and man. In: Lieth H, Whittaker RH (eds) Primary productivity of the biosphere, ecological studies no. 14. Springer, Berlin

Wold S, Martens H, Wold H (1983) The multivariate calibration problem in chemistry solved by the PLS method. In: Kågström B, Ruhe A (eds) Matrix pencils. Springer, Berlin, pp 286–293

Wold S, Sjöström M, Eriksson L (2001) PLS-regression: a basic tool of chemometrics. Chemometr Intell Lab 58:109–130

Wu A, Dijkstra P, Koch GW, Peñuelas J, Hungate BA (2011) Responses of terrestrial ecosystems to temperature and precipitation change: a meta-analysis of experimental manipulation. Glob Change Biol 17:927–942

Xu L, Baldocchi DD (2004) Seasonal variation in carbon dioxide exchange over a Mediterranean annual grassland in California. Agric For Meteorol 123:79–96

Xu X, Sherry RA, Niu S, Li D, Luo Y (2013) Net primary productivity and rain-use efficiency as affected by warming, altered precipitation, and clipping in a mixed-grass prairie. Glob Change Biol 19:2753–2764
Yahdjian L, Sala OE (2002) A rainout shelter design of intercepting different amounts of rainfall. Oecologia 133:95–101
Yahdjian L, Sala OE (2010) Size of precipitation pulses controls nitrogen transformation and losses in an arid Patagonian ecosystem. Ecosystems 13:575–585
Yahdjian L, Sala OE, Austin AT (2006) Differential controls of water input on litter decomposition and nitrogen dynamics in the Patagonian steppe. Ecosystems 9:128–141
Yan L, Chen S, Huang J, Lin G (2010) Differential responses of auto- and heterotrohpic soil respiration to water and nitrogen addition in a semiarid temperate steppe. Glob Change Biol 16:2345–2357
Yan L, Chen S, Huang J, Lin G (2011a) Increasing water and nitrogen availability enhanced net ecosystem CO_2 assimilation of a temperate semiarid steppe. Plant Soil 349:227–240
Yan L, Chen S, Huang J, Lin G (2011b) Water regulated effects of photosynthetic substrate supply on soil respiration in a semiarid steppe. Glob Change Biol 17:1990–2001
Zhou X, Sherry RA, An Y, Wallice LL, Luo Y (2006) Main and interactive effects of warming, clipping, and doubled precipitation on soil CO_2 efflux in a grassland ecosystem. Glob Biogeochem Cy 20, GB1003. doi:10.1029/2005GB002526
Zhou X, Fei S, Sherry R, Luo Y (2012) Root biomass dynamics under experimental warming and doubled precipitation in a tallgrass prairie. Ecosystems 15:542–554

Plant-Mediated Ecosystem Effects of Tropospheric Ozone

Hans J. Weigel, Elke Bergmann, and Jürgen Bender

Contents

Abstract Tropospheric ozone (O_3) is considered as the most significant phytotoxic pollutant in the atmosphere and is already responsible for widespread effects on crops, trees and native plant species. Globally, there is evidence that the background O_3 concentrations are further increasing. Most research has been conducted on plant and tree species of commercial value, but very little is known about the impacts of O_3 on the scale of forest-, agro- or grassland ecosystems. Exposure to elevated O_3 causes oxidative stress, which results in reduced photosynthesis, visible injury, decreased growth and productivity. We present examples showing that impacts of O_3 on vegetation may lead to long-term effects on ecosystem structure and function. Recent experiments have shown that O_3 can cause a shift in plant

H.J. Weigel (✉) • E. Bergmann • J. Bender
Thünen Institute of Biodiversity, Bundesallee 50, 38116 Braunschweig, Germany
e-mail: hans.weigel@ti.bund.de

U. Lüttge, W. Beyschlag (eds.), *Progress in Botany*, Progress in Botany 76,
DOI 10.1007/978-3-319-08807-5_15

species composition and can indirectly affect soil processes. Ozone has also been shown to affect water cycling through its effect on stomata and can alter overall ecosystem productivity.

1 Introduction

During the last 60 years tropospheric or ground-level ozone (O_3) has emerged as an air pollution problem of global dimension with respect to its harmful impacts on human health and vegetation (Akimoto 2003; Royal Society 2008). As a secondary air pollutant O_3 is formed in the troposphere through a number of sunlight driven photochemical reactions involving the main precursor substances: nitrogen mono- and dioxide (NO/NO_2), volatile organic compounds (VOC), methane (CH_4) and carbon monoxide (CO) (Staehelin 2001). These precursors are produced naturally or emitted from anthropogenic activities such as vehicles, power plants, biomass burning and all other forms of combustion.

Terrestrial ecosystems are the major sink for tropospheric O_3 and consequently, vegetation is at particular risk from this pollutant. Ozone enters the plant interior through the stomata and as a strong oxidant O_3 and its breakdown products, respectively, are able to impact plants by altering plant cellular functions and by reducing photosynthesis and changing other important physiological functions. Collectively, this may result in visible leaf injury, growth and biomass reduction and overall inferior plant vigour (Ashmore 2005; Booker et al. 2009; Matyssek et al. 2010a, b). Whether or not these effects at the single plant level have implications or are relevant in an ecosystem context is still a matter of debate (Laurence and Andersen 2003).

In the past four decades O_3 effects have been thoroughly investigated with crops (reviewed by, e.g. Heagle 1989; Fiscus et al. 2005; Booker et al. 2009; Mills and Harmens 2011) and particularly with deciduous and coniferous trees (reviewed e.g. by; Sandermann et al. 1997; Percy et al. 2003a, b; Matyssek et al. 2010a, b, 2013). Other types of natural or semi-natural vegetation have only recently and to a lesser extent received attention (reviewed by, e.g. Fuhrer 1997; Davison and Barnes 1998; Ashmore 2005; Hayes et al. 2007). While the interest related to O_3 effects on crops and commercially relevant trees was mainly driven by concerns about the potential economic losses, the more recent emphasis in assessing its potential effects on ecosystem integrity and related ecosystem functions and services is based on concerns of the potential threats of O_3 to the biodiversity of these habitats, and the long-term, more subtle impacts on ecosystem functions and services such as carbon sequestration, nutrient cycling, water relations and pollination.

While there is a wealth of information on O_3 effects on plant metabolism and plant growth, respectively (e.g. Fiscus et al. 2005; Heath 2008; Booker et al. 2009; Cho et al. 2011), it is the intention of the present contribution to report on the progress that emerged from O_3 effects research with different types of vegetation during the last approx. 15 years with an emphasis on studies that addressed potential implications of this pollutant in an ecosystem context. With this objective we will not address and discuss the large body of information that has emerged during this period with respect to adequate O_3 risk assessments for vegetation, particularly with regard to the progress that has been made in describing phytotoxically relevant "absorbed O_3 doses" by overcoming the concept of O_3 exposure of vegetation (e.g. Matyssek et al. 2013). Rather, we focus on a more qualitative description of potential O_3 effects on plants and ecosystems, respectively, primarily without considering dose–response relationships.

In the following sections we first describe current and future O_3 exposure scenarios and the most common methods by which O_3 effects on vegetation are assessed. We then briefly summarise the current understanding of O_3-induced impairments at the individual plant level that are relevant for the understanding of its ecosystem effects (Fig. 1). Predominately we then address selected examples of how these O_3 effects relate to the ecosystem level, and consider and discuss results that are equally relevant for managed and unmanaged, natural ecosystems.

2 Ozone Levels: Trends and Variation in Space and Time

Naturally occurring O_3 concentrations in the troposphere (ground-level) in the pre-industrial era have been described to range between approx. 5–20 parts per billion (ppb) (Marenco et al. 1994). Since the pre-industrial era the global annual mean background O_3 concentrations have increased considerably to values between approx. >20–45 ppb depending on the geographical location (Vingarzan 2004) with a rate of increase in the annual mean values ranging between 0.1 and 1.0 ppb per year. This increase has been observed over large areas of Europe and North America, and more recently in many countries in Asia (e.g. China, India, Pakistan), South America (e.g. Brazil) and Africa with rapidly emerging industrialisation and hence increasing emissions of precursors of O_3. In these countries, O_3 has reached levels in ambient air which are of concern with respect to vegetation damage and human health effects (Emberson et al. 2003; Royal Society 2008).

The pattern of O_3 exposure of vegetation is changing on a global scale. While in large parts of Western Europe, North America and Japan, a trend of decreasing frequencies of O_3 peak values ("photosmog episodes") can be observed due to control measures on the emission of the precursor compounds, background O_3 values are increasing (Oltmans et al. 2006; Jonson et al. 2006). In the northern hemisphere at mid-latitudes, mean values at background sites have been increasing by 0.5–2 % per year (Derwent 2008). Future changes of the O_3 levels will be determined by the trends of the emissions of the precursors and of temperature and solar radiation. Ongoing global population growth coupled to increasing demands for resources such as land, fossil fuels and polluting activities like energy production, agriculture and transport will lead to enhanced production of natural and

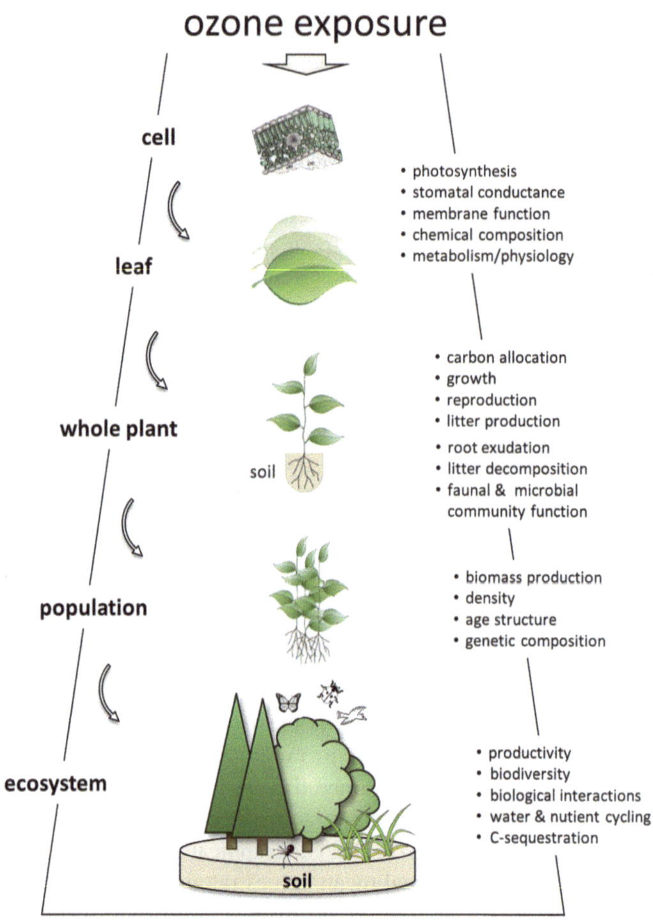

Fig. 1 General scheme of major endpoints that are affected by O_3 exposure at different levels of biological organisation in plants and ecosystems

anthropogenic O_3 precursors like VOC's, NO_x, CH_4 and CO. Higher surface temperatures along with climate change may also trigger the formation of surface O_3 levels. While predictive models, e.g. based on IPCC-SRES global emission scenarios, indicate that background O_3 concentrations will continue to increase at a rate of 0.5–2 % per year in the Northern Hemisphere during the next 100 years and will be in the range of ca. 42–84 ppb by 2100 (Prather et al. 2003; Vingarzan 2004; Jacob and Winner 2009; Fig. 2), recent models predict more moderate increases of O_3 levels until 2050 (Wild et al. 2012). These changes of the global O_3 exposure will be accompanied by other predicted changes in atmospheric chemistry (e.g. increasing atmospheric CO_2 concentrations) and climate which again may modify O_3 effects in the future.

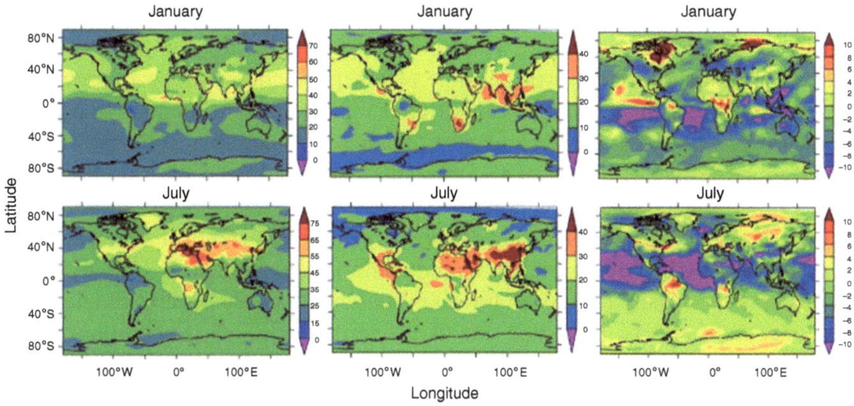

Fig. 2 Modelled surface O_3 (ppb) in January and July from the present-day simulation (*left*) and changes in surface O_3 (ppb) between 2000 and 2100 due to anthropogenic emission changes (*right*). Reprinted from Zeng et al. (2008)

Ozone concentrations influenced by human activities vary significantly with time (diurnally, seasonally, inter-annually) and with geographic location. This variability is of particular relevance for the effects on vegetation, as different vegetation types or developmental stages of plants, respectively, may be exposed to very different levels of O_3 during the course of the year. As O_3 formation is dependent on sunlight and as some of the chemical reactions involved in the O_3 formation in the troposphere are temperature-dependent, O_3 concentrations are particularly high at warm sunny days (Royal Society 2008). This link to the weather conditions also contributes to the inter-annual variation of O_3 concentrations. At least in large parts of Europe, peak O_3 concentrations occur especially in spring and summer. While at low elevation sites O_3 concentrations show diurnal cycles with low concentrations during the night and in the morning and high and peak concentration during the afternoon, high elevation sites mostly do not show such distinct diurnal variation (Stockwell et al. 1997). In general, at a particular location the build-up of phytotoxic O_3 concentrations depends on the local meteorology, the topography and the regional sources of O_3 precursors.

3 Methods to Study Ozone Impacts

The interpretation of plant responses to O_3 to a large extend depends on the methodology that is used to study its impacts. Therefore, a brief description of the most prominent methods in O_3 effects research is given here. Methods to investigate O_3 effects on vegetation can broadly be categorised into methods involving experimental exposure to O_3 and methods where plants are exposed to O_3 in ambient air. Each method has its particular advantages or disadvantages and its usefulness depends on the questions to be addressed and the objectives and budgetary circumstances of the respective study (Manning and Krupa 1992). Experimental techniques to expose single plants, plant communities and segments of ecosystems to modified O_3 concentrations range from controlled environmental chambers, greenhouses, field chambers to open-air O_3 exposure systems. Most of the information of the effects of O_3 on plants is derived from the use of various types of indoor and outdoor chambers.

Laboratory fumigation chambers of various designs (e.g. Heck et al. 1978; Payer et al. 1993) which provide highly reproducible environmental and O_3 exposure conditions have widely been used for assessing visible injury or physiological and biochemical O_3 effects. However, due to different microclimatic conditions in the chambers compared to open air ("chamber effects"), plants often show morphological or physiological differences compared to field-grown plants. Moreover, laboratory chambers are limited in space and mostly only useful for small scale pot or mesocosm studies. To overcome some of these limitations, open-top chambers (OTCs) were developed (Heagle et al. 1973) and have been the most widely used O_3 exposure system up to now (e.g. Heagle et al. 1988; Jäger et al. 1999; Zheng et al. 2013; Oksanen et al. 2013; Burkart et al. 2013). In OTCs, plants can be grown in their natural soil environment, in pots or as artificial model communities (mesocosms). Air either enriched with O_3 or filtered to remove O_3 from ambient air is introduced into the chamber with a blower system. Open-top chambers are best suited for in situ studies with low stature vegetation, e.g. like most crop or grassland species. As with the laboratory fumigation chambers for the interpretation of O_3 effects, the chamber microclimate may interfere with O_3 effects. For example, when used in species-rich systems like in certain grasslands, the differences between the ambient and OTC climate can lead to changes in vegetation structure in the chambers in comparison to chamberless ambient air plots (Grünhage and Jäger 2003).

To allow studies with taller trees, large versions of OTCs have been constructed (Musselman and Hale 1997). According to Kolb and Matyssek (2001) chamber studies with trees cover only a short period of the entire life history of a forest stand; they are thus limited in predicting longer-term ecosystem effects of O_3.

To overcome the various types of "chamber effects" in any kind of enclosure system used for O_3 effect studies and to overcome space limitations and restricted plant root volumes, respectively, chamberless O_3 exposure facilities have been developed (McLeod 1995). The most often used chamberless exposure system for O_3 effect studies is a modification of the circular free air carbon dioxide enrichment (FACE) system (Hendrey et al. 1999; Miglietta et al. 2001), which was modified to dispense O_3 into plant canopies. During the last two decades large-scale FACE-type O_3 exposure systems have been employed in O_3 effects studies with crops like soybean (Morgan et al. 2004), rice and wheat (Tang et al. 2011), and young tree species (Karnosky et al. 1999). A similar custom-designed circular free air O_3 exposure system was used by Volk et al. (2003) in a Swiss grassland system. A free air O_3 fumigation system in mature tree crowns of beech and spruce in Germany was developed by Werner and Fabian (2002) and tested and used by

Matyssek et al. (2010a, b, 2013). A similar system has recently been established in northern Japan addressing potential impacts of O_3 on deciduous oak and white birch (Watanabe et al. 2013). FACE-type O_3 exposure systems require sophisticated infrastructures and can be only used to increase O_3 levels in ambient air. In free air O_3 exposure systems, the coupling between the atmosphere and the plant canopy as well as between the canopy and the respective soil volume largely remains unchanged. Thus, in situ water and nutrient fluxes at the ecosystem level can be investigated. Ideally, FACE type O_3 exposure systems allow O_3 effect research at various hierarchical levels, for example, to link molecular biology with ecophysiological research.

Among the methods of O_3 exposure where there is no manipulation of the O_3 concentration surrounding the plants are ambient gradient studies. Plants or plant community responses are examined along gradients of O_3 concentrations across a landscape or regional transect providing multiple levels of exposure to O_3 that are naturally occurring. Forest tree species and ecosystem responses to O_3 in the USA have been assessed using gradient studies (Winner et al. 1989; McLaughlin et al. 2007a, b). Examples of other methods to assess O_3 effects on plants are the use of protecting chemicals against O_3 stress (Manning et al. 2011), biomonitoring techniques using indicator plants (Manning et al. 2002) and the use of plant growth models (Martin et al. 2001; Hogsett et al. 2008).

4 Ozone Impacts at the Single Plant Level

Ozone impacts on ecosystems result from excessive uptake of O_3 by plants, direct toxicity and cell damage, or from indirect effects mediated by the individual organisms. Although extrapolation of results from studies with individual plants or plant assemblages to the ecosystem level is difficult to make, the knowledge of the processes and mechanisms by which plants are affected by O_3 is an essential prerequisite to understand ecosystem responses to O_3. In this section, we briefly summarise the current understanding of the major mechanisms of O_3 effects at the individual plant scale governing vegetation response to O_3 including uptake, altered physiology such as stomatal functioning, photosynthesis, carbon allocation, reduced growth (above- and below-ground) and reproduction. These individual plant-scale responses to O_3 in the short term may lead to long-term changes in species and genetic composition, changes in water economy and functioning of species communities and, hence, ecosystem structure and function (Sect. 5).

4.1 Deposition of O_3 and Plant Uptake

Vegetation is an important natural sink for O_3. Ozone is transferred from the atmosphere onto plant canopies by turbulent diffusion (deposition), which is

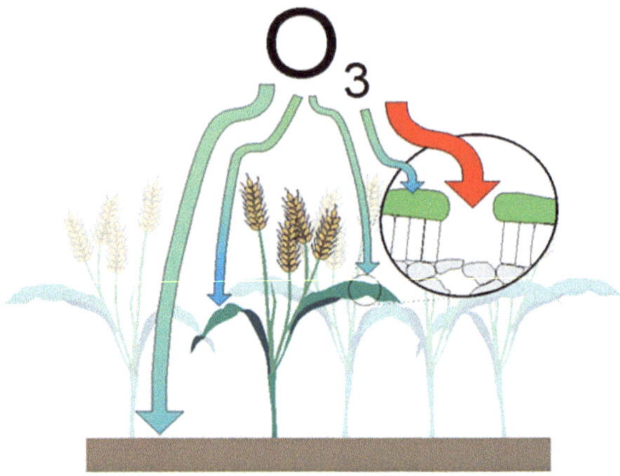

Fig. 3 Major pathways of the transfer of O_3 to terrestrial surfaces. The uptake by vegetation is attributed to both stomatal and non-stomatal (external plant surfaces and soil) deposition. The thickness of the *arrows* denotes the relative importance of the respective pathway. Reprinted with permission from Dr. Lisa Emberson, SEI York, UK

governed by micro-meteorological conditions (radiation, temperature, wind, etc.) and the roughness of the vegetation. The uptake of O_3 by vegetation is attributed to both non-stomatal and stomatal deposition (Fig. 3). Non-stomatal deposition includes deposition to soil, stems, cuticles and other external surfaces. It has long been known (Kerstiens and Lendzian 1989; Massman and Grantz 1995) that penetration of O_3 through the plant cuticle is of minor importance in comparison to the route of uptake through the stomata. This transfer of the gas through the atmosphere (turbulent diffusion) into the plant via molecular diffusion through the stomata is currently considered the key process in relating O_3 exposure to plant responses (Fowler et al. 2009; Fig. 3). Consequently, all environmental factors that modify the stomatal aperture (e.g. temperature, light and soil water conditions, other pollutants, atmospheric CO_2 concentration) and which thus affect leaf gas exchange have an influence on the uptake of O_3 into the plant interior (Fiscus et al. 2005; Fuhrer 2009).

Field measurements of O_3 deposition (flux) in various ecosystems indicate that total dry deposition is largely dominated by stomatal uptake during the most active parts of the growing season, but, at other times of the year and depending on vegetation type and weather conditions, non-stomatal deposition can be larger than stomatal uptake (Cieslik 2004; Cape et al. 2009). For example, continuous multi-year O_3 flux measurements over moorland vegetation in Scotland indicated that over a seasonal scale non-stomatal deposition dominated the overall O_3 flux and represented 70 % of the total O_3 deposition (Fowler et al. 2001). Nunn et al. (2010) compared sap flow measurements—i.e. tree level transpiration—and eddy co-variance approaches—i.e. stand level transpiration—in mixed beech/ spruce stands and found that stomatal O_3 flux amounted to 33 % of the total stand O_3 flux. The concentration of O_3 diminishes as one moves down a canopy to the soil. In addition, gas-phase reactions between O_3 and biogenic volatile organic compounds (BVOCs) and nitric oxide (NO) emitted from the ecosystem contribute to the removal of O_3 from the atmosphere (Fares et al. 2010). Canopy architecture and the density of the foliage may also determine to which O_3 concentration individual leaves are finally exposed. For example, in comparison to dense plant canopies, in open canopies like in widely spaced crop rows leaves inside the canopy will be exposed to similar O_3 concentration as at the canopy surface.

4.2 *Effects on Stomatal Functioning*

There has been long-term and widespread evidence that elevated O_3 levels alter stomatal performance and hence stomatal conductance (g_s) of various plant species (Darrall 1989; Mansfield 1998; Robinson et al. 1998). However, the type of stomatal responses to O_3 exposure is still not fully clear. While high O_3 concentrations appear to reduce g_s, there are variable results under more moderate O_3 levels, i.e. those resembling current ambient conditions (Paoletti and Grulke 2005; Wittig et al. 2007). For example, under chronic exposure conditions O_3 may lead either to enhanced stomatal aperture and a delayed stomatal closure during the night or to a reduction of the stomatal conductance. Moreover, exposure to O_3 has been shown to aggravate the heterogeneous stomatal aperture across the leaf surface, which is known as patchiness (Beyschlag and Eckstein 1998). This increase of stomatal patchiness upon O_3 exposure implies that only the integrated response of groups of stomata will determine the response to O_3 at the larger scales of the total plant and the canopy, respectively (Paoletti and Grulke 2005).

Stomata closure and reduced g_s due to O_3 exposure are often found when measured under steady-state high light conditions. In a recent meta-analysis (Wittig et al. 2007), which compiled results of 73 primary research articles of O_3 effects on photosynthesis and g_s of various tree species, the authors found that the O_3 levels in the atmosphere today suppresses g_s by, on average, 13 % compared to pre-industrial O_3 levels. When ambient background versus elevated O_3 was compared, g_s decreased by 6 % in the elevated O_3 treatment. Evidence for an O_3-induced stomatal closure is also available from recent studies with crops under chamberless O_3 exposure (Morgan et al. 2003; Kitao et al. 2009). Current assumptions of the possible mechanisms that may explain O_3-induced stomatal closure include (1) reduced photosynthesis and increased substomatal CO_2 concentration, (2) direct impact on guard cells, (3) altered calcium homeostasis or (4) altered hormone production (McAinsh et al. 2002; Wittig et al. 2007; Wilkinson and Davies 2010). Overall, this kind of stomatal responses implies that plants are protected from water loss.

However, often the measurements of g_s were carried out under conditions that normally do not prevail in the field in the course of a day, i.e. high light and steady-state vapour pressure deficit. Consequently, when measurements are compared that were carried out under more variable environmental conditions, stomatal aperture was not uniformly decreased by O_3 during the day. Such a "sluggish" stomatal response has long been known (Keller and Häsler 1984) and describes the delay in stomatal response to changing environmental factors relative to controls (Paoletti and Grulke 2010). Sluggish stomatal responses have been observed in O_3 effect studies, e.g. with tree (Wallin and Skarby 1992; Matyssek et al. 1995; Grulke et al. 2007) or grass species (Mills et al. 2009). Thus, if stomata fail to close under low light or water-stressed conditions, water loss may be greater over time. In other situations, it is possible that sluggish stomata may fail to completely open in response to environmental stimuli and result in decreased water loss.

As O_3 may affect plant stomata in relation to the response to other environmental variables like vapour pressure deficit, drought and light (Uddling et al. 2009; McAinsh et al. 2002; Paoletti and Grulke 2010), the resulting stomatal sluggishness may result in increasing g_s and hence increased water use under conditions which normally induce stomatal closure (e.g. drought, high vapour pressure deficit and low light). Such an increase of g_s even under reduced water supply in response to O_3 exposure was observed in several recent studies with grassland species (Mills et al. 2009; Wilkinson and Davies 2010; Hayes et al. 2012). It was suggested that O_3 can prevent stomatal closure under drought by altering the sensitivity of stomata to abscisic acid, a plant hormone stimulating stomatal closure under drought conditions (Wilkinson and Davies 2010; Wilkinson et al. 2012).

There are also recent research results, where O_3 apparently has no effects on stomata at all. In a multi-year FACE-type exposure of soybean to O_3 concentrations predicted by approx. 2050, there were no significant effects on midday g_s, and no effects on instantaneous g_s on 13 of the 15 measurement days (Bernacchi et al. 2006).

Ozone-induced physiological changes, such as reduced leaf area index and accelerated leaf senescence, have also been suggested to have an effect on water-use efficiency of plants. For example, some previous and more recent chamber and field studies have shown that that O_3 exposure is correlated with lower foliar retention (e.g. Karnosky et al. 1996, 2003; Topa et al. 2001).

Overall, there is still a high variability in the results of stomata responses to an O_3 exposure (Table 1). Thus, a better understanding of O_3 effects on leaf stomatal

Table 1 Summary of studies investigating O_3 effects on stomatal functioning in trees, crops and grassland species [after Mills et al. (2013)]

	Total number	No effect	Sluggish control	Increased opening	Stomatal closing
Crops (no. of species)	16	1	2	1	12
Crops (no. of experiments)	22	2	2	1	17
Trees (no. of species)	44	12	4	13	15
Trees (no. of experiments)	60	12	10	17	21
Grasslands (no. of species)	8	2	1	2	3
Grasslands (no. of expts.)	11	2	1	5	3
Total (no. of species)	68	15	7	16	30
Total (no. of experiments)	93	16	13	23	41
Ozone range (ppb) (25th to 75th percentile)		35–80	70–120	50–90	59–100
Mean ozone concentration		59 ppb	91 ppb	67 ppb	89 ppb

functioning remains a challenge, as this type of O_3 impacts on plants may have wider implications for the overall hydrology at the ecosystem level (see Sect. 5.1).

4.3 Physiological Effects

Once O_3 molecules have passed the stomatal pore, its subsequent effects on the plant include reactions with the apoplastic fluid and generation of reactive oxygen species (ROS), effects on the cell membrane structure and function, changes of cell metabolism and cellular events, which finally result in the generation of observable plant responses like chlorotic or necrotic tissue damage, reduced photosynthesis, temporal shifts in the plant's development and losses in productivity (Cho et al. 2011; Dizengremel et al. 2013; Fig. 1).

In the substomatal cavity O_3 rapidly reacts with water which results in the generation of ROS like hydrogen peroxide, singlet oxygen and hydroxyl radicals and with various compounds in the adjacent cell walls or on their outer cell membranes (Iriti and Faoro 2008). Based on this rapid chemical turnover, it has long been assumed that the O_3 concentration in the substomatal cavity is close to zero (Laisk et al. 1989); however, there is no unequivocal evidence for this assumption. According to the present understanding the reaction products of O_3 with the apoplastic fluid and with various biomolecules, respectively, are assumed to interfere with a signalling pathway of the plant cell which is related to cell death and which is triggered by ROS (e.g. Baier et al. 2005; Kangasjarvi et al. 2005; Cho et al. 2011). Apoplastic antioxidants (e.g. ascorbic acid, glutathione), the role of which is to protect cell membranes from a ROS attack, can interfere with O_3 or its reaction products. For example, reaction with the apoplastic acorbate pool seems to be a particular important process for ROS detoxification and is believed to be the first line of defence against O_3 injury, although other defence compounds may also be involved (Fiscus et al. 2005; Fuhrer 2009). Detoxification occurs from both existing antioxidants and those stimulated by O_3 itself. Defence reactions require energy for regeneration of antioxidants, i.e. particularly at prolonged O_3 exposure detoxification capacity may decline due to decreased rates of carbon assimilation and limited available energy (Wieser and Matyssek 2007). In general, cell injury or death of plant tissues occurs when the O_3 uptake exceeds the detoxification capacity. ROS that remain unscavenged can cause a variety of leaf injury symptoms such as necrotic stippling, bronzing, chlorosis or premature senescence.

Visible injury resulting from ambient O_3 pollution has been observed on a wide range of plant species including trees, crops and species of semi-natural vegetation in North-America and in Europe (Flagler 1998; Innes et al. 2001; Mills et al. 2010) and is usually classified as acute or chronic. While acute injury involves the death of the cells and develops within a few hours or days following exposure to high pollutant levels, chronic injury typically develops more slowly within days or weeks following O_3 exposure. While on broad-leaved plants visible injuries include stippling, flecking, surface bleaching, bifacial necrosis, pigmentation (e.g. bronzing) and chlorosis, for conifers visible injury includes chlorotic banding, tip-burn, flecking and chlorotic mottling (Flagler 1998). For both plant types O_3-induced symptoms of premature senescence of leaves and needles, respectively, can be observed. These foliar lesions can vary between and within taxonomic groups and the degree and extent of visible foliar injury development may vary from year to year and site to site. The extent of O_3-induced visible foliar injury is often related to the amount of soil moisture available to the respective plants during the year in which the visible foliar injury is being assessed. As drought conditions generally decreases stomatal conductance and limit the amount of O_3 entering the plant leaf, the result can be less injury. Several studies have shown that dry periods in local areas tend to decrease the incidence and severity of O_3-induced visible foliar injury (Matyssek et al. 2006; Grulke et al. 2003). Therefore, the incidence of visible foliar injury is not always higher in years and areas with higher O_3.

Ozone-induced visible injury is of particular significance when the quality and the marketable value of a plant depend on the appearance of the foliage. Such O_3 damage has been observed on a number of horticultural crops in several countries (Fumagalli et al. 2001; Kostka-Rick et al. 2002; Sheu and Liu 2003).

At chronic O_3 exposure, visible injury is often not observed, but decreased rates of CO_2 assimilation indicate adverse O_3 effects on plant vitality. The response of photosynthesis to O_3 has received much attention in order to explain O_3-induced losses of plant productivity in a wide variety of tree and crop species as well as in grassland and other native plant species (Reich 1987; Morgan et al. 2003; Fiscus et al. 2005; Wittig et al. 2007; Booker et al. 2009; Fuhrer 2009). It may be assumed that plant growth retardation under longer-term O_3 exposure at moderately enhanced concentrations is mostly the result of reduced rates of assimilation at the leaf level, although within-tree alterations of carbon allocation due to disturbed phytohormonal regulation have also been shown to affect tree growth (Winwood et al. 2007; Kitao et al. 2012). For example, a recent analysis of results from different experimental O_3 fumigation studies with tree species covering ambient or near-ambient O_3 concentrations revealed that O_3 levels of approximately 40 ppb can suppress net assimilation rate on average by 11 % compared with pre-industrial (10 ppb) O_3 exposure conditions (Wittig et al. 2007). Summarising 53 O_3 exposure studies with soybean in a meta-analytic approach Morgan et al. (2003) found a ca. 20 % reduction in net assimilation rate due to an average O_3 exposure of 70 ppb.

The impairment of photosynthesis by O_3 has been attributed to effects on the electron transport system (light reactions), a decline in the efficiency of carboxylation (dark reactions), and indirect effects on stomata, i.e. lower rates of diffusion of CO_2 into the leaf interior (Paoletti and Grulke 2005).

Recent studies with plants of natural ecosystems (Scebba et al. 2006), tree species (maple: Calatayud et al. 2007) and crops (tomato: Degl'Innocenti et al. 2007; soybean: Singh et al. 2009), particularly using chlorophyll fluorescence as a measurement tool, have shown that O_3 can alter photosynthetic processes at the level of the electron transport. The effects were connected with decreases in leaf chlorophyll content, reduction in the efficiency of excitation capture, reduced numbers of intact or open photosystem II reaction centres or increases in dissipation of energy through heat.

However, loss of carbon assimilation capacity is mostly considered as the primary reason for a decline of photosynthesis under O_3 exposure. This reduction in carbon assimilation is primarily the result of an O_3-induced decline in the amount and activity of Rubisco (Long and Naidu 2002; Matyssek and Sandermann 2003; Singh et al. 2009). Both, a decrease in Rubisco synthesis and an enhanced degradation of the protein contribute to the measured reduction in its quantity. This impairment of Rubisco is accompanied by a loss of the mRNA coding for the small (rbcS) and large (rbcL) subunits of the enzyme. For example, declines in rbcS mRNA were measured in beech saplings in a free air O_3 exposure system (Olbrich et al. 2009). Effects of O_3 on proteins involved in carbon assimilation have also been observed, as reductions in quantities of the small and large subunit (rbcL) of Rubisco and Rubisco activase were measured in soybean plants exposed to 120 ppb O_3 for 3 days (Ahsan et al. 2010). Similar results were observed with wheat (approx. 70 ppb O_3 for 50 days for 5 h day^{-1}) (Sarkar and Agrawal 2010).

Overall recent research results confirm earlier studies that decreased photosynthesis is commonly observed in plants grown under elevated O_3 concentration. Although many different changes are observed in the photosynthetic apparatus, decreased activity and amount of Rubisco appear to be the prevailing causes of loss of photosynthetic capacity.

4.4 Effects on Growth and Reproduction

Reduced photosynthesis due to an O_3 exposure may finally result in decreased growth rates and reduced overall plant productivity. Along with these effects impaired translocation of assimilates from source (e.g. leaves) to sink (e.g. roots; seeds) organs and early senescence likely contribute to O_3 effects on plant growth and reproduction. In particular, O_3 exposure has been shown to reduce the allocation of carbohydrate from shoots to roots and thus lower the root/shoot biomass ratio (fraction of total biomass in root tissue) (Cooley and Manning 1987), an effect that may have wider implications for below ground processes in the plant's environment (see Sect. 5.2) Several recent meta-analyses have summarised the available information across various types of plants (trees, crops, grassland and native species) and O_3 exposure conditions (Morgan et al. 2003; Grantz et al. 2006; Wittig et al. 2009; Wang and Taub 2010). For example, according to the literature compilation of Grantz et al. (2006), who used a root:shoot allometric coefficient k, which is the relative ratio growth rate of the root and shoot, to describe O_3 effects, O_3 reduced k on average by 5.6 %. Out of 125 observations of changes in k, 55 % yielded a decrease in k. However, about a third of all observations yielded an increase in k. Wang and Taub (2010) examined root mass fraction (i.e. the fraction of root to total biomass) of mostly herbaceous and a few woody species and found that, on average, O_3 reduced biomass allocation by 8.5 %. Wittig et al. (2009)

focused on tree species only and concluded from their analysis that the root-to-shoot ratio indicated a greater sensitivity to O_3 than shoot production. Their results also pointed to a greater sensitivity of angiosperm species in comparison to gymnosperms, which may have wider implications for community persistence (see Sect. 5.3). On the other hand, there are also reports of positive O_3 effects on root biomass production and root-to-shoot ratios especially in tree species (Pregitzer et al. 2008; Matyssek et al. 2010a, b). Overall, although the results of the majority of investigations point to a reduction of carbon allocation to roots, there is still uncertainty as such an effect depends on the respective O_3 concentration, the duration of exposure, the plant species or genotype, respectively, and on modifying factors of other growth variables. For example, already Maurer and Matyssek (1997) pointed out that the outcome of an O_3 effect on the root/shoot ratio of birch trees may be co-determined by the nutritional status of the respective plants.

There has been clear evidence over the last years that exposure to O_3 decreases growth in numerous plants representing important species of agro- and forest ecosystems as well as of different natural ecosystems. Wittig et al. (2009), for example, analysed growth responses of forest species from 263 studies over the past 40 years and found that elevated O_3 concentrations (97 ppb) decreased annual total biomass growth by 11–17 %. The decreased effect at current ambient concentrations as reported in these studies was 7 %. Detrimental O_3 effects on growth and yield of the major global food crop species, such as wheat, rice, soybean and cotton, have repeatedly been described (e.g. Heagle 1989; Morgan et al. 2003; Fiscus et al. 2005; Ashmore 2005; Ainsworth 2008; Booker et al. 2009). Mills et al. (2007) analysed O_3 exposure-response data for 19 agricultural and horticultural crops, respectively, and identified wheat, water melon, pulses, cotton, turnip, tomato, onion, soybean and lettuce as the most O_3-sensitive crops, while, for instance, barley was classified as O_3 resistant. Morgan et al. (2003) calculated an average yield loss of soybean of 24 % compared to charcoal-filtered air with small losses (approx. 8 %) at low (30–60 ppb) and high losses (approx. 35 %) at high (80–120 ppb) O_3 concentrations when they compiled 53 studies which included chamber and open-air O_3 exposure studies. Feng et al. (2008) calculated yield losses of winter and spring wheat ranging from 20 % at 42 ppb (7-h daily average) to 60 % at 153 ppb. According to a review of rice studies by Ainsworth (2008), average rice yields declined by 14 % when exposure to O_3 at a concentration of 62 ppb was compared to charcoal-filtered air. Despite this evidence of negative O_3 effects, one of the most common observations in studies with crop species is that there is considerable genotypic variability in O_3 sensitivity, suggesting that there is potential to breed for O_3 tolerance (Ainsworth et al. 2008; Booker et al. 2009). Production of biomass in grassland or pasture plants can also be negatively influenced by O_3 (Fuhrer 1997, 2009), but for species grown in mixtures other endpoints than biomass growth such as seed output or species composition may be important (see Sect. 5.3).

Studies conducted during the last three decades have repeatedly demonstrated that various stages of reproductive development are clearly sensitive to O_3. A recent meta-analysis of O_3 effects on reproductive growth and development of various plant species indicated that current ambient O_3 concentrations significantly reduced seed number, fruit number and fruit weight, while there was a trend towards increasing flower number and flower weight at elevated O_3 (Leisner and Ainsworth 2012). Negative effects on the reproductive performance in response to O_3 may result from a reduction in plant growth, a decreased reproductive allocation or from direct effects on reproductive structures (Black et al. 2000). Bender et al. (2006a) observed contrasting effects on resource allocation to the vegetative and reproductive organs of 17 herbaceous species that were exposed to different O_3 regimes from the seedling stage to the flowering stage. Although O_3 caused comparable reductions in both vegetative and reproductive growth in the majority of the investigated species, three species (*Chenopodium album*, *Matricaria discoidea*, *Stellaria media*) showed a greater vegetative growth and reduced reproductive allocation. Germinability of the seeds was affected by O_3 such that germination rate was up to 30 % lower in O_3-treated plants compared to control plants (Bender et al. 2006a). Similarly, Darbah et al. (2008) investigated the effects of elevated O_3 on reproductive fitness in paper birch (*Betula papyrifera*) under free air O_3 exposure. Elevated O_3 increased flowering, but decreased seed weight and germination rate. These results suggest that O_3 can significantly affect resource allocation patterns and reproductive fitness which may have significant implications for the establishment and survival of the progeny and hence for plant productivity and composition of plant communities under the influence of O_3 (see Sect. 5.3).

Any impact of O_3 exposure on the timing of flowering may also play an important role in reproductive success, particularly for species in which flowering is closely synchronised with pollinating species (Black et al. 2000; Hayes et al. 2012). However, the impact of O_3 on the timing of flowering varies markedly between species. Such O_3 effects have particularly been investigated in herbaceous species of grassland and ruderal ecosystems. For example, O_3 exposure has been reported to delay flowering in two species (*Campanula rotundifolia* and *Vicia cracca*) of simulated meadow community mesocosms (Rämö et al. 2007). In mesocosms representing "calcareous grassland", O_3 has been found to accelerate the timing of the maximum number of flowers in *Lotus corniculatus* (Hayes et al. 2012). By contrast, Bergmann et al. (1996) showed that the timing of flowering and seed set in 17 wild plant species were not significantly influenced by season-long exposure to 1.5 × ambient O_3 concentration in OTCs. However, O_3-induced changes in flowering timing could have large ecological impacts on plant pollination and the food supply of nectar feeding insects.

4.5 Abiotic and Biotic Factors Modifying O_3 Responses

There are complex interactions between O_3 effects on plants and other abiotic and biotic factors, as O_3 effects may be modified by these factors or O_3 itself may modify plant responses to these other factors. Important abiotic modifiers are temperature, humidity, light, water and nutrient availability, the occurrence of other air and soil pollutants and altered atmospheric chemistry (e.g. CO_2 concentration). On the other hand, biotic factors that interfere with O_3 effects are insect pests and other diseases or pathogens and root microorganisms, resulting in either detrimental biological effects or mutually beneficial relationships (e.g. root nodulation by Rhizobium; mycorrhizal infection). There is a very large body of previous information ranging from controlled environment to field experimentation that has investigated these interactions. Although these modifying factors may be of particular relevance for an assessment of O_3 cause–effect relationships at the ecosystem level, the vast scope of their possible interactions cannot be listed here in detail. Rather, we will provide a brief summary evaluation on important modifying factors of O_3 effects based on recent studies.

Light, temperature and air humidity are prominent abiotic factors that interfere with O_3 effects. Increased light intensity has been claimed to increase the sensitivity to O_3 of light-tolerant species while decreasing that of shade-tolerant species; this assumption has many exceptions (Topa et al. 2001). While previous studies revealed little modifying influence of temperature, some recent field studies have indicated that O_3 impact significantly increases with increased ambient temperature (Mills et al. 2000). On the other hand, there is no new evidence to contradict that O_3 enhances the sensitivity of plants to low temperature stress. It is also known that air humidity enhances the adverse effects of O_3 by affecting stomatal conductance and thereby increasing O_3 flux into the plant. For the current understanding of $O_3 \times$ drought interactions refer to Sect. 4.2.

It has also long been known that the nutritional status of plants can influence its response to O_3; however, the interaction of O_3 with specific nutrients is still contradictory. While some experiments point to higher sensitivity towards O_3 under low nutrient supply, other research results with trees suggest that O_3 and nutrient supply do not interact. With respect to forest and other nutrient poor ecosystems, the co-occurrence of nitrogen (N) deposition and O_3 impacts are of particular concern. Generally, existing information including several more recent studies with tree (Handley and Grulke 2008; Thomas et al. 2006; Watanabe et al. 2007) and pasture species (Bassin et al. 2007b; Volk et al. 2011; Wyness et al. 2011) shows that the interactive effects of N deposition and O_3 vary among species and ecosystems, i.e. there is no consistent information whether N deposition either enhances O_3 toxicity or increases tolerance of plants towards O_3 stress. There are hardly any recent studies on interactions of other air pollutants such as sulphur dioxide or nitrogen oxide with O_3.

Along with the increasing concern about climate change effects on ecosystems during the last decades, research into O_3 interactions particularly with elevated atmospheric CO_2 concentrations [eCO_2] has increased (Fuhrer 2003; Paoletti and Grulke 2005; Lindroth 2010). As [eCO_2] is known to stimulate photosynthesis, to decrease stomatal conductance and mostly to enhance plant growth, while O_3 has negative impacts on photosynthesis and plant growth, interactions between the two gases can be expected. Consistent across different vegetation types and derived from various experimental approaches, there is evidence that [eCO_2] has the potential to mitigate negative effects of O_3, mainly due to a CO_2-induced reduction in stomatal conductance, which reduces O_3 uptake. On the other hand, negative O_3 effects limit positive responses to [eCO_2] in many plants as well (Fiscus et al. 2005). While the $CO_2 \times O_3$ interaction is of little relevance for the current ambient conditions, it may be suggested that the continuing future increase of the CO_2 component of climate change may be ameliorative for the effects of O_3.

Among the interactions of O_3 with biotic factors, plant pathogens and insect pests have repeatedly been studied in various experimental approaches (Percy et al. 2003a, b; Eastburn et al. 2011). While it may be assumed that reduced plant vigour due to O_3 stress can make plants more susceptible to plant pathogens, general predictions of O_3 effects on particular plant–pathogen systems are difficult to make, because the available data for specific pests and diseases remain controversial. With respect to agricultural plants there is no fundamentally new recent information to replace the summary by Manning and von Tiedemann (1995), that increased susceptibility after O_3 exposure can be expected for necrotrophic pathogens, while obligate biotrophic infections tend to be diminished by O_3.

Insects can respond to O_3-induced changes in the plant chemical composition or insect performance is directly affected by O_3. Overall assessments and some more recent studies, respectively (Holopainen 2002; Valkama et al. 2007; Bidart-Bouzat and Imeh-Nathaniel 2008; Lindroth 2010), can be interpreted that an O_3 exposure may increase the likelihood and success of chewing insect attacks. Existing studies on interactions of O_3 with sucking insects such as aphids do not allow to deduce consistent trends. Among the aspects of the many other biotic factors that interact with O_3 or which modify plant O_3 responses, respectively, symbioses with mycorrhizae and plant–plant interactions by competition are partly addressed in Sects. 5.2 and 5.3.

5 O_3 Impacts at the Ecosystem Level

The recent emphasis of the importance of ecosystems for the functioning of the biosphere and their role in providing goods and services to humans has resulted in various attempts to assess the role of O_3 at the ecosystem scale (MEA 2005). However, ecosystem effects of O_3 are difficult to detect and to evaluate, as the different systems vary at spatial and temporal scales. With the exception of two recent multi-year field experiments in a rapidly growing tree plantation (Karnosky et al. 2001) and in an adult forest tree stand (Matyssek et al. 2013), the number of studies at the scale of forest-, agro- or grassland ecosystems especially under exposure to O_3 over several growing seasons is still very limited, which is even more true for other types of vegetation like fens, bogs, etc. The following section will consider selected examples where O_3 effects may be relevant in an ecosystem context (Fig. 4).

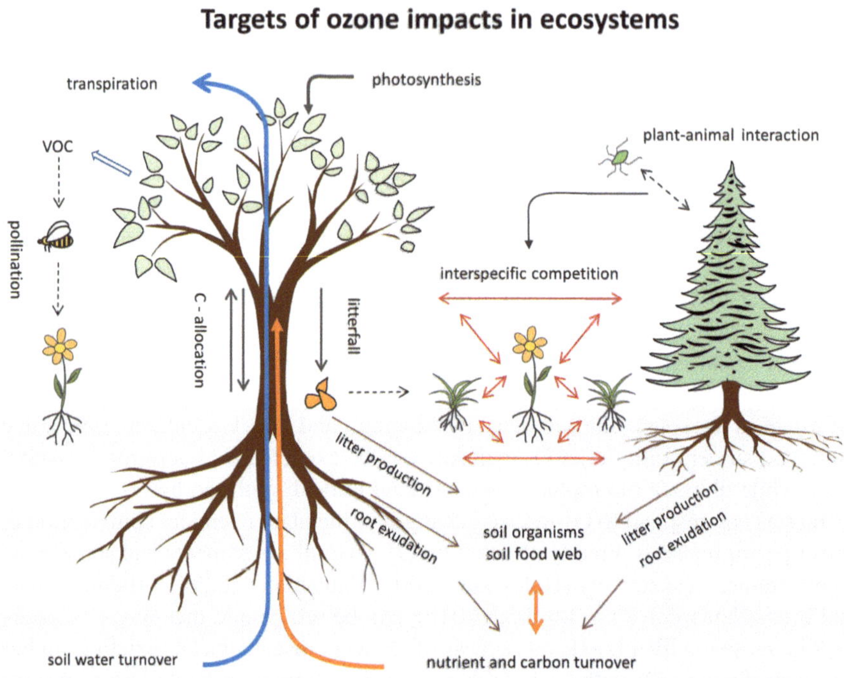

Fig. 4 Schematic representation of structural elements and processes in ecosystems that are potential direct and indirect targets of an O_3 exposure and which are partly described in the text

5.1 Effects on Water Relations and Hydrology

As shown in Sect. 4.1, the main route of entry of O_3 into the plant interior is via stomatal uptake at the leaf level. This role of the stomata has also been proofed at larger scales. For example, previous studies of Duyzer et al. (1995) in forest ecosystems have demonstrated that O_3 deposition is related to g_s at the stand level. However, as O_3 effects on g_s have been variable (Sect. 4.2), there remains uncertainty with respect to O_3 responses of water use at the canopy or stand level.

Investigations with tree species under open-air O_3 exposure conditions using sap flux measurements have shown that stand level water use per unit ground area of aspen clones was not significantly affected by elevated O_3 although the treatment decreased leaf area index by 22 % and basal area by 20 % (Uddling et al. 2008). Uddling et al. (2009) attributed this to an increased leaf area-specific hydraulic conductance. The O_3 effect was assumed to be caused by the sluggish stomatal response (Sect. 4.1), as under O_3 exposure the stomatal closure response to increasing vapour pressure deficit was less sensitive than under the control treatment. Moreover, mid-day leaf water potential was more negative under elevated O_3 compared to controls. The lack of an O_3 effect on stand-level water use may also be caused by a higher proportion of sun leaves in trees under elevated O_3 compared with control trees (Uddling et al. 2008). Also, in an experiment with aspen and birch, Rhea et al. (2010) found that O_3 changed the branch architectural parameters, which may alter tree crown interception of precipitation and thus affect evapotranspiration.

Field studies in a deciduous forest in eastern Tennessee provided some insight into the possible consequences of stomatal sluggishness at the leaf level for ecosystem water cycling (McLaughlin et al. 2007a, b). It was found that daily O_3 levels with maxima ranging from 69 to 82 ppb reduced stem growth by up to 50 % in 1 year with high O_3 levels. The authors suggested that peak hourly O_3 exposures resulted in an increase of the rate of water loss through transpiration as indicated by an increased stem sap flow. Due to the increased canopy water loss water uptake by the trees increased as reflected in reduced soil moisture in the rooting zone. At the landscape level this change in tree water use was assumed to lead to further impacts on the hydrological cycle and O_3 has been reported to contribute to variation in late-season streamflow by as much as 23 % in areas of highest exposure in forested watersheds in Tennessee (Sun et al. 2012). A loss in stomatal sensitivity associated with increased evapotranspiration and reduced streamflow can be expected to increase drought episodes and to have implications for flow-dependent aquatic biota (Sun et al. 2012). Recent studies with several crops and grassland species have also indicated an O_3-induced loss of stomatal sensitivity to drought, light and vapour pressure deficit (Wilkinson et al. 2012; see Sect. 4.2), but the implication of these observations on individual plant species for intact agro- and grassland ecosystems remains unclear.

Although there is no clear pattern of the impact of O_3 on stomatal functioning (see Sect. 4.2), recent ecosystem models that address the larger scale effects of O_3 on water turnover are often based on the assumption that O_3 induces a stomatal closure. For example, in order to assess the interactions of O_3, climate, elevated CO_2 and N limitation on the hydrological cycle in the eastern USA, Felzer et al. (2009) used the terrestrial ecosystem model TEM-Hydro. According to this model elevated CO_2 decreased evapotranspiration by 2–4 % and increased runoff by 3–7 %, as compared to the effects of climate alone. Including O_3 damage and N limitation into the calculations, evapotranspiration was reduced by an additional 4–7 % and runoff was increased by an additional 6–11 %. Hanson et al. (2005) using a stand-level simulation model found a modest 3 % reduction in water use when the O_3 concentration was raised by about 20 ppb above the prevailing ambient level.

5.2 *Below-Ground Effects*

Atmospheric ground level O_3 exposure does not directly affect structural and functional soil properties. However, above- and below-ground processes are interconnected via plant shoot and root communication mediated by the flow of carbon (see Sect. 4.4). Consequently, any O_3-induced alteration of the quantity and quality of carbon supply from either photosynthates or from litter fall to the soil or from an enhanced carbon allocation to above-ground processes due to an O_3-induced stimulation of the production of antioxidants and other chemical compounds for defence and repair processes may exert indirect O_3 effects on the soil system (Cooley and Manning 1987; Andersen 2003; Pregitzer and Talhelm 2013). Moreover, reduced biomass allocation to roots may lower the water availability to plants, which feeds back to stomatal conductance, canopy water flux, hydrology (Sect. 5.1) and nutrient cycling. Examples of O_3 effects on the soil system include changes in litter quality and consequences for various soil biota. Additional effects like alterations of soil carbon formation and of decomposer activities as well as of soil food web structures are not considered here.

Modification in the quality of litter (e.g. soluble sugars, tannins, phenolics, lignin, macro- and micronutrients) released from plants exposed to O_3 have repeatedly been observed. In their review Kasurinen et al. (2007a) concluded from existing information with boreal and temperate forest trees that O_3 effects on litter chemistry were mostly observed only at high O_3 concentrations. For example in an OTC study with birch (*Betula pendula*) clones O_3 slightly increased the content of leaf low molecular weight phenolic compounds, an effect which did not persist over the 3-year study. On the other hand, litter chemistry was not affected by short-term O_3 exposure in a study with beech *Fagus sylvatica* (Schloter et al. 2005). In free air type O_3 exposure experiments, it has been demonstrated that O_3-induced changes in litter quality of *Populus tremuloides* and *Betula papyrifera* communities led to reduced inputs of hemicellulose and lignin (Liu et al. 2005; Meehan et al. 2010) and thus caused a decrease in nutrient flux into soil (Liu et al. 2007). In contrast, Stoelken et al. (2010) detected additional nitrogen incorporation into the soil down to 30 cm resulting from an enhanced nitrogen mobilisation from leaf litter in an O_3 exposure experiment with *Fagus sylvatica* grown in lysimeters.

Ozone effects on soil biota have received more recent attention especially with respect to possible implications for soil biodiversity; however, there is little evidence from experimental studies in real ecosystems with their respective native soil properties. Effects of O_3 on the soil microbial community have been investigated both in the rhizosphere and in the bulk soil of trees, grassland and arable crop species.

In an OTC study with potted plants total bacterial biomass was decreased by O_3 in the soil of the N_2-fixing herbaceous legume *Lathyrus pratensis*, but not in the soil of the grass *Agrostis capillaris* (Manninen et al. 2010). In a multi-year mesocosm study with *Pinus ponderosa*, elevated O_3 tended to increase the ratio of fungal to bacterial biomass (Olszyk et al. 2001) and such an effect was also observed under similar O_3 exposure conditions for blue wildrye (*Elymus glaucus*, Yoshida et al. 2001). For the tree species *Fagus sylvatica*, a shift in the overall community structure of soil microorganisms based on phospholipid fatty acids (PLFA) analysis as a biomarker in response to O_3 (Pritsch et al. 2009) has been found to be associated with a reduction in the potential nutrient turnover (Schloter et al. 2005) and a higher abundance of plant-carbon utilising microbes (Esperschutz et al. 2009).

Aneja et al. (2007) characterised the diversity of microbial communities colonising control and O_3-exposed litter from *Fagus sylvatica/Picea abies* and provided evidence that changed litter quality due to elevated O_3 influenced the structure of litter-colonising microbial communities. In peat-land microcosms (*Eriophorum vaginatum*), Morsky et al. (2008) found an O_3-induced increase in microbial biomass only at the end of a 3-year exposure experiment, a result which supports the conclusion of Kasurinen et al. (2007a) that the onset of microbial responses due to an O_3 impact may take years.

Dohrmann and Tebbe (2005) studied the rhizosphere bacterial community composition of five low-managed grassland species using genetic profiling of PCR amplified 16S rRNA gene sequences based on single-strand conformation polymorphism (SSCP). They found that a 5-week exposure to elevated O_3 did not select for a different bacterial community composition. This was also true if other more O_3 susceptible herbaceous plant species with severe visible O_3 injury were studied (Dohrmann and Tebbe 2006). Also with a grassland system Kanerva et al. (2008) in a 3-year O_3 exposure study provided evidence that elevated O_3 is able to modify the structure of the microbial community in a meadow soil, as bacterial, actinobacterial and fungal PLFA biomass were decreased simultaneously. Again this study points to the fact that long-term observations are necessary to understand the effects of O_3 on the biology of soil processes in ecosystems.

Because of its implication for nutrient acquisition, *mycorrhization* is of high importance for ecosystem function. An ectomycorrhizal community responded to changes in environmental conditions with a change in its total amount of extramatrical mycelium, leading to changes in space occupation, and consequently, to alterations in its capacity to exploit soil resources (Agerer et al. 2012). Several recent studies described effects on mycorrhizal abundance when host trees were exposed to O_3. For example, for *Betula pendula* Kasurinen et al. (2005) found a stimulation of total mycorrhiza infection, Haikio et al. (2009) an increased mycorrhizal status for hybrid aspen (*Populus tremula* L. *x Populus*) and Pritsch et al. (2009) and Grebenc and Kraigher (2007a) a higher total number of mycorrhiza types under OTC and free air O_3 exposure conditions. For ectomycorrhizae collected underneath mature Norway spruce trees at the “Kranzberger Forst” free-air O_3 fumigation site, differences in carbon allocation to the mycorrhizal communities have been shown between the different O_3 treatments by means of differences in the enzyme activity profiles of the ectomycorrhizae communities (Agerer et al. 2012). Moreover, there is past and recent evidence from studies with tree species that O_3 impacts the microbial diversity also in terms of mycorrhizal species composition as shown for *Pinus taeda* (Edwards and Kelly 1992), *Betula pendula* (Kasurinen et al. 2005), *Fagus sylvatica*, (Haberer et al. 2007; Grebenc and Kraigher 2007b) and for an aspen and aspen-birch community in a free air O_3 exposure experiment (Edwards and Zak 2011). On the other hand, data about O_3 effects on the mycorrhization of grassland and other crop species are limited. The mycorrhizal colonisation of blue wildrye (*Elymus glaucus*) was reduced in response to O_3, and it has been demonstrated that this effect depends on the genotype of the grass tested (Yoshida et al. 2001).

Until now few studies have addressed possible implications of plant exposure to O_3 for detritivore invertebrate communities and particularly soil mesofauna composition which are important for ecosystem functioning. In a free air O_3 exposure experiment with temperate forest tree species (aspen and aspen-birch communities) the individual density of soil mites was reduced by nearly 50 % under elevated O_3 conditions, whereas the abundances of collembolans remained unchanged (Loranger et al. 2004). Feeding experiments with litter with altered quality due to previous O_3 exposure of trees were able to demonstrate that growth rates were reduced for the earthworm (Lumbricus terrestris) fed with birch litter (Kasurinen et al. 2007b) or for a collembolean species fed with aspen litter (Meehan et al. 2010). With regard to arable agroecosystems Schrader et al. (2009) observed a decrease in the individual density of enchytraeids, collembolans and soil mites in the rhizosphere of O_3-exposed wheat plants in OTCs and Chang et al. (2011) found a reduction in the abundance and diversity of collembolans associated with cotton plants exposed to O_3. Overall, these few selected examples clearly show that the above-ground impact of O_3 on plant performance may translate into significant secondary below-ground implications in the ecosystem.

5.3 Plant Competition and Community Composition

Driven by concerns about the global changes in biodiversity and the well-known fact that plant growth responses to O_3 vary significantly between species and genotypes, an arising question is whether exposure of vegetation to high levels of O_3 may alter the strength of competitive interactions between different plant species and whether this may lastly result in changes of plant community composition. The way by which elevated levels of O_3 will shape the composition includes a change in the cover or abundance of single plant species, which in turn will change the genetic structure of the community. The challenge to describe O_3 risks on plant communities is the understanding of how competitive interactions may modify growth responses of individual species to O_3 and, conversely, how the impact of O_3 may modify their competitive ability within a plant assemblage. Research on this issue comprises both, experiments with older, established ecosystems and artificially newly created plant communities with the majority of experiments designed to study two-species mixtures or model plant communities under laboratory and field conditions.

Artificial forest communities have been investigated under laboratory and field conditions. Phytotron studies to test O_3 effects on the competition between *Fagus sylvatica* and *Picea abies* revealed that the responses to O_3 strongly depended on the type of competition: although the response to O_3 of *P. abies* was not significantly affected by either intra- or interspecific competition, the competitive ability of this species was scarcely affected by O_3 as indicated by an enhanced above-ground growth of the competing *F. sylvatica* plants (Grams et al. 2002; Kozovits et al. 2005). Under conditions of interspecific competition, *P. abies* was found to be superior in nitrogen acquisition whereas *F. sylvatica* in turn appeared to be nitrogen-limited (Grams and Matyssek 2010; Kozovits et al. 2005). Effects on nutrient efficiency indicate that processes of stress defence due to O_3 exposure trigger a nutrient demand at the expense of above-ground competition (Rodenkirchen et al. 2009). Recently, Grams et al. (2012) demonstrated that the more intense the competition between *F. sylvatica* and *P. abies* is, the stronger the response to other stressors may be modified.

Under free air O_3 exposure conditions, stands of different clones of *Populus tremuloides* (aspen) or mixed stands with either *Betula papyrifera* or *Acer saccharum* were investigated during a 12-year experiment. The growth response of *P. tremuloides* depended on clone and competitive status (Kubiske et al. 2007; McDonald et al. 2002). After 7 years of exposure, O_3 slightly enhanced the rate of conversion of a *P. tremuloides* stand to a *B. pendula* stand (Kubiske et al. 2007), whereas the cumulative nitrogen-acquisition decreased in both species (Zak et al. 2007). When the experiment went on for 12 years, the rank order of nitrogen-acquisition among *P. tremuloides* genotypes was not shifted over time, indicating no change when juvenile trees mature (Zak et al. 2012).

In the above experiment an understory community established which consisted of more than plant 30 species dominated by perennial old field vegetation. Observed effects on total and individual species biomass, N content, and ^{15}N recovery of this understory vegetation could not be related directly to the O_3 treatments but rather reflected the effects on the structure of the overstory community, which is determined by the present tree species and their response to the treatments (Bandeff et al. 2006).

Ozone effects on established forest plant communities have already been assessed in some earlier studies. For example, Nygaard (1994) and Steubing et al. (1989) investigated the responses of understory species growing in an intact conifer or beech forest, respectively, to relatively high O_3 exposures and found high variability between species in O_3 sensitivity. Barbo et al. (1998) examined the response to sub-ambient and enhanced O_3 levels of an early successional plant community associated with *Pinus taeda*. In this study, O_3 exposures caused shifts in the competitive interactions between plants and the abundance of the five most common species was affected already in the first year. The authors concluded that total vegetative cover, vertical density of foliage as well as species richness, diversity and evenness may be at risk by the prevailing ambient O_3 exposures.

A group of beech and spruce trees within a 55-year-old Norway spruce stand (Kranzberger Forst, Germany) has been exposed for 8 years to enhanced O_3 levels using a newly developed free-air O_3 fumigation system (Matyssek et al. 2010a, b, 2013). One of the results showed that individuals of both tree species, spruce and beech, grew faster in mixture than in pure stands reflecting a facilitation of spruce and a reduction in competitiveness of beech. The results of this research project led to the conclusion that increasing levels of O_3 stress may change the pattern of carbon allocation in mixed stands of beech and spruce and the outcome of competition (Pretzsch and Schutze 2009). Actually, Pretzsch et al. (2010) demonstrated a shift in the resource allocation in mature trees caused by exposure to high O_3 levels.

In comparison to studies with forest plants, a much large number of more recent studies on O_3 effects on plant competition are available for grassland communities. This may be due to the fact that a generally accepted outcome of screening experiments with single plants or monocultures is that members of Fabaceae (legumes) with the genus *Trifolium* in particular belong to the most O_3-sensitive plant species, whereas members of the family Poaceae are much less responsive to O_3 (Fuhrer 1997). Experimental approaches to address O_3 effects on interspecific competition between herbaceous plant species have thus focused on grass/clover mixtures.

With respect to artificial grassland communities, more recent experiments with potted plants of artificial grass/clover mixtures confirm earlier results such that markedly negative growth response of the clover to O_3 exposure is observed, whereas the grass species were hardly impaired by O_3 (González-Fernández et al. 2008; Haldemann and Fuhrer 2005; Hayes et al. 2009, 2010a). If the total biomass of the species mixture was largely influenced by the more sensitive component, a decline in total yield of the mixture has been recorded under O_3 exposure (e.g. for *Trifolium repens*/*Lolium perenne*, Hayes et al. 2009; for *Trifolium pratense*/*Trisetum flavescens*, Nussbaum et al. 2000). On the other hand, a range of O_3 exposure experiments resulted in unchanged total yield quantities (e.g. for *Trifolium repens*/*Lolium perenne*, González-Fernández et al. 2008; or grass/alfalfa, Johnson et al. 1996). This result derives from the fact that an O_3-induced decline in the relative yield of clover entailed an increase in the relative yield of the grass component. In addition, it has been argued that altered root/shoot ratios (Haldemann and Fuhrer 2005) and reduced remobilisation of reserves after grazing (Nussbaum et al. 2000) due to the O_3 impact could facilitate the less sensitive species. Thus, O_3 may interact with cutting or grazing by reducing the capacity for regrowth from energy reserves (Ashmore and Ainsworth 1995). It should be mentioned that these recent findings from experiments using potted plants had already been observed with similar tendencies in older field experiments with natural grass/clover communities, where plants were exposed to O_3 under OTC conditions (Blum et al. 1983; Rebbeck et al. 1988; Heagle et al. 1989).

There is also recent evidence that the presence of an interspecific competitor (grass) may affect the response of the clover species to O_3. In two-species mixtures, adverse effects of O_3 on productivity may be enhanced by competition as shown for *Trifolium pratense* in competition with *Trisetum flavescens* (Haldemann and Fuhrer 2005) or mitigated by competition as shown for *Trifolium repens* in competition with *Lolium perenne* (González-Fernández et al. 2008).

Ozone effects on plant competition of two-species mixture have recently also been investigated without legumes. Using a phytometer approach, the response of early season O_3 stress on model communities of wet grassland species (Tonneijck et al. 2004) and ten different extensively managed grassland species (Bender et al. 2002, 2006b) was investigated over three seasons under OTC conditions with moderately enhanced O_3 levels. In the latter studies, for none of the ten species O_3 impacts on growth were detected when grown in monoculture. Target species differed significantly in their competitive ability against the phytometer (*Poa pratensis*) but the experiments did not provide evidence that interspecific competition altered the harmful effects of an early season O_3 stress on aboveground growth. Only for *Veronica chamaedrys*, O_3 was shown to affect its competitive ability against *Poa pratensis* negatively (Bender et al. 2002, 2003). A similar increase in the grass cover ratio under O_3 exposure was demonstrated for *Anthoxanthum odoratum* and *Dactylis glomerata*, respectively, when grown in competition with *Leontodon hispidus* (Hayes et al. 2011).

In previous studies with artificial pasture model plant communities composed of grasses, clover and weeds, the decline in the clover component, *T. repens* and/or *T. pratense* due to the O_3 impact, was associated with a slight increase in the yield of grasses (e.g. field-sown, Fuhrer et al. 1994 and pot-sown, Ashmore et al. 1996). Conversely, in a simulated community representing a typical multi-species UK upland grassland, the grass *Anthoxanthum odoratum* was most affected by an experimental O_3 exposure in terms of aboveground biomass reduction and this effect contributed to a decrease in total community biomass and grass:forb ratio (Hayes et al. 2010b). In Finland, in a study with meadow species in mesocosms, after only 2 years of moderate exposure to O_3, the early season coverage of plant communities was decreased (Rämö et al. 2007); however, the reductions in aboveground biomass were not reflected in changes in the dominance of different functional groups or in the total community root biomass (Rämö et al. 2006).

There are also reports that describe modifying effects of plant competition in grassland communities on the impact of O_3 on plant flowering and visible leaf injury. Among characteristic species of therophytic dehesa grasslands, flower production of *Trifolium cherleri*, *Trifolium subterraneum* and *Trifolium striatum* was suppressed by O_3 in competition with the grass species *Briza maxima*, but there was no interaction between competition and O_3 response (Gimeno et al. 2003). In a complex grassland model community, the timing of flowering and the number of flowers of *Lotus corniculatus* were accelerated by O_3, while a significant reduction in the numbers of flowers with increasing O_3 levels was found for *Campanula rotundifolia*, *Scabiosa columbaria* and *Vicia cracca* (Hayes et al. 2012; Rämö et al. 2007). Such changes in timing and number of flowers could have implications for pollination and the long-term outcome of the whole community development. Reduced proportions of injured leaves on O_3 exposed grasses due to the presence of a competitor have been described. For example, *Trisetum flavescens* responded more strongly in mixture with the low stature species *Centaurea jacea* than in mixture with *Trifolium pratense* (Nussbaum et al. 2000). Similarly, *Leontodon hispidus* exhibited a larger increase in O_3-induced senescence observed in the more open canopy of *A. odoratum* compared to the denser canopy of *D. glomerata* (Hayes et al. 2011). Modification of microclimate and canopy structure are thought to be potential mechanisms that influence the interaction between O_3 responses and competition (Haldemann and Fuhrer 2005; Hayes et al. 2010a).

Investigations on O_3 effects on older, established communities of managed and semi-natural grassland have been done in several earlier and more recent studies (Table 2) and will be discussed here in more detail. For example, Nebel and Fuhrer (1994) classified 31 species according to the appearance of visible injury when

Table 2 Effects of O_3 exposures on species composition in experiments with established grassland communities

Community	Main species	Exposure	Effect on species composition	References
Mesotrophic grassland	*Festuca rubra* *Bromus erectus* *Filipendula vulgaris* *Pimpinella saxifraga* *Arrhenatherum elatius* *Dactylis glomerata* *Holcus lanatus*	OTC	Trends of changes in species composition Shift towards a more calcareous grassland community	Ashmore et al. (1995)
Semi-natural chalk grassland	*Festuca rubra* *Campanula rotundifolia* *Galium verum* *Plantago lanceolata* *Festuca rubra* *Arrhenatherum elatius* *Bromus erecta* *Poa pratensis* *Dactylis glomerata*	OTC	Consistent decline in cover of *F. rubra* *C. rotundifolium* was lost from all ozone treatments Increase in frequency of G. verum Increase in frequency of P. lanceolata	Thwaites et al. (2006)
Low managed grassland Geo-Montani-Nardetum (Alp Flix)	*Festuca violacea Nardus stricta Carex sempervirens Ranunculus villarsii Leontodon helveticus Ligusticum mutellina Potentilla aurea*	Free air	No effects on the abundance of the most frequent species	Bassin et al. (2007b)
Semi-natural grassland Arrhenatheretum elatius-Festuca rubra subcommunity	*Agrostis capillaris* *Festuca rubra* *Poa pratensis* *Veronica chamaedrys* *Trifolium repens* *Plantago lanceolata* *Stellaria graminea*	OTC	No change in species richness Effect on the proportion of grass to forb cover values *T. repens* and *V. chamaedrys* increased in cover *A. capillaris* decreased in cover	Evans and Ashmore (1992)

(continued)

Table 2 (continued)

Community	Main species	Exposure	Effect on species composition	References
Arrhenatherion elatioris Low-to-medium productivity 30-year–old field (Le Mouret)	*Bromus hordeaceus* *Holcus lanatus* *Trisetum flavescens* *Alopecurus pratensis* *Arrhenatherum elatius* *Plantago lanceolata* *Ranunculus friesianus* *Trifolium pratense*	Free-air	Change in fraction of functional groups Legume fraction shows a negative response Negative effects of ozone on grass and legume fraction No response of the forb fraction	Volk et al. (2006) Stampfli and Fuhrer (2010)
Mesotrophic grassland	*Festuca rubra* *Holcus lanantus* *Anthoxanthum odoratum*	Free-air	Influence on the composition of the herb and legume group	Wedlich et al. (2012)

grown in soil blocks of intact semi-natural grassland vegetation and Evans and Ashmore (1992) showed that during a season with relative high O_3 levels total aboveground biomass of a semi-natural grassland community was decreased. More recently, an old, species-rich (53 species) pasture at a mid-elevation site in Switzerland was exposed for seven years to O_3 in a free air exposure system under real field conditions (Volk et al. 2003). For individual growth periods, no relationship between the O_3 exposure level and yield differences was observed. After 5 years a loss in annual dry matter yield of about 23 % was calculated for conditions of moderately elevated O_3 levels (1.5 × ambient air) showing a strong negative response of the yield of the fraction of legumes (Volk et al. 2006) but not of the frequency of legumes at the experimental plots (Stampfli and Fuhrer 2010). In a natural upland mesotrophic grassland in UK, Wedlich et al. (2012) revealed clear evidence for a cumulative effect of moderately elevated O_3 levels (free-air exposure) over time (3 years) on species biomass composition as there was a significant negative effect of O_3 exposure on herb biomass, but not on total grass or legume biomass suggesting that finally O_3 had become the dominant factor influencing species composition within the combined herb and legume component.

Particularly for semi-natural calcareous grassland, a shift in species composition has been recorded which was indicated either by a decline (e.g. of the dominant grass species *Festuca rubra* or *Campanula rotundifolia*) or an increase (*Galium verum* and *Plantago lanceolata*) in cover or frequency of species (Thwaites et al. 2006). Bassin et al. (2007b) concluded that in old, species-rich grassland communities, effects of elevated O_3 on the productivity and floristic composition seem to develop rather slowly, as evidenced from the lack of significant vegetation responses of the sub-alpine grassland community to the elevated O_3 treatment over 7 years (Bassin et al. 2013). With respect to species-specific traits Bassin et al. (2009) suggested that commonly used principles of functional growth analysis do not directly hold under the specific conditions of such plant communities. As a reason, an adaptation to oxidative stress of the alpine species was discussed to account for the low sensitivity in response to the chronic low-level O_3 exposure used in this experiment (Bassin et al. 2013).

In summary, the studies cited above indicate that current and future O_3 concentrations could affect natural and semi-natural grassland communities and point out that detrimental effects on species balance may occur. In clover:grass mixtures a shift in species composition is the predominant effect, favouring the tillering of the grass component, whereas the effect on the total forage yield seems to be determined by the susceptibility of the individual species. Experiments with newly established grassland communities indicate that nitrogen-poor meadows are potentially very sensitive towards an O_3 impact. Mesotrophic grassland communities are characterised by the occurrence of faster growing species which are known to be more susceptible to O_3 than the slower growing calcareous grassland species. The low susceptibility of old, species rich grassland communities to O_3 is linked to specific characteristics of these systems. Low productivity vegetation such as subalpine grassland is mainly composed of species with a stress-tolerant growth strategy, which have been considered relatively unresponsive to O_3 (Bassin et al. 2007a). Probably, the high genetic diversity and the large rooting system, which entails resources to allow repeated establishment of a new photosynthetic canopy are the basis for a large resilience against declining biomass production (Bassin et al. 2007a, b; Volk et al. 2011).

As already shown for grassland communities, any differential O_3 susceptibility between plant species in terms of growth or fitness may alter their competitive interactions. This may also be assumed for plant competition in agro-ecosystems and has been shown for crop–weed interactions. Unfortunately, overall knowledge about this issue is rather scarce. Ozone impacts on competition between crops and the C-4 weed *Cyperus esculentus* have been investigated by Grantz and Shrestha (2005, 2006) and Shrestha and Grantz (2005). Fruit productivity of *Lycopersicon esculentum* in competition with this weed was reduced under low and moderate O_3 levels, whereas the crop responded only to high O_3 levels in the absence of the weed (Shrestha and Grantz 2005). In competition with *Gossypium barbadense*, O_3 impacts were compounded by *C. esculentus* (Grantz and Shrestha 2005) suggesting that high O_3 concentrations appear to increase the competitiveness of the weed with respect to cotton (Grantz and Shrestha 2006). Grantz et al. (2010) thus assumed that it is more likely that the level of threat to agricultural production from *C. esculentus* may increase due to enhanced competition for edaphic resources driven by the O_3 impact. Pfleeger et al. (2010) observed the response of a plant community emerging from a farm soil over several generations. Individuals from some of the species appeared to be diminished in number by the third year, such as *Capsella bursa-pastoris*, *Erodium cicurtarium* and *Spergula arvensis*, while biomass decreased with increasing O_3 exposure. Changes in competitive interactions and community dynamics seemed to be an indirect effect of premature senescence of taller species by altering light availability.

5.4 Ecosystem Productivity

Ozone effects on plant vigour, water relations and soil processes may finally all contribute to altered net primary productivity which is one of the key characteristics of any ecosystem function and service.

Forest productivity is of particular interest not only for timber production but also due to its implications for the global carbon cycle and climate change. Current O_3 levels are considered an important stressor of over 30 % of the world's forests (IPCC 2007; Royal Society 2008) and also constitute a risk for forests in Europe (Ashmore 2005; Matyssek et al. 2008). Such assessments of O_3 effects on forest ecosystem properties are based on experiments and models, but still remain uncertain. For example, most experimental approaches addressing this question were carried out with seedlings or individual young tree species, therefore extrapolation to the results of mature forest stands is limited (Karnosky et al. 2007).

DeMarco et al. (2013) applied a generalised linear/non-linear regression model to assess cause–effect relationships between primary productivity of *Quercus cerris*, *Quercus ilex* and *Fagus sylvatica* and climate and pollutants including O_3 in Italy and concluded that O_3 did not significantly affect net primary productivity. But this conclusion must be viewed with caution, because the authors only considered the external O_3 concentration (as AOT40; accumulated hourly mean O_3 concentration above 40 ppb), i.e. the O_3 exposure in their model rather than the O_3 uptake into the plants, which is toxicologically relevant for any risk assessment. Ollinger et al. (1997) combined leaf-level O_3 response data from O_3 fumigation studies with a forest ecosystem model in order to simulate the effects of ambient O_3 on mature hardwood forests in the northeastern United States. The predicted declines in annual net primary production in this modelling study ranged from 3 to 16 %.

Information on O_3 effects on the productivity of natural and semi-natural vegetation are to a large extent represented by studies on grassland (Bassin et al. 2007a; Fuhrer 2009; see Sect. 5.3). As grasslands comprise a variety of habitats described as meadows and fens, as well as agricultural grassland used for grazing, albeit maintained to conserve species diversity, a general assessment of an O_3 impact on these ecosystem type is difficult. Although O_3 has been reported to decrease productivity in individual grassland species grown in simulated mixtures (see Sect. 5.3), the few experiments with established grassland ecosystems have shown that their net primary production is quite resilient to elevated O_3 (Thwaites et al. 2006; Volk et al. 2011).

Ozone effects on agroecosystem productivity at the field scale and related to this the consequences for regional and global yields and agricultural productivity, respectively, have been assessed by experimental field studies and by regression models using the O_3 dose–response functions derived from these experiments. The most prominent examples of this approach are the previous multi-site field studies in National Crop Loss Assessment Network in the USA (Heagle 1989) and in the European Open-top Chamber Network (Jäger et al. 1992), where various crop species were exposed to O_3 in OTCs. Data from these experiments have been used widely to develop O_3 exposure–response models which again formed the basis to estimate regional or global productivity losses (in terms of crop yields) caused by O_3 (Wang and Mauzerall 2004; van Dingenen et al. 2009). For example, yield losses of important US crops crop species (maize wheat, sorghum, soybean) were calculated to be in the range of approx. 10 % when exposed to an average O_3 concentration below 50 ppb (7-h day^{-1}) or when exposed to O_3 concentrations above 80 ppb (7-h day^{-1}) (Booker et al. 2009). According to Mills et al. (2007) more than 20 % of the European crop production area is at risk for yield losses of about 5 % at current O_3 levels. Also, mostly based on OTC studies considerable yield losses of crops in Asian countries like India, Pakistan (Wahid 2006) and China have been estimated (Cho et al. 2011). Aunand et al. (2000) estimate that yield losses of soybean and wheat may range between 20 and 30 % by 2020 in China. More recently, the only two free air O_3 exposure experiments with crops worldwide have similarly shown that modest enhancements of ambient O_3 concentrations (which ranged between 42 and 62 ppb) resulted in yield losses of 5–18 % for rice (Shi et al. 2009), 15–25 % for soybean (Morgan et al. 2006) and 10–35 % for wheat (Zhu et al. 2011). Overall, the above examples all provide reasonable evidence that the productivity of important agro-ecosystems are at risk from current and future O_3 exposure. From the perspective of the growing global population with ever increasing future needs for food supply, the estimated yield losses of these crops are of concern. However, there remains uncertainty with these estimates as they rely on exposures to the external O_3 concentration rather than on the actual O_3 uptake into the crops.

6 Conclusions

Globally there is widespread evidence that tropospheric O_3 concentrations tend to increase. There is also long-term evidence that O_3 is highly phytotoxic and that vegetation is at particular risk from this pollutant. Driven by concerns about the potential losses in food crop and timber productivity due to O_3 exposures the mode of action of O_3 on individual plant species has been studied intensively during the last decades. Consequently, we now have a reasonable understanding how plant metabolism, physiology and growth vigour is affected by this pollutant and evidence that O_3 exposure causes yield losses of crops and forest trees. More recently, along with overall concerns about the pressures on global ecosystems derived from land use, climate change and overexploitation, etc. emphasis on the importance of ecosystems for the overall functioning of the biosphere has raised new questions about the role of O_3 as an additional threat to that role of terrestrial ecosystems. Here we have highlighted O_3 effects on plant water relations and the possible consequences for the hydrology of whole ecosystems, the possible consequences of an O_3-induced alteration of the carbon transfer between above- and belowground plant parts for soil carbon and soil organisms and the potential role of O_3

as a driver of plant biodiversity in vegetation. It is evident that in comparison to the level of single plants much of the existing information that would allow us to assess O_3 effects at the ecosystem level is still missing and inconsistent.

From the perspective of ecosystem effects this is particularly due to a paucity of adequate research efforts to study O_3 effects at the system level. This is equally true for agro-, forest- and other semi(natural) or grassland ecosystems, albeit these different systems require to consider different time horizons to assess any risk from O_3 stress. While two recent large-scale O_3 experiments with a forest plantation and a mature forest stand applying free air O_3 enrichment techniques have provided important information on the multitude of potential O_3 effects at the system or stand level, we need more of such experiments. This holds true not only with respect to other forest ecosystems at other sites but similarly also for other types of natural or semi-natural vegetation such as, e.g. pastures and grassland, particularly with an increased emphasis on biodiversity issues under the impact of O_3. Also for agroecosystems where O_3 effects have almost always been considered under aspects of food security, the challenge remains to clearly demonstrate at the field level the "true" extent of either direct or indirect O_3 impacts on crop yield and quality. To address the various inconsistencies in the current understanding of O_3 effects at the ecosystem level especially long-term factorial experimental approaches are required that address questions of interactions of O_3 with other environmental factors more systematically. Such efforts should be underpinned by a more mechanistic research trying to better understand the various interactive feedbacks of the components of a particular system under O_3 exposure, e.g. by applying ecophysiological and molecular approaches.

Acknowledgements This work was partly supported by the German Federal Environmental Agency (Umweltbundesamt, FKZ: 3711 63 235).

References

Agerer R, Hartmann A, Pritsch K, Raidl S, Schloter M, Verma R, Weigt R (2012) Plants and their ectomycorrhizosphere: cost and benefit of symbiotic soil organisms. In: Matyssek R, Schnyder H et al (eds) Growth and defence in plants. Springer, Berlin, pp 213–242

Ahsan N, Nanjo Y, Sawada H, Kohno Y, Komatsu S (2010) Ozone stress-induced cell death, hormone biosynthesis and gene expression in Arabidopsis thaliana. Plant J 58:1–12

Ainsworth EA (2008) Rice production in a changing climate: a meta-analysis of responses to elevated carbon dioxide and elevated ozone concentration. Glob Change Biol 14:1642–2650

Ainsworth EA, Rogers A, Leakey ADB (2008) Targets for crop biotechnology in a Future high CO2 and high O3 world. Plant Physiol 147:13–19

Akimoto H (2003) Global air quality and pollution. Science 302:1716–1719

Andersen CP (2003) Source-sink balance and carbon allocation below ground in plants exposed to ozone. New Phytol 157:213–228

Aneja MK, Sharma S, Fleischmann F, Stich S, Heller W, Bahnweg G, Munch JC, Schloter M (2007) Influence of ozone on litter quality and its subsequent effects on the initial structure of colonizing microbial communities. Microb Ecol 54:151–160

Ashmore MR (2005) Assessing the future global impacts on vegetation. Plant Cell Environ 28:949–964

Ashmore MR, Ainsworth N (1995) The effects of ozone and cutting on the species composition of artificial grassland communities. Funct Ecol 9:708–712

Ashmore MR, Thwaites RH, Ainsworth N, Cousins DA, Power SA, Morton AJ (1995) Effects of ozone on calcareous grassland communities. Water Air Soil Pollut 85:1527–1532

Ashmore MR, Power SA, Cousins DA, Ainsworth N (1996) Effects of ozone on native grass and forb species: a comparison of responses of individual plants and artificial communities. In: Kärenlampi L, Skärby L (eds) Critical levels for ozone: testing and finalizing the concepts. UN-ECE Workshop Report, Kuopio, pp 193–197

Aunand K, Berntsen TK, Seip HM (2000) Surface ozone in China and its possible impact on agricultural crops. Ambio 29:294–301

Baier M, Kandlbinder A, Golldack D, Dietz K (2005) Oxidative stress and ozone: perception, signaling and response. Plant Cell Environ 28:1012–1020

Bandeff JM, Pregitzer KS, Loya WM, Holmes WE, Zak DR (2006) Overstory community composition and elevated atmospheric CO_2 and O_3 modify understory biomass production and nitrogen acquisition. Plant Soil 282:251–259

Barbo DN, Chappelka AH, Somers GL, Miller-Goodman MS, Stolte K (1998) Diversity of an early successional plant community as influenced by ozone. New Phytol 138:653–662

Bassin S, Volk M, Fuhrer J (2007a) Factors affecting the ozone sensitivity of temperate European grasslands: an overview. Environ Pollut 146:678–691

Bassin S, Volk M, Suter M, Buchmann N, Fuhrer J (2007b) Nitrogen deposition but not ozone affects productivity and community composition of subalpine grassland after 3 yr of treatment. New Phytol 175:523–534

Bassin S, Werner RA, Sorgel K, Volk M, Buchmann N, Fuhrer J (2009) Effects of combined ozone and nitrogen deposition on the in situ properties of eleven key plant species of a subalpine pasture. Oecologia 158:747–756

Bassin S, Volk M, Fuhrer J (2013) Species composition of subalpine grassland is sensitive to nitrogen deposition, but not to ozone, after seven years of treatment. Ecosystems 16:1105–1117

Bender J, Bergmann E, Dohrmann A, Tebbe CC, Weigel HJ (2002) Impact of ozone on plant competition and structural diversity of rhizosphere microbial communities in grassland mesocosms. Phyton 42:7–12

Bender J, Bergmann E, Weigel HJ (2003) Multi-year experiments on ozone effects on semi-natural vegetation: implications for the development of critical levels. In: Karlsson PE, Selldén G, Pleijel H (eds) Establishing ozone critical levels II. UNECE workshop report. IVL report B 1523, Gothenburg, Sweden, pp 211–217

Bender J, Bergmann E, Weigel HJ (2006a) Responses of biomass production and reproductive development to ozone exposure differ between European wild plant species. Water Air Soil Pollut 176:253–267

Bender J, Muntifering RB, Lin JC, Weigel HJ (2006b) Growth and nutritive quality of Poa pratensis as influenced by ozone and competition. Environ Pollut 142:109–115

Bergmann E, Bender J, Weigel HJ (1996) Effects of chronic ozone stress on growth and reproduction capacity of native herbaceous plants. In: Knoflacher M, Schneider J, Soja G (eds) Exceedances of critical loads and levels: spatial and temporal interpretation of elements in landscape sensitivity to atmospheric pollutants. Federal Ministry for Environment, Youth and Family, Vienna, Austria, pp 177–185

Bernacchi CJ, Leakey ADB, Heady LE, Morgan PB, Dohleman FG, McGrath JM, Gillespie KM, Wittig VE, Rogers A, Long SP, Ort DR (2006) Hourly and seasonal variation in photosynthesis and stomatal conductance of soybean grown at future CO_2 and ozone concentrations for 3 years under fully open-air field conditions. Plant Cell Environ 29:2077–2090

Beyschlag W, Eckstein J (1998) Stomatal patchiness. In: Behnke K, Esser K, Kadereit JW, Lüttge U, Runge M (eds) Progress in botany, vol 60. Springer, Berlin, pp 283–298

Bidart-Bouzat MG, Imeh-Nathaniel A (2008) Global change effects on plant chemical defenses against insect herbivores. J Integr Plant Biol 50:1339–1354
Black VJ, Black CR, Roberts JA, Stewart CA (2000) Impact of ozone on the reproductive development of plants. New Phytol 147:421–447
Blum U, Heagle AS, Burns JC, Linthurst RA (1983) The effects of ozone on fescue-clover forage—regrowth, yield and quality. Environ Exp Bot 23:121–132
Booker F, Muntifering R, McGrath M, Burkey K, Decoteau D, Fiscus E, Manning KS, Chappelka A, Grantz D (2009) The ozone component of global change: potential effects on agricultural and horticultural plant yield, product quality and interactions with invasive species. J Integr Plant Biol 51:337–351
Burkart S, Bender J, Tarkotta B, Faust S, Castagna A, Ranieri A, Weigel HJ (2013) Effects of ozone on leaf senescence, photochemical efficiency and grain yield in two winter wheat cultivars. J Agro Crop Sci 199:275–285
Calatayud V, Cervero J, Sanz MJ (2007) Foliar, physiological and growth responses of four maple species exposed to ozone. Water Air Soil Pollut 185:239–254
Cape JN, Hamilton R, Heal MR (2009) Reactive uptake of ozone at simulated leaf surfaces: implications for 'non-stomatal' ozone flux. Atmos Environ 43:1116–1123
Chang LA, Liu XH, Ge F (2011) Effect of elevated O_3 associated with Bt cotton on the abundance, diversity and community structure of soil Collembola. Appl Soil Ecol 47:45–50
Cho K, Tiwar S, Agrawal SB, Torres NL, Agrawal M, Sarkar A, Shibato J, Agrawal GK, Kubo A, Rakwal R (2011) Troposheric ozone and plants: absorption, responses, and consequences. In: Whitacre DM (ed) Reviews of environmental contamination and toxicology, vol 212, pp 61–111
Cieslik SA (2004) Ozone uptake by various surface types: a comparison between dose and exposure. Atmos Environ 38:2409–2420
Cooley DR, Manning WJ (1987) The impact of ozone on assimilate partitioning in plants: a review. Environ Pollut 47:95–113
Darbah JNT, Kubiske ME, Nelson N, Oksanen E, Vapaavuori E, Karnosky DF (2008) Effects of decadal exposure to interacting elevated CO_2 and/or O_3 on paper birch (*Betula papyrifera*) reproduction. Environ Pollut 155:446–452
Darrall NM (1989) The effect of air pollutants on physiological processes in plants. Plant Cell Environ 12:1–30
Davison AW, Barnes JD (1998) Effects of ozone on wild plants. New Phytol 139:135–151
Degl'Innocenti E, Guidi L, Soldatini GF (2007) Effects of elevated ozone on chlorophyll a fluorescence in symptomatic and asymptomatic leaves of two tomato genotypes. Biol Plant 51:313–321
DeMarco A, Screpanti A, Attore F, Proietti C, Vitale M (2013) Assessing ozone and nitrogen impact on net primary productivity with a generalised non-linear model. Environ Pollut 172:250–263
Derwent R (2008) New directions: prospects for regional ozone in north-west Europe. Atmos Environ 42:1958–1960
Dizengremel P, Jolivet I, Tuzet A, Ranieri A, LeThiec D (2013) Integrative leaf-level phytotoxic ozone dose assessment and forest risk modelling. In: Matyssek R, Clarke N, Cudlin P, Mikkelsen TN, Tuovinen J-P, Wieser G, Paoletti E (eds) Climate change, air pollution and global challenges: understanding and perspectives from forest research. Developments in Environmental Science, vol 13, Elsevier, pp 267–282
Dohrmann AB, Tebbe CC (2005) Effect of elevated tropospheric ozone on the structure of bacterial communities inhabiting the rhizosphere of herbaceous plants native to Germany. Appl Environ Microbiol 71:7750–7758
Dohrmann AB, Tebbe CC (2006) Genetic profiling of bacterial communities from the rhizospheres of ozone damaged Malva sylvestris (Malvaceae). Eur J Soil Biol 42:191–199
Duyzer JH, Weststrate H, Walton S (1995) Exchange of ozone and nitrogen oxides between the atmosphere and coniferous forest. Water Air Soil Pollut 85:2065–2070

Eastburn DM, McElrone AJ, Bilgin DD (2011) Influence of atmospheric and climatic change on plant-pathogen interactions. Plant Pathol 60:54–69

Edwards GS, Kelly JM (1992) Ectomycorrhizal colonization of loblolly pine seedlings during three growing seasons in response to ozone, acidic precipitation, and soil Mg status. Environ Pollut 76:71–77

Edwards IP, Zak DR (2011) Fungal community composition and function after long-term exposure of northern forests to elevated atmospheric CO_2 and tropospheric O_3. Glob Change Biol 17:2184–2195

Emberson LD, Ashmore MR, Murray F (2003) Air pollution impacts on crops and forests: a global assessment. Imperial College Press, London

Esperschutz J, Pritsch K, Gattinger A, Welzl G, Haesler F, Buegger F, Winkler JB, Munch JC, Schloter M (2009) Influence of chronic ozone stress on carbon translocation pattern into rhizosphere microbial communities of beech trees (Fagus sylvatica L.) during a growing season. Plant Soil 323:85–95

Evans PA, Ashmore MR (1992) The effects of ambient air on a semi-natural grassland community. Agric Ecosyst Environ 38:91–97

Fares S, McKay M, Holzinger R, Goldstein AH (2010) Ozone fluxes in a Pinus ponderosa ecosystem are dominated by non-stomatal processes: evidence from long-term continuous measurements. Agric For Meteorol 150:420–431

Felzer BS, Cronin TW, Melillo JM, Kicklighter DW, Schlosser CA (2009) Importance of carbon-nitrogen interactions and ozone on ecosystem hydrology during the 21st century. J Geophys Res 114:G01020

Feng ZZ, Kbayashi K, Ainsworth EA (2008) Impact of elevated ozone concentration on growth, physiology and yield of wheat (*Triticum aestivum* L.): a meta-analysis. Glob Change Biol 14:2696–2708

Fiscus EL, Booker FL, Burkey KO (2005) Crop responses to ozone: uptake, modes of action, carbon assimilation and partitioning. Plant Cell Environ 28:997–1011

Flagler RB (1998) Recognition of air pollution injury to vegetation. Air and Waste Management Association, Pittsburgh

Fowler D, Flechard C, Cape JN, Storeton-West RL, Coyle M (2001) Measurements of ozone deposition to vegetation quantifying the flux, the stomatal and non-stomatal components. Water Air Soil Pollut 130:63–74

Fowler D, Pilegaard K, Sutton MA et al (2009) Atmospheric composition change: ecosystems-atmosphere interactions. Atmos Environ 43:5193–5267

Fuhrer J (1997) Ozone sensitivity of managed pastures. In: Cheremisinoff PN (ed) Ecological issues and environmental impact assessment. Gulf Publishing, Houston, pp 681–706

Fuhrer J (2003) Agroecosystern responses to combinations of elevated CO2, ozone, and global climate change. Agric Ecosyst Environ 97:1–20

Fuhrer J (2009) Ozone risk for crops and pastures in present and future climates. Naturwissenschaften 96:173–194

Fuhrer J, Shariatmadari H, Perler R, Tschannen W, Grub A (1994) Effects of ozone on managed pasture. 2. Yield, species composition, canopy structure, and forage quality. Environ Pollut 86:307–314

Fumagalli I, Gimeno B, Velissariou D, DeTemmerman L, Mills G (2001) Evidence of ozone-induced adverse effects on crops in the Mediterranean region. Atmos Environ 35:2583–2587

Gimeno BS, Bermejo V, Sanz J, de la Torre D, Gil JM (2003) Ozone and competition adversely affect the reproductive ability of three clover species from Iberian rangelands. In: Karlsson PE, Selldén G, Pleijel H (eds) Establishing ozone critical levels II. UNECE workshop report. IVL report B 1523, Gothenburg, Sweden, pp 218–223

González-Fernández I, Bass D, Muntifering R, Mills G, Barnes J (2008) Impacts of ozone pollution on productivity and forage quality of grass/clover swards. Atmos Environ 42:8755–8769

Grams TEE, Matyssek R (2010) Stable isotope signatures reflect competitiveness between trees under changed CO_2/O_3 regimes. Environ Pollut 158:1036–1042

Grams TEE, Kozovits AR, Reiter IM, Winkler JB, Sommerkorn M, Blaschke H, Häberle KH, Matyssek R (2002) Quantifying competitiveness in woody plants. Plant Biol 4:153–158

Grams TEE, Daigo MJ, Winkler JB, Gayler S, Matyssek R (2012) Growth and space use in competitive interactions between juvenile trees. In: Matyssek R, Schnyder H et al (eds) Growth and defence in plants. Springer, Berlin, pp 273–286

Grantz DA, Shrestha A (2005) Ozone reduces crop yields and alters competition with weeds such as yellow nutsedge. Calif Agric 59:137–143

Grantz DA, Shrestha A (2006) Tropospheric ozone and interspecific competition between yellow nutsedge and pima cotton. Crop Sci 46:1879–1889

Grantz DA, Gunn S, Vu HB (2006) O_3 impacts on plant development: a meta-analysis of root/shoot allocation and growth. Plant Cell Environ 29:1193–1209

Grantz DA, Shrestha A, Vu HB (2010) Ozone impacts on assimilation and allocation to reproductive sinks in the vegetatively propagated C-4 weed, yellow nutsedge. Crop Sci 50:246–252

Grebenc T, Kraigher H (2007a) Changes in the community of ectomycorrhizal fungi and increased fine root number under adult beech trees chronically fumigated with double ambient ozone concentration. Plant Biol 9:279–287

Grebenc T, Kraigher H (2007b) Types of ectomycorrhiza of mature beech and spruce at ozone-fumigated and control forest plots. Environ Monit Assess 128:47–59

Grulke NE, Johnson R, Esperanza A, Jones D, Nguyen T, Posch S, Tausz M (2003) Canopy transpiration of Jeffrey pine in mesic and xeric microsites: O3 uptake and injury response. Trees Struct Funct 17:292–298

Grulke N, Neufeld H, Davison A, Roberts M, Chappelka A (2007) Stomatal behavior of ozone-sensitive and -insensitive coneflowers (Rudbeckia laciniata var. digitata) in Great Smoky Mountains National Park. New Phytol 173:100–109

Grünhage L, Jäger HJ (2003) From critical levels to critical loads for ozone: a discussion of a new experimental and modelling approach for establishing flux–response relationships for agricultural crops and native plant species. Environ Pollut 125:99–110

Haberer K, Grebenc T, Alexou M, Gessler A, Kraigher H, Rennenberg H (2007) Effects of long-term free-air ozone fumigation on delta N-15 and total N in Fagus sylvatica and associated mycorrhizal fungi. Plant Biol 9:242–252

Haikio E, Makkonen M, Julkunen-Tiitto R, Sitte J, Freiwald V, Silfver T, Pandey V, Beuker E, Holopainen T, Oksanen E (2009) Performance and secondary chemistry of two hybrid aspen (Populus tremula L. x Populus tremuloides Michx.) clones in long-term elevated ozone exposure. J Chem Ecol 35:664–678

Haldemann C, Fuhrer J (2005) Interactive effects of CO_2 and O_3 on the growth of Trisetum falvescens and Trifolium pratense grown in monoculture or a bi-species mixture. J Crop Improv 13:275–289

Handley T, Grulke NE (2008) Interactive effects of O_3 exposure on California black oak (Quercus kelloggii Newb.) seedlings with and without N amendment. Environ Pollut 156:53–60

Hanson PJ, Wullschleger SD, Norby RJ, Tschaplinski TJ, Gunderson CA (2005) Importance of changing CO_2, temperature, precipitation, and ozone on carbon and water cycles of an upland-oak forest: incorporating experimental results into model simulations. Glob Change Biol 11:1402–1423

Hayes F, Jones MLM, Mills G, Ashmore M (2007) Meta-analysis of the relative sensitivity of semi-natural vegetation species to ozone. Environ Pollut 146:754–762

Hayes F, Mills G, Ashmore M (2009) Effects of ozone on inter- and intra-species competition and photosynthesis in mesocosms of Lolium perenne and Trifolium repens. Environ Pollut 157:208–214

Hayes F, Mills G, Ashmore M (2010a) How much does the presence of a competitor modify the within-canopy distribution of ozone-induced senescence and visible injury? Water Air Soil Pollut 210:265–276

Hayes F, Mills G, Jones L, Ashmore M (2010b) Does a simulated upland grassland community respond to increasing background, peak or accumulated exposure of ozone? Atmos Environ 44:4155–4164

Hayes F, Mills G, Harmens H, Wyness K (2011) Within season and carry-over effects following exposure of grassland species mixtures to increasing background ozone. Environ Pollut 159:2420–2426

Hayes F, Williamson J, Mills G (2012) Ozone pollution affects flower numbers and timing in a simulated BAP priority calcareous grassland. Environ Pollut 163:40–47

Heagle AS (1989) Ozone and crop yield. Annu Rev Phytopathol 27:397–423

Heagle AS, Body DE, Heck WW (1973) An open-top field chamber to assess the impact of air pollution on plants. J Environ Qual 2:365–368

Heagle AS, Kress LW, Temple PJ, Kohut RJ, Miller JE, Heggestad HE (1988) Factors influencing ozone dose-yield response relationships in open-top field chamber studies. In: Heck WW, Taylor OC, Tingey DT (eds) Assessment of crop loss from air pollutants. Elsevier Applied Science, London, pp 141–149

Heagle AS, Rebbeck J, Shafer SR, Blum U, Heck WW (1989) Effects of long-term ozone exposure and soil-moisture deficit on growth of a ladino clover tall fescue pasture. Phytopathology 79:128–136

Heath RL (2008) Modification of the biochemical pathways of plants induced by ozone: What are the varied routes to change? Environ Pollut 155:453–463

Heck WW, Philbeck RB, Dunning JA (1978) A continuous stirred tank reactor (CSTR) system for exposing plants to gaseous air contaminants: principles, specifications, construction, and operation. USDA, ARS-S-81. U.S. Government Printing Office, Washington, DC

Hendrey GR, Ellsworth DS, Lewin KF, Nagy J (1999) A free-air enrichment system for exposing tall forest vegetation to elevated atmospheric CO_2. Glob Change Biol 5:293–309

Hogsett WE, Tingey DT, Lee EH, Beedlow PA, Andersn CP (2008) An approach for evaluating the effectiveness of various ozone Air Quality Standards for protecting trees. Environ Manage 41:937–948

Holopainen JK (2002) Aphid response to elevated ozone and CO_2. Entomol Exp Appl 104:137–142

Innes JL, Skelly JM, Schaub M (2001) Ozone and broadleaved species. A guide to the identification of ozone-induced foliar injury. Haupt, Bern

IPCC (Intergovernmental Panel on Climate Change) (2007) Climate change 2007: synthesis report. Contribution of working groups I, II and II to the fourth assessment report of the IPCC. IPCC, Geneva

Iriti M, Faoro F (2008) Oxidative stress, the paradigm of ozone toxicity in plants and animals. Water Air Soil Pollut 187:285–301

Jacob DJ, Winner DA (2009) Effects of climate change on air quality. Atmos Environ 43:51–63

Jäger HJ, Unsworth MH, De Temmerman L, Mathy P (1992) Effects of air pollution on agricultural crops in Europe—results of the European open-top chamber project. Air Pollution Research Report 46, Commission of the European Communities, Brussels

Jäger HJ, Hertstein U, Fangmeier A (1999) The European stress physiology and climate change experiment—project 1: wheat (ESPACE-wheat): introduction, aims and methodology. Eur J Agron 10:155–162

Johnson BG, Hale BA, Ormrod DP (1996) Carbon dioxide and ozone effects on growth of a legume-grass mixture. J Environ Qual 25:908–916

Jonson JE, Simpson D, Fagerli H, Solberg S (2006) Can we explain the trends in European ozone levels. Atmos Chem Phys 6:51–66

Kanerva T, Palojarvi A, Ramo K, Manninen S (2008) Changes in soil microbial community structure under elevated tropospheric O_3 and CO_2. Soil Biol Biochem 40:2502–2510

Kangasjarvi J, Jaspers P, Kollist H (2005) Signalling and cell death in ozone-exposed plants. Plant Cell Environ 28(1021):1036

Karnosky DF, Gagnon ZE, Dickson RE, Coleman MD, Lee EH, Isebrands JG (1996) Changes in growth, leaf abscission, biomass associated with seasonal tropospheric ozone exposures of Populus tremuloides clones and seedlings. Can J For Res 26:23–37

Karnosky DF, Mankovska B, Percy K, Dickson RE, Podila GK, Sober J, Noormets A, Hendrey G, Coleman MD, Kubiske M, Pregitzer KS, Isebrandts JG (1999) Effects of tropospheric ozone on trembling aspen and interaction with CO2: results from an O3-gradient and a FACE experiment. Water Air Soil Pollut 116:311–322

Karnosky DF, Gielen B, Ceulemans R, Schlesinger WH, Norby RJ, Oksanen E, Matyssek R, Hendrey GR (2001) FACE systems for studying the impacts of greenhouse gases on Forest Ecosystems. In: Karnosky DF, Scarascia-Mugnozza G, Ceulemans R, Innes JL (eds) The impacts of carbon dioxide and other greenhouse gases on forest ecosystems. CABI Press, Wallingford, pp 297–324

Karnosky DF et al (2003) Tropospheric O_3 moderates responses of temperate hardwood forests to elevated CO_2: a synthesis of molecular to ecosystem results from the Aspen FACE project. Funct Ecol 17:289–304

Karnosky DF, Werner H, Holopainen T, Percy K, Oksanen T, Oksanen E, Heerdt C, Fabian P, Nagy J, Heilman W, Cox R, Nelson N, Matyssek R (2007) Free-air exposure systems to scale up ozone research to mature trees. Plant Biol 9:181–190

Kasurinen A, Keinanen MM, Kaipainen S, Nilsson LO, Vapaavuori E, Kontro MH, Holopainen T (2005) Below-ground responses of silver birch trees exposed to elevated CO_2 and O_3 levels during three growing seasons. Glob Change Biol 11:1167–1179

Kasurinen A, Peltonen P, Holopainen JK, Vapaavuori E, Holopainen T (2007a) Leaf litter under changing climate: Will increasing levels of CO_2 and O_3 affect decomposition and nutrient cycling processes? Dynamic Soil, Dynamic Plant. Glob Sci Books 1:58–67

Kasurinen A, Peltonen PA, Julkunen-Tiitto R, Vapaavuori E, Nuutinen V, Holopainen T, Holopainen JK (2007b) Effects of elevated CO_2 and O_3 on leaf litter phenolics and subsequent performance of litter-feeding soil macrofauna. Plant Soil 292:25–43

Keller T, Häsler R (1984) The influence of a fall fumigation with ozone on the stomatal behavior of spruce and fir. Oecologia 64:284–286

Kerstiens G, Lendzian KJ (1989) Interaction between ozone and plant cuticles: I. ozone deposition and permeability. New Phytol 112:13–19

Kitao M, Löw M, Heerdt C, Grams TEE, Häberle KH, Matyssek R (2009) Effects of chronic elevated ozone exposure on gas exchange response of adult beech trees (Fagus sylvatica) as related to the within-canopy light gradient. Environ Pollut 157:537–544

Kitao M, Winkler JB, Loew M, Nunn AJ, Kuptz D, Haeberle K-H, Reiter IM, Matyssek R (2012) How closely does stem growth of adult beech (*Fagus sylvatica*) relate to net carbon gain under experimentally enhanced ozone stress? Environ Pollut 166:108–115

Kolb TE, Matyssek R (2001) Limitations and perspectives about scaling ozone impact in trees. Environ Pollut 115:373–393

Kostka-Rick R, Bender J, Bergmann E, Weigel HJ (2002) Symptoms of ozone-induced foliar injury on horticultural crops. In: Klumpp A, Fomin A, Klumpp G, Ansel W (eds) Bioindication and air quality in European cities. G. Heimbach, Stuttgart, pp 191–196

Kozovits AR, Matyssek R, Blaschke H, Gottlein A, Grams TEE (2005) Competition increasingly dominates the responsiveness of juvenile beech and spruce to elevated CO_2 and/or O_3 concentrations throughout two subsequent growing seasons. Glob Change Biol 11:1387–1401

Kubiske ME, Quinn VS, Marquardt PE, Karnosky DF (2007) Effects of elevated atmospheric CO_2 and/or O_3 on intra- and interspecific competitive ability of aspen. Plant Biol 9:342–355

Laisk A, Kull O, Moldau H (1989) Ozone concentration in leaf intercellular air spaces is close to zero. Plant Physiol 90:1163–1167

Laurence JA, Andersen CP (2003) Ozone and natural systems: understanding exposure, response, and risk. Environ Int 29:155–160

Leisner CP, Ainsworth EA (2012) Quantifying the effects of ozone on plant reproductive growth and development. Glob Change Biol 18:606–616

Lindroth RL (2010) Impacts of elevated atmospheric CO_2 and O_3 on forests: phytochemistry, trophic Interactions, and ecosystem dynamics. J Chem Ecol 36:2–21
Liu LL, King JS, Giardina CP (2005) Effects of elevated concentrations of atmospheric CO_2 and tropospheric O_3 on leaf litter production and chemistry in trembling aspen and paper birch communities. Tree Physiol 25:1511–1522
Liu L, King JS, Giardina CP (2007) Effects of elevated atmospheric CO_2 and tropospheric O_3 on nutrient dynamics: decomposition of leaf litter in trembling aspen and paper birch communities. Plant Soil 299:65–82
Long SP, Naidu SL (2002) Effects of oxidants at the biochemical, cell and physiological level. In: Bell JNB, Treshow M (eds) Air pollution and plant life. Wiley, Chichester, pp 69–88
Loranger GI, Pregitzer KS, King JS (2004) Elevated CO_2 and O_3 concentrations differentially affect selected groups of the fauna in temperate forest soils. Soil Biol Biochem 36:1521–1524
Manninen S, Aaltonen H, Kanerva T, Ramo K, Palojarvi A (2010) Plant and soil microbial biomasses in Agrostis capillaris and Lathyrus pratensis monocultures exposed to elevated O_3 and CO_2 for three growing seasons. Soil Biol Biochem 42:1967–1975
Manning WJ, Krupa SV (1992) Experimental methodology for studying the effects of ozone on crops and trees. In: Lefohn AS (ed) Surface level ozone exposure and their effects on vegetation. Lewis, Chelsea, pp 93–156
Manning WJ, von Tiedemann A (1995) Climate change: potential effects of increased atmospheric carbon dioxide (CO_2), ozone (O_3) and ultraviolet-B (UV-B) radiation on plant diseases. Environ Pollut 88:219–245
Manning WJ, Godzik B, Musselman R (2002) Potential bioindicator plant species for ambient ozone in forested mountain areas of central Europe. Environ Pollut 119:283–290
Manning WJ, Paoletti E, Sandermann H, Ernst D (2011) Ethylenediurea (EDU): a research tool for assessment and verification of the effects of ground level ozone on plants under natural conditions. Environ Pollut 159:3283–3293
Mansfield TA (1998) Stomata and plant water relations: does air pollution create problems? Environ Pollut 101:1–11
Marenco AH, Gouget PN, Pages JP (1994) Evidence of a lon-term increase in tropospheric ozone from Pic Du Midi data series: consequences positive radiative forcing. J Geophys Res 99:166–177
Martin MJ, Host GE, Lenz KE, Isebrands JG (2001) Simulating the growth response of aspen to elevated ozone: a mechanistic approach to scaling a leaf-level model of ozone effects on photosynthesis to a complex canopy architecture. Environ Pollut 115:425–436
Massman WJ, Grantz DA (1995) Estimating canopy conductance to ozone uptake from observations of evapotranspiration at the canopy scale and at the leaf scale. Glob Change Biol 1:183–198
Matyssek R, Sandermann H (2003) Impact of ozone on trees: an ecophysiological perspective. Prog Bot 64:349–403
Matyssek R, Gunthardt-Georg MS, Maurer S, Keller T (1995) Nighttime exposure to ozone reduces whole-plant production in Betula pendula. Tree Physiol 15:159–165
Matyssek R, Le'Thiec D, Low M, Dizengremel P, Nunn AJ, Häberle KH (2006) Interactions between drought and O_3 stress in forest trees. Plant Biol 8:11–17
Matyssek R, Sandermann H, Wieser G, Booker F, Cieslik S, Musselman R, Ernst D (2008) The challenge of making ozone risk assessment for forest trees more mechanistic. Environ Pollut 156:567–582
Matyssek R, Karnosky DF, Wieser G, Percy K, Oksanen E, Grams TEE, Kubiske M, Hanke D, Pretzsch H (2010a) Advances in understanding ozone impact on forest trees: messages from novel phytotron and free-air fumigation studies. Environ Pollut 158:1990–2006
Matyssek R, Wieser G, Ceulemans R, Rennenberg H, Pretzsch H, Haberer K, Low M, Nunn AJ, Werner H, Wipfler P, Oßwald W, Nikolova P, Hanke DE, Kraigher H, Tausz M, Bahnweg G, Kitao M, Dieler J, Sandermann H, Herbinger K, Grebenc T, Blumenrother M, Deckmyn G, Grams TEE, Heerdt C, Leuchner M, Fabian P, Haberle KH (2010b) Enhanced ozone strongly

reduces carbon sink strength of adult beech (Fagus sylvatica): resume from the free-air fumigation study at Kranzberg Forest. Environ Pollut 158:2527–2532
Matyssek R, Clarke N, Cudlin P, Mikkelsen TN, Tuovinen J-P, Wieser G, Paoletti E (eds) (2013) Climate change, air pollution and global challenges: understanding and perspectives from forest research. Developments in environmental science, vol 13, Elsevier, p 622
Maurer S, Matyssek R (1997) Nutrition and the ozone sensitivity of birch (*Betula pendula*), II. Carbon balance, water-use efficiency and nutritional status of the whole plant. Trees 12:11–20
McAinsh MR, Evans NH, Montgomery LT, North KA (2002) Calcium signalling in stomatal responses to pollutants. New Phytol 153:441–447
McDonald EP, Kruger EL, Riemenschneider DE, Isebrands JG (2002) Competitive status influences tree-growth responses to elevated CO_2 and O_3 in aggrading aspen stands. Funct Ecol 16:792–801
McLaughlin SB, Nosal M, Wullschleger SD, Sun G (2007a) Interactive effects of ozone and climate on tree growth and water use in a southern Appalachian forest in the USA. New Phytol 174:109–124
McLaughlin SB, Wullschleger SD, Sun G, Nosal M (2007b) Interactive effects of ozone and climate on water use, soil moisture content and streamflow in a southern Appalachian forest in the USA. New Phytol 174:125–136
McLeod AR (1995) An open-air exposure system for exposure of young trees to sulphur dioxide and ozone. Plant Cell Environ 18:215–225
MEA (Millennium Ecosystem Assessment) (2005) Ecosystems and human well-being: synthesis. Island Press, Washington, DC
Meehan TD, Crossley MS, Lindroth RL (2010) Impacts of elevated CO_2 and O_3 on aspen leaf litter chemistry and earthworm and springtail productivity. Soil Biol Biochem 42:1132–1137
Miglietta F, Peressotti A, Vaccari FP, Zaldei A, deAngelis P, Scarascia-Mugnozza G (2001) Free-air CO_2 enrichment (FACE) of a poplar plantation: the POPFACE fumigation system. New Phytol 150:465–476
Mills G, Harmens H (eds) (2011) Ozone pollution: a hidden threat to food security. Report prepared by the ICP Vegetation. Centre for Ecology and Hydrology, Bangor
Mills G, Ball G, Hayes F, Fuhrer J, Skarby L, Gimeno B, De Temmerman L (2000) Development of a multi-factor model for predicting the effects of ambient ozone on the biomass of white clover. Environ Pollut 109:533–542
Mills G, Buse A, Gimeno B, Bermejo V, Holland M, Emberson L, Pleijel H (2007) A synthesis of AOT40-based response functions and critical levels of ozone for agricultural and horticultural crops. Atmos Environ 41:2630–2643
Mills G, Hayes F, Wilkinson S, Davies WJ (2009) Chronic exposure to increasing background ozone impairs stomatal functioning in grassland species. Glob Change Biol 15:1522–1533
Mills G, Hayes F, Simpson D, Emberson L, Norris D, Harmens H, Büker P (2010) Evidence of widespread effects of ozone on crops and (semi-)natural vegetation in Europe (1990–2006) in relation to AOT40- and flux-based risk maps. Glob Change Biol 17:592–613
Mills G, Wagg S, Harmens H, Bender J, Bergmann E (2013) Effects of ozone on ecological processes and supporting ecosystem services. In: Mills G, Wagg S, Harmens H (eds) Ozone pollution: impacts on ecosystem services and biodiversity. Report prepared by the ICP Vegetation. Centre for Ecology and Hydrology, Bangor, pp 17–23
Morgan PB, Ainsworth EA, Long SP (2003) How does elevated ozone impact soybean? A meta-analysis of photosynthesis, growth and yield. Plant Cell Environ 26:1317–1328
Morgan PB, Bernacchi CJ, Ort DR, Long SP (2004) An in vivo analysis of the effect of season-long open-air elevation of ozone to anticipated 2050 levels on photosynthesis in soybean. Plant Physiol 135:2348–2357
Morgan PB, Mies TA, Bollero GA, Nelson RL, Long SP (2006) Season-long elevation of ozone concentration to projected 2050 levels under fully open-air conditions substantially decreases the growth and production of soybean. New Phytol 170:333–343

Morsky SK, Haapala JK, Rinnan R, Tiiva P, Saarnio S, Silvola J, Holopainen T, Martikainen PJ (2008) Long-term ozone effects on vegetation, microbial community and methane dynamics of boreal peatland microcosms in open-field conditions. Glob Change Biol 14:1891–1903

Musselman R, Hale BA (1997) Methods for controlled and field ozone exposures of forest tree species in North America. In: Sandermann H, Wellburn AR, Heagle A (eds) Forest decline and ozone. Ecological studies, vol 127. Springer, Berlin, pp 277–315

Nebel B, Fuhrer J (1994) Interspecific and intraspecific differences in ozone sensitivity in semi-natural plant communities. Angew Bot 68:116–121

Nunn AJ, Cieslik S, Metzger U, Wieser G, Matyssek R (2010) Combining sap flow and eddy covariance approaches to derive stomatal and non-stomatal O_3 fluxes in a forest stand. Environ Pollut 158:2014–2022

Nussbaum S, Bungener P, Geissmann M, Fuhrer J (2000) Plant-plant interactions and soil moisture might be important in determining ozone impacts on grasslands. New Phytol 147:327–335

Nygaard PH (1994) Effects of ozone on Vaccinium myrtillus, Hylocomium splendens, Pleurozium schreberi and Dicranum polysetum. Rapport fra skogforsk, pp 1–17

Oksanen E, Pandey V, Pandey AK, Keski-Saari S, Kontunen-Soppela S, Sharma C (2013) Impacts of increasing ozone on indian plants. Environ Pollut 177:189–200

Olbrich M, Gerstner E, Welzl G, Winkler JB, Ernst D (2009) Transcript responses in leaves of ozone-treated beech saplings seasons at an outdoor free air model fumigation site over two growing seasons. Plant Soil 323:61–74

Ollinger SV, Aber JD, Reich PB (1997) Simulating ozone effects on forest productivity: interactions among leaf-, canopy-, and stand-level processes. Ecol Appl 7:1237–1251

Olszyk DM, Johnson MG, Phillips DL, Seidler RJ, Tingey DT, Watrud LS (2001) Interactive effects of CO_2 and O_3 on a ponderosa pine plant/litter/soil mesocosm. Environ Pollut 115:447–462

Oltmans SJ, Lefohn AS, Harris JM, Galbally I, Scheel HE, Bodeker G, Brunke E, Claude H, Tarasick D, Johnson BJ, Simmonds P, Shadwick S, Anlauf K, Hayden K, Schmidlin F, Fujimoto T, Akagi K, Meyer C, Nichol S, Davies J, Redondas A, Cuevas E (2006) Long-term changes in tropospheric ozone. Atmos Environ 40:3156–3173

Paoletti E, Grulke NE (2005) Does living in elevated CO_2 ameliorate tree response to ozone? A review on stomatal responses. Environ Pollut 137:483–493

Paoletti E, Grulke NE (2010) Ozone exposure and stomatal sluggishness in different plant physiognomic classes. Environ Pollut 158:2664–2671

Payer HD, Blodow P, Kofferlein M, Lippert M, Schmolke W, Seckmeyer G, Seidlitz HK, Strube D, Thiel S (1993) Controlled environment chambers for experimental studies on plant responses to CO_2 and interactions with pollutants. In: Schulze ED, Mooney HA (eds) Ecosystems research report Nr. 6: Design and execution of experiments on CO_2 enrichment. Commission European Communities, Brussels, pp 127–145

Percy KE, Legge AH, Krupa SV (2003a) Tropospheric ozone: a continuing threat to global forests? In: Karnosky DF, Percy KE, Chappelka AH, Simpson C, Pikkarainen J (eds) Air pollution, global change and forests in the new millennium. Elsevier, Amsterdam, pp 85–118

Percy KE, Mankovska B, Hopkin A, Callan B, Karnosky DF (2003b) Ozone affects leaf surface-pest interactions. In: Karnosky DF, Percy KE, Chappelka AH, Simpson C, Pikkarainen J (eds) Air pollution, global change and forests in the new millennium. Elsevier, Oxford, pp 247–258

Pfleeger TG, Plocher M, Bichel P (2010) Response of pioneer plant communities to elevated ozone exposure. Agric Ecosyst Environ 138:116–126

Prather M, Gauss M, Berntsen T, Isaksen I, Sundet J, Bey I, Brasseur G, Dentener F, Derwent R, Stevenson D, Grenfell L, Hauglustaine D, Horowitz L, Jacob D, Mickley L, Lawrence M, von Kuhlmann R, Muller J, Pitari G, Rogers H, Johnson M, Pyle J, Law K, van Weele M, Wild O (2003) Fresh air in the 21st century. Geophys Res Lett 30:1100–1104

Pregitzer KS, Burton AJ, King JS, Zak DR (2008) Soil respiration, root biomass, and root turnover following long-term exposure of northern forests to elevated atmospheric CO_2 and tropospheric O_3. New Phytol 180:153–161

Pregitzer KS, Talhelm AF (2013) Below ground carbon cycling at Aspen FACE: Dynamic responses to CO2 and O3 in developing forest. In: Matyssek R, Clarke N, Cudlin P, Mikkelsen TN, Tuovinen J-P, Wieser G, Paoletti E (eds) Climate change, air pollution and global challenges: understanding and perspectives from forest research. Developments in Environmental Science, vol 13, Elsevier, pp 209–226

Pretzsch H, Schutze G (2009) Transgressive overyielding in mixed compared with pure stands of Norway spruce and European beech in Central Europe: evidence on stand level and explanation on individual tree level. Eur J For Res 128:183–204

Pretzsch H, Dieler J, Matyssek R, Wipfler P (2010) Tree and stand growth of mature Norway spruce and European beech under long-term ozone fumigation. Environ Pollut 158:1061–1070

Pritsch K, Esperschuetz J, Haesler F, Raidl S, Winkler B, Schloter M (2009) Structure and activities of ectomycorrhizal and microbial communities in the rhizosphere of Fagus sylvatica under ozone and pathogen stress in a lysimeter study. Plant Soil 323:97–109

Rämö K, Kanerva T, Nikula S, Ojanpera K, Manninen S (2006) Influences of elevated ozone and carbon dioxide in growth responses of lowland hay meadow mesocosms. Environ Pollut 144:101–111

Rämö K, Kanerva T, Ojanpera K, Manninen S (2007) Growth onset, senescence, and reproductive development of meadow species in mesocosms exposed to elevated O_3 and CO_2. Environ Pollut 145:850–860

Rebbeck J, Blum U, Heagle AS (1988) Effects of ozone on the regrowth and energy reserves of a ladino clover tall fescue pasture. J Appl Ecol 25:659–681

Reich PB (1987) Quantifying plant response to ozone: a unifying theory. Tree Physiol 3:63–91

Rhea L, King J, Kubiske M, Saliendra N, Teclaw R (2010) Effects of elevated CO_2 and tropospheric O_3 on tree branch growth and implications for hydrologic budgeting. Environ Pollut 158:1079–1087

Robinson M, Heath J, Mansfield T (1998) Disturbances in stomatal behaviour caused by air pollutants. J Exp Bot 49:461–469

Rodenkirchen H, Gottlein A, Kozovits AR, Matyssek R, Grams TEE (2009) Nutrient contents and efficiencies of beech and spruce saplings as influenced by competition and O_3/CO_2 regime. Eur J For Res 128:117–128

Royal Society (2008) Ground-level ozone in the 21st century: future trends, impacts and policy implications. RS policy document 15/08. The Royal Society, London

Sandermann H, Wellburn AR, Heath RL (1997) Forest decline and ozone: a comparison of controlled chamber and field experiments. Ecological studies, vol 127, Springer, Berlin

Sarkar A, Agrawal SB (2010) Elevated ozone and two modern wheat cultivars: an assessment of dose dependent sensitivity with respect to growth, reproductive and yield parameters. Environ Exp Bot 69:328–337

Scebba F, Giuntini D, Castagna A, Soldatini G, Ranieri A (2006) Analysing the impact of ozone on biochemical and physiological variables in plant species belonging to natural ecosystems. Environ Exp Bot 57:89–97

Schloter M, Winkler JB, Aneja M, Koch N, Fleischmann F, Pritsch K, Heller W, Stich S, Grams TEE, Gottlein A, Matyssek R, Munch JC (2005) Short term effects of ozone on the plant-rhizosphere-bulk soil system of young beech trees. Plant Biol 7:728–736

Schrader S, Bender J, Weigel HJ (2009) Ozone exposure of field-grown winter wheat affects soil mesofauna in the rhizosphere. Environ Pollut 157:3357–3362

Sheu BH, Liu CP (2003) Air pollution impacts on vegetation in Taiwan. In: Emberson L, Ashmore M, Murray F (eds) Air pollution impacts on crops and forests: a global assessment. Imperial College Press, London, pp 145–163

Shi GY, Yang LX, Wang YX, Kobayashi K, Zhu JG, Tang HY, Pan ST, Chen T, Liu G, Wang YL (2009) Impact of elevated ozone concentration on yield of four Chinese rice cultivars under fully open-air field conditions. Agric Ecosyst Environ 131:178–184

Shrestha A, Grantz DA (2005) Ozone impacts on competition between tomato and yellow nutsedge: above- and below-ground effects. Crop Sci 45:1587–1595

Singh E, Tiwari S, Agrawal M (2009) Effects of elevated ozone on photosynthesis and stomatal conductance of two soybean varieties: a case study to assess impacts of one component of predicted global climate change. Plant Biol 11:101–108
Staehelin J (2001) Ozone measurements and trends (troposphere). In: Meyers RA (ed) Encyclopedia of physical science and technology, vol 11. Academic, New York, pp 539–561
Stampfli A, Fuhrer J (2010) Spatial heterogeneity confounded ozone-exposure experiment in semi-natural grassland. Oecologia 162:515–522
Steubing L, Fangmeier A, Both R, Frankenfeld M (1989) Effects of SO_2, NO_2, and O_3 on population development and morphological and physiological-parameters of native herb layer species in a beech forest. Environ Pollut 58:281–302
Stockwell WR, Kramm G, Scheel HE, Mohnen VA, Seiler W (1997) Ozone formation, destruction and exposure in Europe and the United States. In: Sandermann H Jr, Wellburn AR, Heath RL (eds) Forest decline and ozone, a comparison of controlled chamber and field experiments. Ecological studies, vol 127. Springer, Berlin, pp 1–38
Stoelken G, Pritsch K, Simon J, Mueller CW, Grams TEE, Esperschuetz J, Gayler S, Buegger F, Brueggemann N, Meier R, Zeller B, Winkler JB, Rennenberg H (2010) Enhanced ozone exposure of European beech (*Fagus sylvatica*) stimulates nitrogen mobilization from leaf litter and nitrogen accumulation in the soil. Plant Biosyst 144:537–546
Sun G, McLaughlin SB, Porter JH, Uddling J, Mulholland PJ, Adams MB, Pederson N (2012) Interactive influences of ozone and climate on streamflow of forested watersheds. Glob Change Biol 18:3395–3409
Tang H, Liua G, Hana Y, Zhu J, Kobayashi K (2011) A system for free-air ozone concentration elevation with rice and wheat: control performance and ozone exposure regime. Atmos Environ 45:6276–6282
Thomas VFD, Braun S, Fluckiger W (2006) Effects of simultaneous ozone exposure and nitrogen loads on carbohydrate concentrations, biomass, growth, and nutrient concentrations of young beech trees (Fagus sylvatica). Environ Pollut 143:341–354
Thwaites RH, Ashmore MR, Morton AJ, Pakeman RJ (2006) The effects of tropospheric ozone on the species dynamics of calcareous grassland. Environ Pollut 144:500–509
Tonneijck AEG, Franzaring J, Brouwer G, Metselaar K, Dueck TA (2004) Does interspecific competition alter effects of early season ozone exposure on plants from wet grasslands? Results of a three-year experiment in open-top chambers. Environ Pollut 131:205–213
Topa MA, Vanderklein DW, Corbin A (2001) Effects of elevated ozone and low light on diurnal and seasonal carbon gain in sugar maple. Plant Cell Environ 24:663–677
Uddling J, Teclaw RM, Kubiske ME, Pregitzer KS, Ellsworth DS (2008) Sap flux in pure aspen and mixed aspen-birch forests exposed to elevated concentrations of carbon dioxide and ozone. Tree Physiol 28:1231–1243
Uddling J, Teclaw RM, Pregitzer KS, Ellsworth DS (2009) Leaf and canopy conductance in aspen and aspen-birch forests under free-air enrichment of carbon dioxide and ozone. Tree Physiol 29:1367–1380
Valkama E, Koricheva J, Oksanen E (2007) Effects of elevated O_3, alone and in combination with elevated CO2, on tree leaf chemistry and insect herbivore performance: a meta-analysis. Glob Change Biol 13:184–201
Van Dingenen R, Dentener FJ, Raes F, Krol MC, Emberson L, Cofala J (2009) The global impact of ozone on agricultural crop yields under current and future air quality legislation. Atmos Environ 43:604–618
Vingarzan R (2004) A review of surface ozone background levels and trends. Atmos Environ 38:3431–3442
Volk M, Geissmann M, Blatter A, Contat F, Fuhrer J (2003) Design and performance of a free-air exposure system to study long-term effects of ozone on grasslands. Atmos Environ 37:1341–1350
Volk M, Bungener P, Contat F, Montani M, Fuhrer J (2006) Grassland yield declined by a quarter in 5 years of free-air ozone fumigation. Glob Change Biol 12:74–83

Volk M, Obrist D, Novak K, Giger R, Bassin S, Fuhrer J (2011) Subalpine grassland carbon dioxide fluxes indicate substantial carbon losses under increased nitrogen deposition, but not at elevated ozone concentration. Glob Change Biol 17:366–376

Wahid A (2006) Influence of atmospheric pollutants on agriculture in developing countries: a case study with three new wheat varieties in Pakistan. Sci Total Environ 371:304–313

Wallin G, Skarby L (1992) The influence of ozone on the stomatal and non-stomatal limitation of photosynthesis in Norway spruce, Picea abies (L.) Karst, exposed to soil moisture deficit. Trees Struct Funct 6:128–136

Wang XP, Mauzerall DL (2004) Characterizing distributions of surface ozone and its impact on grain production in China, Japan and South Korea: 1990 and 2020. Atmos Environ 38:4383–4402

Wang XZ, Taub DR (2010) Interactive effects of elevated carbon dioxide and environmental stresses on root mass fraction in plants: a meta-analytical synthesis using pairwise techniques. Oecologia 163:1–11

Watanabe M, Yamaguchi M, Tabe C, Iwasaki M, Yamashita R, Funada R, Fukami M, Matsumura H, Kohno Y, Izuta T (2007) Influences of nitrogen load on the growth and photosynthetic responses of Quercus serrata seedlings to O_3. Trees Struct Funct 21:421–432

Watanabe M, Hoshika Y, Inada N, Wang X, Mao Q, Koike T (2013) Photosynthetic traits of Siebold's beech and oak saplings grown under free air ozone exposure in northern Japan. Environ Pollut 174:50–56

Wedlich KV, Rintoul N, Peacock S, Cape JN, Coyle M, Toet S, Barnes J, Ashmore M (2012) Effects of ozone on species composition in an upland grassland. Oecologia 168:1137–1146

Werner H, Fabian P (2002) Free-air fumigation of mature trees—a novel system for controlled ozone enrichment in grown-up beech and spruce canopies. Environ Sci Pollut Res 9:117–121

Wieser G, Matyssek R (2007) Linking ozone uptake and defense towards a mechanistic risk assessment for forest trees. New Phytol 174:7–9

Wild O, Fiore AM, Shindell DT, Doherty RM, Collins WJ, Dentener FJ, Schultz MG, Gong S, MacKenzie IA, Zeng G, Hess P, Duncan BN, Bergmann DJ, Szopa S, Jonson JE, Keating TJ, Zuber A (2012) Modelling future changes in surface ozone: a parameterized approach. Atmos Chem Phys 12:2037–2054

Wilkinson S, Davies WJ (2010) Drought, ozone, ABA and ethylene: new insights from cell to plant to community. Plant Cell Environ 33:510–525

Wilkinson S, Mills G, Illidge R, Davies WJ (2012) How is ozone pollution reducing our food supply? J Exp Bot 63:527–536

Winner WE, Lefohn AS, Cotter IS, Greitner CS, Nellessen J, McEvoy LR Jr, Olson RL, Atkinson CJ, Moore LD (1989) Plant responses to elevational gradients of O_3 exposures in Virginia. Proc Natl Acad Sci USA 86:8828–8832

Winwood J, Pate AE, Price J, Hanke DE (2007) Effects of long-term, free-air ozone fumigation on the cytokinin content of mature beech trees. Plant Biol 9:265–278

Wittig VE, Ainsworth EA, Long SP (2007) To what extent do current and projected increases in surface ozone affect photosynthesis and stomatal conductance of trees? A meta-analytic review of the last 3 decades of experiments. Plant Cell Environ 30:1150–1162

Wittig VE, Ainsworth EA, Naidu SL, Karnosky DF, Long SP (2009) Quantifying the impact of current and future tropospheric ozone on tree biomass, growth, physiology and biochemistry: a quantitative meta-analysis. Glob Change Biol 15:396–424

Wyness K, Mills G, Jones L, Barnes JD, Jones DL (2011) Enhanced nitrogen deposition exacerbates the negative effect of increasing background ozone in Dactylis glomerata, but not Ranunculus acris. Environ Pollut 159:2493–2499

Yoshida LC, Gamon JA, Andersen CP (2001) Differences in above- and below-ground responses to ozone between two populations of a perennial grass. Plant Soil 233:203–211

Zak DR, Holmes WE, Pregitzer KS, King JS, Ellsworth DS, Kubiske ME (2007) Belowground competition and the response of developing forest communities to atmospheric CO_2 and O_3. Glob Change Biol 13:2230–2238

Zak DR, Kubiske ME, Pregitzer KS, Burton AJ (2012) Atmospheric CO_2 and O_3 alter competition for soil nitrogen in developing forests. Glob Chang Biol 18:1480–1488

Zeng G, Pyle JA, Young PJ (2008) Impact of climate change on tropospheric ozone and its global budgets. Atmos Chem Phys 8:369–387

Zheng F, Wang X, Zhang W, Hou P, Lu F, Du K, Sun Z (2013) Effects of elevated O_3 exposure on nutrient elements and quality of winter wheat and rice grain in Yangtze River Delta, China. Environ Pollut 179:19–26

Zhu XK, Feng ZZ, Sun TF, Liu XC, Tang HY, Zhu JG, Guo WS, Kobayashi K (2011) Effects of elevated ozone concentration on yield of four Chinese cultivars of winter wheat under fully open-air field conditions. Glob Change Biol 17:2697–2706

MIX
Papier aus verantwortungsvollen Quellen
Paper from responsible sources
FSC® C105338
FSC
www.fsc.org

If you have any concerns about our products,
you can contact us on
ProductSafety@springernature.com

In case Publisher is established outside the EU,
the EU authorized representative is:
Springer Nature Customer Service Center GmbH
Europaplatz 3, 69115 Heidelberg, Germany

Printed by Libri Plureos GmbH
in Hamburg, Germany